Novel Approach to the Study of Noble Metals

Novel Approach to the Study of Noble Metals

Edited by **Peggy Rusk**

NY RESEARCH
PRESS

New York

Published by NY Research Press,
23 West, 55th Street, Suite 816,
New York, NY 10019, USA
www.nyresearchpress.com

Novel Approach to the Study of Noble Metals
Edited by Peggy Rusk

International Standard Book Number: 978-1-63238-351-8 (Hardback)

Printed in China.

Contents

Preface

I am honored to present to you this unique book which encompasses the most up-to-date data in the field. I was extremely pleased to get this opportunity of editing the work of experts from across the globe. I have also written papers in this field and researched the various aspects revolving around the progress of the discipline. I have tried to unify my knowledge along with that of stalwarts from every corner of the world, to produce a text which not only benefits the readers but also facilitates the growth of the field.

This book aims to demystify noble metals. This book provides a wide spectrum of observations into the optical principle, resource, production, nanoscience and nanotechnology of noble metals. It also deals with the enhanced applications of noble metals in various fields. This book will be useful for scientific researchers and professionals interested in this field.

Finally, I would like to thank all the contributing authors for their valuable time and contributions. This book would not have been possible without their efforts. I would also like to thank my friends and family for their constant support.

Editor

Part 1

Principles

1

Dielectric Screening Properties and Many Body Effects in Molten Salts

Shigeki Matsunaga[1], Takahiro Koishi[2], Masatoshi Saito[3] and Shigeru Tamaki[3]
[1]*Nagaoka National College of Technology*
[2]*University of Fukui Graduate School of Engineering*
[3]*Niigata University*
Japan

1. Introduction

The dielectric screening properties in molten salts have been a matter of particular interest. There is a close relation between the dielectric screening and the charge fluctuation of the constituents of the substances. The charge-charge structure factor, or $S_{ZZ}(q)$ (Hansen & McDonald, 1986), indicates the charge fluctuation in a molten salt, which can be easily obtained by the linear transformation from their partial structure factors. The dielectric screening function, or $\varepsilon(q)$, of a mono-valent molten salt has been represented by $S_{ZZ}(q)$ as follows,

$$1/\varepsilon(q) = 1 - \{4\pi e^2 \beta n\, S_{zz}(q)/q^2\} \tag{1}$$

where $\beta = 1/k_B T$ and n is the number density of constituent ions (Hansen & McDonald, 1986).

On the other hand, the partial structure factors of various molten salts have been experimentally obtained by applying a combination of different diffraction methods, *e.g.* X-ray and neutron diffraction, to the same molten salts. Therefore, $S_{ZZ}(q)$ of molten salts are obtainable by experiment from their partial structure factors (Saito et al, 1999). Several attempts have been made at deriving the dielectric screening functions from the experimental structure factors and Eq. (1). However, the obtained results indicate a *negative* sign in the small q region. This fact shows the difficulty in the appropriate explanation of the results, because $\varepsilon(q)$ should be *positive* in the meaning of the potential screening. In this situation, we have proposed the *new* equation of $\varepsilon(q)$ and $S_{ZZ}(q)$ in molten salts which formula is different from Eq.(1). In order to test the new theory, we have applied it to molten alkali-halides (Koishi et al, 2007) and noble metal halides and their mixtures (Matsunaga et al, 2007, 2008, 2011).

In this book chapter, firstly, we summarize the re-examination of the theory, and the derivation of the alternative expression of the dielectric function. Then, we wish to show the application of the theory to the molten salts; alkali halides NaCl, RbBr, noble metal halides AgBr, CuBr, and mixture AgBr-AgI. AgBr-CuBr system will be treated in the context of the many body effect. The extension of the theory to the polarizable ion model will also been discussed with the application to CuI and CsAu melt.

2. Theoretical background of dielectric screening

The theory of the dielectric screening effect in molten salt has been re-examined in relation with the charge-charge structure factor in our previous work (Koishi et al, 2007; Matsunaga et al, 2007, 2008, 2011). In this chapter, for the readers' benefit, we summarize the theoretical back ground of the charge-charge structure factor firstly, then the derivation of the new expression of the dielectric function in relation with the dielectric screening.

2.1 Charge-charge structure factors in molten salts

We consider a typical binary molten salt system $A_\nu B_\mu$. N_ν and N_μ stand for the number of ion species + and -. With the total number of particles N, the concentration of species ν is expressed as $x_\nu = N_\nu/N = 0.5$. The densities of particles are $n^\nu = n^\mu = n^0 = n/2$ and $n = N/V$, where V is the volume of the system. The effective charges of ions are $|z^\nu| = |z^\mu| = z$. The charge-charge structure factor, $S_{ZZ}(q)$, defined by Hansen-McDonald is useful representation for charged fluid, which is written as follows (Hansen & McDonald, 1986),

$$S_{ZZ}(q) = <\rho_q^Z \rho_q^Z> / N = \Sigma_\nu \Sigma_\mu z_\nu z_\mu S_{\nu\mu}(q) \quad (2)$$

where $S_{\nu\mu}(q)$ is a partial structure factor, *i.e.* the Fourier transformation of the pair distribution function of ion ν around ion μ in r-space, $g_{\nu\mu}(r)$. $S_{\nu\mu}(q)$ is defined as,

$$S_{\nu\mu}(q) = x_\nu \delta_{\nu\mu} + n x_\nu x_\mu \int_0^\infty (\sin qr / qr)\{(g_{\nu\mu}(r) - 1)\} 4r^2 dr \quad (3)$$

$S_{\nu\mu}(q)$ and $g_{\nu\mu}(r)$ directly reflect the ionic configuration and are easily obtainable from neutron diffraction experiment or molecular dynamics simulation.

Using $g_{\nu\mu}(r)$ (ν,μ= +,-), $S_{zz}(q)$ is given by,

$$S_{zz}(q) = 1 + (n/2)\int_0^\infty (\sin qr / qr)\{(g_{++}(r) - 1) + (g_{--}(r) - 1) - 2(g_{+-}(r) - 1)\} 4\pi r^2 dr \quad (4)$$

On the other hand, the concentration-concentration fluctuation in r-space $g_{cc}(r)$ defined by Bhatia-Thornton representation is written as follow (Bhatia & Thornton, 1970),

$$\begin{aligned} g_{cc}(r) &= x_\nu^2 x_\mu^2 [(g_{++}(r) - 1) + (g_{--}(r) - 1) - 2(g_{+-}(r) - 1)] \\ &= (1/16)\left[(g_{++}(r) - 1) + (g_{--}(r) - 1) - 2(g_{+-}(r) - 1)\right] \end{aligned} \quad (5)$$

Corresponding to $g_{cc}(r)$, the concentration-concentration structure factor $S_{cc}(q)$ is defined as follows,

$$\begin{aligned} S_{cc}(q) &= x_\nu x_\mu + n\int_0^\infty (\sin qr / qr) g_{cc}(r) 4\pi r^2 dr \\ &= 1/4 + n\int_0^\infty (\sin qr / qr) g_{cc}(r) 4\pi r^2 dr \end{aligned} \quad (6)$$

Inserting (5) into (6) and using (4), we have,

$$S_{zz}(q) = 4\,S_{cc}(q) \tag{7}$$

Using the formulae obtained in this section, the dielectric screening functions in terms of $S_{zz}(q)$ will be shown in the next section.

2.2 Dielectric screening function expressed in terms of $S_{zz}(q)$

In order to derive the dielectric screening function, we will consider the same molten salt system considered in the preceding section. If a charge $e\rho_{ext}(\mathbf{r})$ were introduced from outside into this system at the position $\mathbf{r}$, then there occurs an induced charge $e\rho_{ind}(\mathbf{r})$ at its position. Therefore we have the following Poisson equations in the electrostatic unit,

$$\mathrm{div}\boldsymbol{D}(\mathbf{r}) = 4\pi e\rho_{ext}(\mathbf{r}) \tag{8}$$

$$\mathrm{div}\boldsymbol{E}(\mathbf{r}) = 4\pi e\{\rho_{ext}(\mathbf{r}) + \rho_{ind}(\mathbf{r})\} \tag{9}$$

where $\boldsymbol{D}$ ($\mathbf{r}$) and $E(\mathbf{r})$ the electric displacement and the electric field, respectively.

The electric potential $\varphi(\mathbf{r})$ due to this external charge density is given by

$$E(\mathbf{r}) = -\mathrm{grad}\varphi(\mathbf{r}) \tag{10}$$

Using (9) we have

$$\nabla^2\varphi(\mathbf{r}) = -\{4\pi e\rho_{ext}(\mathbf{r}) / \varepsilon\} \tag{11}$$

$$= -4\pi e\{\rho_{ext}(\mathbf{r}) + \rho_{ind}(\mathbf{r})\} = -4\pi e\rho_{ext}(\mathbf{r})[1 + \{\rho_{ind}(\mathbf{r}) / \rho_{ext}(\mathbf{r})\}] \tag{12}$$

where ε is the dielectric constant, but it may be extended to an isotropic r-dependent term written as $\varepsilon(r)$.

In their text book, Hansen and McDonald took partly $\varepsilon(r) = 1$ for the dielectric function in molten salts and obtained the formula expressed as shown in Eq. (1).

Now we can express each quantity $\varepsilon(\mathbf{r})$, $\varphi(\mathbf{r})$ and $\rho_{ext}(\mathbf{r})$ in this equation as the following Fourier representations,

$$\varepsilon(\mathbf{r}) = \Sigma_q \underline{\varepsilon}(\mathbf{q})e^{i\mathbf{q}.\mathbf{r}},\ \varphi(\mathbf{r}) = \Sigma_q \underline{\varphi}(\mathbf{q})e^{i\mathbf{q}.\mathbf{r}},\ \rho_{ext}(\mathbf{r}) = \Sigma_q \underline{\rho}_{ext}(\mathbf{q})e^{i\mathbf{q}.\mathbf{r}}$$

and

$$\rho_{ind}(\mathbf{r}) = \Sigma_q \underline{\rho}_{ind}(\mathbf{q})e^{i\mathbf{q}.\mathbf{r}} \tag{13}$$

Putting some of these into equation (11) and taking $\mathbf{r} = 0$, we have

$$q^2\varepsilon(q)\underline{\varphi}(\mathbf{q}) = 4\pi e\underline{\rho}_{ext}(\mathbf{q}) \tag{14}$$

where $\underline{A}(\mathbf{q}) = \{A_x(\mathbf{q}), A_y(\mathbf{q}), A_z(\mathbf{q})\}$. And we assume that $\underline{\varepsilon}(\mathbf{q})$ is isotropic, which is expressed as $\varepsilon(q)$.

Since $e\rho_{ind}(\mathbf{r})$ is a charge fluctuation by the insertion of the external charge $e\rho_{ext}(\mathbf{r})$, it might be expressed in the following form, by using the linear charge response function $\chi_{zz}(q)$,

$$\underline{\rho}_{ind}(\mathbf{q}) = \chi_{zz}(q) e \underline{\varphi}(\mathbf{q}) \tag{15}$$

Based on the fluctuation dissipation theorem, $\chi_{zz}(q)$ is expressed in terms of $S_{zz}(q)$ as follows,

$$\chi_{zz}(q) = -\beta n S_{zz}(q) \tag{16}$$

Compare equations (11) and (12), the inverse dielectric function $1/\varepsilon(q)$ is expressed as follows,

$$1/\varepsilon(q) = FT\{\mathrm{div}\boldsymbol{E}(\mathbf{r})\} / FT\{\mathrm{div}\boldsymbol{D}(\mathbf{r})\} = \{\mathbf{q}\cdot\underline{\boldsymbol{E}}(q)\} / \{\mathbf{q}\cdot\underline{\boldsymbol{D}}(q)\} = 1 + \{\underline{\rho}_{ind}(\mathbf{q}) / \underline{\rho}_{ext}(\mathbf{q})\} \tag{17}$$

where FT means the Fourier component.

Putting (14) and (15) into (17), and using (16), we have

$$1/\varepsilon(q) = 1 + \{4\pi e^2 \chi_{zz}(q) / \varepsilon(q) q^2\} = 1 - \{4\pi e^2 \beta n S_{zz}(q) / \varepsilon(q) q^2\} \tag{18}$$

or

$$1/\varepsilon(q) = 1/\{1 + (\kappa_s^2 / q^2) S_{zz}(q)\} \tag{19-a}$$

where

$$\kappa_s^2 = 4\pi e^2 \beta n \tag{19-b}$$

Equation (19-a) is resulted from an isotropic configuration of surrounded ions, which condition can be satisfied by a symmetric configuration in the short range region in the molten salt that consists of almost perfectly ionized ions such as in molten NaCl and CsCl. If the ions' configuration deviates extremely from an isotropic one, which can be, for instance, seen in molten CuI in which the oscillation of $g^{Cu\text{-}I}(r)$ for the distance r coincides basically with that of $g^{Cu\text{-}Cu}(r)$ (Waseda et al.,2000), then the dielectric screening may include some anisotropic effect and the application of equation (19-a) becomes insufficient. We will discuss this point in the later section. A rather simplified revision for such a case is the insertion of a parameter δ into equation (19-a) defined as,

$$1/\varepsilon(q) = 1/\{1 + \delta(\kappa_s^2 / q^2) S_{zz}(q)\} \tag{20}$$

Magnitude of δ may be in the range of $0<\delta\leq 1$ and the condition $\delta=1$ may occur in a well symmetric configuration. In the region of higher values of $q > 10A^{-1}$, we have usually $S_{zz}(q) \simeq 1$ and then

$$1/\varepsilon(q) = 1/\{1 + \delta(\kappa_s^2 / q^2)\} \tag{21}$$

On the other hand, the inverse dielectric function in the long wavelength limit is already well known by the theory of classical one component plasma (Hansen & McDonald, 1986; March & Tosi, 1976). Starting from the continuity equation relating to the charge and taking Fourier transforms under the condition of long wavelength limit, the relation between $\rho_{ind}(q)$ and $\rho_{ext}(q)$ is expressed as follows,

$$\rho_{ind}(q) = -\rho_{ext}(q) / \{1 + (q^2 / \lambda_s^2)\} \tag{22}$$

where

$$\lambda_s^2 = 4\pi e^2 n^2 \chi_T \tag{23}$$

here χ_T being the isothermal compressibility and $n = 2n_0 = n^+ + n^-$. If an external point charge ze is put at the position r, then we have the well-known Thomas-Fermi type screening potential by using equation (22), which indicates that the inverse dielectric function near the long wavelength limit is written as follows,

$$1 / \varepsilon(q) = 1 / \{1 + (\lambda_s^2 / q^2)\} \tag{24}$$

It is interesting that the inverse dielectric function $1/\varepsilon(q)$ in its higher q-region and very lower q-region is expressed by a similar form as shown in equations (21) and (24), although their screening constants are quite different. In the intermediate region, $1/\varepsilon(q)$ exhibits an oscillating behavior due to the effect of $S_{zz}(q)$, as a logical consequence.

Using (6) and (7), $S_{zz}(q)$ is given by

$$S_{zz}(q) = \sum_{\upsilon} \sum_{\mu} z_{\upsilon} z_{\mu} S_{\upsilon\mu}(q) = 1 + 8n \int_0^{\infty} \frac{\sin qr}{qr} g_{cc}(r) 4\pi r^2 dr \tag{25}$$

where we took that the ionic charges are equal to the ionic valences as $z^+ =- z^- = z = 1$. Therefore $1/\varepsilon(q)$ is converted to

$$\frac{1}{\varepsilon(q)} = \frac{1}{1 + \delta(\kappa_s^2 / q^2)\{1 + 8n \int_0^{\infty} (\sin qr / qr) g_{cc}(r) 4\pi r^2 dr\}} \tag{26}$$

Under the assumption of $\delta = 1$, equation (19) is useful for deriving the inverse dielectric function, $1/\varepsilon(q)$, from experimental results for the partial structure factors and also equation (25) is applicable for deriving $1/\varepsilon(q)$ by a computer simulation.

It is apparent that equation (24) is equal to zero at q=0. And therefore, the inverse dielectric function $1/\varepsilon(q)$ has the following relation for any positive values of $S_{zz}(q)$,

$$0 \leqq 1 / \varepsilon(q) \leqq 1 \tag{27}$$

At this point, it should be stressed that the validity of Eq.(20) and Eq.(21) is also confirmed by this fact in the meaning of screening; $1/\varepsilon(q)$ remains *positive* for all positive q value. It is therefore possible to derive the inverse dielectric function $1/\varepsilon(q)$ if $S_{zz}(q)$ or

$g_{cc}(r)$ are known by either the experimental method or computer simulation under the assumption of $\delta= 1$. In the following sections we will show several results for $1/\varepsilon(q)$ by using experimental $S_{zz}(q)$ and simulated $g_{cc}(r)$.

2.3 Screening for Coulomb potential in a molten salt

In a molten salt having the charges of $z^+ =- z^- = 1$, the screened potential between a cation and an anion, $\varphi^{+-}{}_{sc}(r)$, may be divided into two parts as follows,

$$\varphi^{+-}{}_{sc}(r) = \varphi^{+-}{}_{sc\ rep}(r) + \varphi^{+-}{}_{sc\ at}(r) \tag{28}$$

Here $\varphi^{+-}{}_{sc\ rep}(r)$ is the repulsive potential influenced by a small amount of screening effect in the short range distance between cation and anion, and $\varphi^{+-}{}_{sc\ at}(r)$ is the attractive screened potential in the long range.

The utilization of equation (19-a) for the repulsive potential seems not to be a good and enough approximation as the screening effect, because the application of linear response theory for a highly nonlinear functional form cannot give any good approximation. Therefore, equation (16) is a poor approximation for the repulsive potential because of its highly nonlinear functional form, and therefore the screening effect for the direct repulsive potential should be treated in a different way. A simple consideration for the screening effect for the repulsive potential is the introduction of parametric multiplier, hereafter taken as α, which can be multiplied to the well-known formula of the repulsive potential as a multiplicand. Then the screened repulsive potential may be written as follows,

$$\varphi^{+-}{}_{sc\ rep}(r) = \alpha\varphi^{+-}{}_{rep}(r) \tag{29}$$

where $\varphi^{+-}{}_{rep}(r)$ is the bare repulsive potential. On the other hand, the attractive screened potential $\varphi^{+-}{}_{sc\ at}(r)$ is effectively screened by the existence of other ions, although its bare-potential may be mainly ascribed to the form of Coulomb interacting potential. Let put an ion of positive point charge at the origin. Then another ion of negative point charge in the region of attractive Coulomb potential at *r* feels the following potential,

$$\varphi^{+-}{}_{at}(r) = -e^2 / r \tag{30}$$

Therefore, the screened attractive potential in q-space is written as follows,

$$\varphi^{+-}{}_{sc\ at}(q) = -\ 4\pi e^2 / \varepsilon(q)q^2 \tag{31}$$

The numerical result for $\varphi^{+-}{}_{sc\ at}(r)$ is given by the inverse Fourier transformation $FT\{\varphi^{+-}{}_{sc\ at}(q)\}$. The repulsive potential $\varphi^{+-}{}_{rep}(r)$ which is often approximated by either the Born-Mayer type potential or the inverse function of r^n ($n \leq 12$). That is,

$$\varphi^{+-}{}_{rep}(r) = A\cdot\exp(-br) \quad \text{or} \quad \varphi^{+-}{}_{rep}(r) = B / r^n \quad (n\leq 12) \tag{32}$$

where A, b, B and n are constants. Equation (28) is therefore converted to

$$\varphi^{+-}{}_{sc}(r) = \alpha\ A\exp(-br) + FT\{\varphi^{+-}{}_{sc\ at}(q)\} \tag{33-a}$$

or

$$\varphi^{+-}_{sc}(r)=\alpha(B/r^n)+FT\{\varphi^{+-}_{sc\,at}(q)\} \tag{33-b}$$

In the region of $S_{zz}(q)\sim1$, equation (21) is applicable. If δ=1, then we have,

$$\varphi^{+-}_{sc\,at}(q)=-(4\pi e^2/q^2)\{q^2/(q^2+\kappa_s^{\ 2})\}=-\,4\pi e^2/(q^2+\kappa_s^{\ 2}) \tag{34}$$

This equation is easily converted to the r-dependent expression as

$$\varphi^{+-}_{sc\,at}(r)=-(e^2/r)\exp(-\kappa_s r) \tag{35}$$

The screening parameter κ_s is exactly equal to the inverse of the Debye screening length. Using (32) and (35), the effective potential of mean force between cation and anion in the region of $S_{zz}(q)\sim1$ is then expressed as,

$$\varphi^{+-}_{sc}(r)=\alpha\ A\cdot\exp(-br)-(e^2/r)\exp(-\kappa_s r) \tag{36-a}$$

or

$$\varphi^{+-}_{sc}(r)=\alpha\ B/r^n-(e^2/r)\exp(-\kappa_s r) \tag{36-b}$$

It is emphasized that the inverse dielectric function $1/\varepsilon(q)$ as the screening effect can be multiplied onto the bare attractive inter-ionic potential between two ions in the q-space, and the screening parameter α is also multiplied onto the repulsive potential in r-space, in order to obtain the effective screened potential

2.4 Equivalency between the screened pair potential and the potential of mean force

We have obtained an asymptotic form for the screened attractive potential, as shown in Eq. (35). The factor $\exp(-\kappa_s r)$ is familiar in the elementary Debye-Hückel theory and the corresponding pair distribution function $g_{+-}(r)$ is given by,

$$g_{+-}(r)=\exp\{\frac{\beta e^2}{r}\exp(-\kappa_s r)\} \tag{37}$$

In this section, we will prove that this asymptotic form of the screened attractive potential $\varphi^{+-}{}_{sc\,at}(r)$ and the potential of mean force $U^{+-}(r)$ defined by the formula of

$$g_{+-}(r)=\exp\{-U^{+-}(r)/k_BT\} \tag{37-a}$$

are equal to each other at the long distance of r, by using the Ornstein-Zernike equation. Putting a cation 1 at the origin and an anion 2 at the position r, and furthermore the third ion 3 at the position r', then we have an approximate Ornstein-Zernike equation as follows,

$$-\beta U^{+-}(r)=-\beta\varphi^{+-}(r)+n\int d\boldsymbol{r}\ c_{+-}(|\boldsymbol{r}-\boldsymbol{r}'|)h_{--}(\boldsymbol{r}') \tag{38}$$

where $c_{+-}(r)$ is the direct correlation function between cation and anion, and $h_{--}(r')$ is equal to $(g_{--}(r')-1)$. Hence we have inferred that the third ion represented by 3 is mainly another neighboring anion located around the cation at the origin, because the ion 3 is located at a closer position to the origin, which means that its sign is negative. By Fourier transformation, Eq.(38) is converted to the following form,

$$U^{+-}(q) = \varphi^{+-}(q)\left[1 - \frac{c_{+-}(q)h_{--}(q)}{\beta n\varphi^{+-}(q)}\right] \tag{39}$$

On the other hand, it is well known that the direct correlation function $c_{+-}(r)$ has a approximate form for large *r*, as follows,

$$c_{+-}(r) \cong -\beta\varphi^{+-}(r) \tag{40}$$

The Fourier transformation of Eq.(40) is immediately written as,

$$c_{+-}(q) \cong -n\beta\varphi^{+-}(q) \tag{41}$$

Putting (41) into (39), we have

$$U^{+-}(q) \cong \varphi^{+-}(q)\{1 + h_{--}(q)\} = \varphi^{+-}(q)S_{--}(q) \tag{42}$$

Apart from a quantitative disagreement with either the experimental result or simulated one, $S_{--}(q)$ may be analogically expressed as follows, by using the random phase approximation (March and Tosi, 1976),

$$S_{--}(q) = \frac{1}{1+n\beta\varphi^{--}(q)} \tag{43}$$

Here the pair potential between anions in its Fourier transformation $\varphi^{--}(q)$ is also approximately expressed as $4\pi e^2/q^2$, and then we have,

$$U^{+-}(q) = -\frac{4\pi e^2}{q^2+\kappa_S^2} \tag{44}$$

And finally we have,

$$U^{+-}(q) = -\frac{e^2}{r}\exp(-\kappa_S r) \tag{45}$$

Therefore, $\varphi^{+-}_{sc\ at}(r)$ described in Eq. (35) and the potential mean force $U^{+-}(r)$ are equal to each other at the long distance of *r*. The expression of $\varphi^{+-}_{sc\ at}(r)$ is certainly not rigorous, and therefore there is inevitably a minor discrepancy between $\varphi^{+-}_{sc\ at}(r)$ and $U^{+-}(r)$ in the numerical agreement. However, in an approximate sense, both functions are equivalent. In fact, the screened potential between the centered cation 1 and neighboring anion 2 locating at an appropriate distance, is expressed in terms of the form of direct interacting potential multiplied by the inverse dielectric function $1/\varepsilon(q)$. The factor $1/\varepsilon(q)$ is evidently given by the force acting on the centered cation from all other neighboring ions represented by 3. Therefore, the screened potential is physically equivalent to the potential of mean force.

2.5 Deviation from Nernst-Einstein relation

In this section, we wish to deal with a practical application of the potential of mean force $U^{+-}(r)$ in molten salts. The deviation, Δ, from the Nernst-Einstein relation in nomovalent molten salts is defined as follows,

$$\sigma = \sigma^{+} + \sigma^{-} = \left(ne^{2} / k_{B}T\right)\left(D^{+} + D^{-}\right)(1 - \Delta) \tag{46}$$

The partial conductivities for cation and anion, σ^+ and σ^-, and the diffusion constants D^+ and D^- were already derived before (Koishi & Tamaki, 2005). Taking these results, we have

$$\Delta = 1 - (1 - \Delta_{BR})\left[\left(2 / \alpha^{0}\right) / \left(1 / \alpha^{+} + 1 / \alpha^{-}\right)\right] \tag{47}$$

where

$$\alpha^{0} = n\int_{0}^{\infty}\left\{\partial^{2}\varphi^{+-} / \partial r^{2} + (2 / r)\left(\partial\varphi^{+-} / \partial r\right)\right\} g_{+-}(r) 4\pi r^{2} dr \tag{48}$$

$$\alpha^{+} = (n / 2)\int_{0}^{\infty}[\left\{\partial^{2}\varphi^{++} / \partial r^{2} + 2 / r\left(\partial\varphi^{++} / \partial r\right)\right\} g_{++}(r) + \\ +2\left\{\partial^{2}\varphi^{+-} / \partial r^{2} + 2 / r\left(\partial\varphi^{+-} / \partial r\right)\right\} g_{+-}(r)] 4\pi r^{2} dr \tag{49}$$

$$\alpha^{-} = (n / 2)\int_{0}^{\infty}[\left\{\partial^{2}\varphi^{--} / \partial r^{2} + 2 / r\left(\partial\varphi^{--} / \partial r\right)\right\} g_{--}(r) + \\ +2\left\{\partial^{2}\varphi^{+-} / \partial r^{2} + 2 / r\left(\partial\varphi^{+-} / \partial r\right)\right\} g_{+-}(r)] 4\pi r^{2} dr \tag{50}$$

and

$$\Delta_{BR} = (4\pi n / 3k_{B}T)\int_{d}^{\infty}\left(\partial\varphi^{+-}(r) / \partial r\right) g_{+-}(r) r^{3} dr \tag{51}$$

where $\varphi^{ij}(r)$ is an effective inter-ionic potential between ions i and j. $g_{ij}(r)$ is the partial pair distribution function between ions i and j as defined in the previous section. In this equation, we can approximate that d is equal to the hard-core contact distance between cation and anion, if both ionic sizes are rigid. However, this assumption may not be always valid for the heavier atoms. Here, we will use the first maximum position of $g_{+-}(r)$ as the distance of ionic contact. The quantity Δ_{BR} is essentially caused by an asymmetric distribution of the surrounded ions around the centered ion under an applied external field; it was obtained by Berne and Rice (Berne & Rice, 1964), and recently certified by the present authors in a different way (Koishi & Tamaki, 2005). Since it was found that values of α^0, α^+ and α^- are numerically close to one another, we can take $\Delta \sim \Delta_{BR}$ as the deviation from Nernst-Eistein relation in a molten salt.

The effective potential $\varphi^{+-}(r)$ shown in Eq.(48) is equal to the potential of mean force acting on the ion at the origin from other ion of opposite sign located at the distance r, and therefore it is possible to use the potential of mean force $U^{+-}(r)$, instead of $\varphi^{+-}(r)$. Therefore, equation Eq.(51) is converted to,

$$\Delta_{BR} = (4\pi n / 3k_{B}T)\int_{d}^{\infty}\left(\partial U^{+-}(r) / \partial r\right) g_{+-}(r) r^{3} dr = -(4\pi n / 3)\int_{d}^{\infty}(\partial g_{+-}(r) / \partial r)\, r^{3} dr \tag{52}$$

We can basically calculate the deviation from Nernst-Einstein relation, by using Eq.(52), if $g_{+-}(r)$ is experimentally obtained. However, because of the presence of the r^3 term in the integrand, convergence at large r is slow and full information on the approach of $g_{+-}(r)$ to unity is necessary. At the present stage, $g_{+-}(r)$'s in molten salts being satisfactory this requirement seem to be rare.

3. Application of the theory to the real systems

In this section, the new theory of the dielectric screening is applied to the real systems to test the validity of the theory utilizing the structure factors obtained by simulation or experiment. The theory is applied to molten alkali-halides (Koishi et al., 2007), and noble metal halides (Matsunaga et al, 2007, 2008, 2011). We summarize the results in the following subsections. Application to molten AuCs is also mentioned.

3.1 Inverse dielectric function in molten NaCl obtained by computer simulation

It is not necessary to apply the dielectric function for the computer simulation, because the simulation procedure itself involves automatically the screening behaviors and only the utilization of appropriate bare inter-ionic potentials is required, although the ion's charges are not always equal to their valence ones. In fact, the effective charge z* used in computer simulation is sometimes smaller than the ion's valence number because of a partly covalent character of the constituents. In the cases of MD or Monte-Carlo simulations, the effective charge in molten CuI is taken to be 0.6 (Waseda et al., 2000). In this section, we derive the charge-charge correlation function Szz(q) and the inverse dielectric function $1/\varepsilon(q)$ of molten NaCl by using MD simulation. The well-known Tosi-Fumi potentials are used as the bare inter-ionic potentials. The simulated $g_{++}(r)$, $g_{--}(r)$ and $g_{+-}(r)$ in the range of r < 8Å agreed with the experimental results (Edwards et al., 1975) and indicate a good isotropic configuration suggesting $\delta = 1$ and these quantities were inserted into Eq.(4). The obtained Szz(q) of molten NaCl is shown in Fig.1, which agreed with that shown in the literature (Hansen & McDomald,1986). Putting this Szz(q) into (19-a), we have $1/\varepsilon(q)$ in molten NaCl as shown in Fig.1. Using the obtained $1/\varepsilon(q)$ and taking α=0.5 as a trial screening factor for the repulsive potential and (33-a), the estimated screened potential between cation and anion is shown in Fig.2. So far the obtained screened potential is, more or less, close to the mean force $U^{+-}(r)$. A difference between $\varphi^{+-}_{sc}(r)$ and $U^{+-}(r)$, which are shown in Fig.2, may be caused by several reasons. One of them is that the inverse dielectric function is only applicable for a gradual slope of attractive potential and the screening for the repulsive part is restricted to a qualitative way as multiplying the parameter α. This treatment may diminish more or less, the exactness of the result. The other reason is that we have used Tosi-Fumi potential for MD simulation in order to obtain $g_{+-}(r)$. Even though the obtained $g_{+-}(r)$ by this simulation is close to the experimental one in the range r < 8Å, we are not sure of the possibility of the agreement in the range of the distance beyond 8Å, because the decaying tendency of $g_{+-}(r)$ obtained by simulation is slow, indicating a remarkable oscillation up to r = 15Å. Usually it is hard to obtain any visible oscilation in the experimental $g_{+-}(r)$ beyond the distance of r ~ 10Å. It is known that a Monte-Carlo simulation using the deformed dipole model (Gartrell-mills & McGreevy, 1989), which makes it possible to have a more rapid decaying for $g_{+-}(r)$, gives better agreement for the experimental results of molten CsCl (Locke et al., 1985). These facts suggest that the second and third peaks of $U^{+-}(r)$ obtained from simulated $g_{+-}(r)$ should be diminished to some extent in the range of $r > 10$ Å. At the present stage, the obtained $\varphi^{+-}_{sc}(r)$ is inevitably somewhat numerically different from $U^{+-}(r)$. By inserting the calculated $\varphi^{+-}_{sc}(r)$ instead of $U^{+-}(r)$, however, the obtained $g_{+-}(r)$ is semi-quantitatively close to the simulated one, as shown in Fig.2.

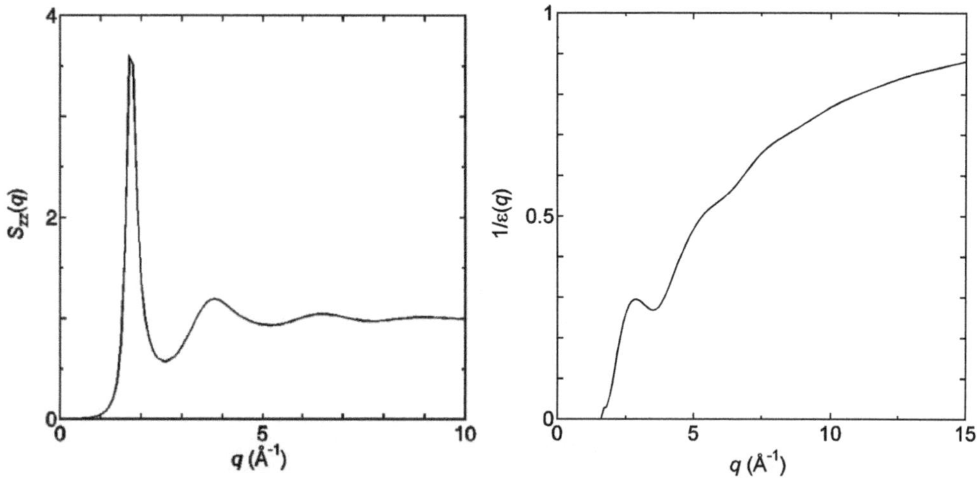

Fig. 1. $S_{ZZ}(q)$ of molten NaCl obtained by MD (left), $1/\varepsilon(q)$ of molten NaCl (right),

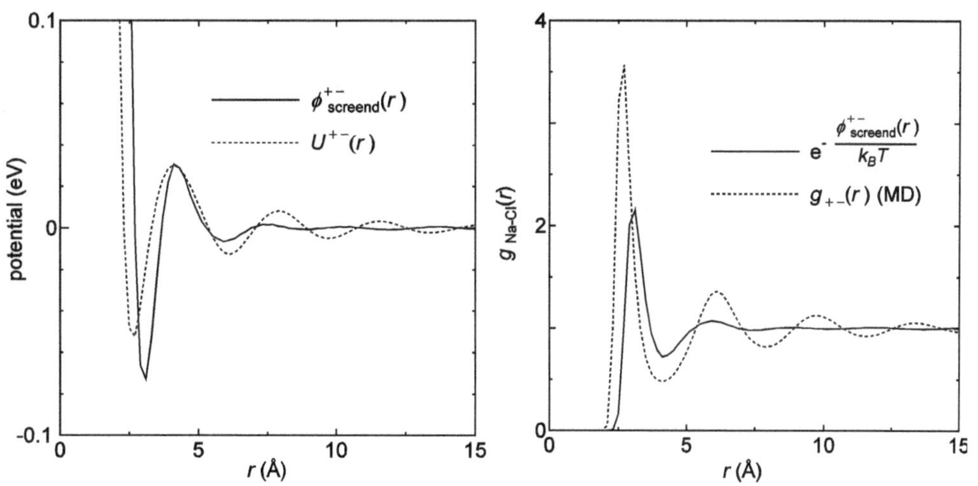

Fig. 2. $\varphi^{+-}_{sc}(r)$ and $U^{+-}(r)$ (left), $g_{+-}(r)$ and $\exp(-\varphi^{+-}_{sc}(r)/k_BT)$ (right) of molten NaCl.

3.2 Inverse dielectric function for molten RbBr obtained from experimental $S_{\nu\mu}(q)$

In this section we will show $1/\varepsilon(q)$ of molten RbBr, using the experimental partial structure factors. The experimental data for $S_{\nu\mu}(q)$ are adopted from the articles observed by Saito et al.(1999). In Fig.3, the experimentally obtained $S_{zz}(q)$ and the corresponding

$1/\varepsilon(q)$ are shown. As seen in these figures, the larger $S_{zz}(q)$ rather than unity yields a relatively small $1/\varepsilon(q)$. This fact is physically plausible, because a large charge fluctuation at the position of r around a centered ion located at the origin indicates a larger distribution of ions of opposite sign at r, which naturally causes a large screening in comparison with that of an averaged distribution of ions.

Using available repulsive potential for molten RbBr (Saito et al., 1999), we have calculated the screened potential by putting the appropriate $1/\varepsilon(q)$ into equation (33-a). The results are shown in Fig.4 under the condition of α=1. As seen in Fig.4, the screened inter-ionic potential between Rb^+ and Br^- ions, φ^{+-}_{sc} (r) has deeper minimum in comparison with the potential of mean force, U^{+-}(r), contrasting with the tendency of molten NaCl starting from the partial structure factors obtained by MD simulation. The most likely reason for this discrepancy seems to be an experimental uncertainty for partial structure factors in their small-q regions. This uncertainty gives, in due course, a numerical error in the inverse dielectric function $1/\varepsilon(q)$ in the very small-q region, even though its magnitude is relatively small. On Fourier transformation, such a small uncertainty gives a remarkably large change in the screened inter-ionic potential. In fact, an artificial modification for the curve of $1/\varepsilon(q)$ in only the small-q region makes it possible to obtain a result similar to U^{+-}(r).

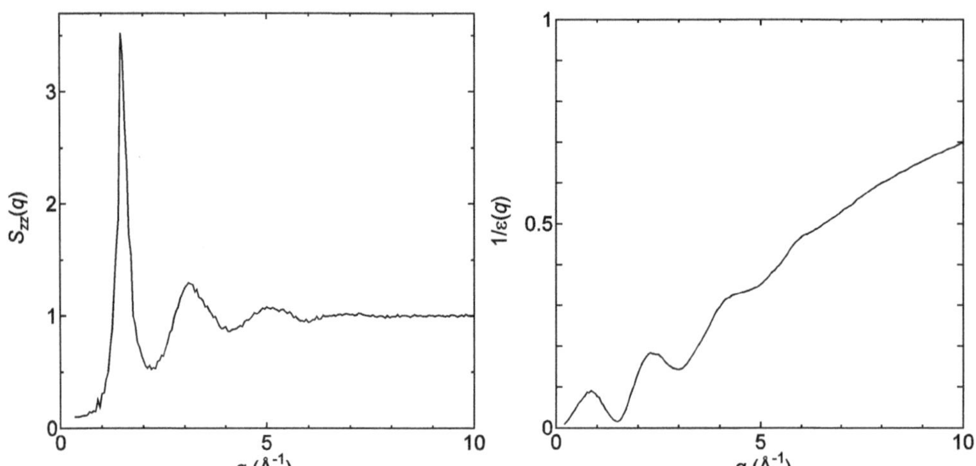

Fig. 3. Experimental Szz(q) (left), and $1/\varepsilon(q)$ (right) of molten RbBr.

3.3 Dielectric screening properties in molten noble metal halide and their mixtures

In this subsection, we focuse on molten noble-metal halides, AgBr and CuBr (Matsunaga et al., 2008), and mixture AgI-AgBr system (Matsunaga et al., 2007). It is well known that the noble metal halides are the typical example of the superionic conductors, which is a group of substances that exhibits high values of ionic conductivity, while they are still in solid phase. The mechanism of high ionic conductivity in the superionic conductors has been widely investigated because of their novel physical behaviour and technological importance, *e.g.* solid state batteries, fuel cells, optical devices (see for example, Chandra, 1981). CuBr shows superionic conduction in its alpha phase, where copper ions statistically distribute around the bromine bcc cubic, and copper ions can move one site to another. However, AgBr has a rock salt structure in its solid state and does not exhibit superionic conduction until its melting temperature, though AgBr is also one of the noble metal halides. We have investigated the structural and transport properties of noble metal halide mixtures by molecular dynamics simulations (Matsunaga, 2003; Matsunaga & Madden, 2004; Matsunaga, 2005). In this subsection, we apply the new theory to molten AgBr, CuBr , and AgI-AgBr to examine the dielectric screening effect, because the screened potentials are the fundamental subject for the transport properties.

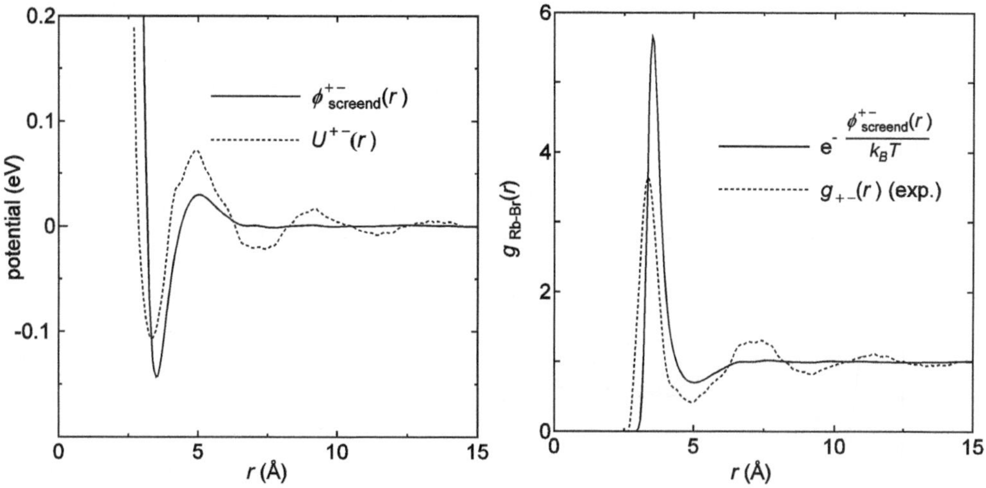

Fig. 4. $\varphi^{+-}_{sc}(r)$ and $U^{+-}(r)$ (left), $g_{+-}(r)$ and $\exp(-\varphi^{+-}_{sc}(r)/k_BT)$ (right) of molten RbBr.

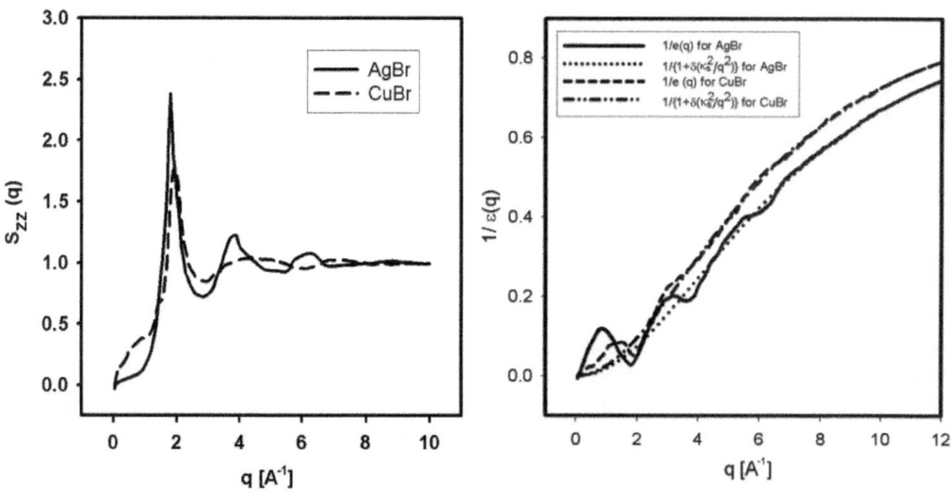

Fig. 5. $S_{ZZ}(q)$ (left), and $1/\varepsilon(q)$ with $1/(1+\delta(\kappa_s^2/q^2))$ for AgBr and CuBr (right).

As mentioned before, Eq.(19-a) or Eq.(20) is useful to derive $1/\varepsilon(q)$ from the experimental partial structure factors. Eq.(3) is applicable to deriving $1/\varepsilon(q)$ from $g_{+-}(r)$ by computer simulation. For examples of $1/\varepsilon(q)$ for binary ionic melts, we will show the results of $1/\varepsilon(q)$ using experimental $S_{ZZ}(q)$ for AgBr and CuBr (Saito et al., 1997; Saito et.al., 1999). As mentioned previously, $S_{ZZ}(q)$ is related to structure factors and pair distribution functions by Eq.(2). In this case, the summation in Eq.(2) is taken for ion species Ag or Cu and Br. The obtained $S_{ZZ}(q)$ for molten AgBr at 753K and CuBr at 810K are shown in Fig.5. We can see the significant first peaks in $S_{ZZ}(q)$ at about 1.8Å^{-1} and 2.0 Å^{-1} for molten AgBr and CuBr, respectively. These peaks correspond to the oscillations of partial structure factors (Saito et al., 1997; Saito et.al., 1999). By substituting these $S_{ZZ}(q)$ into Eq.(20), we can obtain the inverse dielectric function $1/\varepsilon(q)$. The calculated results are shown in Fig.5 with the calculated curve by Eq.(21) for comparison. It is clearly recognized that $1/\varepsilon(q)$ satisfies the condition $0 \leq 1/\varepsilon(q) \leq 1$. We can find the oscillatory features and the significant minimum at around 1.8 Å^{-1} and 2.0 Å^{-1} for AgBr and CuBr, respectively. These features are yielded by the form of $S_{ZZ}(q)$'s which have large maximum and oscillation. These results should be attributed to the effect that an ion in molten salts is estimated to be surrounded by ions of opposite sign, which yields the large charge fluctuation and screening effect.

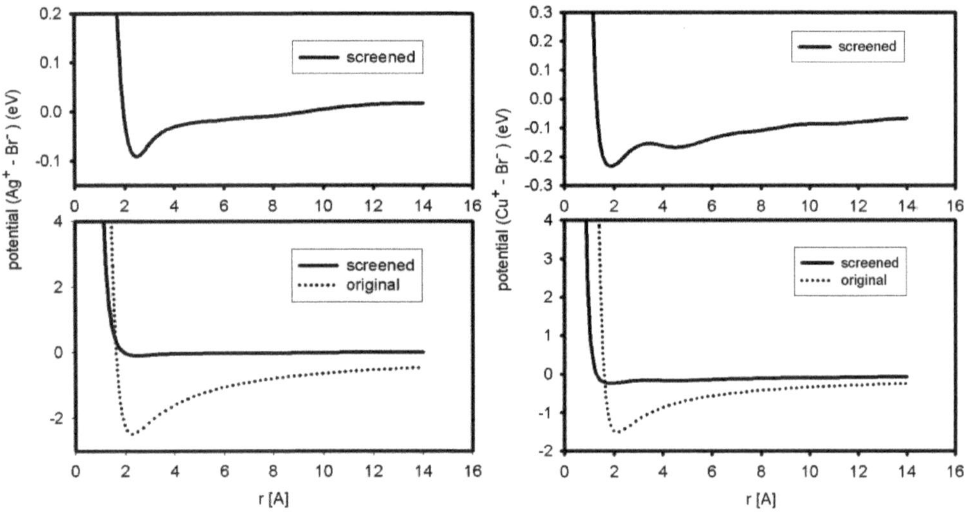

Fig. 6. Screened potential for Ag-Br (left) and Cu-Br (right) with original one (bottom). Screened potentials are also shown with different scale to show their oscillatory feature (top).

From above consideration, it seems interesting to obtain the screened potential between anion and cation, Ag^+ or Cu^+ and Br^- in molten AgBr and CuBr. Besides, the experimental pair distribution functions between cation and anion, *i.e.* $g_{BrAg}(r)$ and $g_{BrCu}(r)$ are obviously different (Saito et al., 1997; Saito et.al., 1999). We adopt the potential by Rahman, Vashishta and Parrinello (RVP) (Parrinello et al., 1983), as bare inter-ionic interaction. The RVP type potential sets for i and j ions are written as,

$$V_{ij}(r) = H_{ij} / r^{n_{ij}} + z_i z_j e^2 / r - P_{ij} / r^4 \tag{53}$$

where the third term is charge-dipole interactions. The adopted parameters are taken from literature (Tasseven et al., 1997; Stafford et al., 1990). The screened potentials of molten AgBr and CuBr are obtained by the procedure described in the previous sections. The results are shown in Fig.6. In the calculation, we adopt the value of δ as 0.50 and 0.40, and the value of α as 0.1 and 0.05 for AgBr and CuBr, respectively. The reduction of the depth of potential by screening effect is obviously recognized in Fig.6. The characteristic oscillations are found in the screened potentials especially in molten CuBr, which may be caused by the charge fluctuation in the distribution of ions.

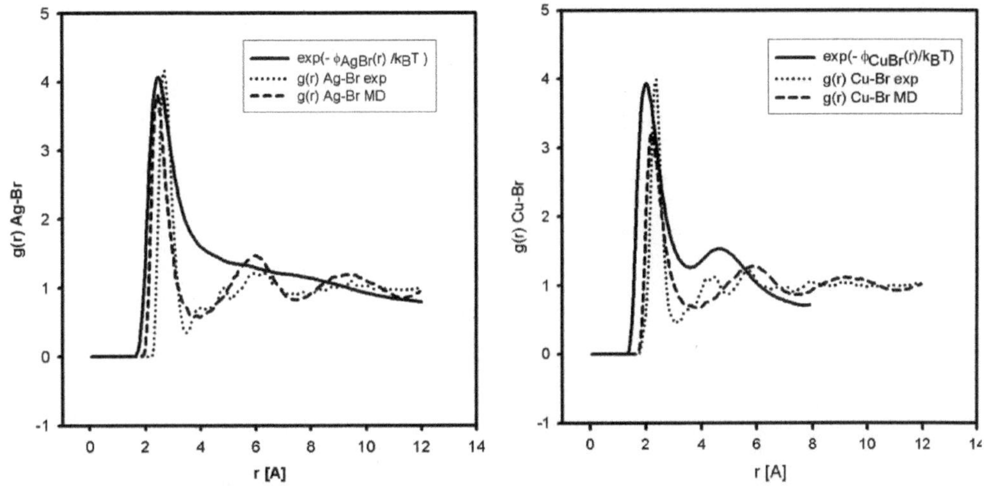

Fig. 7. $_{ij}$(r) for AgBr (left) and CuBr (right) calculated using screened potential with that by experiment and by MD.

To confirm the results of screened potentials between cations and anions, we calculate the pair distribution function $g_{+-}(r)$ by the potential of mean force $U^{+-}(r)$, which is defined by Eq.(37-a), where we can take the screened potential $\varphi^{+-}_{sc}(r)$ as $U^{+-}(r)$, because Eq.(37-a) is originally suggested by weak interacting materials. In the previous section, we have proved that $\varphi^{+-}_{sc}(r)$ and $U^{+-}(r)$ are mathematically equivalent, being our argument based on the Ornstein-Zernike equation (Koishi et al., 2007). $g_{+-}(r)$ are obtained by inserting the calculated $\varphi^{+-}_{sc}(r)$ instead of $U^{+-}(r)$ into Eq.(37-a). They are shown in Fig.7, together with $g_{+-}(r)$'s obtained by experiment and molecular dynamics (MD) simulation. The first peaks of $g_{+-}(r)$ almost agree well, though the oscillations in larger r region are not satisfactory reproduced. This fact may suggest that the screening effect is considerable especially in the small r region, *i.e.* about the distance to the first nearest neighbors. In other words, the first nearest neighbor ions have the main contribution to the screening effects in molten salts.

Next, we apply the theory to molten AgI-AgBr as an example of a pseudo-binary system, in which ions are thought to be in considerable order. It is well known that silver iodide is one of the superionic conductors in its alpha phase, i.e. α-AgI, in which silver ions statistically distribute around the iodine bcc cubic. On the other hand, AgBr has a rock salt structure in its solid state and dose not exhibit high conduction. We have investigated the effect of dissolution of AgBr into AgI on the structural and transport properties in superionic and molten phases by molecular dynamics simulations (Matsunaga, 2005).

Accordingly it is interesting to discuss the dielectric screening effects in this system. We wish to show the result for $1/\varepsilon(q)$ by using $S_{ZZ}(q)$ obtained by simulation. Though the consideration to this point is for the equi-charged binary salts, the theory is easily extendable to the equi-charged pseudo binary molten salts, e.g. AgI-AgBr, and Eq.(19-a) is also applicable to them. As mentioned in the previous section, $S_{ZZ}(q)$ is related to structure factors and pair distribution functions as Eq.(2), $S_{ZZ}(q) = \Sigma_\nu \Sigma_\mu z_\nu z_\mu S_{\nu\mu}(q)$, where summation is taken for species Ag, Br and I. We derive $1/\varepsilon(q)$ from the structure factors obtained by molecular dynamics simulation. The procedure of the simulation is essentially same as the previous work (Matsunaga, 2005). The RVP type pair potential sets are used, which parameters are listed in the same reference. The I-Br interaction is estimated by the appropriate combination rule. The MD calculations are performed for $Ag(Br_xI_{1-x})$ (x=0.20) at 903K.

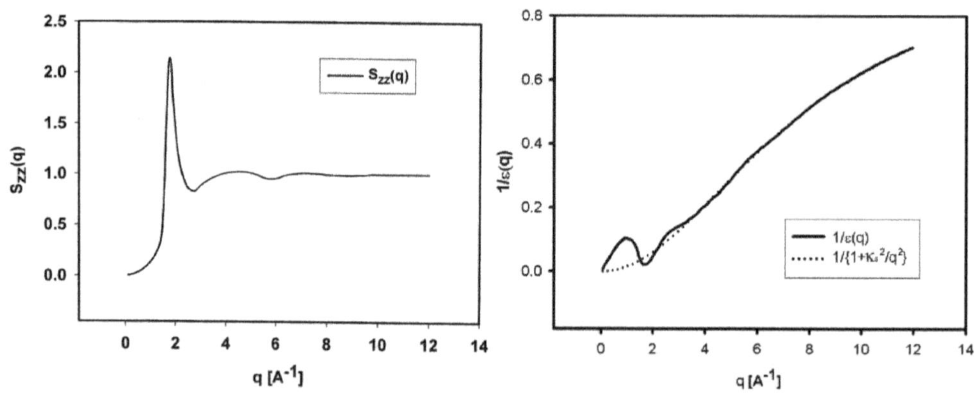

Fig. 8. $S_{ZZ}(q)$ (left), $1/\varepsilon(q)$ and $1/\{1+\kappa_s^2/q^2\}$ (right) for $Ag(Br_xI_{1-x})$ (x=0.20) at 903K.

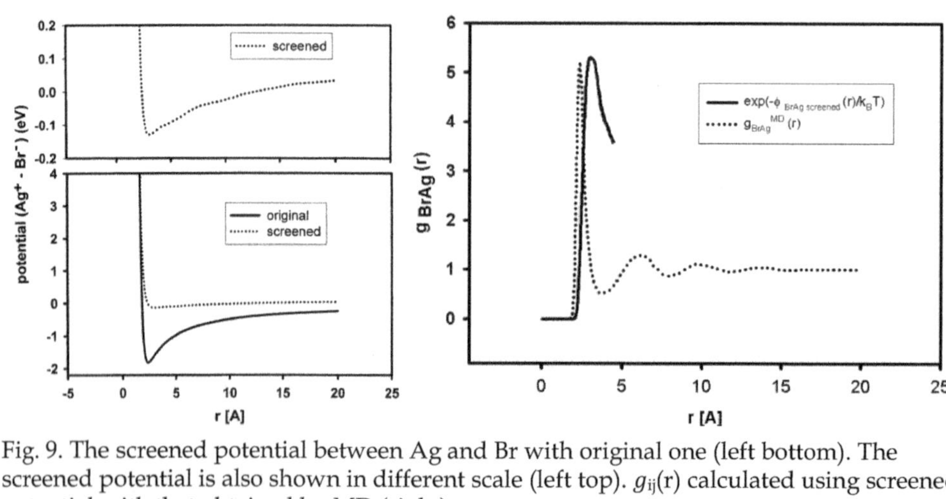

Fig. 9. The screened potential between Ag and Br with original one (left bottom). The screened potential is also shown in different scale (left top). $g_{ij}(r)$ calculated using screened potential with that obtained by MD (right).

The obtained $S_{ZZ}(q)$ at 903K is shown in Fig.8. We can see a significant peak in $S_{ZZ}(q)$ at about 1.7Å^{-1} and a plateau around 4.0Å^{-1} corresponding to the oscillations of partial structure factors (Matsunaga, 2005). We have the inverse dielectric function $1/\varepsilon(q)$ by putting obtained $S_{ZZ}(q)$ into Eq.(19-a), which is shown in Fg.8 with that calculated by Eq.(21) for $\delta=1$ for comparison. It is clearly recognized that $1/\varepsilon(q)$ satisfies the condition Eq.(27). We can see the significant minimum at around 1.7Å^{-1} corresponding the large maximum of $S_{ZZ}(q)$, which shows the large charge fluctuation. This fact may be attributed to the effect that an ion at the origin is supposed to be surrounded by ions of opposite sign at position r, which may yield the large charge fluctuation and screening effect.

The discrepancy in the larger r region between $g_{+-}(r)$'s are attributed to the difference between $\varphi^{+-}{}_{sc}(r)$ and $U^{+-}(r)$, which may be caused by several reasons, in addition to a margin of calculation error in Fourier transformation. One reason is that only the Coulomb attractive potential is used to derive $\varphi^{+-}{}_{sc}(r)$ instead the attractive part of RVP type potential $- P_{ij}/r^4$ which is used for simulation. Another possible reason is that the non-symmetric structure in solid phase remains in molten phase, which may obstruct the screening effect in molten phase. In spite of above facts, however, it may be recognized that the calculated result for $\varphi^{+-}{}_{sc}(r)$ so as to carry on the above treatment is physically significant, and the obtained results are obviously affected by screening effect.

3.4 Cation-cation interaction in noble metal halide mixtures

In the superionic phase of noble metal halides, the distributions of cations have been studied in detail. These studies have been executed mainly in the systems with one kind of cation, *i.e.* Cu^+ or Ag^+, that is supposed to be distributed mainly around the tetrahedral 12(d) site of halogen bcc lattice. However, according to the recent experimental study by NMR and X-ray diffraction, cations Cu^+ and Ag^+ are distributed mainly around the octahedral 6(b) site in the superionic phase of $(Ag_xCu_{1-x})I$, which is contradictory to the generally accepted view (Endo et al., 1999). Besides the interest on the basis of physical chemistry, there would be an advantage of practical application, because the pseudo-binary superionic conductor may lower the transition temperature to the superionic phase. These facts prompt us to investigate the superionic phase of noble metal halide mixture with two kinds of mobile cations. In the previous study (Matsunaga, 2009), we have carried out the molecular dynamics (MD) simulation to examine the structural properties of $(Ag_xCu_{1-x})Br$ for $x < 0.5$ in its superionic phase as an example of the superionic conductor with two kinds of cations, Cu^+ and Ag^+. The MD results were quite noteworthy; the different distributions between Ag^+ and Cu^+ ions in superionic phase were detected. For $x_{Ag} = 0.10$, Cu ions are mainly distributed around the tetrahedral 12(d) site. On the other hand, the significant distribution of Ag ions around the octahedral 6(b) site can be seen. For $x_{Ag} = 0.40$, however, Cu ions are mainly distributed around the octahedral 6(b) site. Ag ions are also distributed around the octahedral 6(b) site, moreover their distribution is more enhanced than those in $x_{Ag} = 0.10$ (Matsunaga, 2009a). These results suggest that there is a concentration dependence in cation distribution for $(Ag_xCu_{1-x})Br$, which may affect the dynamical and thermodynamic properties of this system.

In this subchapter, considering the circumstances mentioned above, as a continuous work (Matsunaga, 2011), we investigate the dynamical and thermodynamic properties of the superionic and molten phases of $(Ag_xCu_{1-x})Br$ for $x < 0.5$ by MD. The essential procedure of MD simulations is same as our previous works (Matsunaga, 2009a, 2009b; Matsunaga &

Tamaki, 2008a, 2008b). *RVP* type pair potentials are used. The used potential parameters are taken from literature (Stafford et al.,1990; Tasseven et al., 1997) The interactions between cations are obtained using the adequate combination rule. The used potential parameter sets are listed in the Table 1 in the previous paper (Matsunaga, 2009a).

The concentration-concentration (C-C) structure factor is quite effective to clarify the structural features of the molten binary system, which have been obtained by the experiments (Bhatia & Thornton, 1970; Matsunaga et al., 1983). The C-C structure factor has also been extended to the multi component systems (Bhatia & Ratti,1977). In this subsection, the C-C structure factors will be evaluated to examine how the structural features in the superionic phase of the system $(Ag_{1-x}Cu_{0x})Br$ will be observed in the molten phase. The procedure to obtain the multi component C-C structure factor is briefly summarized as follows. The partial structure factor defined by Faber and Ziman is expressed as,

$$a_{ij}(q) = 1 + \frac{N}{V}\int\{g_{ij}(q) - 1\}e^{iq\cdot r}d^3r \quad (54)$$

where $g_{ij}(r)$ is the pair distribution function, and *i* and *j* stand for the species of ions. The multi component C-C structure factor, $Sc_ic_j(q)$, is defined as,

$$S_{c_ic_j(q)} = \frac{1}{2}N\langle c_i^*(\boldsymbol{q})c_j(\boldsymbol{q}) + c_i(\boldsymbol{q})c_j^*(\boldsymbol{q})\rangle \quad (55)$$

$Sc_ic_j(q)$ is related to the Faber-Ziman type structure factor $a_{ij}(q)$ as,

$$Sc_ic_j(q) = c_ic_j\left(a_{ij}(q) - \sum_{\xi=1}^{v} c_\xi\left(a_{i\xi}(q) + a_{j\xi}(q)\right) + \sum\sum_{\xi,\zeta=1}^{v} c_\xi c_\zeta a_{\xi\zeta}(q)\right) + c_i\delta_{i,j} - c_ic_j \quad (56)$$

where ξ and ζ also stand for the species of ions. v is the number of the species. In the case of the *'ideal'* mixture, $Sc_ic_j(q)$'s for the long wave length limit, $q \to 0$, are expressed as,

$$S_{c_ic_i}^{id}(0) = c_i(1 - c_i), \qquad S_{c_ic_j}^{id}(0) = -c_ic_j \qquad i \neq j \quad (57)$$

The obtained normalized $Sc_ic_j(q)$'s divided by $|\, S^{id}c_ic_j(0)\, |$ in the molten phase 76K above the liquidus are shown in Fig.10. The difference between $Sc_{Ag}c_{Ag}(q)$ and $Sc_{Cu}c_{Cu}(q)$ can be seen in both figures. In Fig.10 for $(Ag_{0.1}Cu_{0.9})Br$, the profile $Sc_{Ag}c_{Ag}(q) \simeq 1$ shows the weak interaction between Ag ions, thought the interaction between Cu ions and between Cu and Ag ions seems to be significant in $Sc_{Cu}c_{Cu}(q)$ and $Sc_{Ag}c_{Cu}(q)$. The peaks in $Sc_{Ag}c_{Ag}(q)$ can be seen at about 3.6 and 1.8Å^{-1}; in $Sc_{Cu}c_{Cu}(q)$ at 4.9 and 1.8 Å^{-1}; in $Sc_{Ag}c_{Cu}(q)$ at 3.6 and 1.8 Å^{-1} corresponding to the first and the second nearest neighbors, respectively. In Fig.10 for $(Ag_{0.4}Cu_{0.6})Br$, the interaction between Ag ions seems to be somewhat enhanced in $Sc_{Ag}c_{Ag}(q)$, though the interaction between Cu ions is decreased. The interaction between Ag and Cu ions is still significant in $Sc_{Ag}c_{Cu}(q)$. The peaks in $Sc_{Ag}c_{Ag}(q)$ can be seen at about 3.7 and 1.8 Å^{-1}; in $Sc_{Cu}c_{Cu}(q)$ at 5.0 and 1.9 Å^{-1}; in $Sc_{Ag}c_{Cu}(q)$ at 3.8 and 1.9 Å^{-1} corresponding to the first and the second nearest neighbors, respectively.

The expression of $Sc_ic_j(q)$ for the long wave length limit, $q \to 0$, is related to the correlation between the fluctuations Δc_i and Δc_j as follows:

$$Sc_ic_j(0) = N < \Delta c_i \Delta c_j > \quad (58)$$

which has a physical meaning of an order parameter in the molten state (Bhatia & Ratti, 1977; Bhatia & Thornton, 1970). $S_{c_ic_j}(0)$ is related to the thermodynamic quantities. In a ternary system, e.g. $S_{c_1c_2}(0)$ is expressed as

$$S_{c_1c_2}(0) = -c_1c_2\left\{1 + \frac{c_3(\omega_{12}-\omega_{13}-\omega_{23})}{k_BT}\right\}/D \tag{59}$$

where D is the function of ω_{ij}. ω_{ij}'s are the pairwise interchange energies which are related to the excess free energy G^E as follows:

$$G^E = N\sum\sum_{\alpha<\beta} c_\alpha c_\beta \omega_{\alpha\beta} \tag{60}$$

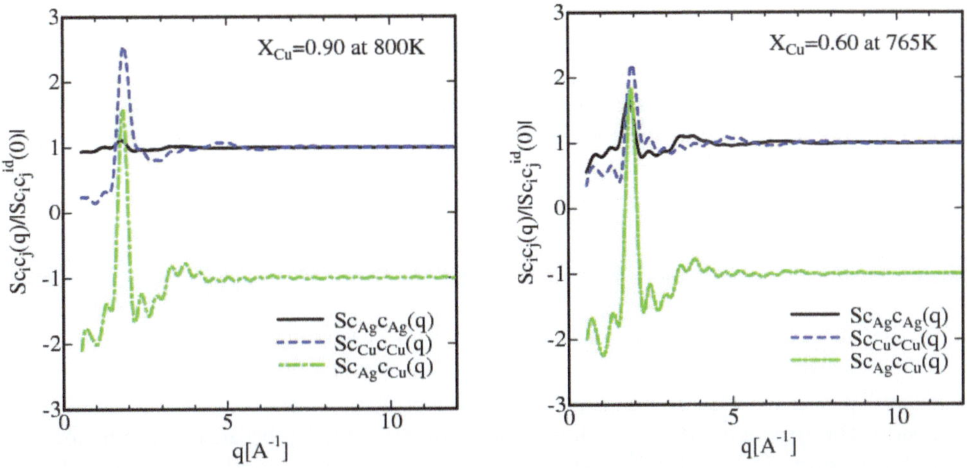

Fig. 10. $S_{cicj}(q)/|S_{cicj}^{id}(0)|$ for $(Ag_{1-x}Cu_x)Br$, x_{Cu}=0.9 at 800K (left), x_{Cu}=0.60 at 765K (right).

Though G^E for molten $(Ag_xCu_{1-x})Br$ is not available as far as we know, the expected values from the MD result in the superionic phase are $|S_{c_{Ag}c_{Cu}}(0)| > |S^{id}_{c_{Ag}c_{Cu}}(0)|$ for x_{Cu} =0.90, $|S_{c_{Ag}c_{Cu}}(0)| \leq |S^{id}_{c_{Ag}c_{Cu}}(0)|$ for x_{Cu} =0.60. In other words, the phase separation tendency of Ag and Cu ions for x_{Cu} =0.90 , and the ordering tendency for x_{Cu} =0.60. This fact suggests that there is the concentration dependence of the interaction between ions, ω_{ij}, in molten phase.

As stated so far, there is a concentration dependence in the dynamical and thermodynamic properties as well as the structural feature in the superionic and molten phases for $(Ag_xCu_{1-x})Br$, which suggests the many body effects and the short range interaction in the molten ternary system.

4. Polarizable ion model and dielectric screening effect

To understand qualitatively and quantitatively the properties of molten salts such as the static and dynamical behaviour, theoretical and molecular dynamics simulation studies are indispensable. For this purpose, the main problem is how to define the appropriate pair-potentials for the system of one's study. Needless to say, the properties of molten salts are, more or less, related with those of the solid phase. There are various ionic crystals; alkali-halides with rock salt structure which has ionic bond; zinc-blend (ZnS) type in which covalent and ionic bond coexist; superionic conductor of which a typical representative is AgI; etc. The constant efforts have been made for about 80 years to study the inter ionic potentials for these ionic crystals of various ionisation (Matsunaga & Tamaki, 2011). In this study, we will show how the theory of screened potentials is modified in connection with the recent development. The new theory will be applied to copper iodide for example. The numerical results will be discussed in comparison with those of experiment and molecular dynamics (MD) simulation.

4.1 Polarizable ion model

In recent years, studies have been developed on the viewpoint of the charge distribution change in a certain ion caused by approaching of surrounding ions. In 1996, Wilson et al. succeeded to perform MD to reproduce the experimental results of molten AgCl structure using new potentials introducing ion polarization. In 2006, Bitrián and Trullàs proposed the similar polarizable ion model to perform MD simulation in molten AgBr. In this case, Vashishta-Rahman (VR) type potential is adopted as the rigid ion model potential (Vashishta and Rahman, 1978), which is familiar in the study of super ionic conductor, expressed as follows,

$$\varphi_{ij}(r) = \frac{z_i z_j e^2}{r} + \frac{H}{r^n} - \frac{C_{ij}}{r^6} - \frac{P_{ij}}{r^4} \equiv \varphi_{ij}^0(r) - \frac{P_{ij}}{r^4} \tag{61}$$

where, n=7 power term is adopted for the repulsive potential in the second term on the right hand side. The dipole-quadrupole interaction term is neglected, because it is smaller than the last two terms in Eq.(61). The feature of the polarizable ion model is that the charge-dipole interaction or the last term in Eq.(61), P_{ij}/r^4, is expressed as a function of *r*, which may be the origin of its name '*polarizable*'. In the polarizable ion model, introducing the dumping function $f_{ab}(r)$ which varies from 0 to 1, the dipole of ion *a* by the surrounding ion *b* is expressed as follows,

$$p_a = \alpha_a \{1 - f_{ab}(r)\}\{(z_b e / r^2) + (2\alpha_b z_a e / r^5)\} / \{1 - (4\alpha_a \alpha_b / r^6)\} \tag{62}$$

which means $p_a(r)$ diverges at $r_c^6 = (4\alpha_a\alpha_b)$, where α_i is the polarizability of ion i. This singular point is called *the polarization catastrophe distance*. This divergence is ascribed to the fact that the iteration has been done disregarding that r_{ij} is required much longer than the distance between the positive and negative dipole charges. Therefore, strictly speaking, the equations of polarizable model are valid only in the case when the polarizability of the ion *a* or *b* is almost zero. In the molten state of Ag or Cu halide including super ionic conductor, e.g. AgI, AgBr, Ag_2Se, Ag_2Te, CuBr etc., the dipole moments of cations are considerably smaller than those of anions, hence $r_c^6 \sim 0$ (see, e.g. tables in Kittel,1996). Consequently, the condition $\alpha_+ \sim 0$ would be strictly applicable to the pair potentials $\varphi_{+-}(r)$ and $\varphi_{++}(r)$. From the considerations based on these facts, the polarizable ion model has been applied to AgBr and AgCl (Bitrián and Trullàs, 2006 ; Wilson et al., 1996).

4.2 Evaluation of polarizable ion model in the previous section

The theory that Bitrián and Trullàs (2006) have developed is effective to revise the rigid ion model to some extent. However, as we have pointed out in the previous section, such dispersion called *the polarization catastrophe* is ascribed to the fact that the iteration has been done disregarding that the inter ionic distance r_{ij} is required to be much longer than the distance between the positive and negative dipole charges. In fact, for example, the polarization catastrophe distance in molten NaCl estimated from the electron polarizabilities of Na and Cl ions is 0.93A. Of course the actual negative increasing of potential to the catastrophe would be relaxed to some extent, as the short range damping factor $f_{ab}(r)$ rapidly approaches to 1. In spite of these facts, it is quite doubtful whether that is the effective theoretical value when *r* approaches to the nearest ionic distance.

Because the $f_{ab}(r)$ is used as a parameter, the following *a prior* expression seems to be more preferable, as,

$$\varphi_{ab}(r) = \varphi_{ab}{}^{0}(r) - P_{ab} / r^4 = \varphi_{ab}{}^{0}(r) - (1/2)\{1 - f_{ab}(r)\}^2 (\alpha_a z_b{}^2 e^2 + \alpha_b z_a{}^2 e^2) / r^4 \quad (63)$$

Though $f_{ab}(r)$ in Eq.(63) is different from those Wilson et al. (1996), or Bitrián and Trullàs (2006) have defined, the function which satisfies the following condition will be appropriate; when *r* approaches to 0, then $f_{ab}(r)$ approaches to 1; when *r* increases (about twice of the inter ionic distance), $f_{ab}(r)$ approaches to 0. In any case, by setting $f_{ab}(r)$ as Eq.(63), the repulsive part of the potential will be more relaxed by taking into account the last term in Eq.(63), the charge dipole interaction, in the range where the repulsive potential is effective.

4.3 Polarization and dielectric constant

To discuss the screened inter ionic potential by the dielectric function, we will briefly summarize the relation between the polarization and the dielectric constant. Suppose that the local electric field at the position of the ion is E_{loc}, the contribution to the polarizability is ordinary classified to the following three types; (*a*) the electron distribution change of the atomic core which occurs inside the ion. It has been discussed in the previous sections; (*b*) the contribution from ions, which occurs from the relative configuration of other neighbouring ions; (*c*) if the ion has the permanent electric dipole, its contribution should be considered, especially in the case that the ion consists of plural atoms. However, it is unnecessary to be taken into account in the case of molten alkali chloride or molten carbonate.

In the solid phase, if the frequency analysis is applied to the local electric field E_{loc}, (c) vanishes at about the micro wave length; next (*b*) disappears at the infrared region; (*a*) lasts to the end. In this discussion, we pay attention to (*a*) and (*b*).

According to the linear response theory, the relation between the electron polarizability discussed in the previous sections and the dielectric constant corresponding to it is expressed as,

$$1 - (1/\varepsilon(q)) = 4p\alpha(q) \quad (64)$$

Therefore, it is possible to obtain the effective inter ionic potentials in molten salts by the calculation of the term *(a)* and *(b)* using the polarization or the dielectric constant.

4.4 Our final form of inter-ionic potentials in molten salts

The dielectric function of the attractive part of the inter ionic potential of molten salts have been derived (Koishi et al., 2007; Matsunaga et al. 2007, 2008), and the theory have been extended to the polarizable model (Matsunaga and Tamaki, 2011). We briefly describe the procedure as follows. The Fourier component of the Coulomb potential in $\varphi_{ij}^0(r)$, i.e. $FT[z_i z_j e^2/r]=4\pi z_i z_j e^2/q^2$ is screened by the dielectric function, as,

$$\{\text{the FT result of the screened Coulomb term in } \varphi_{ij}^{\ 0}(r)\} = 4\pi z_i z_j e^2 / q^2 \varepsilon(q) \tag{65}$$

As will be described in the next section, $\varepsilon(q)$ is related to the charge-charge structure factor $Szz(q)$. In other words, $\varepsilon(q)$ is introduced to the pair potential as, so to say, a factor of the many body effect, or the environmental factor. Adding the repulsive term $(\zeta H_{ij}/r^n)$ and the dipole-dipole interaction term $(-\xi C_{ij}/r^6)$ in $\varphi_{ij}^0(r)$ with the simplified reduction constants ζ and ξ to (65), the promise inter-ionic potentials in *a-b* binary molten salt screened by the many-body force are expressed as,

$$\begin{aligned}\varphi_{ij}(r) \equiv \text{Inv.FT}[4\pi z_i z_j e^2 / q^2 e(q)] + \zeta H_{ij} / r^n - (1/2)\{1 - f_{ij}(r)\}^2 (\alpha_i z_j e^2 + \alpha_j z_i e^2) / \\ / r^4 - \xi C_{ij} / r^6 \qquad (i,j = a,\ b)\end{aligned} \tag{66}$$

where Inv.FT means the inverse Fourier transformation. The repulsive term in (66) can be replaced by the exponentially decaying functions.

Provided that the following function Eq.(67) is adopted as the decaying factor $f_{ab}(r)$, the desirable condition will be satisfied; as r approaches to 0, then $f_{ab}(r)$ approaches to 1; as r increases, then $f_{ab}(r)$ approaches to 0.

$$f_{ab}(r) = 2 / \{\exp(+k_{ab} r) + \exp(-k_{ab} r)\} \tag{67}$$

where k_{ab} is a variable parameter, e.g. if $f_{+-}(r) \sim 0.2$ is required when the inter ionic distance is about 3 A, the estimated value is $k_{+-} \sim 0.77\ A^{-1}$. Regarding $\varepsilon(q)$, the detailed discussion can been found in our previous works (Koishi et al., 2007; Matsunaga et al. 2007, 2008). As we have stated so far, the polarizable ion model in the short-range distance has been adopted in Eq.(66) where the repulsive potential (in both case $1/r^n$ function form and the exponentially decaying function) is dominant. On the other hand, the screening contribution is adopted in the long-range distance where the Coulomb attractive potential is dominant. Thus the plausible inter ionic potential can be obtained, as it was, the polarized and screened inter-ionic potentials model (PSIPM). We will demonstrate the example in the next section.

4.5 Application to molten CuI

In this section, we will show the example of application of the new theory to molten CuI. In molten CuI, the anomaly approach in Cu-Cu distribution is known, which is not reproduced by the classical MD (Waseda et al., 2000). Firstly, the outline of the theory of the dielectric screening that we have developed in the previous section is briefly described as follows for the readers' benefit (Koishi et al., 2007; Matsunaga et al. 2007, 2008). As stated before, the Coulomb potential is screened by the dielectric constant $1/\varepsilon(q)$. $\varepsilon(q)$ is expressed using the charge-charge structure factor, $S_{ZZ}(q)$, which is defined as follows (Hansen and McDonald, 1986),

$$S_{ZZ}(q) = < r_q^{\ Z} r_{-q}^{\ Z} > /N = \Sigma_i \Sigma_j z_i z_j S_{ij}(q) \tag{68}$$

where z_i and z_j are the effective charge of ions. $S_{ij}(q)$ is the partial structure factor, i.e. the Fourier transformation of the pair distribution function of ion i and j, $g_{ij}(r)$, which is expressed as follows,

$$S_{ij}(q) = x_i\delta_{ij} + nx_ix_j \int_0^\infty (\sin qr/qr)\{g_{ij}(r) - 1\}4\pi r^2 dr \tag{69}$$

The relation between $1/\varepsilon(q)$ and $S_{ZZ}(q)$ is expressed as,

$$\frac{1}{\varepsilon(q)} = \frac{1}{1+\delta\left(\kappa^2/q^2\right)S_{ZZ}(q)} \tag{70}$$

where $\kappa_s^2 = 4\pi e^2\beta n$, $\beta = 1/k_BT$, and n is the number density of constituent ions. The parameter δ is introduced to reduce the screening effect by the structure anisotropy as seen in molten noble metal halide. The value of δ is supposed to be in the range of $0 < \delta \leq 1$. The condition $\delta = 1$ may occur in a well-symmetric configuration like molten alkali halide. In the region of high values of $q > 10A^{-1}$, $S_{ZZ}(q) \cong 1$ is valid, hence Eq.(70) will be written as,

$$\frac{1}{\varepsilon_i(q)} = \frac{1}{1+\delta\left(\kappa^2/q^2\right)} \tag{71}$$

As we have describes in the previous section, the screened potentials can be obtained using $\varepsilon(q)$. In our previous work, to confirm the validity of the obtained potential, we have calculated the pair distribution function by the potential mean force (Hansen and McDonald, 1986), which is expressed as,

$$g_{ij}(r) = exp\left[-U^{ij}(r)/k_BT\right] \tag{72}$$

However, the screened potential $\varphi_{ij}(r)$ with the small reduction parameters multiplied on the repulsive term was used for $U^{ij}(r)$, and the calculation was restricted to the cation and anion pair.

On this stage, we try to make a slight modification of the interpretation of the theory. We consider two types of screened potentials; the first one, $\varphi_{ij}^S(r)$, is the screened potential that we have discussed in the previous section, which is affected by the structure through $S_{ZZ}(q)$; the second one, $\varphi_{ij}^I(r)$, is so to say, the 'ideal' screened potential with $\varepsilon_i(q)$ in Eq.(71), which expresses the effect from the uniform back ground. The potentials are written as,

$$\varphi_{ij}^S(r) \equiv \text{Inv.FT}[4\pi z_i z_j e^2 / q^2\varepsilon(q)] + \zeta^S H_{ij} / r^n - (1/2)\{1 - f_{ij}(r)\}^2(\alpha_i z_j e^2 + \alpha_j z_i e^2) / \\ /r^4 - \xi^S C_{ij} / r^6 \qquad (i,j = a, b) \tag{73}$$

$$\varphi_{ij}^{I}(r) \equiv \text{Inv.FT}[4\pi z_i z_j e^2 / q^2 \varepsilon_i(q)] + \zeta^I H_{ij} / r^n - (1/2)(\alpha_i z_j e^2 + \alpha_j z_i e^2) / \\ /r^4 - \xi^I C_{ij} / r^6 \qquad (i,j = a, b) \tag{74}$$

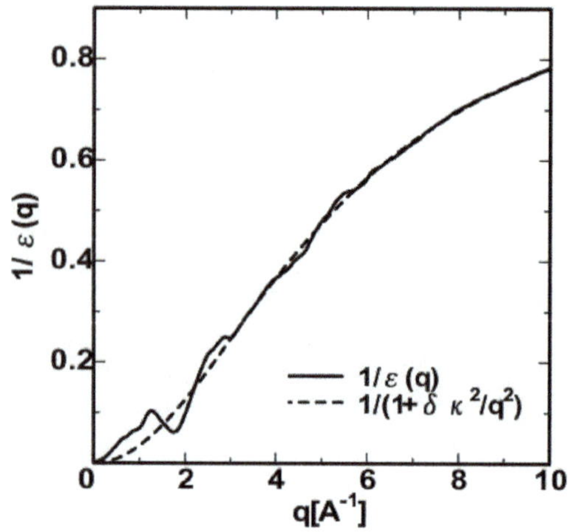

Fig. 11. $1/\varepsilon(q)$ of molten CuI with $1/(1+\delta(\kappa_s^2/q^2))$.

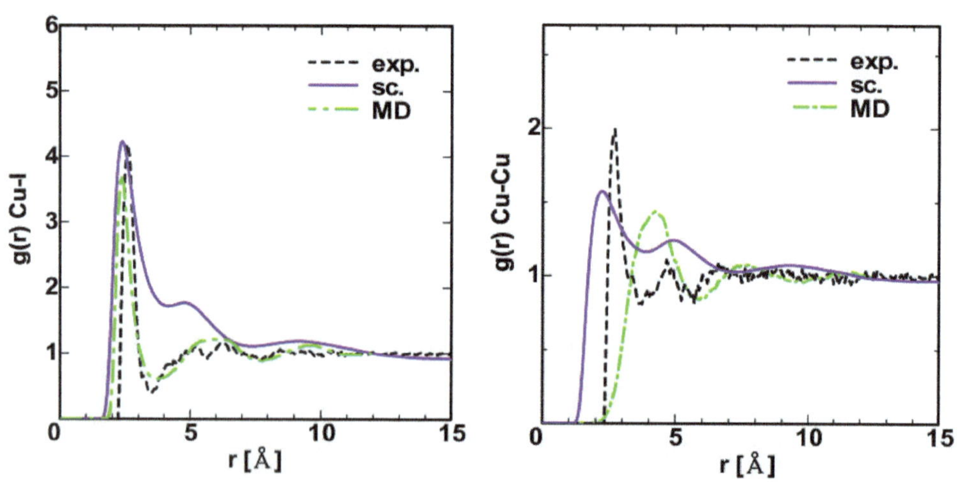

Fig. 12. $g_{CuI}(r)$ (left) and $g_{CuCu}(r)$ (right) calculated using the screened potential with that obtained by experiment and by MD.

We take $U^{ij}(r)$ for the potential of mean force in Eq.(72) as the deviation from the ideal screening, as,

$$U^{ij}(r) \equiv -\Delta\varphi_{ij}(r) = -|\varphi_{ij}^{S}(r) - \varphi_{ij}^{I}(r)| \tag{75}$$

This interpretation seems to be appropriate because the effective mean force is originally defined for the weak interaction between particles, which corresponds to $\Delta\varphi_{ij}(r)$ in this case. In the numerical calculation, we adopt the value of δ in Eq.(70) and Eq.(71) as 0.4, $\Delta\zeta = |\zeta^S - \zeta^I| = 0.12$. C_{CuCu} and C_{CuI} are 0 in Eq.(73) and Eq.(74).

The obtained $g_{+-}(r)$ and $g_{++}(r)$ are shown in Fig.12 with $g_{ij}(r)$ obtained by experiment (Waseda et al., 2000), and MD using the original VR type potentials (Stafford et al., 1990). The first peak of $g_{+-}(r)$ and the phase of oscillations almost agree well, though the second peak height is not satisfactory reproduced. The first peak position of $g_{++}(r)$ obtained by the experiment suggests the anomalous closer distribution of Cu ions, which has been reproduced by the screened potentials and the effective mean force to some extent, except their peak heights and the oscillation phase of the third peak. On the other hand, the first peak of $g_{++}(r)$ obtained by MD obviously disagrees with the experiment. The discrepancy between $g_{ij}(r)$ by the experiment and that obtained by the screened potential may be caused by several reasons, besides a margin of error in the Fourier transformation. One of the expected reasons is that the non-symmetric structure in solid phase remains in molten phase, which may obstruct the screening effect in molten phase. Despite these facts, however, it may be suggested that the treatment described so far is physically significant, and the screening effect has the obvious contribution to the obtained results.

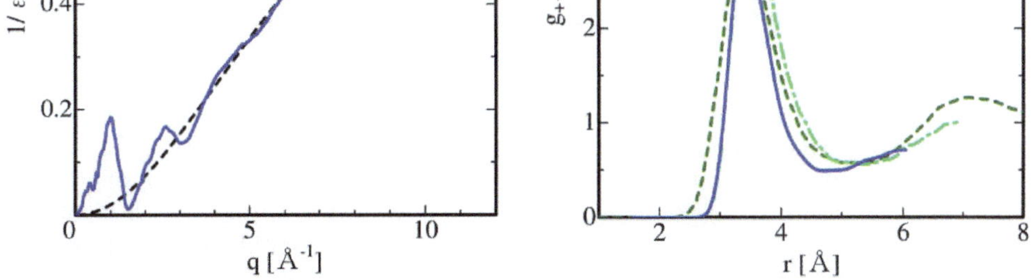

Fig. 13. 1/ε(q) of molten CsAu with $1/(1+\delta(\kappa_s^2/q^2))$ (left). $g_{+-}(r)$ calucaed using the screened potential with that obtained by MD and QMD.

4.6 Application to molten CsAu

Finally, we wish to show the application to molten CsAu system. Molten CsAu is a typical example which exibits a transition from a metallic to non-metallic state around a stoichiometric composition in molten phase, though both components are metallic conductors. Molten CsAu at the equiatomic composition exhibits a very low conductivity like molten salts (Schmutzler et al., 1976). These characteristics have been interpreted that a chemical short range order in the molten state are formed, in other words the molten CsAu alloy is composed of Cs^+ and Au^-. Structural measurements for molten CsAu systems using the neutron diffraction were performed (Marchin et al., 1980). Molten CsAu alloys have some structural characteristics of the short range order indicating a typical ionic compound. But partial structure factors have not yet been obtained experimentally.

Many attempts have been made to investigate the three dimensional configurations of the molten CsAu alloys by considering the partial structure factors and the pair distribution functions. The charged hard sphere model with the mean spherical approximation is mainly used for this purpose (Evans & Telo da Gama,1980; Holzhey et al., 1982). Hoshino indicated that the partial structure factors and the entropy of mixing depend strongly on the amount of charge transfer (K. Hoshino, 1984). Both of their consequences showed that the partial structure factors and the pair distribution functions of molten CsAu alloy have the same characteristics as those observed in molten alkali halides. However, the pair distribution functions obtained by the Monte Carlo simulation are very different from those obtained by the charged hard sphere model; Au^- - Au^- distance indicates the most closest approach (Costa Cabral et al., 1992). In this situation, we performed molecular dynamics simulation using the Tosi-Fumi type potentials to examine the structural and transport properties in molten CsAu (Matsunaga, 2000, 2001). The obtained structure of molten CsAu shows the similar characteristics to CsCl structure. On the other hand, the recent *ab initio* calculation in molten CsAu shows the good agreement in the pair distribution function of Cs-Au, $g_{CsAu}(r)$, with the molecular dynamics result, however there are discrepancies in $g_{AuAu}(r)$ and $g_{CsCs}(r)$ (Charpentier and Clérouin, 2008). This result may suggest the existence of the anomalous inter-ionic interactions in molten CsAu like molten CuI. This fact prompts us to examine the dielectric screening effect in molten AuCs. The calculation procedure is similar to that described before. The obtained calculation results are shown in Fig.13. It can be seen in Fig.13 that $g_{CsAu}(r)$s obtained by three different methods agrees well to some extent.

5. Conclusion

In this book chapter, we have re-examined the theory of the dielectric screening, and summarized the derivation of the new expression of the dielectric function. The extension of the theory to the potential mean force, and the deviation from the Nernst-Einstein relation has also been discussed. Then, we have reviewed the application studies of the new theory to the binary and ternary molten salts, i.e. alkali halides NaCl, RbBr, noble metal halides AgBr, CuBr, and AgBr-AgI mixture. The concentration-concentration fluctuation for the multi-component systems $S_{CiCj}(q)$ (Bhatia & Ratti, 1977) has also been discussed in relation with the many body effect and the short range order. We have applied the theory to the poralizable ion model. The new expression of the potential which enables to avoide the polarization catastrophe has been proposed. The model has been applied to molten CuI and CsAu systems.

6. Acknowledgment

One of the authors (SM) is grateful to the Ministry of Education, Science and Culture for Financial support of Grant-in-Aid for Science Research. One of the authors (ST) expresses his thanks to Professors S. Takeno and M. Kusakabe of Niigata Institute of Technology for their helpful instruction for a mathematical treatment. He also wishes to express his cordial thanks to Professor W. H. Young for fruitful comments on this subject.

7. References

Berne, B. & Rice, S. A. (1964). On the Kinetic Theory of Dense Fluids. XVI. The Ideal Ionic Melt. *Journal of Chemical Physics.,Vol.* 40, Issue 5, (March 1964), pp. 1347-1362; DOI:10.1063/1.1725318

Bhatia, A. B. & Ratti, V. K. (1977). Number-Concentration structure factors and their long wavelength limit in multicomponent fluid mixtures. *Physics and Chemistry of Liquids*, Vol.6, Issue 3, (1977), pp.201-213, DOI:10.1080/00319107708084140

Bhatia, A. B. & Thornton, D. E. (1970). Structural Aspect of the Electrical Resistivity of Binary Alloys. *Physical Review B*, Vol.2, (October 1970), pp.3004-3012, DOI:10.1103/PhysRevB.2.3004

Bitrián, V. & Trullàs, J. (2006). Molecular Dynamics Study of Polarizable Ion Models for Molten AgBr. *Journal of Physical Chemistry*, Vol.110, Issue 14 (2006) pp.7490–7499, DOI: 10.1021/jp056818u

Chandra, S. (1981). *Superionic Solids: Principles and Applications*, Elsevier Science Ltd, ISBN: 0444860398, Amsterdam

Charpentier, N. & Clérouin, J. (2008). *Ab initio* simulations of the liquid alloy Au-Cs. Physical Review B, Vol.78, Issue 10, (September 2008), pp.100202-1-100202-2, DOI:10.1103/PhysRevB.78.100202

Costa Cabral, B. J.; Natalia, M.; Cordeiro, D. S. & Telo da Gama, M. M. (1991). The structure of molten CsAu: ab initio and Monte Carlo study. *Journal of Physics: Condensed Matter*, Vol.3, No.29, (July 1991), pp.5615-5620, DOI:10.1088/0953-8984/3/29/014

Edwards, F. G.; Enderby, J. E.; Howe, R. A. & Page, D. I. (1975). The structure of molten sodium chloride. *Journal of Physics C: Solid State Physic*, Vol.8, No.21, (November 1975), pp.3483-3490, DOI:10.1088/0022-3719/8/21/018

Endo, K.; Ida, T.; Kimura, J.; Mizuno, M.; Suhara, M. & Kihara, K. (1999). Structural analysis of $Ag_xCu_{1-x}I$ ($0.75 \leqslant x < 1.00$) in the superionic phase using solid NMR and X-ray diffraction methods. *Chemical Physics Letters*, Vol.308, Issues 5-6, (July 1999), pp.390-396, DOI:10.1016/S0009-2614(99)00631-4

Evans, R. &Telo Da Gama, M. M. (1980). Structural evidence that molten CsAu is ionic. *Philosophical Magazine Part B*, Vol.41, Issue 3, (1980), pp.351-356, DOI:10.1080/13642818008245391

Gartrell-mills, P. R. & McGreevy, R. L. (1989). Monte-Carlo Simulation of Molten CsCl Using a 'Deformation Dipole' Polarisable Ion Potential. *Molecular Simulation*,Vol.2, Issue 3, (1989), pp. 209-216, DOI:10.1080/08927028908031369

Hansen, J.P. &McDonald, I.R. (1986). *Theory of Simple Liquids* 2nd ed, Academic Press, New York.

Holzhey, C.; Brouers, F.; Franz, J. R. & Schirmacher,W. (1982). The theory of metal-non-metal transitions in gold alkali liquid alloys. *Journal of Physics F: Metal Physics,* Vol.12, No.11, (November 1982), pp.2601-2610, DOI:10.1088/0305-4608/12/11/018

Kittel, C. (1996) *Introduction to Solid State Physics 7th Edition*, John Wiley & Sons, Inc., New York, ISBN: 9780471111818

Koishi, T. & Tamaki, S. (2005). A theory of transport properties in molten salts. *Journal of Chemical Physics.* Vol.123, Issue 19, (November 2005), pp.194501-1 - 194501-11, DOI: 10.1063/1.2102901

Koishi, T.; Saito, M.; Matsunaga, S. & Tamaki, S. (2007). Dielectric screening properties in molten salts. *Phys. Chem. Liq.* Vol.45, No.2, (April 2007), pp.181-196, ISSN 0031-9104

Locke, J.; Messoloras, S.; Stewart, R. J.; McGreevy, R. L. & Mitchell, E. W. J. (1985). The structure of molten CsCl. *Philosophical Magazine Part B*,Vol.51, Issue 3, (1985), pp. 301-315, DOI:10.1080/13642818508240576

March, N.H. & Tosi, M. P. (1976). *Atomic Dynamics in Liquids.* Macmillan, London, ISBN: 0333112784

Martin, W.; Freyland, W.; Lamparter, P. & Steeb, S. (1980). Structure and Density of Gold-Cesium-Melts. II. Neutron Diffraction with Molten Gold-Cesium-alloys. *Physics and Chemistry of Liquids,* Vol.10, Issue 1, (1980), pp. 61-76, DOI:10.1080/00319108008078457

Matsunaga, S.; Ishiguro, T. & Tamaki, S. (1983). Thermodynamic properties of liquid Na-Pb alloys. *Journal of Physics F: Metal Physics,* Vol.13, No.3, (March 1983), pp.587-596, DOI:10.1088/0305-4608/13/3/009

Matsunaga, S. (2000). Structural Study on Liquid Au–Cs Alloys by Computer Simulations. *Journal of the Physical Society of Japan*, Vol.69, No.6, (June 2000), pp.1712-1716, DOI: 10.1143/JPSJ.69.1712

Matsunaga, S. (2001). Transport Properties in Liquid AuCs Alloy. *Journal of the Physical Society of Japan,* Vol.70, No.12, (December 2001) pp.3591-3595, DOI: 10.1143/JPSJ.70.3591

Matsunaga, S. (2003). Structural Properties in Ag3SI: A Molecular Dynamics Study of Superionic and Molten Phases. *Journal of the Physical Society of Japan*, Vol.72, No.6, (June 2003), pp.1396-1402, DOI: 10.1143/JPSJ.72.1396

Matsunaga, S. & Madden, P.A. (2004). Structural and transport properties in the Ag3SI system: a molecular dynamics study of alpha, beta and molten phases. *Journal of Physics: Condensed Matter*, Vol.16, (January 2004), pp. 181-194, DOI:10.1088/0953-8984/16/3/001

Matsunaga, S. (2005). Structural and transport influence of dissolving AgBr into AgI in super ionic and molten phases by molecular dynamics simulations. *Solid State Ionics*, Vol.176, Issues 23-24, (July 2005), pp.1929-1940, DOI:10.1016/j.ssi.2005.04.045

Matsunaga, S.; Saito, M.; Koishi, T. & Tamaki, S. (2007). Dielectric screening effects in molten AgI-AgBr system. *Molecular Simulation*, Vol.33, Nos 1-2, (January–February 2007), pp.153-158, ISSN 0892-7022

Matsunaga, S.; Saito, M.; Koishi, T. & Tamaki, S. (2008). Dielectric screening properties in molten noble-metal halides. *Journal of Alloys and Compounds*, Vol.452, (March 2008), pp.182-187, DOI:10.1016/j.jallcom.2006.12.161

Matsunaga, S. & Tamaki, S. (2008). Premelting phenomena in ionic crystals. *Journal of Physics: Condensed Matter*, Vol.20, No.11, (March 2008), pp.114116-1-114116-9, DOI:10.1088/0953-8984/20/11/114116

Matsunaga, S. & Tamaki, S. (2008). Hetero-phase fluctuations in the pre-melting region in ionic crystals. *The European Physical Journal B*, Vol.63, No.4, (June 2008), pp.417-424, DOI: 10.1140/epjb/e2008-00245-3

Matsunaga, S. (2009). Structural features of superionic phase in AgBr-CuBr system by molecular dynamics simulation. *Journal of Physics: Conference Series*, Vol.144, No.1, (January 2009), pp.012011-1-012011-4, DOI:10.1088/1742-6596/144/1/012011

Matsunaga, S. (2009). Structure and Atomic Dynamics of Silver Halide Mixtures. *Progress of Theoretical Physics Supplement*, No. 178 (2009) pp. 113-119, DOI:10.1143/PTPS.178.113

Matsunaga, S. & Tamaki, S. (2011). On the inter ionic potentials. *EPI Web of Conferences*, (May 2011), Vol.15, pp.02010-p.1 -02010-p.6, DOI: 10.1051/epjconf/20111502010

Parrinello, M.; Rahman, A. & Vashishta, P. (1983). Structural Transitions in Superionic Conductors. *Physical Review Letters*, Vol.50, Issue 14, (April 1983), pp.1073–1076, DOI:10.1103/PhysRevLett.50.1073

Saito, M.; Park, C.; Omote, K.; Sugiyama K. & Waseda Y. (1997). Partial Structural Functions of Molten CuBr Estimated from the Anomalous X-Ray Scattering Measurements. *Journal of the Physical Society of Japan*, Vol.66, No. 3, (March 1997), pp. 633-640, DOI: 10.1143/JPSJ.66.633

Saito, M.; Kang, S.; Sugiyama, K. & Waseda, Y. (1999). Partial Structural Functions of Molten AgBr Estimated from the Anomalous X-Ray Scattering Data Coupled with Neutron Diffraction. *Journal of the Physical Society of Japan*, Vol.68, No.6, (June 1999), pp. 1932–1938, ISSN 0031-9015

Schmutzler, R. W.; Hoshino, H.; Fischer, R. & Hensel, F. (1976). Nonelectronic Electrical Transport in Liquid CsAu. *Berichte der Bunsengesellschaft für Physikalische Chemie*, Vol.80, Issue 2, (February 1976), pp.107–113, DOI: 10.1002/bbpc.19760800203

Stafford, A. J.; Silbert, M.; Trullàs, J. & Giró, A. (1990). Potentials and correlation functions for the copper halide and silver iodide melts. I. Static correlations. *Journal of Physics: Condensed Matter*, Vol.2, No.31, (August 1990), pp.6631-6642, DOI:10.1088/0953-8984/2/31/016

Tasseven, Ç.; Trullàs, J.; Alcaraz, O.; Silbert, M. & Giró, A. (1997). Static structure and ionic transport in molten AgBr and AgCl. *Journal of Chemical Physics*. Vol. 106, Issue 17, (May 1997), pp.7286-7294 , doi:10.1063/1.473690

Vashishta, P. & Rahman, A. (1978). Ionic Motion in α-AgI. *Physical Review Letters*, Vol.40, Issue 20, (May 1978), pp. 1337–1340, DOI:10.1103/PhysRevLett.40.1337

Waseda, Y.; Kang, S.; Sugiyama, K.; Kimura, M. & Saito, M. (2000). Partial structural functions of molten copper halides CuX (X = Br, I) estimated from the anomalous x-ray scattering measurements. *Journal of Physics: Condensed Matter*, Vol.12, (February 2000), A195-A202, DOI:10.1088/0953-8984/12/8A/323

Wilson, M.; Madden, P. A. & Costa-Cabral, B. J. (1996). Quadrupole Polarization in Simulations of Ionic Systems: Application to AgCl. *the Journal of Physical Chemistry*, Vol.100, Issue 4, (January 1996), pp 1227–1237, DOI: 10.1021/jp9512319

Part 2

Resource

2

Pb Isotope Signatures of Polymetallic (Au-Cu-Zn) Deposits of the SW Amazonian Craton and Their Relation to Crustal Evolution

Mauro C. Geraldes
Department of Geology, Rio de Janeiro State University
Rio de Janeiro-RJ
Brazil

1. Introduction

The distribution of ore deposits through geological time allows defining the correlation between magmatic, metamorphic and tectonic events and mineralizing processes. Among geochronologic methods, Pb isotope geochemistry is a powerful tool in helping to solve problems of metallogenesis, because they provide informations about the origin of the fluids responsible for the metal concentration (Changkakoti et al., 1986; Crocetti et al., 1988; Deloule et al., 1989; Kerrich, 1989; Kerrich, 1991; Wilton, 1991). As the isotopic ratios yielded a time-integred record of the U/Pb and Th/Pb ratios of the sources in which the lead developed, they also shed light upon the time-dependent geochemical behavior of these elements within the lithosphere (Cumming and Richards, 1975; Doe and Zartman, 1979; Zartman and Doe, 1981; Zartman and Haines, 1988). This work, based mainly on Pb isotopes constitutes a preliminary attempt to correlate the Proterozoic crustal evolution with metalogenetic processes in the southwestern Amazonian craton.

Reported works using U-Pb and Sm-Nd methods in the SW Amazonian craton had led to better understanding about the framework of the Proterozoic terranes (Teixeira et al., 1989, Tassinari and Macambira, 1999). The craton has been divided (Figure 1) into two major domains: the Archean nuclei in which is included the Central Amazonian Province; and the Proterozoic Provinces, represented by the Maroni-Itacaiúnas (ca. 2.2 Ga), the Ventuari-Tapajós (1.95-1.80 Ga), the Rio Negro-Juruena 1.79-1.52 Ga), the Rondonian-San Ignácio (1.51-1.34 Ga), and the youngest Sunsás-Aguapeí (1.24-1.00 Ga).

These provinces have been divided into orogenic belts. In this way, the 1.79-1.74 Ga Alto Jauru and the 1.58-1.52 Ga Cachoeirinha magmatic arcs (Van Schmus et al., 1999; Geraldes et al., 2001, respectivelly) represent the Rio Negro/Juruena Province. The Rondonian/San Ignácio Province is marked by important events involving magmatic arc settings and continental collision processes between 1.51 Ga and 1.34 Ga. These comprise the 1.51-1.48 Ga Rio Alegre, the 1.45-1.42 Ga Santa Helena and the 1.42-1.32 Ga San Ignácio arcs (Matos et al., 2004; Geraldes et al., 2001; Geraldes et al., 2004 respectively). Finally, the youngest Sunsás/Aguapeí Province comprises sequences deposited during basin tectonic (1.1 Ga Nova Brasilândia and 1.0 Ga Aguapeí Group; Rizzotto et al., 1999; Geraldes et al., 1997, respectively) and magmatic products (1.0 Ga. Sunsás; Litherland et al., 1986).

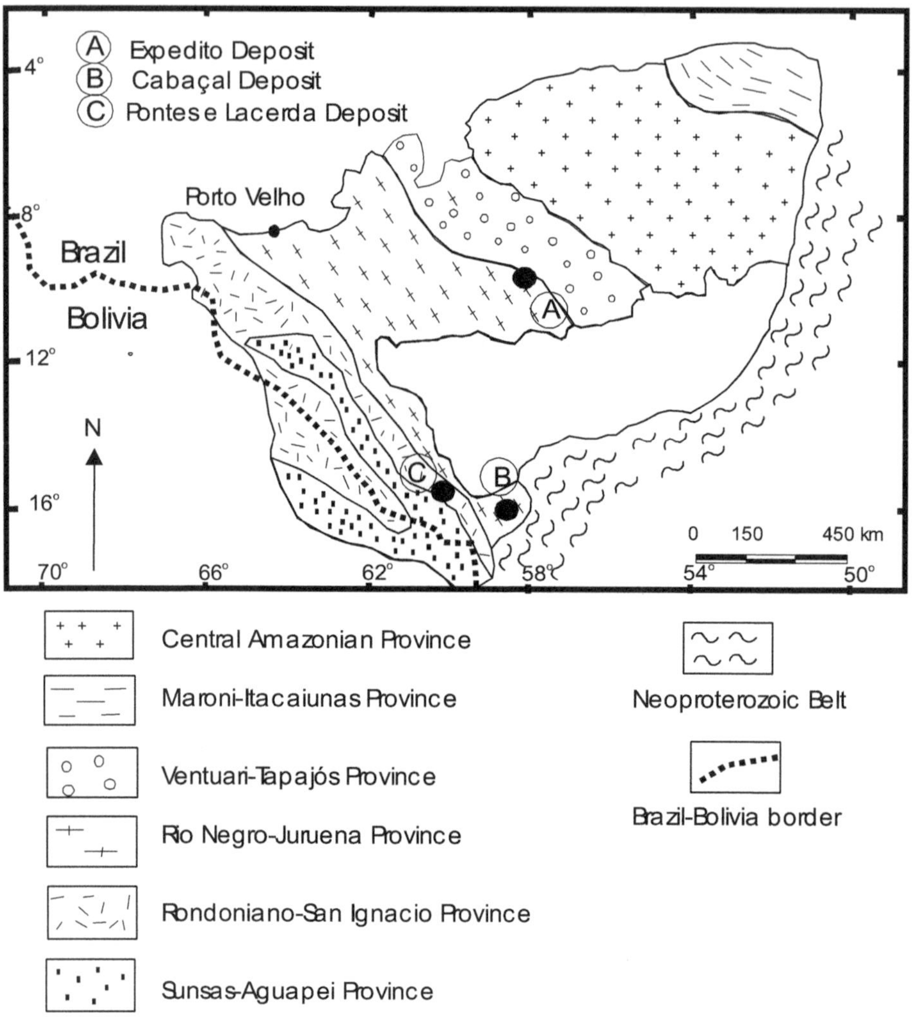

Fig. 1. Mineral deposits location and Geochronological Provinces of the southern sector of the Amazonian craton. The Cabaçal, Expedito and Pontes e Lacerda deposits are located in the map.

The study of SW part of the Amazonian craton Au, Cu and Sn deposits suggests a strong correlation between the time period of the tectonic events and the formation age of mineral concentrations of economic importance (Tassinari and Melito, 1994) and include polymetallic veins, magmatic and VMS types (Souza, 1988; Silva and Rizzoto, 1994; Dardenne and Schobbenhaus, 2000). Paleoproterozoic terranes contain the Moriru Au deposit (related to felsic 1796-1773 Ma volcanic rocks), the Expedito Cu-Au deposit (comprised of a thick pile of 1762-1755 Ma acidic to intermediate volcanic rocks; Pinho et al., 2001) the Cabaçal Zn-Au ore deposit (hosted by ca. 1750 Ma felsic volcanic and volcanoclastic rocks; Pinho et al., 1997; Toledo, 1997). Mesoproterozoic terranes contain the Puquio Norte Au deposit (hosted in Mesoproterozoic greenstone belt; Sáens, 2002); the Rondônia Tin Province (comprised of bimodal intraplate rapakivi suites); and the Au deposits of Pontes e Lacerda (related to the occurrence of a 927-908 Ma NW-SE striking ductile shear zone). The Cachoeirinha, Santa Helena, Rio Alegre, Nova Brasilândia and Sunsás orogens have no associated mineral deposits reported up to now. The only three (Cabaçal, Expedito and Pontes e Lacerda) dated deposits are discussed below.

2. Analytic techniques

The analysis here reported were carried out during the last 10 years as result of several field and laboratorial works. The samples from the three deposits (Cabaçal, Expedito and Pontes e Lacerda) were prepared and analysed at Geochronological Research Center of University of São Paulo, Brazil. Initially the samples were crushed and sift at 10 mesh and the sulfide crystals were separated from quartz by hand-picking in binocular microscopy. 0,2 g of clean crystals was digest in 9 N HCl during two hours (with heating) and the solute was collected. The Pb was purified using collums of exchance resin in HBr and loaded on rhenium filament using the standart silica gelphosphoric acid technique and analysed in a VG 354 solid-source mass spectrometer at 1,250 °C. Reproducibility of 0,24, 0,32 and 0,36 per mil (1σ) for $^{206}Pb/^{204}Pb$, $^{207}Pb/^{204}Pb$ and $^{208}Pb/^{204}Pb$ ratio, respectively, and the mass fractionation level of 1,3 per mil per mass unit difference was determined from repeated measurements. Detais of laboratoty procedures may be found in Tassinari et al., (1990), Iyer et al., (1992) and Tassinari and Cavalcanti (1994).

The international standarts of Pb isotopes used during the laboratory were presented in Table 1, with their respective results.

standarts	$^{206}Pb/^{204}Pb$	$^{207}Pb/^{204}Pb$	$^{208}Pb/^{204}Pb$
NBS981	16,9371	15,49175	36,7213
NBS982	36,7390	17,19971	36,7449

Table 1. Pb isotope analysis of International standards results used in CPGeo-USP.

3. The Cabaçal Au-Zn deposit

The Cabaçal gold deposit is located in the SW Amazonian craton, Mato Grosso State, Brazil, where the Alto Jauru orogenic rocks (U-Pb ages from 1790 Ma to 1744 Ma) and Cachoeirinha orogenic rocks (U-Pb ages from 1580 Ma to 1520 Ma) occur (Pinho et al., 1997; Toledo, 1997; Geraldes et al., 2002; Tassinari et al., 2000). The mineralization is hosted within felsic volcanic and volcanoclastic rock of the Alto Jauru rocks. Detailed petrologic and geochemical investigations (Pinho, 1997) indicate that gold deposition is associated with metamorphic fluids migrating along regional shear zones. Hydrothermal solutions in the Cabaçal deposit originated sericitic, biotitic, and chloritic alteration zones. The ore zones are irregularly-shaped, presenting ondulating outlines in general coincident with the principal foliation plane. The ore planes plunge towards in two different directions: SSW, with variable dip and SSE, dipping about 20°. Sericite and chlorite are common minerals alterations related to the concentration of sulfide minerals. Alterations consists of an inner, chloritized core surrounded by an intermediate biotitic zone and a sericitic zone. Sericitic and chloritic alterations commonly occur associated with volcanic-hosted massive sulfide ores.

Sample	Mineral	206/204	207/204	208/204	ref.
Cabaçal Deposit					
CB-01-BC	pyrite	15,461	15,227	35,154	1
CB-01-BX	pyrite	16,245	15,318	35,924	1
CB-04	pyrite	15,531	15,410	35,633	1
CB-05	carbonate	20,463	15,777	39,833	1
CB-07	pyrite	21,208	15,999	39,936	1
CB-09X	chalcopyrite	18,664	15,709	37,975	1
Expedito Deposit					
F16/219	galena	15,861	15,414	35,575	2
F16/257	galena	15,835	15,44	35,685	2
F25/207	galena	16,004	15,652	36,302	2
513	galena	15,955	15,492	35,833	2
514	galena	16,057	15,731	36,607	2
515	galena	15,846	15,432	35,692	2
516	galena	15,304	15,592	36,202	2
517	galena	15,407	15,396	36,603	2
34123	K-felds	16,736	15,459	36,34	2
Pontes e Lacerda					
ONÇA I (a)	galena	15,606	17,585	36,646	3
ONÇA I (2)	galena	15,499	17,675	36,375	3
ONÇA I (c)	galena	15,523	17,677	36,440	3
ONÇA I (d)	galena	15,645	17,768	36,767	3
ONÇA II (a)	galena	15,522	17,584	36,463	3
ONÇA II (b)	galena	15,528	17,657	36,403	3
ONÇAII (c)	galena	15,638	17,660	36,760	3
ONÇA II (d)	galena	15,574	17,666	36,521	3
ONÇA III(a)	galena	15,562	17,498	36,551	3
ONÇA III(b)	galena	15,539	17,734	36,434	3
ONÇA III(c)	galena	15,539	17,668	36,433	3
ONÇA III (d)	galena	15,681	17,934	36,897	3
PL-Au-O2	Gold	18,007	15,674	37,257	1
PL-Mg 01	Magnetite	21,250	15,827	38,393	1
PL-202	pyrite	20,665	15,668	38,687	1
PL-202	pyrite	20,514	15,763	38,601	1
PL-202	pyrite	29,815	16,321	38,542	1
PL-202	pyrite	20,484	15,692	38,134	1
PL-207 res.	pyrite	29,856	16,303	45,090	1
PL-207 L1	pyrite	21,381	15,553	39,336	1
PL-207 L2	pyrite	27,662	16,131	44,807	1
PL-207 L3	pyrite	21,566	15,758	40,272	1
PL-207 L4	pyrite	22,714	15,791	42,443	1
PL-207 L5	pyrite	28,826	16,279	44,136	1
PL-208 res.	pyrite	20,074	15,678	39,069	1
PL-208 L1	pyrite	18,351	15,554	38,075	1
PL-208 L2	pyrite	20,665	15,723	40,222	1
PL-208 L3	pyrite	21,869	15,817	41,399	1
PL-208 L4	pyrite	26,239	16,069	49,555	1
PL-208 L5	pyrite	23,574	15,979	41,029	1

References: (1) this work; (2) Neder (2002); and (3) Geraldes et al (1997).

Table 2. Pb isotopes data form Cabaçal, Expedito and Pontes e Lacerda deposits.

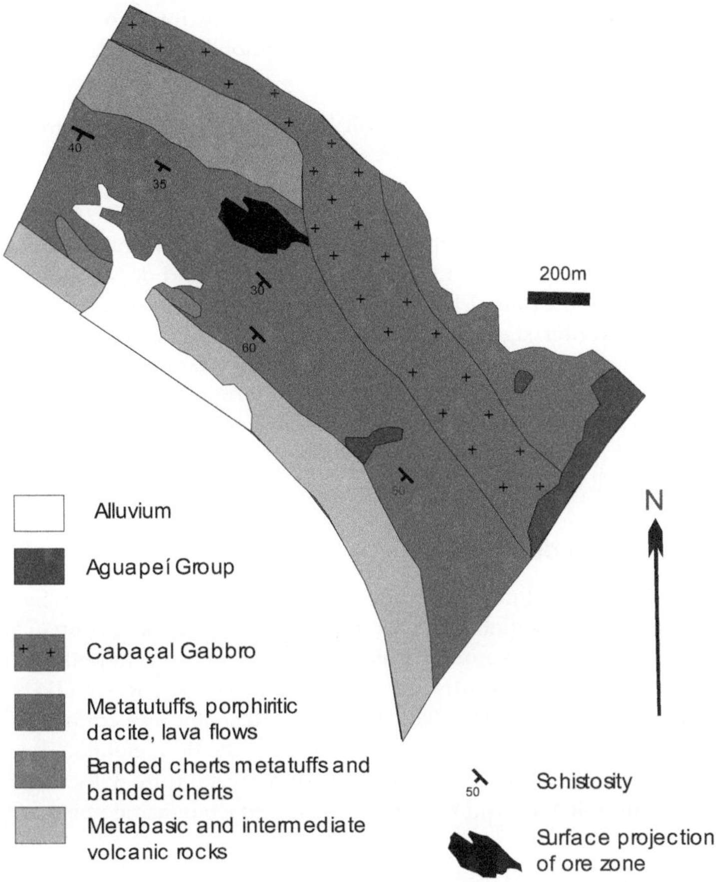

Fig. 2. Local geologic map of the Cabaçal deposit. Adapted from Pinho, (1996).

The ore is represented by concentrations of Cu-Au and Zn-Pb-Cu-Au, where sulfide and selenide minerals, native Bi, and Au-Ag and Au- Bi alloys are important phases. The ore occurs as disseminated, banded, veined, brecciated and massive types. The banded ore shows continuous laminae of sulfide minerals coincident with foliation, and is usually associated with banded tuff, chert and the upper part of the chlorotized zone, suggesting a volcanogenic origin. The veined ore is widespread in all over mineralized area and veins are composed chiefly of milky quartz but carbonate veins also occur. The brecciated ore is represented by fragments of chloritic volcanic rocks, banded cherts, metatuffs, and quartz veins set in a matrix of sulfide minerals. Chalcopyrite is the most common sulfide, followed by pyrite, sphalerite, pyrrhotite, and visible native gold and Bi-Te minerals are common. The massive ore is restricted to the chlorite zone, comprised of chalcopyrite, pyrrhotite and lesser pyrite. Sericites from the hydrothermal zones were dated by the $^{40}Ar/^{39}Ar$ method, using laser step-heating dating in single grains (Geraldes et al., 2003). One sample is from a bore hole 107m deep and yielded a plateau age of 1521.3 ± 1.3 Ma. Another sample is 36.6 m deep, and yielded a plateau age of 1510.4 ± 1.2 Ga. The Ar cooling ages indicate a regional heating during Cachoeirinha suite (U-Pb of 1560-1520 Ma) intrusion which may play an important role in the genesis of the ore deposit due remobilization processes of the original volcanogenic metal concentration.

Pb-Pb signatures for the Cabaçal gold deposit indicate two sources (Geraldes et a., 2003): one more radiogenic ($^{206}Pb/^{204}Pb$ from 15.941 to 16.600 and $^{207}Pb/^{204}Pb$ from 15.527 to 15.600; and $^{208}Pb/^{204}Pb$ from 35.549 to 35.630), and other less radiogenic ($^{206}Pb/^{204}Pb$ from 15.650 to 15.843 and $^{207}Pb/^{204}Pb$ from 15.318 to 15.376 and and $^{208}Pb/^{204}Pb$ from 35.324 to 35.469). These discordant Pb isotope signature may suggest that more than two Pb components were involved in the formation of the Cabaçal gold deposit. The less radiogenc group may indicate a contribution Pb from volcanic-plutonic host-rocks. According to Dean and Carr, (1982) the similar signature between ore and magmatic country rocks may be interpreted as co-magmatic and coeval origin. The second group of Pb values is characterized by strong radiogenic Pb and may be originated from a source that was external to the main volcanic-related hydrothermal systems. Similar Pb signature has been reported by Relvas, et al., (2001) in Iberian Pyrite Belt and interpreted as multiple source of fluids for the ore deposit origin.

4. The Expedito Cu-Au deposit

The Expedito Cu-Zn deposit (located 20 km northern from Aripuanã city) occurs within a thick pile of acidic to intermediate volcanic rocks of the Uatumã Group. These rocks are believed to be related to the Mesoproterozoic intracontinental rift, hypothesis supported by the 1762 Ma dacitic volcanics, exhalative sediments and a co-genetic granitic intrusion about 1755 Ma (SHRIMP U-Pb zircon age; Neder et al., 2002). The unmetamorphosed volcanic rocks are interlayed with chemical and epiclastic sediments. Base metal and gold are hosted in the transition from the fine to coarse ash, lapilli and crystal tuffs with minor intercalation of sericitized feldspatic siltstone and rhyolitic, dacitic and rhyodacitic rocks .

The ore is hosted by dacitic lapilli and crystal tuff interlayered with massive dacitic porphyritic flows, carbonate and chert layers. The deposit consists of several discordant and discontinuous lenses of massive and disseminated pyrrhotite, pyrite, sphalerite, galena, chalcopyrite and arsenopyrite. The deposit is enveloped by a hydrothermal alteration halo consisting of chlorite, biotite and carbonate zones and it is interpreted of volcanic origin according to Neder et al. (2002a). $^{40}Ar/^{39}Ar$ data determined in amphibole and biotite from alteration halo over volcanic rocks yielded 1580-1560 Ma (Neder et al., 2003). Zircon U-Pb dating reported by Rizzotto et a., (2002) for intrusive granites of Aripuanã region yielded 1538 ± 7 Ma, suggesting a partial resetting of the Ar was results of heating caused by the intrusive bodies. These geochronologic data indicate an important remobilization that occurred in the volcanic rocks, which may be responsible for Au concentration in shear zones and peculiar (amphibole) hydrothermal alteration.

Pb isotopic compositions of Expedito deposit were determined from sulfides and the values of $^{207}Pb/^{204}Pb$ range from 15.731 and 15.396; $^{206}Pb/^{204}Pb$ values range from 15.304 to 16.057; $^{208}Pb/^{204}Pb$ values range from 35.575 to 36.607 (Neder et al., 2002b). The Pb data plotted in the Stacey and Krammer (1975) lead growth curve yields a model age abound 1.75 Ga which is roughly coeval to the 1762-1755 Ma U-Pb SHRIMP ages of the volcanic and plutonic rocks of the Aripuanã metalogenetic district.

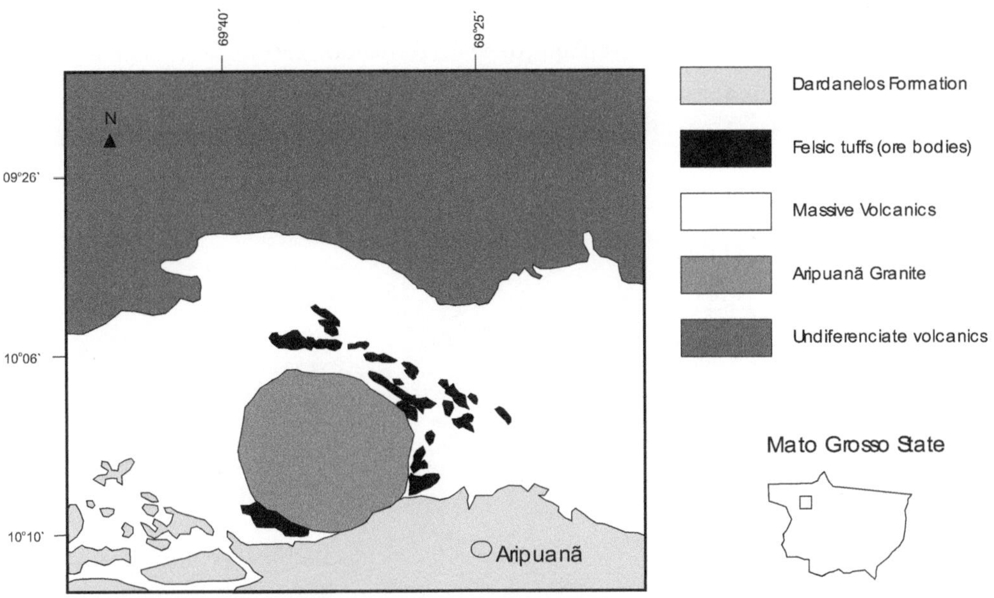

Fig. 3. Geologic map with ores bodies of Expedito deposit. Adapted of Neder (2002).

5. The Au deposits of Pontes e Lacerda region

Gold deposits occur at more than twenty localities in the Pontes e Lacerda region situated between the Guaporé and Jauru rivers in the SW part of Amazonian Craton, State of Mato Grosso, Brazil (Souza, 1988; Saes et al., 1991; and Silva & Rizzoto, 1994; Saes et al., 1994; Geraldes et al., 1997). In the Figure 4 are plotted the following deposits: Ribeiro, Onça, Japonês, Marinho, Lavrinha, Ernesto, Cantina, Pombinha, Nene, João Cumprido and Maraboa. They are distributed along a NW striking, 40 km wide and more than 200 km long shear belt originated during the Middle Proterozoic Aguapeí-Sunsás tectonic event. Mining companies active in the region estimate Ore reserves in ca. 18 t Au.

The sedimentary rocks of the Aguapeí Group include, in the region, metasandstones, metaconglomerates and subordinated metassiltites of the Fortuna Formation overlaid by an intermediate sequence of metapelites (phyllites, low grade siltstone and mudstone, and metapsamite) of the Vale da Promissão Formation, (Figueiredo et al., 1974; Souza & Hildred, 1980) all correlated to the Bolivian Sunsás Group referred to the Middle Proterozoic (1300-950 Ma) by Litherland et al. (1986). Saes & Fragoso Cesar (1994) considered the Aguapeí rock associations and structures as indicative of rifting geological environment which basin had correlation with Sunsas basin in Bolivia.

The Au deposits of Pontes e Lacerda region are related to the occurrence of a NW-SE striking ductile shear zone along which sedimentary rocks of the Aguapeí Group. Tectonics involved oblique overthrusting (which led to formation of recumbent folds and thrusts (pathways for the mineralizing fluids), upright folds and faults with dominant strike-slip component. These unconfomities are potential sites for mineralization as in the main exploted deposits reported: São Vicente deposit (Scabora and Duarte, 1998), Lavrinha deposit (Costa Neto, 1996) and Pau-a-Pique deposit (Fernandes, 1999).

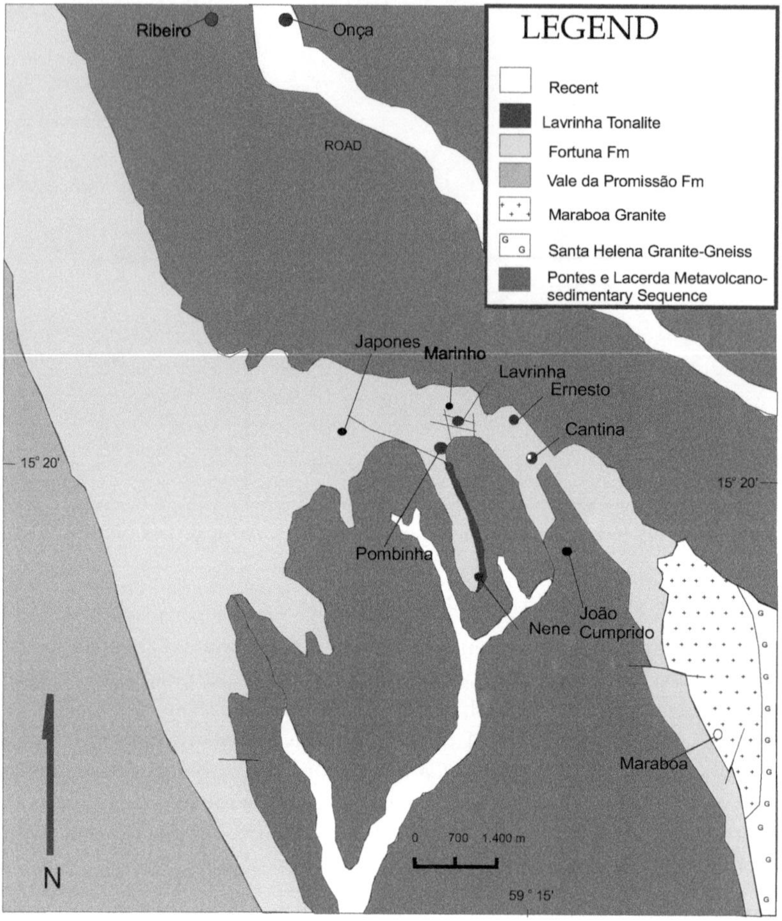

Fig. 4. Geologic map of the Pontes e Lacerda main ore deposits. Adapted from Geraldes et al. (1997).

Most of the gold deposits lay along the tectonic contact between the 1.51-1.49 Ga Rio Alegre metavolcanic-metassedimentary rocks and the Aguapei metassedimentary rocks. Secondarily, some gold deposits are hosted by clastic sedimentary rocks (base unit of the Aguapeí group), schists and granitoids (1,44-1,42 Ga Santa Helena suite). Disseminated and vein controlled mineralization are commonly found in volcanic host-rocks whereas sedimentary rock or granite-hosted deposits are mainly formed by veins. The ore consists of quartz, pyrite and gold, and the hydrothermal alteration zones contain quartz, sericite, pyrite, magnetite (altered to hematite), chalcopyrite, galena and sphalerite. $^{40}Ar/^{39}Ar$ ages obtained in sericite of the hydrothermal veins showed ages from 908.1 ± 0.9 Ma to 927 ± 1 Ma for mineralization (Fernandes et al., 2003; Paulo et al., 2005).

The sulfides from the Pontes e Lacerda Au deposits present values of 208Pb/^{204}Pb from 36.760 to 36.433 and ^{206}Pb/^{204}Pb from 17.734 and 17.498 and 207Pb/^{204}Pb from 15.638 and 15.499 (Geraldes et al., 1997). When the Pb isotope composition are plotted in the ^{207}Pb / ^{204}Pb *versus* 206Pb / ^{204}Pb diagram, they form a step linear trend. The variation in ^{206}Pb / ^{204}Pb and for ^{207}Pb / ^{204}Pb values is much greater than analytical error and must reflect heterogeneity of the source region and/or the hydrothermal fluids forming of these mineralized veins. Its possible to suggest that the ore deposit was deposited by hydrothermal solutions with influence of ultramafic basement which deep solutions originate during the regional metamorphism represented by Aguapei thrusting event.

6. Discussion

Pb isotope determination in ore-forming minerals is particularly useful when they can be directly combined with country rocks and can potentially constrains the crustal or mantle reservoirs that are sampled by the ore-forming system. Pb isotope source tracing relies on a number of assumptions, including: (1) measured or calculated initials ratios correspond to the isotopic composition of the ore-forming hydrothermal system; (2) fluid signatures accurately reflect the isotopic composition of the rock reservoir(s) sampled by the hydrothermal system, and (3) contemporaneous isotopic ratios of all possible reservoirs are known (Kerrich, 1991). In many instances theses assumptions cannot be critically evaluated within the constraints imposed by the geological boundary conditions (Bursnall et al., 1989). Moreover Pb many studies have shown that Pb isotope of ore deposits may be a mixture of hydrothermal contributions (Deloule et al. 1989) and Pb indigenous to the contiguous host rocks (Croceti et al. 1988).

For basement rocks it is usual to analyse k-feldspar. This mineral is enriched in Pb and depleted in uranium and thorium and contains Pb concentration up to several hundreds parts per million while its uranium and thorium content is usually less than 1 ppm. As a result, the bulk isotopic composition of Pb in k-feldspar is practically constant (similarly to galena in mineral deposits), thus maintaining the record of the lead originally incorporated from the melt.

Pb initial isotopic compositions reported in the literature carried out in K-Feldspars are now compiled to allow a correlation with the Pb signature of the mineral deposits discussed above. The Pb results of the followings units are available in the literature: 1.55 Ga rocks of Cachoeirinha suite; ca. 1.8 Ga rocks of the Alto Jauru terrane; ca. 1.75 Ga Aripuanã volcanic rocks; 1.45 Ga rocks of the Santa Helena batholith (granites and augen gneisses). The results are also plotted on a 207Pb/204Pb versus 206Pb/204Pb evolutionary diagram (Figure 5) along with Stacey and Kramers (1975) second-stage evolution curve together with Pb isotope results of the mineral deposits.

Reported Pb isotopic compositions (Geraldes et al., 2001) from the 1.8 Ga Alto Jauru rocks (k-feldspar leaching), including the 1.55 Ga Cachoeirinha suite, plot at ca. 1.55-1.30 Ga on a Stacey and Kramers (1975) second stage evolution curve. The results suggest that Pb in the Alto Jauru terrane rocks evolved along this growth curve from 1.8 to 1.30 Ga, at which time regional deformation and metamorphism caused the Pb to be rehomogenized and the Pb compositions in the K-feldspar have been unchanged since then due to absence of U in the feldspar and absence of a significant, younger metamorphic event. Sr/Sr results have two signatures: concordant carbonate veins yielded values from 0.705 to 0.7029 and discordant carbonate veins yielded values from 0.7144 to 0.7119, also suggesting two sources or remobilization.

These studies added to the Cabaçal deposit Pb isotope data reveal that mineralizing solutions may be originated during the Alto Jauru orogen (1.79-1.74 Ga) where a subduction process generated juvenile magmas in an island arc setting. The metal carried out by the hydrothermal solutions probably were deposited along ductile shear zones synchronously to the calc-alkaline magmatism. These studies reveal that Alto Jauru greenstone belt later on underwent to an important remobilization process, represented by the intrusion of the Cachoeirinha suite (1,66-1,62 Ga) which cooling ages are recorded by $^{40}Ar/^{39}Ar$ about 1,56 Ga. With the available data is not possible to define if the metal concentration was only related to the late evolution of the Alto Jauru orogen, recorded by the 1724 ± 30 Ma U-Pb SHRIMP age, or linked to the Cachoeirinha orogen, recorded by the 1.58-1.52 Ga (U-Pb in zircon ages) and 1,56 Ga ($^{40}Ar/^{39}Ar$ cooling ages in biotite).

Only one analysis of Pb isotopes in k-feldspar from Aripuanã region is reported, yielding values of ^{208}Pb / ^{204}Pb of 36.340, ^{206}Pb / ^{204}Pb of 16.739 and 207Pb / ^{204}Pb of 15.459, similar to the Alto Jauru k-feldspar signature. This datum added to the Pb isotopic signature of the sulfides may suggest important metalogenetic implications for Expedito Au-Cu deposit. Lead isotopes composition of sulfides is consistent with upper crust fluid and metal sources. Based upon the Pb isotopic results and the reported U-Pb zircon SHRIMP data for metadacitic rock which results yielded 1762 ± 6 Ma, interpreted as crystallization age, we may suggests that the Expedito massive sulphide deposit probably originated in volcanic environment, and may be included as VMS type deposit.

Pb isotope data from Pontes e Lacerda deposits basement is limited to the Santa Helena suite rocks. The results show a complex behavior and present a large range of isotopic compositions. The Santa Helena rocks have crystallization ages of ca. 1.44-1.42 Ga, but they are strongly deformed (milonitic) with a probable age of deformation about 0.95 Ga, during formation of the Aguapeí thrust belt and coeval to the 0.93 Ga (U-Pb in zircon) Guapé suite. Thus, the linear array formed by the data plot (Figure 5) probably represents growth of radiogenic Pb from 1.45 Ga to 0.95 Ga. These granites had a large range Pb isotope values and their Pb was rehomogenized at 0.95 Ga, with no subsequent growth of radiogeneic Pb in the K-feldspars since that time.

Concluding, in Cabaçal and Expedito deposits there is a strong correlation between country rocks and ore minerals Pb isotopes signatures. Moreover, Cabaçal deposit Pb isotopes suggest that including ultrabasic, basic and felsic volcanics added to gabbros and tonalities of Alto Jauru greenstone belt probably were formed in a magmatic setting. Shear zones (chlorite, sericite and biotite) with intense hydrothermal solutions percolation were the responsibe for the metal deposition. The cratonic volcanosedimentary sequences of the Expedito deposit also had formed thermal flux resulting in hydrothermal solutions percolation and and metal deposition. In the case of Pontes e Lacerda gold deposits, the ore minerals formation is correlated to the hydrothermal fluids percolation during the tectonic event (Aguapei Thrust). The results here presented define regional exploration constraints for the Alto Jauru, Aripuanã and Pontes e Lacerda region of the SW Amazonian craton.

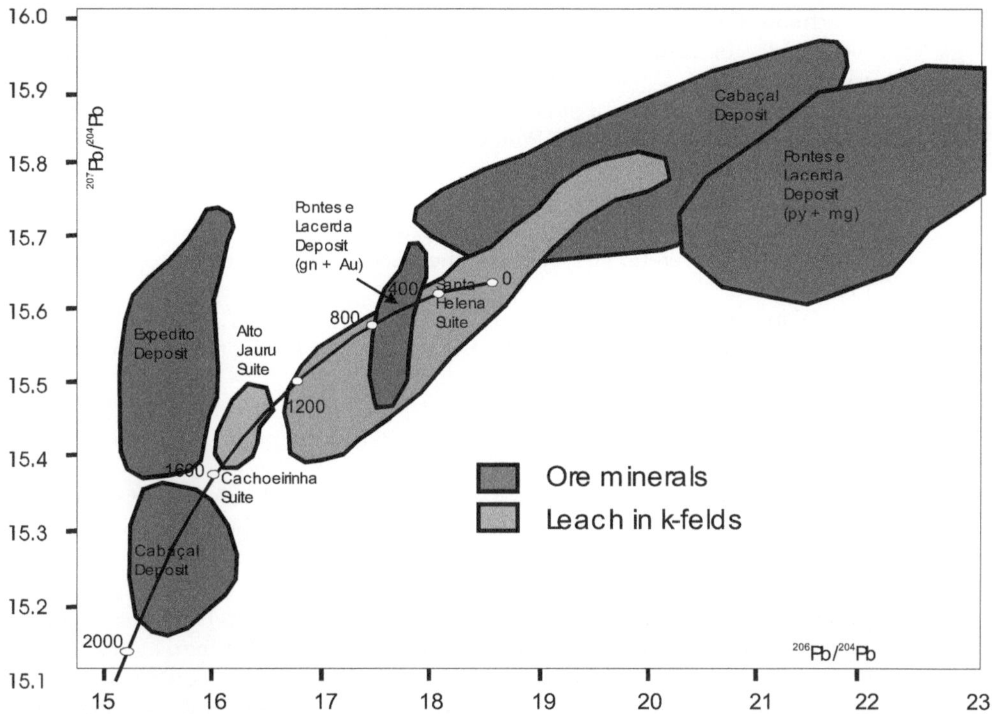

Fig. 5. ^{207}Pb / ^{204}Pb *versus* ^{206}Pb / ^{204}Pb diagram for mineral deposits and rocks from SW Amazonia craton

7. Acknowledgments

The author acknowledges the Brazilian National Research Council (CNPq) grant 301539/2005-7

8. References

Deloule, E; Gariepy, C.; Dupré, B. (1989) Metallogenesis of the Abitibi greenstone belt of Canada: a contribuition from analysis of trace lead in sulphide minerals. Can J. Earth Sci. Letters. 26: 2529-2540.

Changkakoti, A.D.K.; Krstic, D.; Gray, J.; Morton, R. (1986) Pb and Sr isotope composition of hydrothermal mineral from the Great Rear Lake Silver Deposit, N.W.T. (Canada). Econ. Geol. 81:. 739-743.

Crocetti, C. A.; Holland, H.D.; e Mekena, L.W. (1988) Isotopic composition of lead in galena from the Viburnum Trend, Missouri. Econ. Geol. 83: 355-376.

Cumming, G.L. e Richards, J.R. 1975 Ore lead isotope ratios in a continuously changing earth. Eart. and Plan. Sci. Lett. 28: 155-171.

Dean, J.A.; Carr, G.R (1982) Distinguishing plutonic and volcanic genese mineralization in the Lachlan fold belt of eastern Australia using Pb isotopes. Intern. Congr. Geol. Abstracts, p.25.

Doe, B.R. e Zartman, R.E. 1979 Plumbotectonics 1, the Phanerozoic. *In:* H.L. Barnes (ed.), Geochemistry of Hydrotermal Ore Deposits. Holt, Rihehart and Winstons, New York.p.107-135.

Geraldes, M.C., Figueiredo, B.R., Tassinari, C.C.G., Ebert, H.D., 1997. Middle Proterozoic vein-hosted gold deposits in the Pontes e Lacerda region, southwestern Amazonian Craton, Brazil. International Geology Review 39, 438-448

Geraldes, M.C., Teixeira, W., Heilbron, M., 2004. Lithospheric versus asthenospheric source of the SW Amazonian craton A-type granites: the role of the Paleo- and Mesoproterozoic accretionary belts for their coeval continental suite. Episodes 27 (3) 185-189.

Geraldes, M.C., Van Schmus, W.R., Condie, K.C., Bell, S., Teixeira, W., Babinski, M., 2001. Proterozoic geologic evolution of the SW part of the Amazonian Craton in Mato Grosso state, Brazil: Precambrian Research 111, 91-128.

Iyer, S.S.; Hoefs, J.; Krouse, H.R. (1992) Sulfur and lead isotope geochemistry of galenas from Bambuí Group, Minas Gerais, Brazil- impllications for ore genesis. Econ. geol. 87: 437-444.]

Kerrich, R. (1989) Geochemical evidence on source of fluids and solutes for shear zone hosted mesothermal gold deposit. *In* J.T. Burshal (ed.), Mineralization and Shear Zones. GAC. Short Courses Notes. 567: 207-218.

Kerrich, R. (1991) Radiogenic isotope systems to mineral deposits. *In* L. Heaman e J.N. Ludden (eds): Short Course Handbook on Applications of Isotope Systems to Probems in Geology. Mineralogical Association of Canada, Toronto.

Relvas, J.M.R.S., Tassinari, C.C.G., Munhá, J. and Barriga, F.J.A.S. 2001 Multiple sources for the ore-forming fluids I the Neveds Corvo VHMS deposit of the Iberian Pyrite Belt (Portugal): strontium, neodymium and lead isotope evidence. Mineralium Deposita. 36:416-427.

Tassinari, C.C.G. e Cavalcanti, M. (1994) Sr and Pb isotope evidence for the origin of skarn, sulphide and fluor mineralization related to Itaoca granitoid, Brazil. 7 Congresso Geológico Chileno. vol. II. p.1488-1490.

Tassinari, C.C.G.; Babour, A.P., Elias R.D.; Sato,K. 1990 Aplicações dos isótopos de Pb e Sr na determinação da natureza das fontes das mineralizações de chumbo do Vale do Ribeira-SP e PR . 36 Cong. Bras. de Geol. vol. 3 p. 1254-1266.

Tassinari. C.C.G. e Melito, K. (1994) The time-bound characteristics of gold depósits in Brazil and their tectonic impllications. Comunicaciones. 45: 45-55.

Teixeira, W.; Tassinari, C.C.G.; Cordani,U.G.; Kawashita, K. (1989) A Review of the Geochronology of the Amazonian Craton: Tectonic Implications. Prec. Res. 42: 213-227.]

Toledo, F.H. (1996) Mineralização e alteração hidrotermal do depósito de ouro do Cabaçal, Mato Grosso. Dissertação de Mestrado. UNICAMP. São paulo. (em preparação).

Wilton, D.H.C. (1991) Metallogenic and tectonic implication of Pb isotope data for galena separetes from Labrador central mineral belt. Econ. Geol. 86: 1721-1736.

Zartman, R. E. e Doe, B.R. 1981 Plumbotectonics - The Model. Tectonophysics, 75: 135-162.

Zartman, R.E. e Haines, S.M. 1988 The plumbotectonics model for Pb isotopic systematics among major terrestrial resevoirs - a case for bi-direcional transport. Geochim. et Cosm. Acta. 52: 1327-1339.

3

New Technology for Recovery of Gold and Silver by Pressure Cyanidation Leaching and Electrocoagulation

José R. Parga[1], Jesús L. Valenzuela[2] and José A. Díaz[1]
[1]*Department of Metallurgy and Materials Science Institute Technology of Saltillo, Saltillo Coahuila*
[2]*Department of Chemical Engineering and Metallurgy University of Sonora, Hermosillo, Sonora*
México

1. Introduction

The present chapter describes a new technology of pressure oxidation/cyanidation leaching for the dissolution of gold and silver and the recovery of the precious metals by using the electrochemical process of Electrocoagulation (EC). The novel method demonstrates that the oxidation and dissolution of gold and silver in alkaline cyanide solution can be conducted simultaneously in the same reactor in less than 90 minutes with a recovery that exceeds 96%. Then, the pregnant cyanide solution with gold and silver is sent for recovery of precious metals by using a very promising electrochemical technique (EC) that does not require high concentrations of silver and gold in cyanide solutions.

Gold is classed as a noble metal because of its inertness to most chemical reactions under ordinary conditions. At the present time, cyanidation has superseded all previous leaching processes, particularly chlorination, because of its ability to effectively and economically treat ores containing as little as 1-3 g/ton gold. Cyanidation processes are especially suitable for treatment of gold/silver-bearing sulphidic materials. Gold cyanidation has been reported to involve the chemical reactions shown in Eqs. (1) and (2). Silver is accomplished in the same fashion (Senanayake, 2008; Parga et al., 2007).

$$2Au + 4NaCN + O_2 + 2H_2O \rightarrow 2\,Na[Au(CN)_2] + 2NaOH + H_2O_2 \quad (1)$$

$$2Au + 4NaCN + H_2O_2 \rightarrow 2\,Na[Au(CN)_2] + 2NaOH \quad (2)$$

The Equation (1) proposed by Elsner is stoichiometrically correct but does not describe the cathodic reactions associated with the dissolution. The stoichiometry of the process shows that 4 moles of cyanide are needed for each mole of oxygen used. At room temperature and std. atmospheric pressure, approximately 8.2 mg of oxygen are present in one liter of water. This corresponds to 0.27×10^{-3} mol/L. Accordingly, the sodium cyanide concentration (molecular weight of NaCN = 49) should be equal to $4 \times 0.27 \times 10^{-3} \times 49 = 0.05$ g/L or approximately 0.01%. This was confirmed in practice at room temperature by a very dilute solution of NaCN of 0.01% - 0.5% for ores, and for concentrates rich in gold and silver of 0.5 % -5 % (Parga et al., 2007). Details of this electrochemical reaction have received considerable attention and under certain circumstances the reaction is limited by the coupled diffusion of CN^- and O_2 to the gold surface. Lime or sodium hydroxide (caustic) is added to keep the system at an alkaline pH of 10-11. This protective alkalinity is required to counteract the generation of acid during cyanidation, thereby preventing cyanide degradation and the formation of the deadly HCN gas.

Gold and silver ores are classified as refractory when a significant portion of the precious metals cannot be extracted by the conventional cyanidation process. The refractoriness may be of a physical or chemical nature. The former type is usually due to sub-microscopic particles of gold being locked within mineral particles, for example in sulphides or silicates. Refractory gold-silver ores contain precious metals locked up in a matrix of pyrite and/or arsenopyrite (Parga et al., 2007). Such ores are not amenable to cyanidation. To liberate the gold-silver from the sulfide matrix and render it accessible to cyanidation. It is a common practice to subject them to a preoxidation in order to enhance the recovery of the precious metals. Although effective, these techniques result in the production of large quantities of environmental hazardous substances such as sulphur dioxide gas or sulfuric acid, which can cause difficulties in compliance with environmental pollution regulations. After review of the literature (Parga et al., 2007), it was shown that roasting is currently the most cost effective means of oxidizing refractory pyrite and arsenopyrite concentrates to produce a product amenable to the cyanidation process. However, the major drawback of roasting is that it produces large quantities of SO_2 gas which is released into atmosphere and this is not acceptable. The sulphur dioxide must be collected, or an alternative technology used. One such alternative process that has previously been used successfully around the world, is that of pressure aqueous pre-oxidation of the sulphide minerals. Pressure oxidation has been the most popular method for the treatment of refractory gold concentrates (see Table 1); This process utilizes oxygen or air at high pressures and temperatures to oxidize an aqueous slurry of the ore or concentrate to produce hematite, iron sulphates and considerable quantities of free sulphuric acid. Because of this, before the cyanidation process excessive lime or caustic soda must be used in order to the pH of the pulp to 10 or 11.

This process, which is used in many commercial plants around the world, involves the recovery of the precious metals by oxidation pretreatment followed by traditional cyanidation. By consequence the saving in time for the gold and silver dissolution from ore or concentrates is limited by the conventional cyanidation leaching step, a process which requires 48 to 72 hours. It should be remembered that the pressure aqueous pre-oxidation technology is not used to leach the gold but only to make the cyanide ions accessible to the gold in the host mineral.

2. Extraction of gold and silver from the argentopyrite / argentite ore

2.1 Pressure oxidation / cyanidation chemistry

The oxidation of gold and silver is a prerequisite for its dissolution in the alkaline cyanide lixiviant. Pyrite along with arsenopyrite, argentopyrite, sphalerite and covellite are the most common host minerals of gold, silver and electrum. It is important from a process optimization standpoint to understand the behavior of each of these minerals during alkaline pressure oxidation/cyanidation which leads to dissolution/destruction and subsequent liberation of gold and silver which would then be available for the cyanide ions. Pressure oxidation of pyrite and argentopyrite at low temperature (70^{0}C) and oxygen pressure (60 lb/in^2) involves reactions yielding ferrous ion, sulfate ion, and elemental sulfur as products (Anderson and Nordwick, 1996; Chander and Briceño, 1987).

The primary reactions are:

$$2FeS_2 + 7O_2 + 2H_2O \rightarrow 2Fe^{2+} + 4SO_4^{2-} + 4H^+ \quad (3)$$

$$FeS_2 + 2H^+ \rightarrow Fe^{2+} + H_2S + S^0 \quad (4)$$

Plant Location	Feed	Medium	Capacity T/d	Operation Data		
				T (°C)	P (psi)	t (min)
McLaughlin, California USA	Ore	Acid	2700	180	401	90
São Bento Brazil	Concentrate	Acid	240	190	230	90
Barrick Mercur Utah, USA	Ore	Alkaline	790	225	473	70
Getchell USA	Ore	Acid	3000	210	420	130
Goldstrike Nevada, USA	Ore	Acid	1360	225	439	75
Goldstrike Nevada, USA	Ore	Acid	5450	225	400	75
Porgera, Papua New Guinea	Concentrate	Acid	1350	190	258	180
Campbell Canada	Concentrate	Acid	70	195	305	120
Barrik USA	Ore	Acid	2000	220	450	100
Goldstrike Nevada, USA	Ore	Acid	11580	200	200	140
Santa Fe Pacific Gold Nevada, USA	Ore	Acid	4000	225	460	50
Sunshine Mining & Co Idaho, USA	Ore	Acid	1000	120	90	60
Porgera, Papua New Guinea	Concentrate	Acid	2700	190	250	150

Table 1. Commercial plants using pressure aqueous pre-oxidation (Adams M.D., 2005; Parga et al. 2007).

Ferrous ions produced by reaction (3, 4) are subsequently oxidized to ferric ions.

$$2Fe^{2+} + 1/2O_2 + 2H^+ \rightarrow 2Fe^{3+} + H_2O \quad (5)$$

And contribute to the leaching of sulfides only as a source of ferric ions. The ferric ions can also contribute to the oxidation of argentopyrite, argentite, pyrite, pyrrhotite, sphalerite and chalcopyrite:

$$AgFe_2S_3 + Fe^{3+} \rightarrow Ag^+ + 3Fe^{2+} + 2S^{2-} + S^0 \quad (6)$$

$$Ag_2S + 2Fe^{3+} \rightarrow 2Ag^+ + 2Fe^{2+} + S^0 \quad (7)$$

$$FeS_2 + 8Fe^{3+} + 4H_2O \rightarrow 9Fe^{2+} + 8H^+ + S^0 + SO_4^{2-} \quad (8)$$

$$Fe_7S_8 + 32H_2O + 62Fe^{3+} \rightarrow 69Fe^{2+} + 64H^+ + 8SO_4^{2-} \quad (9)$$

$$ZnS + 2Fe^{3+} \rightarrow Zn^{2+} + 2Fe^{2+} + S^0 \quad (10)$$

$$CuFeS_2 + 4Fe^{3+} \rightarrow Cu^{2+} + 5Fe^{2+} + 2S^0 \quad (11)$$

Of course silver has been shown to activate the oxidation and dissociation of chalcopyrite because a porous sulfur layer is formed (Parga et al., 2007). Then, elemental sulfur may also be further oxidized to sulfate by oxygen or by ferric sulfate:

$$2S^0 + 3O_2 + 2H_2O \rightarrow 4H^+ + 2SO_4^{2-} \quad (12)$$

$$S^0 + 6Fe^{3+} + 4H_2O \rightarrow 6Fe^{2+} + 8H^+ + SO_4^{2-} \quad (13)$$

This results in the formation of a porous, but nonprotective, elemental sulfur layer, thus allowing cyanide and dissolved oxygen to access to the previously locked gold, silver and electrum. It was thought that by maintaining high concentrations of CN^- ions in the pressure leach reactor it would be possible to complex gold and silver (Eq. 14) as they became liberated and thus achieve both objectives, i.e. decomposition of refractory minerals and leaching the Au/Ag with the cyanide ions, simultaneously in a single stage. Also, two of the main advantages of this cyanidation process are the selectivity of free cyanide for gold and silver dissolution and the extremely high stability of the cyanide complex as illustrated in Table 2.

Reaction	K	Reaction	K
$Au^+ + 2CN^- = Au(CN)_2^-$	$10^{38.3}$–$10^{39.3}$	$Ag_2S = 2Ag^+ + S^{2-}$	$10^{-53.5*f}$
$Au(CN)_2^- = AuCN + CN^-$	$10^{-7\,a}$	$H^+ + S^{2-} = HS^-$	10^{17m}
$Au^+ + 2HS^- = Au(HS)_2^-$	$10^{30.1\,b}$	$HS^- + H^+ = H_2S$	10^{7}
$Au^+ + HS^- = Au(HS)$	$10^{24.5\,b}$	$2S + HS^- + OH^- = S_2S^{2-} + H_2O$	$10^{2.2}$
$Au^+ + 2OH^- = Au(OH)_2^-$	$10^{22.0\,c}$	$3S + HS^- + OH^- = S_3S^{2-} + H_2O$	$10^{3.9}$
$Au^+ + OH^- = Au(OH)$	$10^{20.6\,c}$–$10^{10.2\,d}$	$4S + HS^- + OH^- = S_4S^{2-} + H_2O$	$10^{4.6}$
$Au^+ + OH^- + CN^- = Au(OH)(CN)^-$	$10^{23.3\,e}$	$5S + HS^- + OH^- = S_5S^{2-} + H_2O$	$10^{4.6}$
$Au^+ + 2CH_3CN = Au(CH_3CN)_2^*$	$10^{1.6\,n}$	$6S + HS^- + OH^- = S_6S^{2-} + H_2O$	$10^{2.3}$
$Au_2S = 2Au^+ + S^{2-}$	$10^{-72.8*f}$	$Cu^{2+} + 4CN^- = Cu(CN)_4^{2-}$	10^{25}
$Ag^+ + 2CN^- = Ag(CN)_2^-$	$10^{20.1}$	$Fe^{3+} + 6CN^- = Fe(CN)_6^{3-}$	$10^{43.6}$
$Ag^+ + 3CN^- = Ag(CN)_3^{2-}$	$10^{21.4}$–$10^{21.8}$	$3Ag^+ + Fe(CN)_6^{3-} = Ag_3Fe(CN)_6(s)$	$10^{18.2}$
$Ag^+ + 2OH^- = Ag(OH)_2^-$	$10^{3.6}$–$10^{4.2}$	$Au^+ + 2S_2O_3^{2-} = Au(S_2O_3)_2^{3-}$	10^{26}
$Ag^+ + OH^- = AgOH$	$10^{2.3}$–$10^{3.9}$	$Ag^+ + S_2O_3^{2-} = AgS_2O_3^-$	$10^{8.80}$
$Ag^+ + OH^- + CN^- = Ag(OH)(CN)^-$	$10^{12.8}$–$10^{13.2}$	$Ag^+ + 2S_2O_3^{2-} = Ag(S_2O_3)_2^{3-}$	$10^{13.7}$
$AgCN = Ag^+ + CN^-$	$10^{-15.7*}$	$Ag^+ + 3S_2O_3^{2-} = Ag(S_2O_3)_3^{5-}$	$10^{14.2}$

Table 2. Equilibrium constants (Marsden and House, 1960).

For this argentopyrite concentrate, both gold and silver extractions achieved were in fact greater than 96%.

Since silver in Mexican ores occurs as argentite the cyanidation and sulfide oxidation reaction are as follows:

$$Ag_2S + 2nCN^- \rightarrow 2Ag(CN)^{(n-1)^-}_n + S^{2-} \qquad (14)$$

$$Ag_2S + 4CN^- + 0.5H_2O + O_2 \rightarrow 2Ag(CN)_2^- + 0.5\, S_2O_3^{2-} + OH \qquad (15)$$

$$S^{2-} + CN^- + 1/2O_2 + H_2O \rightarrow SCN^- + 2OH \qquad (16)$$

$$2S^{2-} + 2O_2 + H_2O \rightarrow S_2O_3^{2-} + 2OH^- \qquad (17)$$

$$S_2O_3^{2-} + 2OH^- + 2O_2 \rightarrow 2SO_4^{2-} + H_2O \qquad (18)$$

According to the above equations, sulfide ions are oxidized to thiosulphate and can contribute with the dissolution of gold. Also Graham et al. (Graham and James, 1995) showed that an alkaline or near neutral solution of thiosulphate dissolved gold metal slowly in the presence of a mild oxidant. The dissolution of the gold can be written as in Equation (19), where oxygen is the oxidant and thiosulphate is the ligand.

$$4\,Au + 8S_2O_3^{2-} + O_2 + 2H_2O \rightarrow 4[Au(S_2O_3)_2]^{3-} + 4OH^- \qquad (19)$$

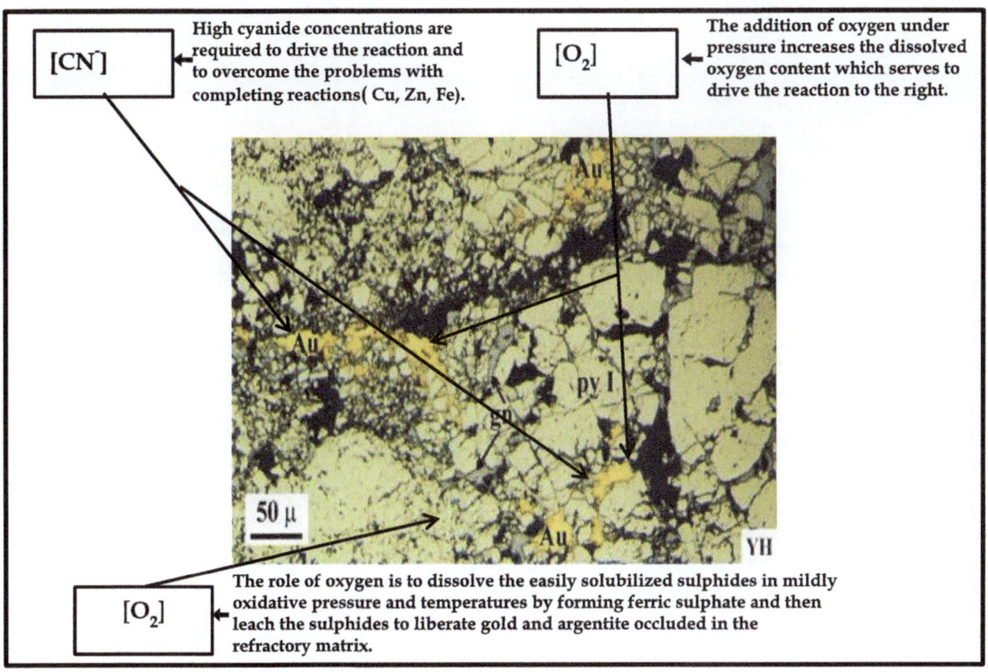

Fig. 1. Schematic mechanism of gold and silver leaching.

Better extractions of gold with thiosulphate are achieved when elevated temperatures (e.g. 65 °C) are used (Wan et al.,1993). The proposed leaching reaction mechanism for gold and silver dissolution in relatively mildly operating conditions of 70 °C and 60 psi oxygen pressure is described in Figure 1. The role of oxygen gas is to dissolve the easily solubilized sulphides in mildly oxidative pressure and temperature by forming sulphate and metals (e.g. Fe^{2+} to Fe^{3+}) to liberate gold occluded in the refractory matrix.

2.2 Experimental details

Mineralogical analysis indicates that the sulfide minerals for the Bacís Mining Co. concentrate include pyrite, argentopyrite, pyrrhotite, arsenopyrite, chalcopyrite, covellite, hematite and magnetite. The non-opaque minerals were quartz, calcite, apatite, gypsum, fluorite and barite. Also, several photomicrographs indicate occlusion and/or dissemination of micron-size gold, silver, electrum particles in the sulfide minerals such as argentopyrite, pyrite, sphalerite, arsenopyrite and quartz (Figure 2). The compositions are listed in Table 3.

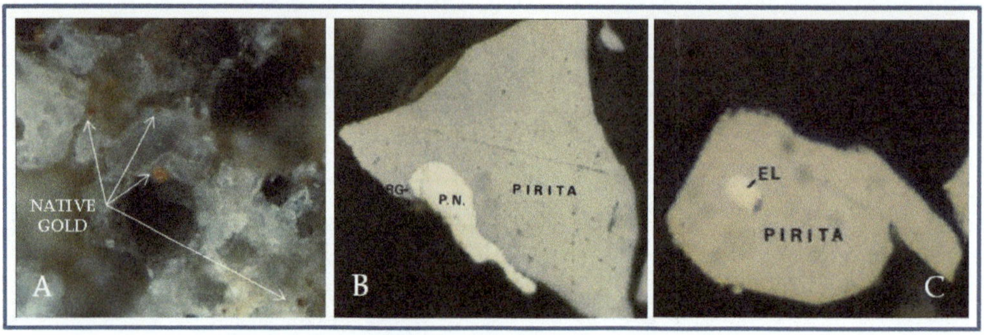

Fig. 2. Photomicrograph, Native gold, size 5 µm , Argentite (ARG) occluded in pyrite, size 60 µm, Electrum (EL) occluded in pyrite, size 10 µm.

	gr./ton		%					
	Au	Ag	Pb	Zn	Cu	Fe	As	S
Concentrate	87.09	12320	2.6	3.8	0.5	29.2	0.15	32
Ore	4.12	289	0.5	0.4	0.04	3.67	0.1	3.7

Table 3. Analytical data for the composite samples.

2.2.1 Leaching experiments

Experiments were carried out in a four-liter Parr autoclave assembled with impeller, thermowell, pressure gauge, gas inlet and outlet pipes were used for simultaneous oxidation and cyanidation in the same autoclave. Samples were ground to 84% minus 40 µm or finer, pulped with fresh tap water. Depending on test requirements, variables such as NaCN, reaction time, temperature, particle size, pH and pulp density were set and the leaching experiments were undertaken.

2.3 Results and discussion

2.3.1 Ambient condition

Preliminary batch testing had indicated that at ambient conditions (25°C, 1atm.) direct cyanide leaching of the concentrate with air gave poor silver and gold recoveries. The results obtained are shown in Figures 3 and 4.

2.3.2 Effect of the autoclave retention time

Figure 5 shows that there is an optimum leach time to achieve both maximum gold and silver extractions. The optimum leach time appears to lie close to 60 minutes at 80°C.

Increase in time beyond this value leads to chemical degradation of gold and silver cyanocomplexes, due to the hydrogen ions produced by the sulfide oxidation. Also for prolonged leaching time there is significant loss in cyanide, since the oxygen present oxidizes the cyanide to cyanate and ultimately to ammonia and carbon dioxide.

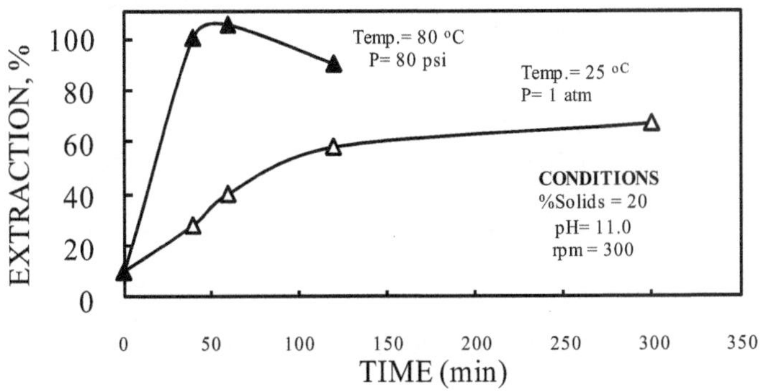

Fig. 3. Comparison of silver extraction at ambient conditions and high pressure.

2.3.3 Effect of cyanide concentration

Concentrations of 0.6 - 1.1 wt. % sodium cyanide were used to leach the concentrate in the autoclave. When the concentration of sodium cyanide was less than 0.6%, gold and silver extraction could not be completed optimally in a single stage of autoclave leaching. Increasing the cyanide concentration resulted in an increase in the gold and silver extraction. The results obtained are shown in Figure 6.

2.3.4 Effect of temperature

The concentrate was leached as 20-wt.% solids slurry with 1 wt.% of NaCN for 60 minutes. Temperature was varied from 60 to 200°C. As the temperature was increased, the gold and silver extractions decreased, since at temperatures higher than 80°C, the oxidation of cyanide ions is too rapid resulting in the formation of CO_2 gas and ammonium ion. Cyanide is then not available for complexetion with the gold and silver. See Figure 7.

$$CNO^- + H_2O + 2H^+ \rightarrow CO_2 + NH_4^+ \qquad (20)$$

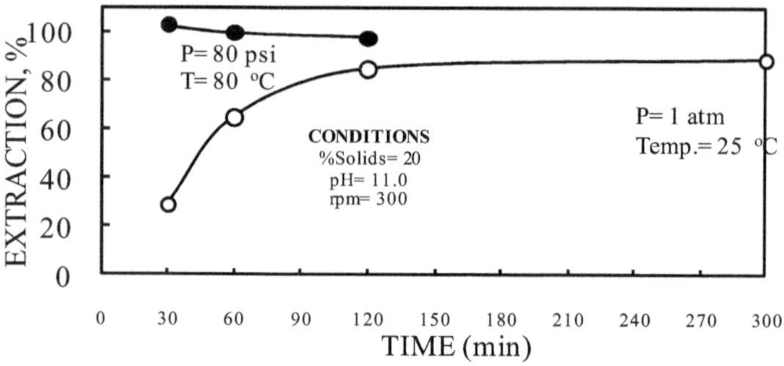

Fig. 4. Comparison of gold extraction at ambient conditions and high pressure.

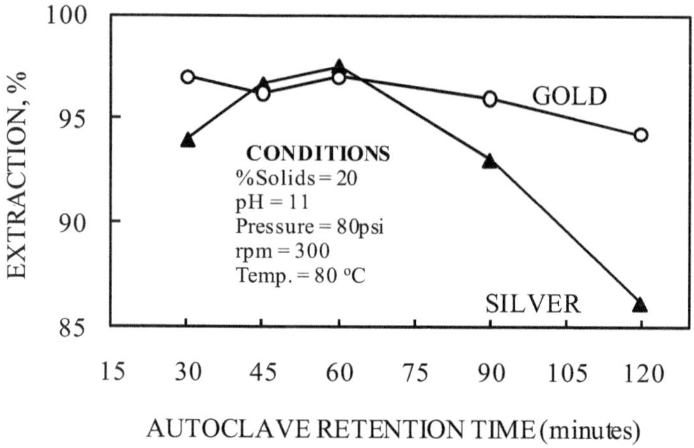

Fig. 5. Effect of autoclave retention time on gold and silver extraction.

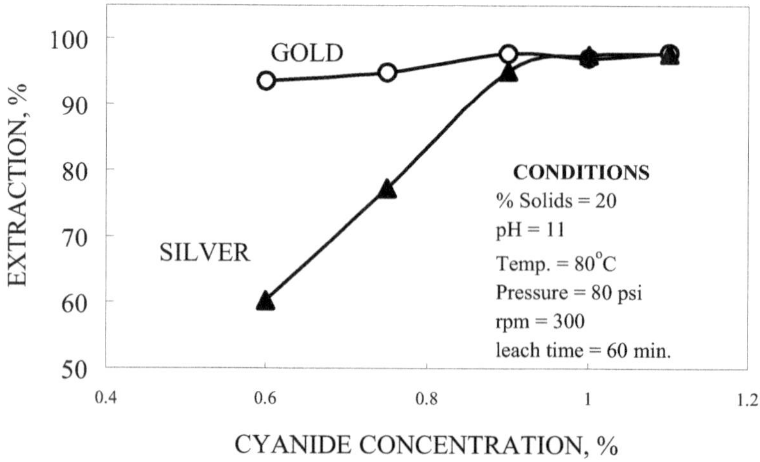

Fig. 6. Effect of the cyanide concentration on gold and silver extraction

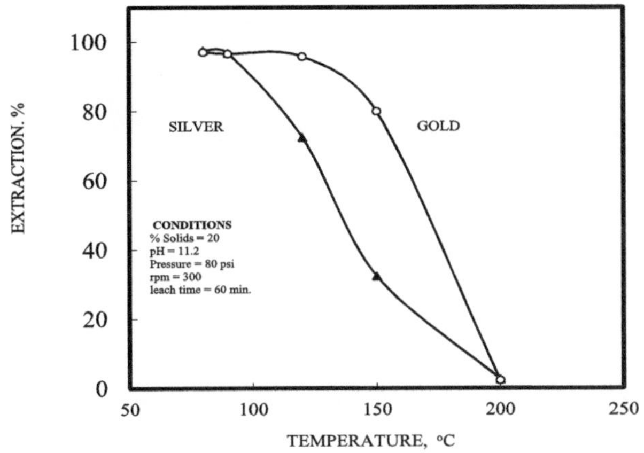

Fig. 7. Effect of temperature on gold and silver extraction.

2.3.5 Effect of pH

A pH range of 10.5-13, was examinated for the extraction of gold and silver. In Figure 8, it can be seen that at a pH of 11.5 or above, optimum extractions could not be maintained.

This behavior is because in alkaline-oxidizing medium the pyrite and pyrrhotite in the concentrate competes with gold and silver for consumption of oxygen, and this mechanism may be responsible for the decrease in extraction.

$$2FeS_2 + 7O_2 + 2H_2O \rightarrow 2Fe^{2+} + 4H^+ + 4SO_4^{2-} \quad (21)$$

$$FeS + 2O_2 \rightarrow Fe^{2+} + SO_4^{2-} \quad (22)$$

Eventually the ferrous sulfate formed is converted to the stable, insoluble ferric hydroxide.

$$FeSO_4 + Ca(OH)_2 \rightarrow Fe(OH)_2 + CaSO_4\downarrow \quad (23)$$

$$2Fe(OH)_2 + 1/2O_2 + H_2O \rightarrow 2Fe(OH)_3\downarrow \quad (24)$$

The decrease in the extraction of gold and silver with increasing pH may also be due to a decrease in the formation of ferric sulfate which is the reagent that contributes to the oxidation of sulfide minerals which occlude gold and silver. Also Deitz (Deitz and Halpern, 1953) found that at high pH values between 12 to 13, the silver surface was coated with a whitish film of CaO_2 and thus the leaching process was retarded.

2.3.6 Effect of pulp density

The effect of varying the pulp density in the range of 15-30% solids is shown in Figure 9.

Gold and silver extractions gradually decreased as the pulp density increased apparently due to the increased apparent viscosity of the slurry which impedes good oxygen dispersion in the system, and also for the excessive acid generation.

2.3.7 Plant operation

The optimum conditions as delineated in the laboratory study (Table 4) were used to study the extraction of gold and silver during a continuous one-month plant campaign. A total of 600 tons of concentrate was processed. Recoveries for both gold and silver averaged approximately 96%. These plant data are shown in Figure 10.

On the basis of the results described above the plant tests were in good agreement with the results obtained in the laboratory autoclave and relative to any comparable process, low temperature, low pressure and with these conditions cheaper autoclaves can be used. Thomas shows (Thomas et al., 1992) that a 1400 t/d acid pressure oxidation circuit might cost C$30 million while the corresponding non-acid circuit might cost only C$15 million.

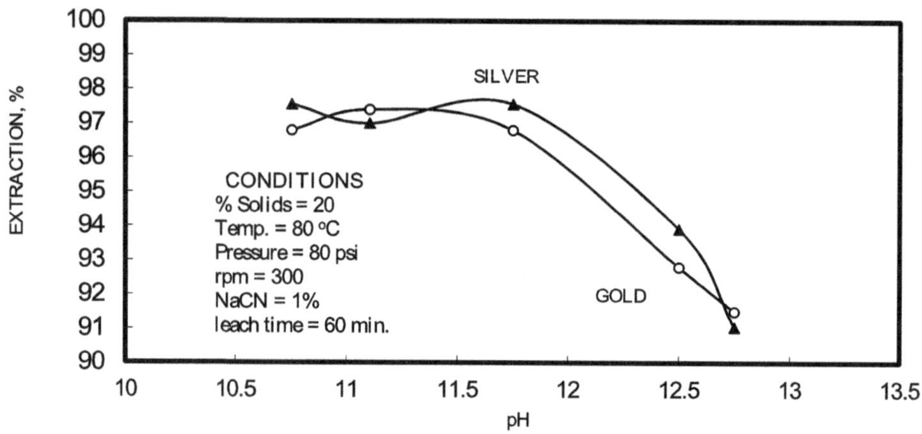

Fig. 8. Effect of pH on gold and silver extraction.

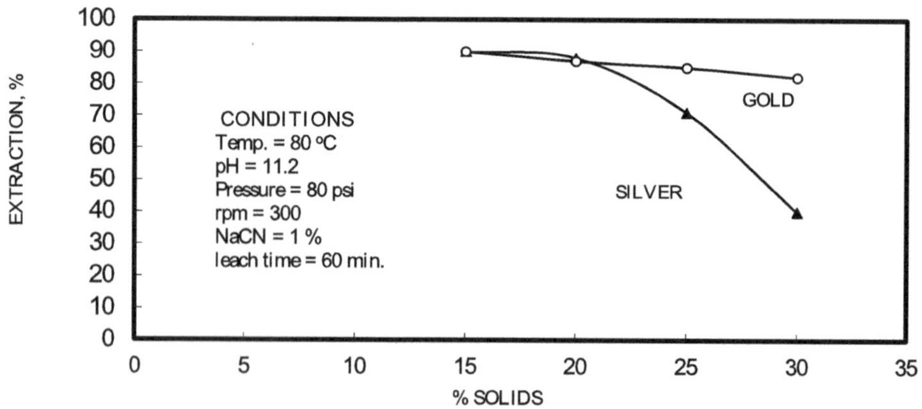

Fig. 9. Effect of percent solids in the extraction of gold and silver.

Solid % = 20	Cyanidation time = 60 min.
Temperature = 80°C	NaCN = 1%
Pressure = 5.6 kg./cm² (psi)	pH = 10.7
rpm = 300	PbO = 100gr./ton

Table 4. Optimum parameters for the autoclave

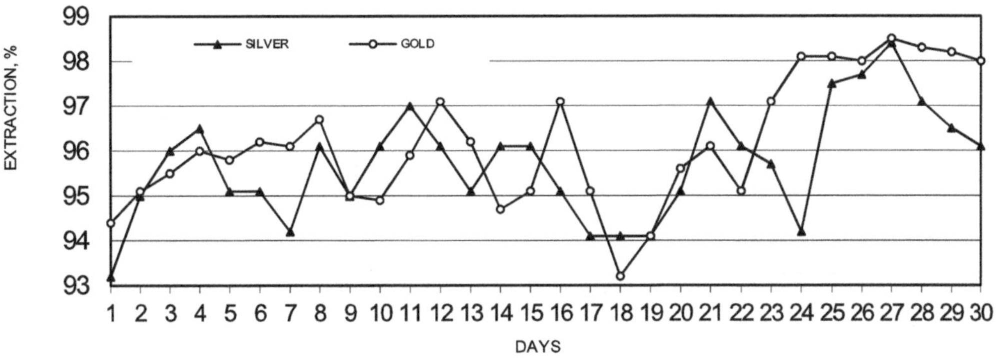

Fig. 10. Result from continuous plant operation.

3. Electrocoagulation process alternative for silver and gold recovery

After the extraction of gold and silver from the argentopyrite/argentite ore, the next step is the recovery of the precious metals from the cyanidation process by which gold and silver are recovered from their ores, and it is recognized that the Carbon in Pulp, the Merrill-Crowe (Metal displacements), the Ion Exchange and Electrowinning processes are used for concentration and purification of gold and silver from cyanide solutions. Each recovery method has advantages and disadvantages. Process selection depends on the specific conditions for a particular operation and the facilities already available. The Merrill-Crowe method had been the preferred process for many years for recovery silver and gold from high reach solutions. Only recently, in the past years, has the carbon adsorption process become popular for recovering gold from large volumes of low grade pregnant leach solutions that contain mainly gold. Other processes, Ion-Exchange Resins and Solvent Extraction, have recently been reviewed as an alternative for gold and silver recovery from alkaline cyanide solutions, (Aguayo et al., 2007). Commercially available resins were unable to compete with activated carbon due to poor selectivity, mechanical breakdown of the beads and the requirement for complex elution, generation of HCN and regeneration process.

Among several available options for recovery of precious metals from cyanide solutions, Electrocoagulation (EC) is a very promising electrochemical technique that does not require high concentrations of silver and gold in cyanide solutions in order to recovery them. Also, literature review showed that the potential of EC as an alternative to traditional treatment recovery of precious metals (silver and gold cyanide) has not yet been exploited. Advantages and disadvantages for the different processes are presented in Table 5, along with those of the EC (Mollah et al., 2004; Parga et al., 2005). EC has been proposed since before the turn of the 20th century. A plant was built in London in 1889

Method	Advantages	Disadvantages
Merrill-Crowe	• Lower capital and operating costs. • Handles solutions containing high silver and gold content. • It is highly efficient (99.5%). • Also can treat high-grade solutions produced by carbon elution. • Alternative to Electrowinning. Well known technology.	• The pregnant solution need clarification and deoxygenating. • Low concentrations of metals, increases amount of zinc • Depends on the pH and concentration of the free cyanide. • The precipitate contain cyanides like copper and arsenic. • The precipitate spend one week in the filter press.
Adsorption with Activated Carbon	• Does not require pre-treatment of pregnant solution. • Not dependent on the concentration of metals. • Large specific surface. • The pulp needs no clarification.	• Fouled carbon needs to be regenerated by heating. • Large carbon inventory. • The pregnant solution has to go through 5 or 6 columns. • High operating costs.
Ion Exchange Resins	• Does not need: washing, revitalization or heat treatment. • High abrasion resistance in tanks of adsorption. • High selectivity.	• High cost of the process. • Lower loading capacity. • Royalty payments. • The resin must be regenerated in acid medium.
Electrocoagulation	• Low residence time (minutes). • Does not use chemicals. • Handles solutions containing lower or high silver and gold contents. • Energy costs per m^3 of pregnant solution are lower than conventional treatment systems.	• Sacrificial anode must be placed periodically. • Precise initial pH control. • New technology. • The product is high in iron. • The sludge need to be leaching with sulfuric acid.

Table 5. Advantages and disadvantages of methods for recovery of gold and silver (Emamjomeh et al. 2004; Mollah et al., 2009; Parga et al., 2007).

for the treatment of sewage mixing it with seawater and electrolyzing it. In 1906, EC was first patented (Parga et al., 2005) and used to treat bilge water from ships. In 1909, in the United States J.T. Harries received a patent for wastewater treatment by electrolysis using sacrificial aluminum and iron anodes (Vik et al., 1984). Matteson (Matteson et al., 1995), described a device of the 1940's, the "electronic coagulator" which electrochemically dissolved aluminum (from the anode) into solution, reacting this with the hydroxyl ion (from the cathode) to form aluminum hydroxide. The hydroxide flocculates and coagulates the suspended solids and thereby purifies water. A similar process was used in Britain in 1956 in which iron electrodes were used to treat river water (Holt et al., 2005).

3.1 Electrocoagulation fundamentals

The electrochemical phenomenon of Electrocoagulation has been employed previously for treating many types of wastewater with varying degrees of success. This electrochemical method of contaminant removal requires smaller quantities of salt addition to increase the conductivity of the solution (typically, an aqueous electrolye), and the maintenance and operation of the EC cells are relatively simple. Electrocoagulation processes offer significant potential for removing ionic species from solution, particularly heavy metals (Mollah et al., 2004; Parga et al., 2007). Operating conditions are highly dependent on the chemical composition and properties of the aqueous medium, specifically, conductivity and pH. Other important process variables, such as particle size, type of electrodes and retention time between electrodes, electrode spacing and chemical-constituent concentrations influence the operating process-parameters (Parga et al., 2007). The fundamental operating-principle is that cations produced electrolytically from iron and/or aluminum anodes (by oxidation) provide for the coagulation of contaminants contained in the aqueous electrolyte. Thus, the (sacrificial) metal-anodes provide a continuous supply of polyvalent metal cations in the vicinity of the anode. These cations participate in the coagulation process by neutralizing the negatively-charged ions (anions) that are transported toward the anode by electrophoresis. In a continuous-flow EC system, the production of polyvalent cations from the oxidation of the sacrificial anodes (Fe and/or Al) and the evolution of electrolysis gases (H_2 at the cathode and O_2 at the anode) are directly proportional to the current (charge) supplied according to Faraday's Law of Electrolysis. The evolved gases enhance the flocculation of the coagulant species.

A schematic of the electrocoagulation process for recovery gold and silver is shown in Figure 11. The gas bubbles produced by electrolysis convey the gold and silver species to the top (free-surface) of the electrolyte where it is concentrated, collected and removed. The removal mechanisms in EC may involve oxidation, reduction, decomposition, deposition, coagulation, absorption, adsorption, precipitation and flotation.

However, it is the reaction involving the metal ions that enhance the formation of the coagulant. The metal cations react with the OH- ions produced at the cathode during the evolution of hydrogen, to yield both soluble and insoluble hydroxides that will react with or adsorb pollutants, respectively, from the solution and also contribute to coagulation by neutralizing the negatively charged colloidal particles that may be present at neutral or alkaline pH. This enables the particles to approach closely and agglomerate under the influence of Van der Waals attractive forces. Depending on the pH range, the electrode reactions that have been proposed to describe EC mechanisms for the production of $H_{2(g)}$, OH^- (cathode) and H^+ and $O2_{(g)}$ (anode) are (Moreno et al., 2009):

i. pH <4

Anode:

$$Fe^{2+} + 2e^- = Fe^0 \qquad (25)$$

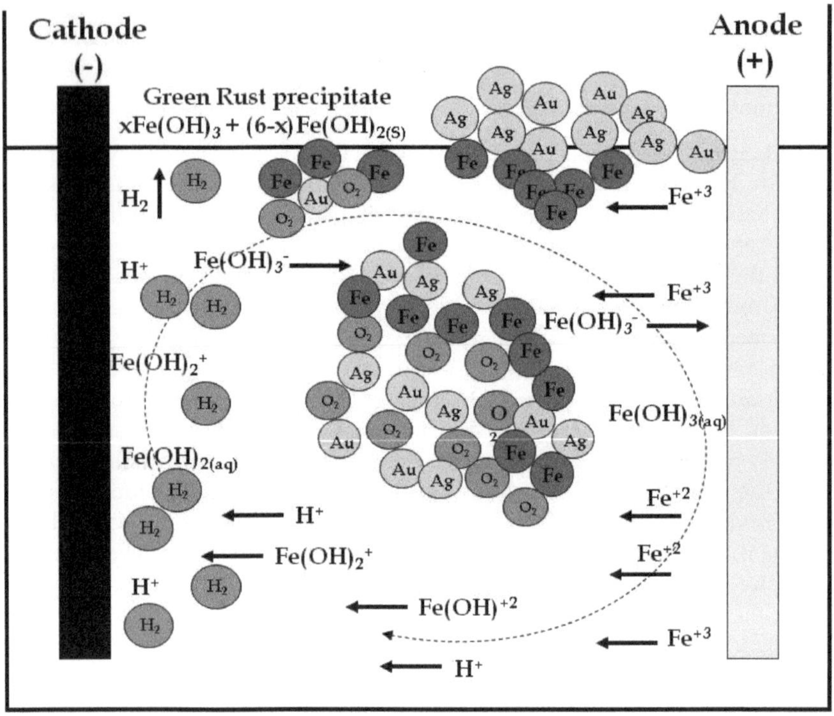

Fig. 11. An illustration of the EC mechanism (arrow indicate the migration of ions, the H_2 evolution and the formation of green rust).

Cathode:

$$2H^{+} + 2\,e^{-} \rightarrow H_{2(g)}\uparrow \qquad (26)$$

ii. pH 4 <pH< 7:

Anode: as before, reactions (25) and (26). Furthermore, iron also undergoes hydrolysis:

$$Fe + 6H_2O \rightarrow Fe(H_2O)_4(OH)_{2\ (aq)} + 2H^{+} + 2e^{-} \qquad (27)$$

$$Fe + 6H_2O \rightarrow Fe(H_2O)_3(OH)_{3\ (aq)} + 3H^{+} + 3e^{-} \qquad (28)$$

Fe(III) hydroxide begins to precipitate as a floc with yellowish color.

$$Fe(H_2O)_3(OH)_{3\ (aq)} \rightarrow Fe(H_2O)_3(OH)_{3(s)} \qquad (29)$$

"Rust" can also be formed.

$$2Fe(H_2O)_3(OH)_3 \leftrightarrow Fe_2O_3\,(H_2O)_6 \qquad (30)$$

Cathode:

$$2H^{+} + 2\,e^{-} \rightarrow H_{2(g)}\uparrow \qquad (31)$$

Additional hydrogen evolution takes place, but $[H^+]$ now comes from iron hydrolysis.

iii. pH 6 <pH< 9:

Anode: reactions (25) and (26). Precipitation of Fe(III) hydroxide occurs according to Reaction (30) simultaneous with Fe(II) hydroxide precipitation whereby a dark-green floc is produced.

$$Fe(H_2O)_4(OH)_{2\,(aq)} \rightarrow Fe(H_2O)_4(OH)_{2\,(s)} \quad (32)$$

The minimum solubility of iron hydroxides, $Fe_x(OH)_y$, occurs in the pH range of 7–8. EC flocs are formed due to the polymerization of iron oxyhydroxides. Formation of rust (dehydrated hydroxides) occurs according to the following:

$$2Fe(OH)_3 \leftrightarrow Fe_2O_3 + 3H_2O \text{ (hematite, maghemite)} \quad (33)$$

$$Fe(OH)_2 \leftrightarrow FeO + H_2O \quad (34)$$

$$2Fe(OH)_3 + Fe(OH)_2 \leftrightarrow Fe_3O_4 + 4H_2O \text{ (magnetite)} \quad (35)$$

$$Fe(OH)_3 \leftrightarrow FeO(OH) + H_2O \text{ (goethite, lepidocrocite)} \quad (36)$$

The species hematite, maghemite, rust magnetite, lepidocrocite and goethite have been identified as EC products by (Parga et al. 2005).

Cathode:

$$H_2O + 2e^- = H_{2(g)} + OH^- \quad (37)$$

Overall reactions are:

$$Fe + 6H_2O \rightarrow Fe(H_2O)_4(OH)_{2\,(s)} + H_{2(g)}\uparrow \quad (38)$$

$$Fe + 6H_2O \rightarrow Fe(H_2O)_3(OH)_{3\,(s)} + 1\tfrac{1}{2}\,H_{2(g)}\uparrow \quad (39)$$

The concentrations of the various species within the cell are not uniform; in addition the concentration of species such as iron and hydronium ion are also time dependent. Typically, the EC process employs bipolar electrodes (Parga et. al. 2005). It has been demonstrated that with this configuration where the electrodes are connected in series, and consequently low current-densities are present, iron (or aluminum) coagulant is produced more effectively, at higher rates and more economically compared to chemical coagulation (Parga et. al. 2005).

Also, in the electrocoagulation cell for the high voltage we produce a very strongly oxidizing environment around the anode and this is suitable for destroying strong cyanide solutions (greater than 1000 ppm) and is a direct oxidation of the cyanide ion at the anode to cyanate ion which is further decomposed to carbon dioxide and nitrogen, ammonium, and carbonate or oxalate ions according to the pH (Hwang et. al., 1987). The reactions are as follows:

In strong alkaline solution (pH=12):

$$CN^- + 2OH^- \rightarrow CNO^- + H_2O + 2e \quad (40)$$

$$CNO^- + 2OH^- \rightarrow CO_2 + 1/2N_2 + H_2O + 3e \quad (41)$$

In neutral and weak alkaline solution (pH = 7.0-11.7):

$$2CN \rightarrow C_2N_2 \quad (42)$$

$$C_2N_2 + 2OH^- \rightarrow CNO^- + CN^- + H_2O \quad (43)$$

$$CNO^- + 2H_2O \rightarrow NH^+_4 + CO^{2-}_3 \quad (44)$$

In weak acidic solution (pH = 5.2-6.8):

$$C_2N_2 + 4H_2O \rightarrow C_2O^{2-}_4 + 2\,NH^+_4 \quad (45)$$

3.2 Experimental details

The EC experiments were performed with a Fisher magnetic stirrer and a 400ml beaker size reactor equipped with two carbon steel electrodes (6 cm x 3 cm) that were 5 mm apart. As a source of current and voltage a universal AC/DC adaptor was used. pH was measured with a VWR scientific 8005 pH meter and electrodes were properly scrubbed and rinsed prior to each experiments to ensure a clean surface free from passive oxide layers . Gold and silver adsorption onto iron hydroxide species was investigated with pregnant cyanide solutions provided by Bacis S.A. de C.V mining group (13.25 mg L^{-1} Au, 1357 mg L^{-1} Ag, 200 mg L^{-1} free CN^- and 1400 mg L^{-1} total CN^- and pH of 8). Analysis were performed by ICP/Atomic Emission Spectrometry (Perkin Elmer 3100). The conductivity of pregnant solutions was adjusted by adding one gram of NaCl per liter (Fisher, 99.8% A.C.S. Certified, lot #995007).To identify and characterize the iron species in the solid products, formed during the EC process for the removal of gold and silver using iron electrodes, X-ray diffraction (XRD) (Phillips model X-PERT), FT-IR analysis were carried out by Thermonicolate FTIR spectrometer and Scanning Electron Microscope (SEM / EDX) (FEI Quanta 2000, Oxford Instruments) were used.

Analysis of Au and Ag were conducted to the Bacis solution, by AES. EC was run at 15 Volts (DC) and the corresponding current was of 0.1 A. EC was run for five minutes, and a sample was taken every minute in order to determinate the removal efficiency for Au and Ag. Solutions and solids from the EC process were separated by filtration through cellulose filter paper. The sludge from the EC was dried either in an oven or under vacuum at room temperature and characterized. The experimental set-up is presented in Figure 12. The current and voltage during the EC process were measured and recorded, using Cen-Tech multimeters. The pH values of the solution before and after EC were measured with a VWR scientific 8005 pH meter.

The resulting sludge of iron hydroxide gel precipitate with Au/Ag is filtered. Then, this rich sludge is treated in an acid leach step with sulfuric acid under oxidizing conditions caused by the addition of air. The conditions of this acid leach are such that the major portion of iron and copper are leached into solution with gold and silver remaining in the residue. The resultant residue from the filtration step is gold and silver, as well as little iron (10 % Au, 80 % Ag and 5 % Fe) suitable for further refining by conventional commercial method.

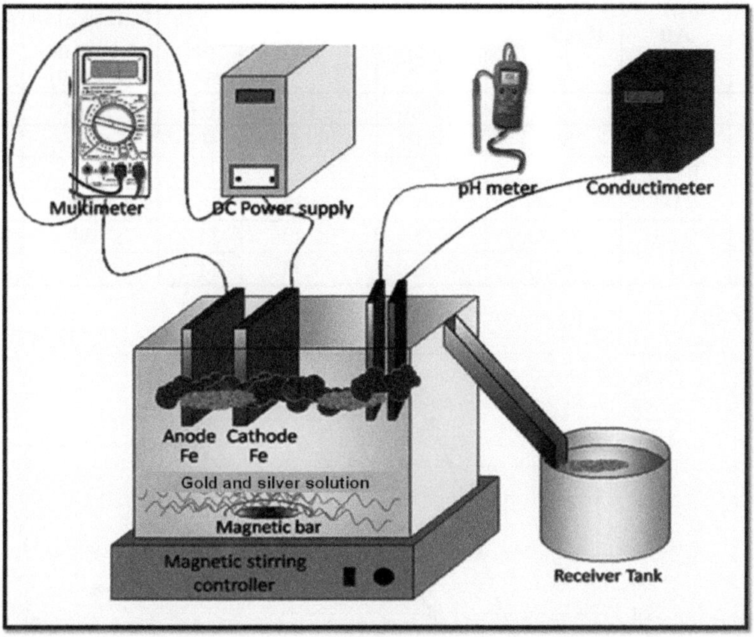

Fig. 12. Experimental set-up of the EC process for the removal of Ag and Au.

3.3 Results and discussion

With the optimal parameters for cell work of EC, they were tested for removal efficiency of gold and silver, the table 6 shows the results when making treatment of the solution containing gold and silver and also the removal efficiency of cyanide, as shown in Figure 13. A maximum in gold and silver recovery were achieved at 5 minutes of treatment and as the arithmetic average of five replications; achieved an efficiency of 99.24% Au to 99.93% Ag for both, with a standard deviation of 0.26 in gold and 0.06 in silver. Also, this studies shown a very efficient recovery in the range of 2 to 3 minutes for gold and 1 to 2 for silver, this occurs in the pH range from 9 to 11 approximately, which coincides with the production of the magnetic iron, Fe_3O_4, which has magnetic properties that accelerates the process of adsorption of metals, the adsorption rate is then physically, because it is caused by the magnetic forces of the magnetite into gold and silver these forces without altering their chemical composition. Also, it is likely that the electrocoagulation cell is oxidizing the Au and Ag cyanide complexes and converting them to a less-soluble form that is captured by the iron hydroxide gel. The high voltage in the EC cell around the anode destroys some of the cyanide.

About the same studies, Figure 14 shows graphically the evolution of pH during the operation time, there was an increase in pH of the solution which is attributed to the evolution of hydrogen at the cathode which is accompanied by alkalinization of the aqueous solution. The final effect is the oxidation suffered by the water coupled with the generation of hydroxyl ions generated during EC.

EC residence time (min)	Au (mg L-1)	Recovery (%)	Ag (mg L-1)	Recovery (%)	pH	Removal of Cyanide (mg L-1)	Removal of Cyanide (%)
0	13.25	0	1357.0	0	8.0	1400	0
1	12.50	5.66	1240.0	8.62	9.2	1050	25
2	10.50	20.37	219.5	83.82	9.5	870	38
3	1.00	92.45	9.0	99.33	10.7	750	46
4	0.50	96.22	7.0	99.48	11.2	400	71
5	0.10	99.24	0.9	99.93	11.5	210	85

Table 6. Recovery of gold and silver by EC.

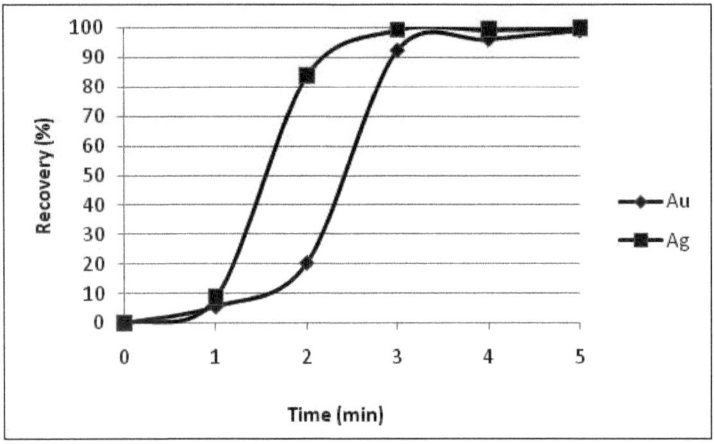

Fig. 13. Gold and silver recoveries from Bacis cyanide solutions.

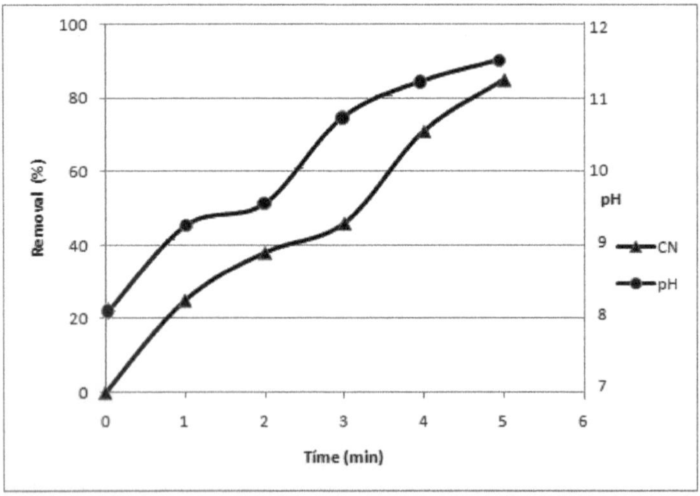

Fig. 14. Variation of pH vs. EC residence time and Cyanide removal vs time.

This results shown that in the EC cell with the iron electrodes decompose the content of cyanide in the pregnant reach solution of gold and silver from initial cyanide content of 1400 to 210 mg L^{-1}. It was found that (Tamura et al., 1974) the anodic oxidation of cyanide is proportional to the alkalinity of the electrolyte and consistent with the following mechanism:

$$CN^- + 2OH \rightarrow CNO^- + H_2O + 2e^- \quad (46)$$

$$2CNO^- + 4OH^- \rightarrow 2CO_2 + N_2 + 2H_2O + 6e \quad (47)$$

$$CNO^- + 2H_2O \rightarrow NH_4 + CO^-_3 \quad (48)$$

3.3.1 Product characterization

X-ray Diffraction Analysis. Diffraction patterns of flocs collected from the experiment with gold and silver, (the sample were ground to a fine powder and loaded into a sample holder) were obtained with a diffracted X-PERT Phillips meters equipped with a vertical goniometer, with a range of analysis 2Θ 10 ° to 70 °. The source of X-rays has a copper anode, whose radiation is filtered with a graphite monochromator (λ = 1.541838Å) with scan rate of 0.02 ° and a duration of 10 seconds per count. The X-Ray Diffractometer is controlled by a Gateway 2000 computer, by PC-APD 2.0 with software for Windows.

Figure 15 shows the ray diffraction pattern of the flocs recovered from the sample of gold and silver, respectively 13.25 mg/L and 1357 mg/L, initial pH 8, 5 minutes of treatment, 0.1 amperes and 15 volts. The species identified were magnetite, lepidocrocite, goehtite, silver and copper hexacyanoferrate.

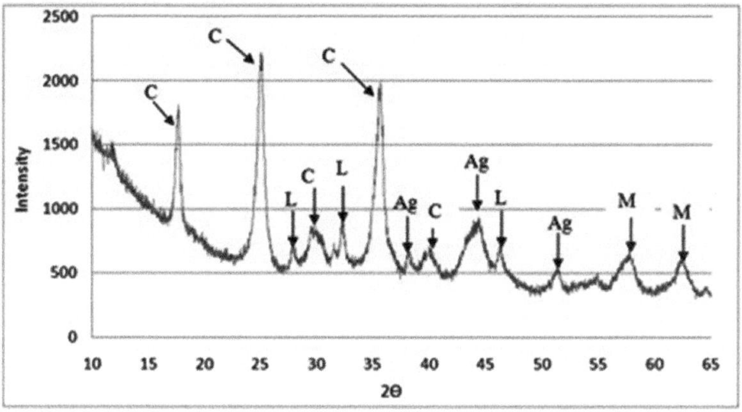

Fig. 15. X-ray diffractogram of solids obtained in the recovery of gold and silver. C: Cu_2Fe (CN) $62H_2O$ A: Silver, M: Magnetite and L: Lepidocrocite.

Fourier Transform Infrared Spectroscopy. FT-IR analysis were carried out by Thermonicolate FTIR spectrometer and OMNIC software using potassium bromide pellets (sample: KBr =1: 50). The spectra were usually recorded in the range of 4000-400 cm^{-1} with 2 cm^{-1} resolution. 64 scans were collected for each specimen. Figure 16 shows the FT-IR spectrum of the by-product. Infrared analysis of iron electrode by-product showed OH stretching at 3738 and 3447 cm^{-1}, hydroxyl bending and γ`(OH) water bending vibration or overtones of hydroxyl bending around 1637 cm^{-1} . Bands for lepidocrocite phase showed up at 1120, 1023, and 745 cm^{-1}. Magnetite (Fe_3O_4 or $Fe_{3-x}O_4$) band at 575 $cm^{-1}$and Fe-O vibration band is seen at 469 cm^{-1}. For details of FT-IR analysis see Table 7.

XRD analyses also confirmed the presence of these species detected by FT-IR.

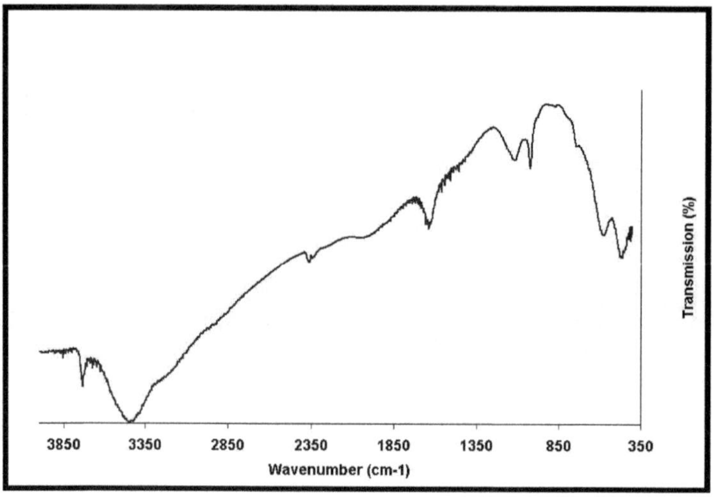

Fig. 16. FT-IR spectrum of iron electrode by-product.

Electrode Material	Type of Vibrations	Vibration wavenumbers (cm^{-1})	Vibration Range (cm^{-1})
Iron	OH stretching	3738	3689-3787
		3447	3550-3000
	Hydroxyl bending	1637	1572-1813
	γ`(OH) water bending	1637	1572-1813
	Overtones of hydroxyl bending	1637	1572-1813
	Magnetite (Fe_3O_4 or $Fe_{3-x}O_4$)	575	526-840
	Fe-O	469	416-510
	Lepidocrocite	1120	1090-1245
		1023	923-1057
		745	730-790

Table 7. FT-IR vibrations and their corresponding wave numbers and region for the bands observed for the EC-byproduct

Scanning Electron Microscopy (SEM/EDAX). Figure 17 shows SEM images and EDAX of silver adsorbed on iron species. These SEM and EDAX results show that the surfaces of these iron oxide/oxyhydroxide particles were coated with a layer of silver. It is worth clarifying that, given the low concentration of gold it was impossible to locate any nanoparticle of it.

Transmission Mössbauer Spectroscopy. Figure 18 shows the spectrum obtained from the EC silver, gold and iron solid product from 1350 mg/L of silver and 13 mg/L of gold cyanide solutions at pH = 11.0, Mössbauer Spectra for each sample was obtained on a ±15 mm/s velocity scale, which allows observation of wide magnetic hyperfine spectra expected from iron oxide compounds. The spectrum consists of a doublet magnetic spectrum, which is probably due to fine particles of iron oxides (non-stoichiometric magnetite) or iron hydroxides (Lepidocrocite, Goethite, etc.).From the analysis of these techniques the in-situ generated small fine particles of iron- oxide/oxyhydroxides in the EC process are: non-stoichiometric magnetite, goethite and iron hydroxide oxide.

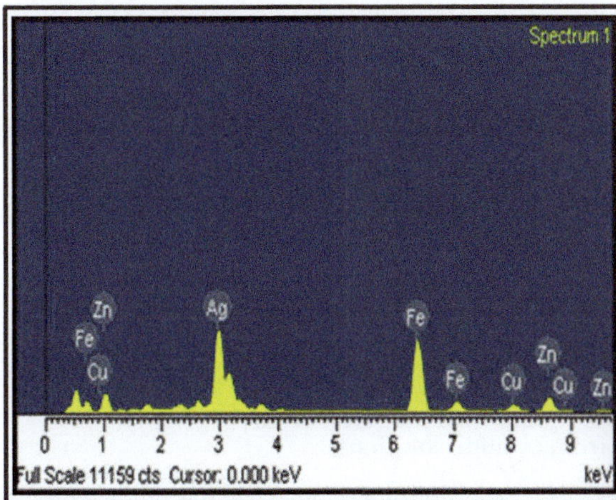

Element	Wt%	At%
Ag	43.17	29.25
Fe	36.78	48.15
Zn	14.50	16.21
Cu	5.55	6.38

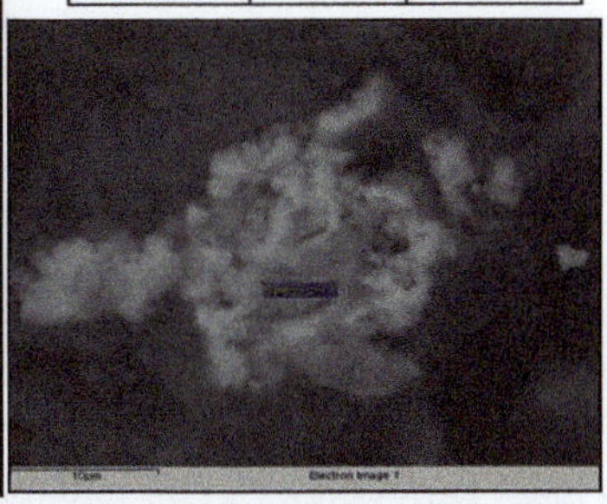

Fig. 17. Chemical composition of solid product as determined by EDX, which shows the presence of silver in the particle of iron.

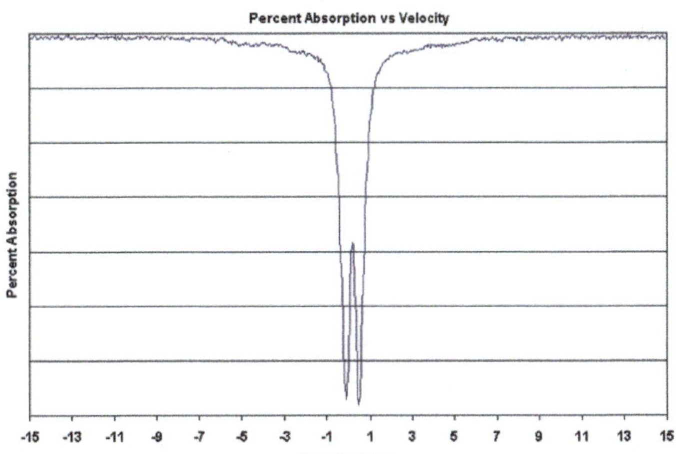

Fig. 18. Mössbauer spectrum with silver and gold at pH=10.5, indicating the presence of magnetite species.

4. Conclusions

This study has been very valuable in identifying that, gold and silver values are associated with argentopyrite, pyrite, sphalerite and chalcopyrite in the Bacís concentrate. The kinetics of the direct pressure oxidation/cyanidation was found to be strongly dependent on particle size, concentration of sodium cyanide, oxygen pressure, temperature and pH. Single stage direct pressure oxidation/cyanidation, has proven to be effective in treating argentopyrite refractory gold and silver concentrates from the Bacís mining and processing operations, for both gold and silver it was found that the precious metals recovery exceeded 95%.

Also in this process, because of the short leaching time the inventory of gold and silver is reduced. The relatively mild operating conditions of 80 °C and 80 psi oxygen pressure offer distinct advantages. For example, low cost materials of construction can be utilized for the autoclave. Finally the pressure oxidation cyanidation process is flexible and can accommodate gold and silver ore of different mineralogical composition and origin.

In addition it has been shown that Eletrocoagulation is an interesting process for the recovery of gold and silver from the cyanide leach solution that is yet to be fully realized. EC comprises complex chemical and physical processes involving many surface and interfacial phenomena. Also, the results of this study suggest that EC produces magnetic particles of magnetite and amorphous iron oxyhydroxides, and that this process can be used to remove gold and silver cyanide ions. The results of this study indicate that silver and gold can be successfully adsorbed on iron species produced by the Electrocoagulation process. So EC may be used to recover gold and silver from cyanide solutions.

The X-Ray Diffraction, FT-IR analysis and Scanning Electronic Microscopy techniques demonstrate that the formed species are of magnetic type, like lepidocrocite and magnetite, and amorphous iron oxyhydroxide which adsorbed the silver and gold particles on his surface due to the electrostatic attraction between both metals.

The 99.5% of gold and silver were removed in the experimental EC reactor, and it was achieved in 5 minutes or less with a current efficiency of 99.7%. Finally, the high voltage in the EC cell around the anode destroys some of the cyanide and this process can be accelerated in the presence of copper ions.

5. Acknowledgment

The author thanks the management of Bacís Mining Co., Williams Mining Co., CONACYT and DEGEST for the support and permission to publish this chapter and appreciation is extended to Professor Jan D. Miller of University of Utah and Gerard P. Martins of Colorado School of Mines for their interest in this research.

6. References

Adams, M.D. (2005). "Advances in gold ore processing" Elsevier Inc. 346-369, ISBN 9780444517302, Amsterdam Netherlands.

Aguayo, S., Valenzuela, J.L., Parga, J.R. & Lewis, R.G. (2007). "Continuous laboratory gold solvent extraction from cyanide solutions using LIX79 reagent". *Chem. Eng. Technol.*, 30, 1532-1536.

Anderson, C.G. & Nordwick, S.M. (1996). "Pretreatment using alkaline sulfide leaching and nitrogen species catalyzed pressure oxidation on a refractory gold concentrate". *EPD Congress the Mineral and Materials Society.*, 323-341.

Chander, S. & Briceño, A. (1987). "Kinetics of pyrite oxidation". *Minerals and Metallurgical Processing.*, 8, 170-176.

Deitz, G.A. & Halpern, J. (1953). "Reaction of silver with aqueous solutions of cyanide and oxygen". *Journal of Metals.*, 5, 1109-1116.

Emamjomeh, M.M. & Sivakumar, M. (2009). "Review of pollutants removed by electrocoagulation and electrocoagulation /flotation processes". *Int. J. Environ. Manage.*, 90, 1663-1679.

Graham, J.S. & James, T.W. (1995). "Cyanide and other lixiviant leaching systems for gold with some practical applications". *Mineral Processing and Extractive Metallurgy Review*, 14, 193-247.

Holt, P., Barton, G. & Mitchell, C. (2005). "The future of EC as a localized water treatment technology". *Chemosphere* 9, 13, 335-367.

Hwang, J. Y., Wang, Y. Y. & Wan, C.C., (1987). "Electrolytic oxidation of cuprocyanide electroplating wastewaters under different pH conditions", *J. Appl. Electrochem*, 17, 684-694.

Marsden J. O. & House, C.L. (2006). "The chemistry of gold extraction", Second edition, *Society of Mining, Metallurgy and Exploration,* 237-238, ISBN 9780873352406. Littleton, Colorado, USA.

Matteson, M. & Dobson, R, (1995). "Electrocoagulation and separation of aqueous suspensions of ultrafine particles". *Colloids and Surfaces A: Physicochemical and Engineering Aspects,* 104, 1, 101-109.

Mollah, M., Morkovsky, P., Gomez, J., Parga, J.R., & Cocke, D. (2004). "Fundamentals, present and future perspectives of electrocoagulation". *J. Hazard. Mat.*, B1, 14, 199-210.

Moreno, H., Cocke, D. L., Gomes, J. A. G., Morkovsky, P., Parga, J. R., Peterson, E. & Garcia, C. (2009). "Electrochemical reactions for electrocoagulation using iron electrodes". *Ind. Eng. Chem. Res.*, 48, 2275–2280.

Parga, J.R., Cocke, D.L., Valenzuela, J.L., Kesmez, M., Gomes, J.A.G. & Valverde, V. (2005). "As removal by EC technology in the Comarca Lagunera Mexico". *Journal of Hazardous Materials*, 124, 1-3, 247-254.

Parga, J.R., Valenzuela, J.L & Cepeda, F. (2007). "Pressure cyanide leaching for precious metals recovery". *Journal of Metals*, 10, 43-47.

Senanayake, G. (2008). "A Review of effects of silver, lead, sulfide and carbonaceous matter on gold cyanidation and mechanistic interpretation", *Journal Science Direct*, 2008, 90, 46-73.

Tamura, H., Arikado, T., Yoneyama, H. & Matsuda, Y. (1974). "Anodic oxidation of potassium cyanide on platinum electrode", *Electrochim.Acta,* 19, 273.

Thomas, K.G, Burns, D. & Hill, A.R. (1992). "Autoclaving technology and barrick gold", *(Paper prepared for the International Precious Metals Institute, Reno, Nevada)*, 11.

Vik, E.A., Carlson, D.A., Eikun, A.S. & Gjessing, E.T. (1984). "Electrocoagulation of potable water". *Water Research,* 18, 11, 1355-1360.

Wan, R.Y., Le Vier, M. & Miller, J.D. (1993). "Research and development activities for the recovery of gold from non-cyanide solutions". *Hydrometallurgy fundamentals, Technology & Innovations. (Society for Mining, Metallurgy & Exploration),* 415-436.

4

Distribution of Precious Metals During the Reducing Pyrometallurgical Processes of Complex Copper Materials

Leandro A. Voisin
University of Chile, Department of Mining Engineering
Chile

1. Introduction

The name precious metal, PM, is given to gold, silver, and the six platinum group metals, PGM, platinum, palladium, rhodium, iridium, osmium and ruthenium. These metals have a long history and close relationship with mankind, and they have exerted considerable influence on the development of society because of their special characteristic and specific properties required mainly for technology under an increasing tendency of their prices.

The variety of processes used in the recovery and refining of precious metals can be classified into two major process categories which are based on the differences in the raw material sources, in this way, we can mentioned the primary category related to those concentrates obtained from mined platinum bearing copper-nickel sulfide ores and the secondary category related to raw material sources such as recycled industrial products, spent catalyst, electronic scrap, spent electrolytes, and jewellery scrap.

The recovery method chosen for precious metals depends on the physical form of the source, its precious metal content, and the nature of the other elements present into the source. Among the primary category sources, precious metals are mainly presented as minor elements associated with base metal sulphide concentrates such as copper, lead and nickel, including arsenates, tellurides or antimonates, in fact, it is a common operational practice to use gold and silver siliceous ores as slag-forming fluxing agent in pyrometallurgy processes.

When base metals are produced by pyrometallurgical processes from primary category source ores, precious metals resulted to be concentrated into the electrolytic refining anode slimes, this situation is mainly because they have a heavy specific gravity and a general stronger affinity to metal-sulphide or metal phases than that to oxide phases or slag. According to this, an important primary source of precious metals corresponds to that derived from anodes slimes, a by-product of primary copper and/or nickel electro refining stage.

In recent times, the smelting of copper concentrates with higher content of impurities such as arsenic and antimony has increased and, unfortunately, their concentrations tend to be high in the intermediate products such as matte, white metal and blister copper. This causes a serious problem in controlling the quality of the final product cathode copper and also on the distribution of their associated precious metals during the overall process. Therefore, under these conditions, the distribution of precious metals and the elimination of impurities before the electro-refining of crude copper are of great concern in the copper production.

This chapter provides essential thermodynamic information, which predicts the distribution of precious elements during the pyrometallurgical treatment, under reducing conditions, of complex copper materials rich in impurities of arsenic and antimony at 1473K which is a common temperature for copper smelting processes. The thermodynamic studies involved in this chapter are related, for example, with the smelting of high impurities-copper sulphide concentrates or with the pyrometallurgical treatment of copper scrap, smelting dust or other kind of materials containing copper as base component, high content of impurities and precious metals as minor elements.

2. Thermodynamic concepts

In the conventional pyrometallurgical process for the production of metallic copper from copper sulfides, the terms related to the melted condensed phases as slag, matte, blister and anodic copper are well known, however, when the concentrations of impurities such as arsenic and/or antimony are high and under reducing conditions, an additional condensed phase called speiss usually appears in the process. Thus, before continuing with the main topic it is important to clarify some thermodynamic terms.

2.1 Formation of speiss in the processing of ores and treatment of secondaries

The speiss phase is basically composed of arsenides and antimonides of iron, copper, nickel and cobalt. Its formation occasionally occurs during treatment of intermediate products of non-ferrous reducing smelters such as sludge, flue dust and dross with high arsenic or antimony concentration in addition to metal bullion, matte and slag phases. A Simplified flowsheet of copper smelter showing the main elimination route of arsenic and antimony from smelting system is shown in Figs. 1. Speiss is considered to have more metallic properties than matte, its density being higher than matte and lower than the metal phase (Fuwa & Otani, 1980) and forms a separated phase between metal and matte in the furnace.

During the practical non-ferrous smelting operations, speiss formation is usually avoided by maintaining the arsenic and antimony contents in the feed materials to low concentration levels. High arsenic and antimony sources are usually blended with other feed materials to maintain low levels of arsenic and antimony in the blended charge, which prevents speiss formation in the main process stream.

Arsenic and antimony contained in the charge are removed to some degree in the slag phase during smelting, while a major portion is volatilized to the flue dust due to the high vapour pressures of sulfidic, oxidic and metallic species of these elements (Hulgren at al., 1973). Arsenic and antimony dissolved in the main process stream are usually eliminated in subsequent processing steps. Sometimes, according to the process, arsenic and antimony form metal compounds with the other base metals components and removed as dross from the system. In addition to arsenic and antimony removed in the flue dust and dross, base metal values such as copper, iron, lead, cobalt and nickel as well as precious metals such as platinum, gold and silver are also distributed in these intermediate products.

The reduction smelting process utilizing an electric furnace is commonly used for recovering these metal values. During treatment of these intermediate products a slag phase, matte phase and metallic bullion phase are formed. When the concentration of arsenic or antimony is significantly high, an additional speiss phase is also equilibrated with slag, matte and metallic bullion phases.

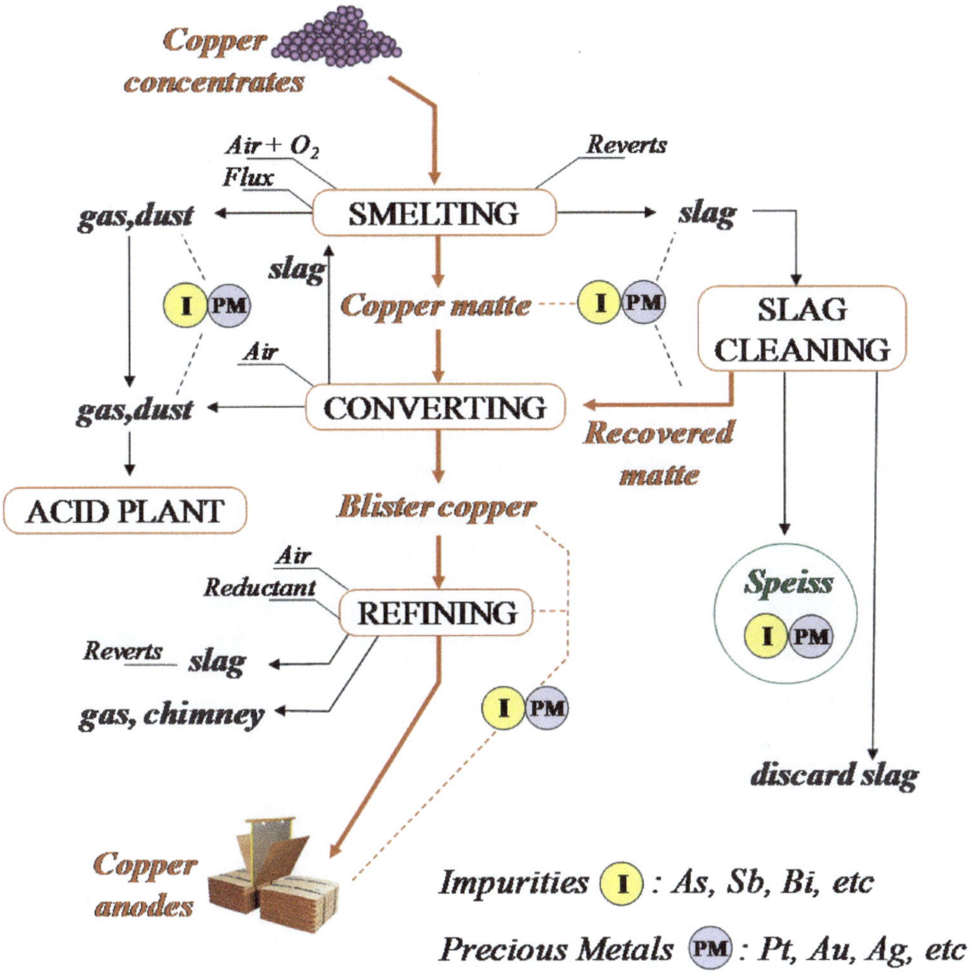

Fig. 1. Simplified copper smelter flowsheet showing the main distributions routes of detrimental impurities and precious metals from the smelting system

All the results presented in this chapter were experimentally obtained. The quenching method combined with the metallographic observation and chemical analyses using combustion-infrared spectrometry for carbon, electron probe micro analysis (EPMA) and inductively coupled plasma spectrometry (ICP) was used to determine the phase relations and the distribution of the precious metals.

2.2 Distribution ratio of precious metals between melted phases

The distribution ratio of a precious metal, X, between two condensed immiscible phases, A_1 and A_2 in a thermodynamic system, $L_X^{A_1/A_2}$, is defined as follows.

$$L_X^{A_1/A_2} = [\text{mass \% X}]A_1 \;/\; \langle \text{mass \% X} \rangle A_2 \tag{1}$$

where [ ]A_1 and < >A_2 indicate the A_1 and the A_2 phase, respectively. Assuming that the denominator of the equation corresponds to the heavier phase of the system, if there were three condensed phases in equilibrium in the system, and then we will have two distribution ratios, both with a common numerator that will correspond to the less dense phase in the system. By the definition, the element X will be concentrated in the heavier phase when the value of distribution ratio is less than unity.

3. Precious metals behaviour in the Cu-Fe-As/Sb-C systems

Sulfide ores containing bituminous coal reserves sometimes constitute a feed for copper smelting (Parviainene & Fugleberg, 1980; Dobrzanski & Kozminski, 2003). The presence of coal in a shaft furnace makes iron excessively dissolved in the melt. In a ladle adjacent to the furnace, an iron base alloy containing a considerable amount of arsenic solidifies and accumulates as a furnace residue, which is generally called "speiss". Several other metals such as copper, cobalt, nickel, silver, gold and platinum are dissolved in the residue and the recovery of these valuable metals has offered a challenging subject.

In recent years, the content of impurities of arsenic and antimony in the sulfide concentrates of non-ferrous metals tends to rise. This results in the formation of matte, slag and flue dust or dross with a considerably high content of these impurities in the nonferrous smelting processes. The speiss may be also made when these intermediate products are treated in a strongly reducing condition where the metallic iron is formed. Therefore, the behaviour of arsenic, antimony and precious metals in the speiss is of importance for treating the sulfide concentrates and by-products with the high content of impurities.

The Cu-Fe-As and the Cu-Fe-Sb ternary systems are a base for the speiss phase related to the production of copper and the treatment of by-products. According to a literature, the Cu-Fe system saturated with carbon (Chang et al., 1979) makes a miscibility gap at considerably low temperatures of less than 1500K, which is composed of the liquid copper phase with very small contents of iron and carbon, and the liquid Fe-C alloy with about 7 mass % copper. This phase separation will be useful for developing a new recovery process to treat the Cu-Fe-As/Sb base speiss, in which the less valuable iron is to be removed into the iron-rich phase, while the valuable copper and other precious metals are enriched in the copper-rich phase.

Information on the phase relations and the distribution of precious metals of silver and platinum in the miscibility gap of the Cu-Fe-As and Cu-Fe-Sb systems saturated with carbon were investigated by the author at 1473 K (Voisin et al., 2004, 2005) by a quenching method.

3.1 Phase relations in the miscibility gap of the Cu-Fe-As-C and Cu-Fe-Sb-C systems

The phase relations for the mass % ratio of Cu/Fe of 1/1 are illustrated in Fig. 2 (a) and 2 (b), respectively, in the N_C and N_{Sb} or N_{As}.

When antimony is added to the Fe-Cu binary system saturated with carbon, antimony and copper form speiss phase, while carbon and iron form an iron-rich phase. As shown in Fig. 2(a), antimony is distributed almost completely in the speiss phase and dissolves very few in the iron-rich phase, while carbon presents an opposite distribution. This behaviour of antimony can be explained from the fact that the activity of antimony shows a more negative deviation from Raoult`s law in the Cu-Sb system than in the Fe-Sb system.

On the other hand, when arsenic is added to the Fe-Cu binary system saturated with carbon, arsenic and copper form preferably speiss phase, while carbon and iron form an iron-rich phase, which also contains arsenic with smaller grade. As shown in Fig. 2(b), arsenic is distributed mostly in the speiss phase and dissolves to less extent in the iron-rich phase, while carbon is distributed almost completely in the iron-rich phase and decreases with increasing arsenic content in the same phase. This behaviour of arsenic can be explained from the fact that the activity of arsenic shows a more negative deviation from Raoult`s law in the Cu-As system than in the Fe-As system.

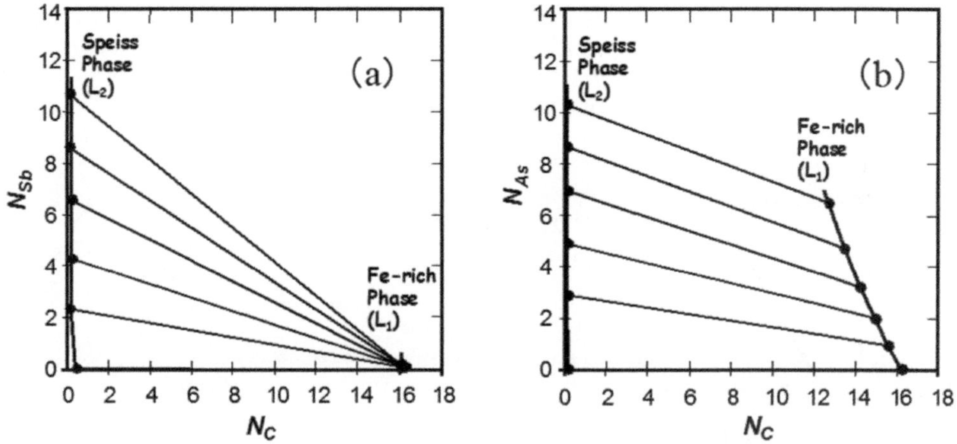

Fig. 2. (a) Relation Between N_{Sb} and N_C in Fe-rich and Speiss Phases in the Cu-Fe-Sb System Saturated with Carbon at 1473K, mass% Fe/Cu = 1; (b) Relation Between N_{As} and N_C in Fe-rich and Speiss Phases in the Cu-Fe-As System Saturated with Carbon at 1473K, mass% Fe/Cu = 1

To represent the whole systems would require quaternary diagrams, however, since the solubility of carbon in liquid copper at the experimental temperature is quite low, the compositions may be expressed with the adjusted calculation result on the pseudo-ternary diagrams in which iron and carbon are regarded as one constituent. The phase relations in the Cu-(Fe+C)-Sb and Cu-(Fe+C)-As pseudo ternary diagrams are illustrated in Figs. 3 and 4, respectively.

It is noticed in Fig. 3 that there is a miscibility gap designated by ($L_1 + L_2$) over a wide range of concentrations, and that the solubility of copper in the Fe-rich phase L_1 and the solubility of iron in the speiss phase L_2 are quite small. The miscibility gap reported in the Cu-Fe-Sb system at 1423 K (Lee & Itagaki, 1986) is shown with the dotted lines in the figure and the result obtained in the present study is very close to their result in shape though with the mutual solubility increased in the whole range studied.

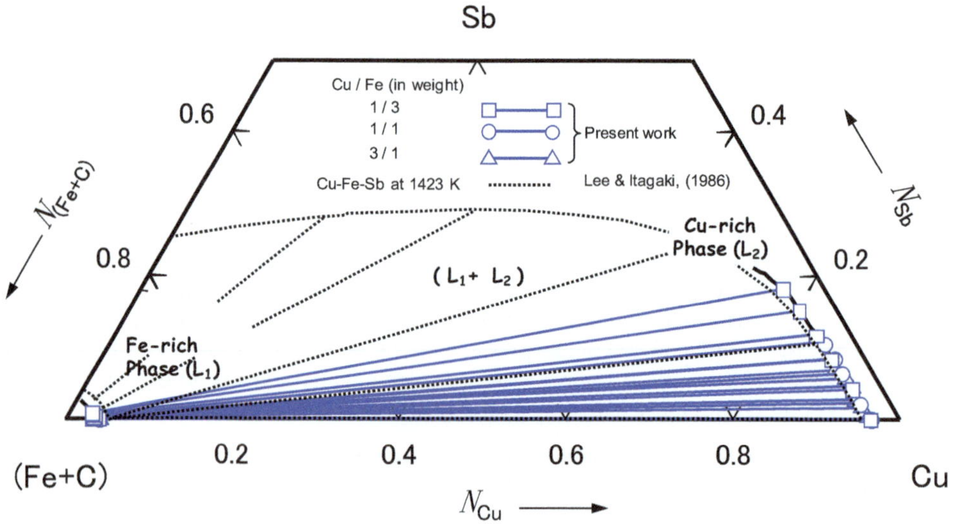

Fig. 3. Phase Relations in the Cu-(Fe+C)-Sb Pseudo Ternary System at 1473 K

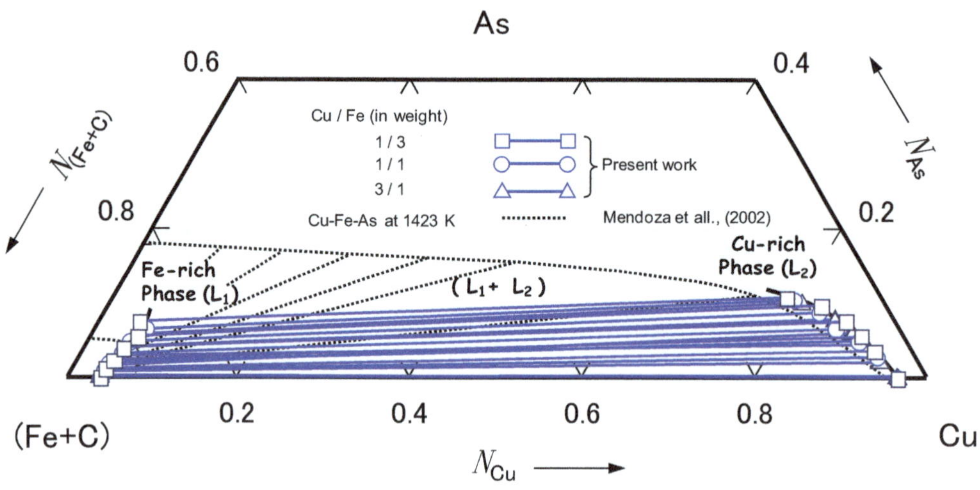

Fig. 4. Phase Relations in the Cu-(Fe+C)-As Pseudo Ternary System at 1473 K

In the Cu-(Fe+C) pseudo binary system the obtained mutual solubility of N_{Fe} = 0.036 in the speiss phase L_2 and of N_{Cu} = 0.039 in the iron-rich phase L_1 differ considerably from the reported values (Massalski, 1990) of N_{Fe} = 0.06 and of N_{Cu} = 0.09 and also from the reported values in the system without carbon (Lee & Itagaki, 1986) of N_{Fe} = 0.05 and of N_{Cu} = 0.08 respectively in the Cu-Fe binary system. This behaviour can be explained by the presence of carbon in the system.

On the other hand, it is noted in Fig. 4 that the Cu-(Fe+C)-As pseudo ternary also exhibits a miscibility gap over a wide range of concentrations, nevertheless, the mutual solubility of iron in the speiss phase and of copper in the iron-rich phase increase with increasing arsenic content. This behaviour can be explained from the fact that the activity of iron as well as copper decreases with increasing activity of arsenic.

The miscibility gap reported in the Cu-Fe-As system (Mendoza et al., 2002) is shown with the dotted lines in the figure and the obtained tendency in the present study is close to their result though the solubility of iron is increased in the whole composition of speiss investigated, which extends above 12 at% As. This behaviour can be also explained by the presence of carbon in the system.

3.2 Distribution of silver and platinum in the miscibility gap of the Cu-Fe-As-C and Cu-Fe-Sb-C systems

Adjusting the equation (1) to the corresponding systems, we will have:

$$L_X^{Fe/Sp} = [\text{mass } \% \text{ X}]Fe \,/\, <\text{mass } \% \text{ X}>Sp \quad (2)$$

where [ ]Fe and < >Sp are the iron-rich and speiss phases, respectively. Then, when the value of the distribution coefficient is larger than unity, the concentration of the element in the iron-rich phase is higher than that in the speiss phase.

Fig. 5 shows the distribution coefficients of silver and platinum in relation to the antimony content in the charge in the Cu-Fe-Sb ternary system saturated with carbon at 1473 K (Voisin et al., 2005). It is clearly demonstrated that the distribution coefficient of platinum is slightly larger than unity, thus it is dissolves more easily in the iron-rich phase while silver dissolves mostly in the speiss phase.

On the other hand, Fig. 6 shows the distribution coefficient of silver and platinum in relation to the arsenic content in the charge in the Cu-Fe-As ternary system saturated with carbon at 1473 K (Voisin et al., 2004). The distribution coefficient of platinum is almost constant at about 0.8 against the arsenic content. It is noted that silver dissolves mostly in the speiss phase, though its distribution coefficient increasing with increasing arsenic in the charge. The distribution of a precious metal between two phases would mainly depend on the affinity between the precious minor element and component elements of the two phases.

As the activity coefficient of antimony (Hino & Toguri, 1987) and arsenic (Hino & Toguri, 1986), are very small in the speiss, a large amount of these elements are concentrated into the speiss. From this fact, antimony and arsenic in nonferrous ores can be concentrated and fixed into the speiss and then be changed to a harmless substance which can be discarded safely. In this case, dissolution of metals in the speiss has to be minimized. On the contrary, there is a way to recover valuable metals from the speiss containing those metals as much as possible. For such metallurgical use of the speiss, it is necessary to know various properties of the speiss.

Since, in the Cu-Fe-Sb system saturated with carbon, platinum distributed in an iron-rich phase containing very few amounts of antimony together with copper around 4.5% in weight, these metals can be recovered by sulphuric leaching. In the Cu-Fe-As system saturated with carbon, arsenic could be concentrated and fixed into the iron-rich phase, and later on to be discarded. In case that cobalt and nickel are not of interest in the process. On the other hand, for both systems the speiss phase of copper, which contains platinum and the entirety of silver, should be treated later on.

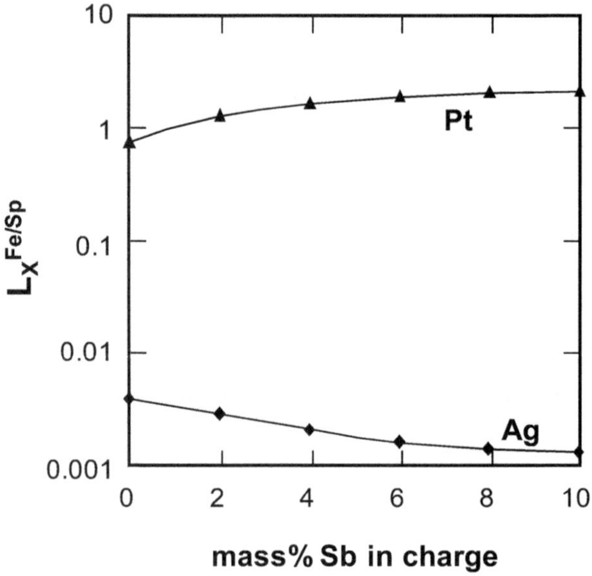

Fig. 5. Distribution Coefficient of Precious Metals in Relation to Mass% Antimony in the Charge in the Cu-Fe-Sb System Saturated with Carbon at 1473K.

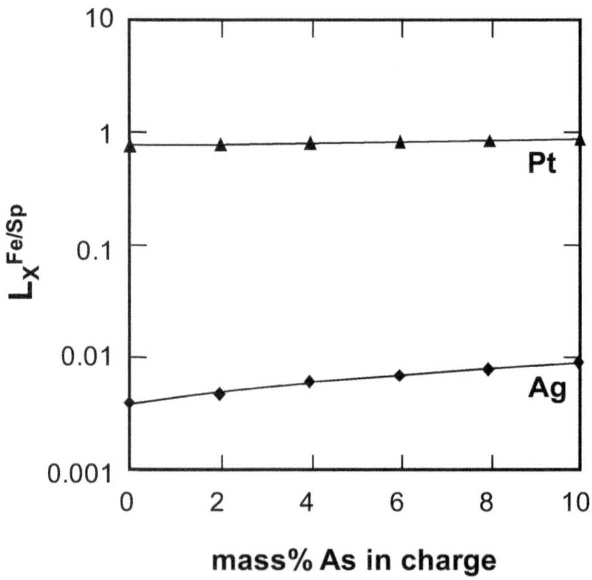

Fig. 6. Distribution Coefficient of Precious Metals in Relation to Mass% Arsenic in the Charge in the Cu-Fe-As System Saturated with Carbon at 1473K.

4. Precious metals behaviour in the Cu-Fe-C-S-As/Sb systems

In the previous section, we studied the behaviour of some precious metals in the Cu-Fe-C-As/Sb systems, which is useful to recover them from high impurities-iron-copper sources like scraps or flue dust by the addition of carbon or to understand their distribution during the formation of speiss in reductive smelting of the copper concentrate.

During the latter process like a blast furnace operation using coke as fuel and reducer or in processing the copper concentrate which contains coal in the bulk (Dobrzanski & Kozminski, 2003), also a matte phase with deficient sulfur apart from the tie line connecting between FeS and Cu_2S in the Cu-Fe-S ternary diagram is often produced in equilibrium with the speiss.

Then, it is also of interest in smelting the copper concentrate in a very strongly reducing condition to know the behaviour of precious metals between the matte and the furnace residue or the speiss. In this respect, the phase relations and distribution of precious metals in the Cu-Fe-S-As-C (Voisin & Itagaki, 2006) and Cu-Fe-S-Sb-C (Voisin et al., 2008) systems are of great importance in the extreme case of reducing smelting. Additionally, the study of those systems will provide the fundamental information on a new process proposed by the author to eliminate arsenic and antimony from the impurities-rich matte produced from the copper concentrate containing significant amount of those elements.

It is considered in the proposed process that the matte is to be reduced with a given amount of molten pig iron at about 1473 K to produce a molten iron base alloy (speiss) with a considerable amount of impurities (arsenic or antimony), but with very few amounts of precious metals such as silver and gold, which can be expected to be discarded as a harmless deposit in the yard. Hence, the phase relations and the distribution ratios of such precious metals as silver, gold and platinum in the Cu-Fe-S and Cu-Fe-S-As/Sb systems saturated with carbon at 1473 K are presented in this section.

4.1 Phase relations in the Cu-Fe-S-C system

Three different zones were observed in the Cu-Fe-S system saturated with carbon at 1473 K, as shown in Fig. 7. Zone I is delimited by points a, b, c and d where the iron-rich alloy and matte phases are in equilibrium. Zone II is delimited by points c, d and e where a further copper-rich alloy is equilibrated with the iron-rich alloy and matte phases. Zone III is delimited by points c, e, f and g where the copper-rich alloy and matte phases are in equilibrium.

In the first zone, the slope of the tie lines shifts to the iron side since the presence of carbon in the system makes the iron fusible at the experimental temperature. In zone II, a large miscibility gap is observed. The obtained matte composition with a copper content (matte grade) of 63 mass % and also the sulfur content in the liquid copper-rich alloy are in a good agreement with those reported in the Cu-Fe-S system at the same temperature (Mendoza et al., 2002a, 2002b) though the solubility of iron in the copper-rich alloy is larger than the reported value.

In zone III, the obtained solubility limits of sulfur for the copper-rich alloy and matte phases in the Cu-S-C system are in a good agreement with the reported data because the content of carbon in both phases is very small. Furthermore, the data obtained for the Cu-Fe-C system at 1473 K (Voisin et al. 2004) showed in the previous section 3, were also plotted in Fig. 7 and the results are in a good projection toward the obtained points e and d in the copper-rich and iron-rich alloys, respectively, in the Cu-Fe-S system saturated with carbon. The Fig. 8 shows the results in the corresponding quaternary system.

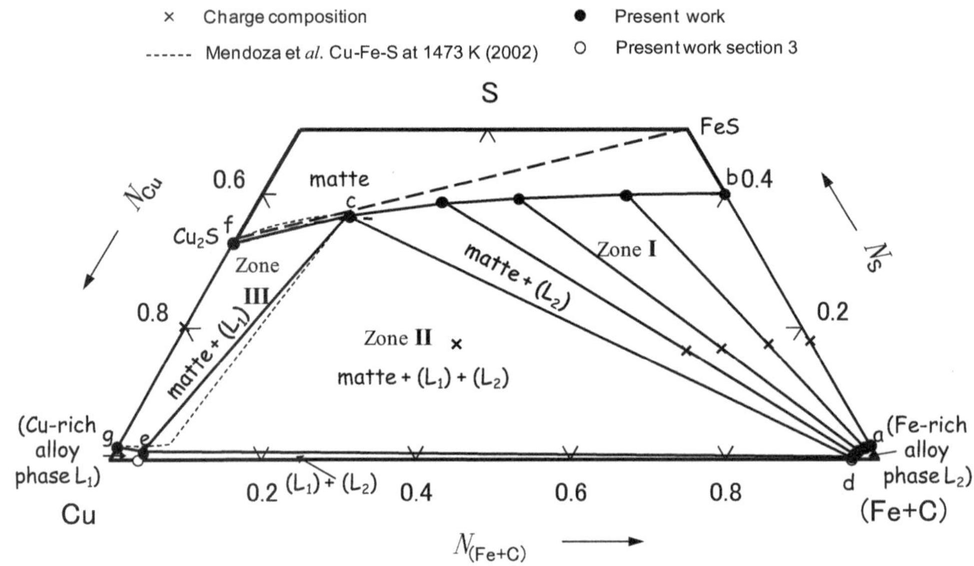

Fig. 7. Phase relations in the Cu-(Fe+C)-S pseudo ternary system at 1473 K

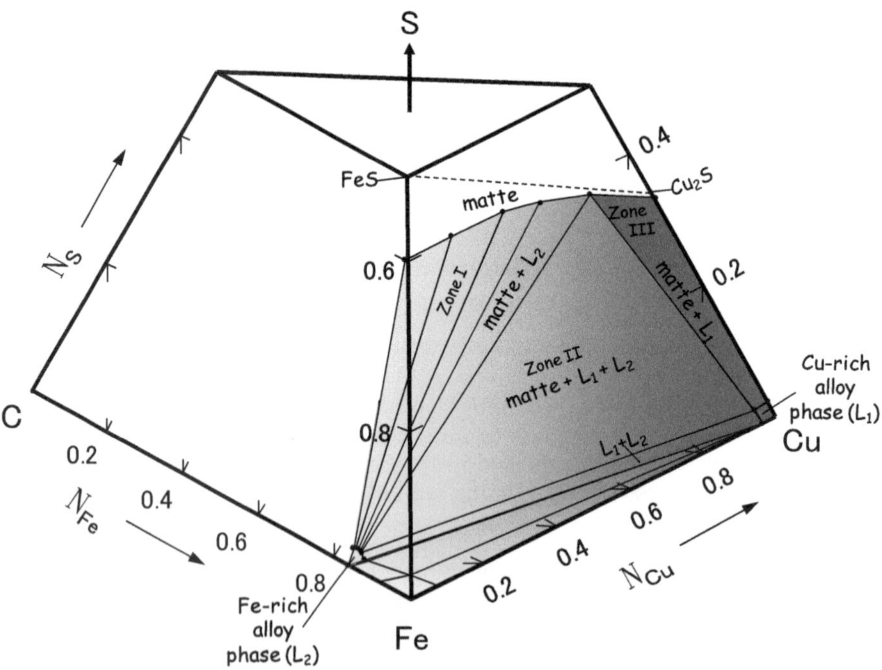

Fig. 8. Phase relations in the Cu-Fe-S-C quaternary system at 1473 K

Since the distribution of precious metals was studied under the presence of impurities of arsenic and antimony into the zone II forming a matte phase in equilibrium with two speiss phases, it is important to clarify the behavior of those impurities in Cu-Fe-S-C system previous to show that for precious metals between the three equilibrated phases.

4.2 Phase relations in the Cu-Fe-S-As-C system

The phase relations in the Cu-Fe-S-As-C system for zone II are shown in Figs. 9-a and -b for the iron-rich alloy and matte phases and for the copper-rich alloy and matte phases, respectively, in relation to the mole fractions of sulfur (N_S) and arsenic (N_{As}). The corresponding compositions are listed in Table 1.

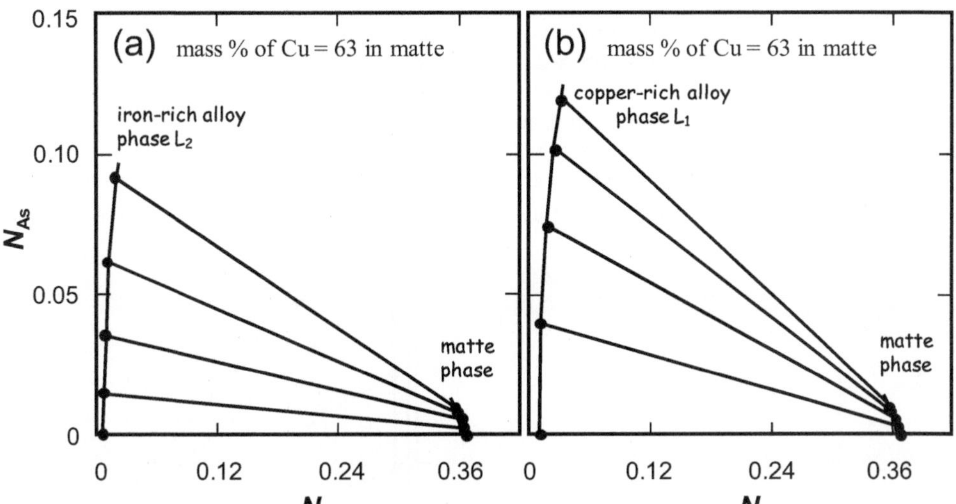

Fig. 9. Relation between N_{As} and N_S in the iron-rich or copper rich alloy and matte phases in the Cu-Fe-S-As systems saturated with carbon at 1473 K (zone II)

mass% in copper-rich alloy phase (L_1)					mass% in iron-rich alloy phase (L_2)					mass% in matte phase				
Fe	S	As	Cu	C	Fe	S	As	Cu	C	Fe	S	As	Cu	C
3.41	0.58	0	95.99	0.021	91.59	0.38	0	4.07	3.96	13.85	23.16	0	62.95	0.047
5.51	0.64	4.69	89.14	0.027	88.88	0.45	2.22	4.90	3.55	14.61	23.00	0.37	61.96	0.044
8.11	0.98	8.80	82.08	0.032	85.58	0.56	5.23	5.31	3.33	14.92	22.84	0.82	61.39	0.034
11.91	1.38	12.06	74.61	0.039	81.39	0.68	8.85	6.38	2.70	15.54	22.43	1.20	60.80	0.027
16.56	1.71	14.27	67.42	0.042	71.58	1.12	12.62	12.77	1.92	16.57	22.29	1.40	59.73	0.016
mass %Cu / %Fe / %S in charge = 27.6 / 62.9 / 9.5														

Table 1. Phase equilibrium composition of the copper-rich alloy, iron-rich alloy and matte phases in the Cu-Fe-S-As system saturated with carbon at 1473 K (zone II)

When arsenic is added to the Cu-Fe-S-C system in zone II, it is preferentially enriched in the metallic copper-rich and iron-rich alloys with a small amount of sulfur, while sulfur, copper and iron form the matte phase with a smaller amount of arsenic when compared with that in the metallic phases. With increasing arsenic content in the charge, the sulfur content decreases in the matte phase and increases in the metallic phases.

Furthermore, as listed in Table 1, with increasing arsenic content, the carbon content in the iron-rich alloy decreases, while that in the copper-rich alloy and matte phases increases and decreases, respectively, though they are negligibly small. The copper content in the copper-rich alloy phase decreases with increasing arsenic content, while the iron content increases. On the contrary, the copper content in the iron-rich alloy phase increases with increasing arsenic content while the iron content decreases.

4.3 Phase relations in the Cu-Fe-S-Sb-C system

The phase relations in the Cu-Fe-S-Sb-C system for zone II are shown in Figs. 10-a and -b for the iron-rich alloy and matte phases and for the copper-rich alloy and matte phases, respectively, in relation to the mole fractions of sulfur (N_S) and arsenic (N_{Sb}). The corresponding compositions are listed in Table 2.

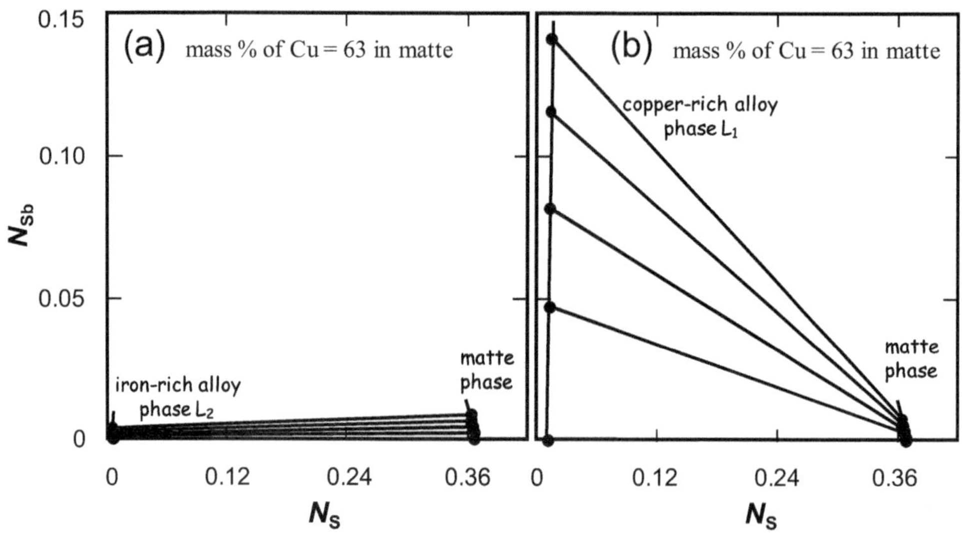

Fig. 10. Relation between N_{Sb} and N_S in the iron-rich or copper rich alloy and matte phases in the Cu-Fe-S-Sb systems saturated with carbon at 1473 K (zone II)

When antimony is added to the Cu-Fe-S-C system in zone II, it is preferentially enriched in the copper-rich alloy with a small amount of sulfur, while the most sulfur, copper and iron form the matte phase with a smaller amount of antimony when compared with that in the copper-rich alloy phase. The iron-rich alloy phase contains very few antimony even less than the matte phase and almost the entirely of carbon and a copper content between 2.7 and 4 mass %.

With increasing antimony content in the charge, the sulfur content decreases in the matte and iron-rich alloy phases and increases in the copper-alloy rich phase. Furthermore, as listed in Table 2, with increasing antimony content, the carbon content in the iron-rich alloy is almost constant at about 4 mass % while that slightly decreases in the copper-rich alloy and matte phases though they are negligibly small. The copper content decreases, while the iron content increases with increasing antimony content in the alloy phases.

mass% in copper-rich alloy phase (L_1)					mass% in iron-rich alloy phase (L_2)					mass% in matte phase				
Fe	S	Sb	Cu	C	Fe	S	Sb	Cu	C	Fe	S	Sb	Cu	C
3.41	0.58	0	95.99	0.021	91.59	0.38	0	4.07	3.96	13.85	23.16	0	62.95	0.047
3.50	0.59	8.98	86.90	0.020	91.76	0.31	0.33	3.65	3.95	14.90	23.26	0.52	61.37	0.045
3.84	0.61	15.06	80.47	0.017	91.84	0.30	0.52	3.38	3.96	15.62	22.93	1.07	60.34	0.038
4.06	0.62	20.61	74.69	0.016	92.12	0.31	0.74	2.88	3.96	18.46	22.85	1.55	57.10	0.036
5.42	0.66	24.66	69.25	0.015	92.01	0.32	1.05	2.67	3.96	20.37	22.80	2.10	54.70	0.033

mass %Cu / %Fe / %S in charge = 27.6 / 62.9 / 9.5

Table 2. Phase equilibrium composition of the copper-rich alloy, iron-rich alloy and matte phases in the Cu-Fe-S-Sb system saturated with carbon at 1473 K (zone II)

4.4 Distribution of silver, gold and platinum in the miscibility gap of the Cu-Fe-S-As-C and Cu-Fe-S-Sb-C systems

Adjusting the equation (1) to the distribution ratio of a precious metal X between the matte and copper-rich alloy or iron-rich alloy phases in the Cu-Fe-S-As or in the Cu-Fe-S-Sb system saturated with carbon, $L_X{}^{m/L}$, is defined by

$$L_X{}^{m/L} = [\text{mass } \% \text{ X}]\text{m} \ / \ <\text{mass } \% \text{ X}>\text{L} \tag{3}$$

where [ ]m and < >L indicate the matte and the iron-rich alloy or copper-rich alloy phases, respectively. By definition, the element X will be concentrated in the alloy phase when the value of distribution ratio is less than unity. Hence, a larger value of $L_X{}^{m/L}$ corresponding to the iron-rich alloy phase and a smaller one corresponding to the copper-rich alloy phase are preferable when a process for treating the by-products containing arsenic or antimony is considered, in which the precious metals will be recovered into the copper-rich alloy phase, while the less valuable iron eliminated into the iron-rich alloy phase.

The distribution ratios of silver, gold and platinum between the matte and copper-rich alloy (L_1) or iron-rich alloy (L_2) phases are shown in Fig. 11 in relation to the arsenic content while in Fig. 12 in relation to the antimony content in the matte for the Cu-Fe-S-As and Cu-Fe-S-Sb system saturated with carbon, respectively.

As shown in Fig. 11, the distribution ratios of silver, gold and platinum between the matte and copper-rich alloy (L_1) phases increase with increasing content of arsenic in the matte phase. It is noted that the distribution ratio at a given arsenic content in the matte decreases in the order of silver, platinum and gold. $L_{Ag}{}^{m/L_1}$ presents values of less than 1, while $L_{Au}{}^{m/L_1}$ and $L_{Pt}{}^{m/L_1}$ very small values around 0.001 in the whole range of arsenic content.

It is shown in the same figure that the distribution ratio between the matte and iron-rich alloy (L_2) phases at a given arsenic content in the matte decreases in the order of silver, gold and platinum. $L_{Ag}{}^{m/L_2}$ is almost constant at about 12 while $L_{Au}{}^{m/L_2}$ abruptly decreases and $L_{Pt}{}^{m/L_2}$ increases against the arsenic content in the matte.

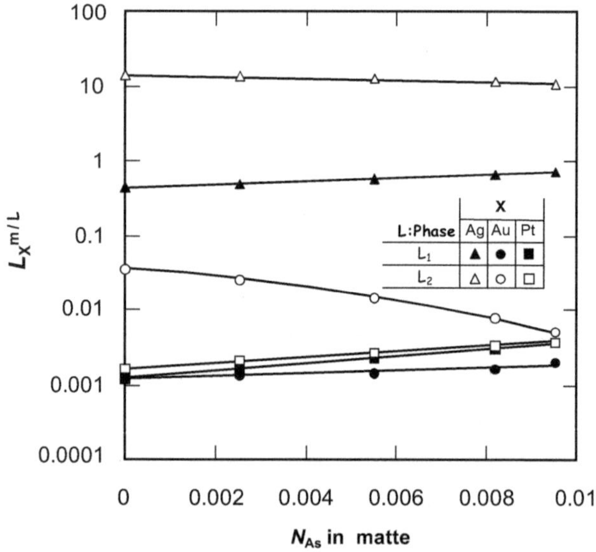

Fig. 11. Distribution ratios of silver, gold and platinum in relation to N_{As} in the matte in the Cu-Fe-S-As system saturated with carbon at 1473 K.

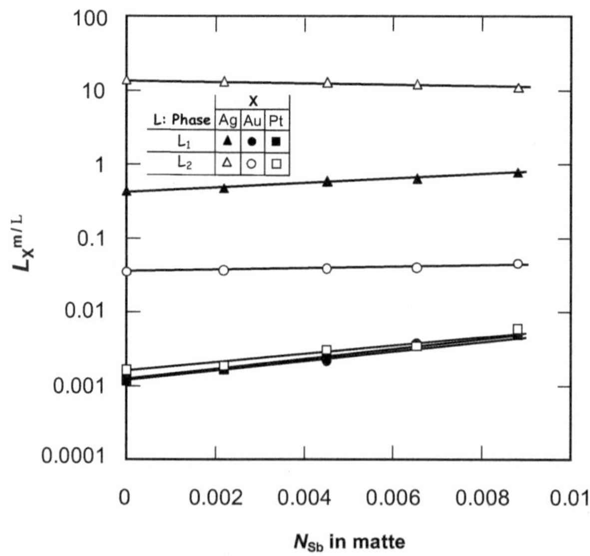

Fig. 12. Distribution ratios of silver, gold and platinum in relation to N_{Sb} in the matte in the Cu-Fe-S-Sb system saturated with carbon at 1473 K.

On the other hand, as shown in Fig. 12, the distribution ratios of those precious metals between the matte and copper-rich alloy (L_1) phases increase with increasing content of antimony in the matte phase. It is noted that the distribution ratio at a given antimony content in the matte decreases in the order of silver, platinum and gold. L_{Ag}^{m/L_1} presents values of less than 1, while L_{Au}^{m/L_1} and L_{Pt}^{m/L_1} very small values between 0.01 and 0.001 in the whole range of antimony content.

It is shown in the same figure that the distribution ratio between the matte and iron-rich alloy (L_2) phases at given antimony content in the matte decreases in the order of silver, gold and platinum. L_{Ag}^{m/L_2} and L_{Au}^{m/L_2} are almost constant at about 12 and 0.04, respectively, while L_{Pt}^{m/L_2} increases from 0.0015 to 0.007 against the antimony content in the matte.

Finally, we can determine the distribution ratio of silver and platinum between two metallic-rich alloy phases in the zone II where they are also in equilibrium with matte in the Cu-Fe-S-As or in the Cu-Fe-S-Sb system saturated with carbon and compare them to the results obtained in section 3 for those precious metals where there is not presence of sulfur and there are only two metallic-rich alloy phases in equilibrium. According to this, we need to adjust the equation (1) as follow:

$$L_X^{L_1/L_2} = [\text{mass \% X}]L_1 \,/\, \langle\text{mass \% X}\rangle L_2 \qquad (4)$$

where [ ]L_1 and < >L_2 indicate the copper-rich alloy and the iron-rich alloy phase, respectively. The distribution ratios of silver and platinum are shown in Figs. 13 and 14 for the Cu-Fe-S-As-C and Cu-Fe-S-Sb-C systems in relation to the arsenic and antimony content in the copper-rich alloy phase, respectively, the results are shown together with those obtained in section 3 in the corresponding system without the presence of sulfur.

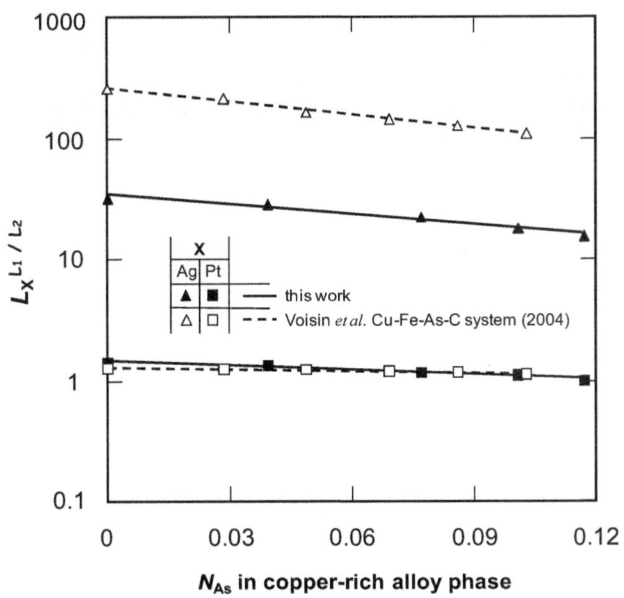

Fig. 13. Distribution ratios of silver and platinum between the copper-rich alloy (L_1) and iron-rich (L_2) alloy phases in relation to N_{As} in the copper-rich alloy phase in the Cu-Fe-S-As system saturated with carbon at 1473 K

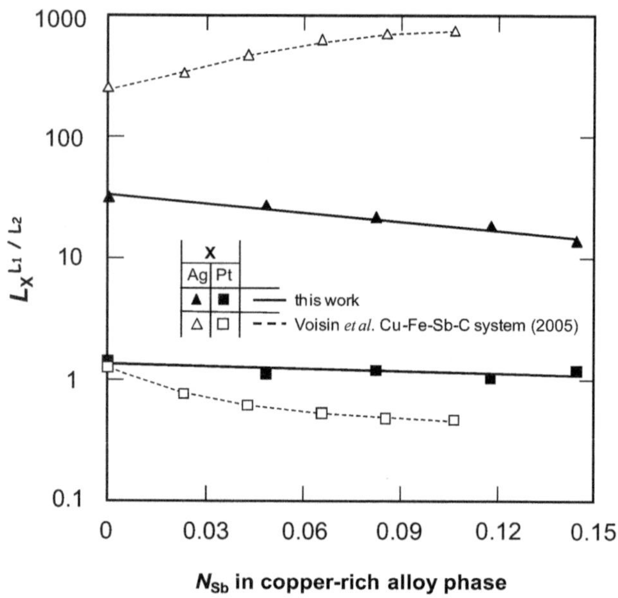

Fig. 14. Distribution ratios of silver and platinum between the copper-rich alloy (L_1) and iron-rich (L_2) alloy phases in relation to N_{Sb} in the copper-rich alloy phase in the Cu-Fe-S-Sb system saturated with carbon at 1473 K

As shown in Figs. 13 and 14, the distribution ratios of silver and platinum between the copper-rich alloy (L_1) and iron-rich alloy (L_2) phases in the Cu-Fe-S-As-C and Cu-Fe-S-Sb-C systems decrease with increasing arsenic and antimony content, respectively, in the copper-rich alloy phase. $L_{Ag}{}^{L_1/L_2}$ is larger than 10 while $L_{Pt}{}^{L_1/L_2}$ close to unity in the whole range of impurity content.

The distribution ratios of silver and platinum in the Cu-Fe-As-C and Cu-Fe-Sb-C systems obtained in section 3 are also shown in the corresponding Figs. 13 and 14, respectively, with broken lines. In Fig, 13, it is noteworthy that, in terms of the magnitude and the dependency against the arsenic content in the copper-rich alloy phase, $L_{Pt}{}^{L_1/L_2}$ is very similar while $L_{Ag}{}^{L_1/L_2}$ is about 10 times larger than that for the Cu-Fe-S-As-C system. On the other hand, in Fig. 14, $L_{Ag}{}^{L_1/L_2}$ presents an increasing trend and it is about 10 times larger and $L_{Pt}{}^{L_1/L_2}$ is similar but decreases more drastically than those for the Cu-Fe-S-Sb-C system.

5. Treatment of complex copper matte by reducing condition

Based on the experimental results showed in this chapter, the material balances were evaluated for the treatment of complex copper matte according to the proposed process by the author.

It was supposed in the calculation that 100 kg of copper matte with a matte grade of 70% Cu containing arsenic and antimony with 0.5 mass % for each one, and precious metals of silver, gold and platinum with each 0.01 mass % is treated at 1473 K by adding different amounts of pig-iron at 1473 K (iron saturated with about 4.3 mass % of carbon) of 10, 20, 30, 40 and 50 kg. Then, these values were converted to the dimensionless ones of α = 0.1, 0.2, 0.3, 0.4 and 0.5 with α = (amount of added pig-iron) / (amount of initial copper matte (100 kg)). According to the results obtained by the author, metallic copper-rich and iron-rich alloys together with a sulfide matte phase containing about 63 % of copper are in equilibrium in these conditions.

In the calculation, the loss of arsenic, antimony and sulfur by volatilization was neglected because the summation of equilibrium partial pressures of predominant As and As_2 for arsenic, Sb, Sb_2 and Sb_4 for antimony and S_2 for sulfur gas species over the corresponding alloys and matte at 1473 K are very small at less 10 (Voisin et al., 2004), 0.1 (Voisin et al., 2005) and 2 Pa (Zekeri et al., 2000), respectively. The schematic diagram with the corresponding charge compositions is shown in Fig. 15, while the compositions of iron, copper and sulfur for the different levels of α are represented in the Cu-(Fe+C)-S pseudo ternary system and shown in Fig. 16.

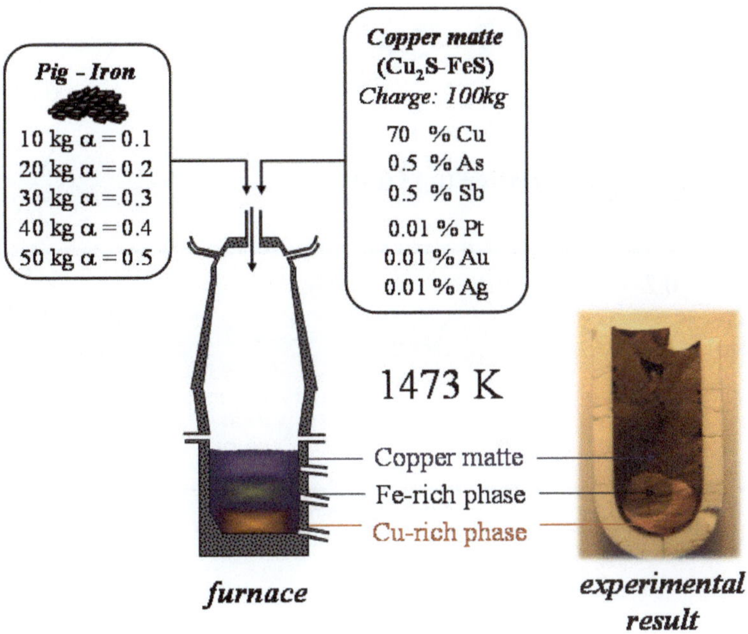

Fig. 15. Schematic diagram with the corresponding charge compositions for the treatment of complex copper matte at 1473 K

The calculated results are listed in Tables 3, 4 and 5 representing the fractional distribution (%) of all the elements for the copper-rich alloy, iron-rich alloy and matte phases, respectively. Furthermore, the fractional distribution, between the three phases and the mass % in the treated matte phase are shown against α in Figs. 17, 18 and 19 for copper, arsenic and antimony as impurities and silver, gold and platinum as precious metals, respectively. Since carbon and sulfur are mostly distributed in the iron-rich alloy and matte phases, respectively, their results are not shown in those figures.

According with the calculation, 0.065 correspond to the minimum value of α where the three phases start appearing when the pig-iron is added to the charged matte. This means that at least 6.5 kg of pig-iron must be added to the system for treating 100 kg of matte. It is indicated in Tables 3, 4 and 5 that the amount of copper-rich alloy phase decreases from 11.6 to 9.2 kg, the amount of iron-rich alloy phase increases from 3.8 to 45.7 kg and the amount of matte phase is almost constant at about 94.5 kg with increasing value α, respectively.

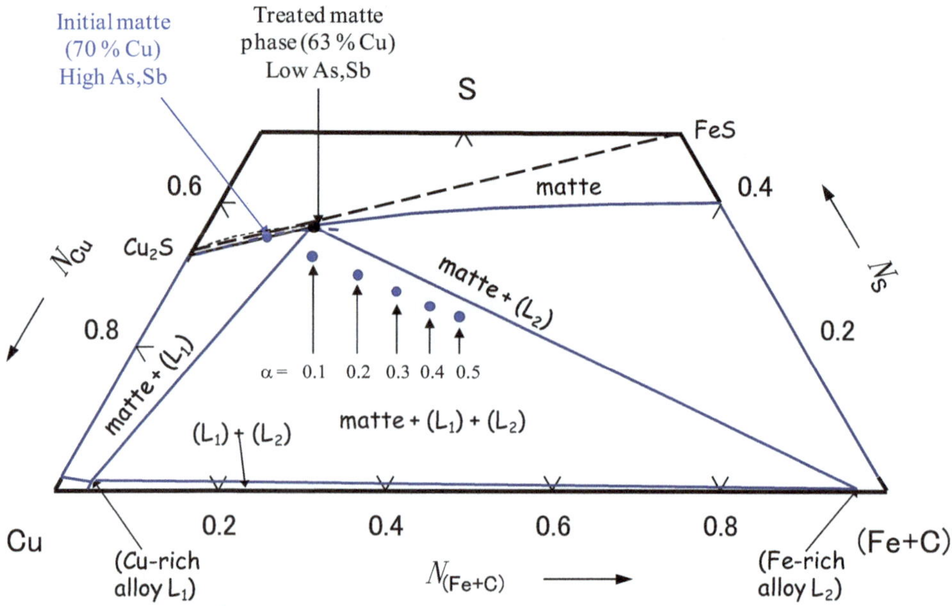

Fig. 16. Phase relations in the Cu-(Fe+C)-S pseudo ternary system at 1473 K

element	fractional distribution (%)				
Cu	15.0	13.6	13.3	12.7	11.9
Fe	2.49	1.44	1.03	0.780	0.604
As	60.1	47.3	39.8	33.3	28.1
Sb	70.0	65.6	62.9	59.9	56.4
C	1.26	0.368	0.220	0.146	0.107
S	0.291	0.265	0.258	0.247	0.232
Ag	20.7	19.1	18.6	18.0	17.0
Au	97.4	93.0	89.3	85.1	81.0
Pt	78.8	48.0	34.9	26.1	20.5
total (kg)	11.6	10.6	10.3	9.8	9.2
pig-iron in charge (kg)	10	20	30	40	50
α	0.1	0.2	0.3	0.4	0.5

Table 3. Material balances for the copper-rich alloy phase in the treatment of 100 kg of complex copper matte at 1473 K

element	fractional distribution (%)				
Cu	0.24	0.89	1.50	2.19	2.81
Fe	20.6	49.6	61.3	70.8	75.1
As	7.50	24.5	35.7	45.6	52.7
Sb	1.18	4.61	7.64	11.1	14.3
C	76.1	92.3	95.3	96.8	97.5
S	0.069	0.259	0.437	0.639	0.820
Ag	0.234	0.890	1.49	2.19	2.81
Au	1.49	5.82	9.53	13.7	17.8
Pt	20.3	51.4	64.6	73.6	79.2
total (kg)	3.83	14.5	24.4	35.6	45.7
pig-iron in charge (kg)	10	20	30	40	50
α	0.1	0.2	0.3	0.4	0.5

Table 4. Material balances for the iron-rich alloy phase in the treatment of 100 kg of complex copper matte at 1473 K

element	fractional distribution (%)				
Cu	84.8	85.5	85.2	85.1	85.2
Fe	76.9	49.0	37.6	28.4	24.3
As	32.4	28.2	24.4	21.2	19.2
Sb	28.8	29.8	29.4	29.0	29.3
C	22.7	7.31	4.49	3.09	2.44
S	99.6	99.5	99.3	99.1	98.9
Ag	79.1	80.0	79.9	79.8	80.2
Au	1.10	1.17	1.16	1.15	1.18
Pt	0.891	0.591	0.437	0.334	0.280
total (kg)	94.3	94.8	95.1	94.4	94.9
pig-iron in charge (kg)	10	20	30	40	50
α	0.1	0.2	0.3	0.4	0.5

Table 5. Material balances for the matte phase in the treatment of 100 kg of complex copper matte at 1473 K

As is shown in Fig. 17, with increasing value of α, the fractional distribution of copper decreases from 15 to 11.9 %, increases from 0.24 to 2.81 % and is almost constant at about 85 % in the copper-rich alloy, iron-rich alloy and matte phases, respectively, and the matte grade (% Cu) of treated matte is about 63 %. For the iron, the fractional distribution decreases from 2.5 to 0.6 % and from 76.9 to 24.3 % in the copper-rich alloy and matte phases, respectively, while it increases from 20.6 to 75.1 in the iron-rich alloy phase. The sulfur is almost completely distributed in the matte phase while the carbon in the iron-rich alloy phase.

For the impurities, as is shown in Fig. 18, with increasing value of α, the fractional distribution of arsenic decreases from 60.1 to 28.1%, and from 32.4 to 19.2 % in the copper rich alloy and matte phases, respectively, and increases from 7.5 to 52.7 % in the iron-rich alloy phase. For the antimony, the fractional distribution decreases from 70 to 56.4 %, increases from 1.18 to 14.3 % and is almost constant at about 29 % in the copper-rich alloy, iron-rich alloy and matte phases, respectively. The mass % of arsenic and antimony in the treated matte for the value of α close to 0.1 are 0.18 and 0.15 respectively.

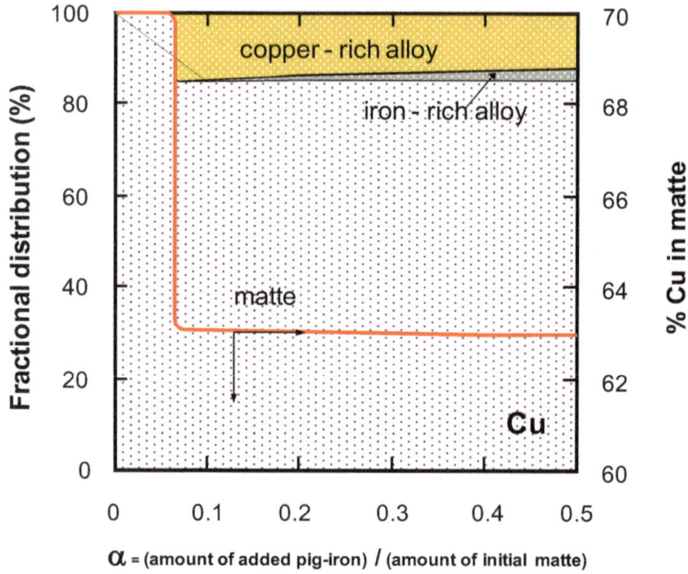

Fig. 17. Fractional distribution of copper and mass % of copper in matte against α during the reducing treatment of complex copper matte at 1473 K

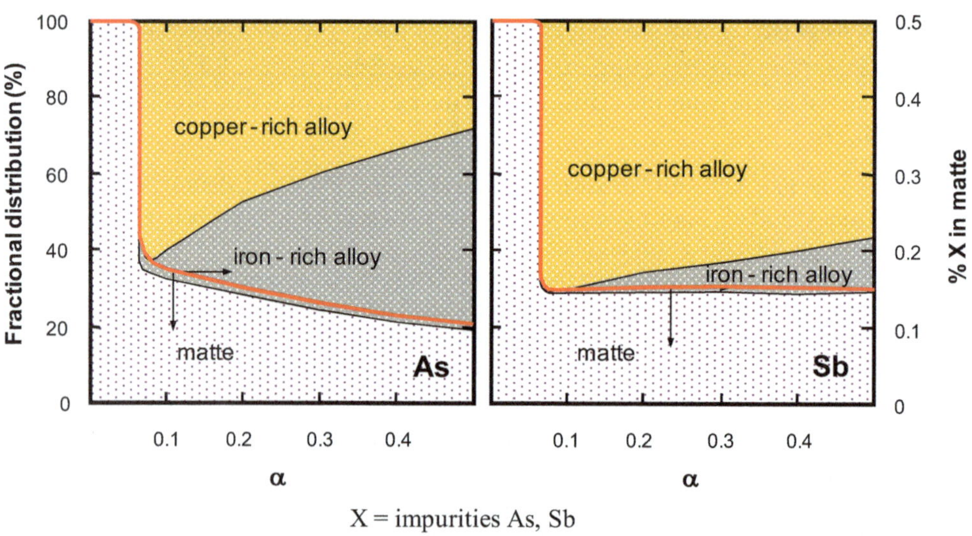

Fig. 18. Fractional distribution of impurities and their mass % in matte against α during the reducing treatment of complex copper matte at 1473 K

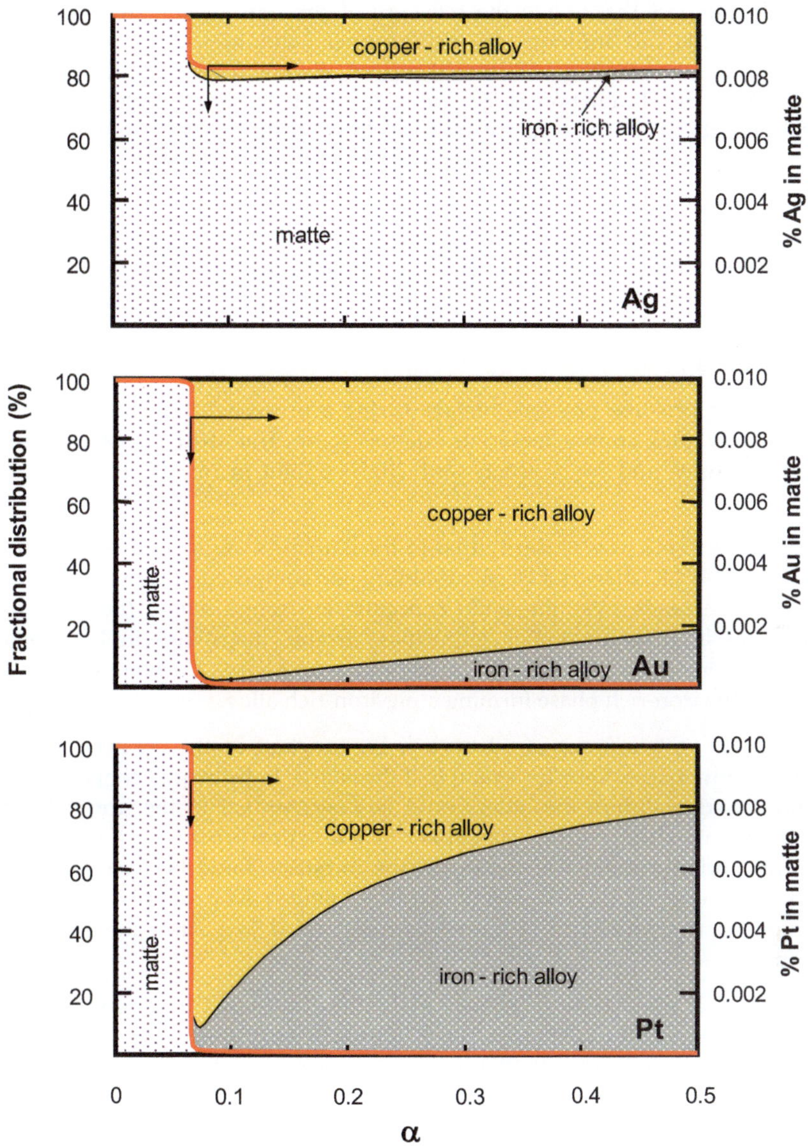

Fig. 19. Fractional distribution of precious metals and their mass % in matte against α during the reducing treatment of complex copper matte at 1473 K

For the precious metals, as is shown in Fig. 19, with increasing value of α, the fractional distribution of platinum abruptly decreases and those of silver and gold slightly decrease in the copper-rich alloy phase and those in the iron-rich alloy phase present an opposite behaviour, while, in the matte phase, the fractional distribution of platinum slightly decreases and those of silver and gold are almost constant.

These results suggest that, when the recovery of precious metals and the elimination of detrimental arsenic and antimony from the copper matte are considered by means of the phase separation, the recovery of valuable copper, silver and gold into the copper-rich alloy phase as well as the elimination of less valuable iron together with the detrimental arsenic into the iron-rich alloy phase might be feasible even though the proportion of valuable platinum lost in the iron-rich phase is considerably large.

The copper-rich alloy may be further treated in a pyro- or hydro-metallurgical process to extract silver, gold and copper; the iron-rich alloy may be harmlessly discarded if the content of platinum in the initial charge is small, while the matte phase might be continuously processed in the following converting stage.

6. Conclusions

The distribution of some precious metals during a proposed reducing pyrometallurgical treatment of complex copper materials such as scrap, flue dust, furnace residue, sulfide matte, concentrates and by-products with high content of impurities was study in this chapter.

According to section 3, it was clarify that in the Cu-Fe-As and Cu-Fe-Sb systems saturated with carbon at 1473 K, there is a large miscibility gap composed of copper-rich and iron-rich phases, which extends over the wide concentration range where the mutual solubility between them is very small. For impurities, arsenic distributes in both phases while antimony preferentially distributes in copper-rich phase. Carbon almost completely distributes in the iron-rich phase forming a pig-iron-rich alloy.

As for precious metals, most of silver will be enriched in the copper-rich phase while platinum in both phases. Thus for example if the source to be treated contains high amounts of antimony and platinum the latter could be concentrated in the iron-rich phase to be recovered it along with the dissolved copper by a leaching process, or if the source contains high arsenic and silver but low platinum, the impurity could be fixed into the iron-rich phase and subsequently be more safely discarded while silver could be concentrated into a smaller volume of copper-rich phase and be recovered in a subsequent hydro- or pyro-metallurgical process.

In section 4, thermodynamic equilibrium in the Cu-Fe-S, Cu-Fe-S-As and Cu-Fe-S-Sb systems saturated with carbon at 1473 K were determined, they presents a large miscibility gap, named zone II in this study, where two metallic-rich alloy phases of copper and iron are in equilibrium with a matte phase which contains around 63 mass % of copper. Arsenic has strong preference to the metal alloy phases, while antimony preferentially distributed in the copper-rich alloy phase. Carbon is distributed almost completely in the iron-rich alloy phase.

The zone II in the Cu-Fe-S-As and Cu-Fe-S-Sb systems saturated with carbon at 1473 K represents the thermodynamic basis for studying the treatment of complex copper mattes containing impurities and precious metals. Here, in section 5, the behaviour of silver, gold and platinum between the three equilibrated phases was investigated and also based on the experimental results material balances were evaluated for a proposed process by the author.

For the treated complex copper mattes, the contents of impurities may be considerably decreased. The elimination of arsenic from the matte increases with increasing pig-iron amount in the charge, while that of antimony is almost independent. According to the results by using the equilibrium phase separation, the recovery of valuable copper, silver and gold into the copper-rich alloy phase as well as the elimination of less valuable iron together with the detrimental arsenic into the iron-rich alloy phase might be feasible.

The copper-rich alloy may be further treated by some hydro- or pyro-metallurgical process to extract silver, gold and copper; the iron-rich alloy may be harmlessly discarded if the content of platinum in the initial charge is small, while the matte phase might be continuously processed in the following converting stage.

7. References

Chang, Y. Newmann, J. Mikula, A. & Goldberg, D. (1979). *INCRA Series on the Metallurgy of Copper - Phase Diagrams and Thermodynamic Properties of Ternary Copper - Metal Systems,* International Copper Research Association, Milwaukee, Wisconsin, USA.

Dobrzanski, J. & Kozminski, W. (2003). *Proceedings of Copper 2003,* Vol. IV, Santiago, Chile, Nov-Dec 2003.

Fuwa, T. & Otani, M. (1980). *Non ferrous Extractive Metallurgy,* Japan Institute of Metals, Japan.

Hino, M. & Toguri, J. (1986). Arsenic Activities in Molten Copper and Copper Sulfide Melts. *Met. Trans. B,* Vol 17B, No. 4, (755-761).

Hino, M. & Toguri, J. (1987). Antimony Activities in Copper Mattes. *Met. Trans. B,* Vol 18B, No. 1, (189-194).

Hultgren, R. Desai, P. D. Hawkins, D. T. Gleiser, M. Kelley, K. K. & Wagman, D. D. (1973). *Selected Values of the Thermodynamic Properties of the Elements,* ASM, USA.

Lee, Y. & Itagaki, K. (1986). Distribution Equilibria of Various Slag Systems with Cu-Fe-Sb Ternary Speiss. J. *Min. Mater. Inst. Japan,* 102, (591-595).

Massalski, T.B. (1990). *Binary Alloy Phase Diagrams* (Second Edition), Vol. 2, ASM International, Materials Park, Ohio.

Mendoza, D. Hino, M. & Itagaki, K. (2002). Phase Relations and Activity of Antimony in Cu-Fe-S-Sb System at 1473 K. *Mater. Trans. JIM,* 43, No. 5, (1166-1172).

Mendoza, D. Hino, M. & Itagaki, K. (2002). Distribution Equilibria between Cu-Fe-As Ternary Speiss and Slag Phases. *J. Min. Mater. Process. Inst. Japan,* 118, No. 3.4, (197-201).

Parviainene, A. & Fugleberg, S. (1980). *Proceedings of Cobalt 80,* CIM, Edmonton, Canada.

Voisin, L. Henao, H. & Itagaki, K. (2004). Phase Relations and Distribution of Some Minor Elements in Cu-Fe-As System Saturated with Carbon at 1473 K. *Mater. Trans., JIM,* 45, No. 9, (2851-2856).

Voisin, L. Henao, H. & Itagaki, K. (2005). Phase Relations and Distribution of Some Minor Elements in Cu-Fe-Sb System Saturated with Carbon at 1473 K. *Mater. Trans., JIM,* 46, No. 1, (74-79).

Voisin, L. & Itagaki, K. (2006). Phase Relations, Activities and Minor Element Distribution in Cu-Fe-S and Cu-Fe-S-As Systems Saturated with Carbon at 1473 K. *Mater. Trans., JIM,* 47, No. 12, (2963-2971).

Voisin, L. Okura, T. & Itagaki, K. (2008). Phase Relations, Activities and Precious Metal Distribution in the Cu-Fe-S-Sb Systems Saturated with Carbon at 1200 °C. *Mater. Trans., JIM*, 49, No. 6, (1311-1319).

Zakeri. A, Hino, M. & Itagaki, K. (2000). Silver Activity in Matte and Copper Two Phase Region of Cu-Fe-S System. *J. Min. Mater. Process. Inst. Japan*, 17, (43-60).

Part 3

Novel Fabrication and Application

5

Synthesis and Characterization of Noble Metal Nanowires

Dragos-Pinzaru Oana-Georgiana,
Dumitru-Daniel Herea and Horia Chiriac
National Institute of Research & Development for Technical Physics
Romania

1. Introduction

In the ten last years the nanomaterials science and technology have represented one of the most attractive interdisciplinary science researches. The growing interest for the nanoscience domain resides in potential applications in physics, chemistry, biology and electronics. Nowadays, the research in the nanomaterials field takes advantage from important funding since they are the basis for the development of new technologies, devices and systems. Bibliographical data present many synthesis methods of simple and/or multilayered nanowires such as: photochemical synthesis (Kim et al., 2002), catalytical synthesis (Huang et al., 2002), vapour-liquid-solid growing (Björk et al., 2002), electrochemical deposition (Yu et al., 1997; Inguanta et al., 2009; Xu & Wang, 2008).

The preparation of nanowires by electrochemical deposition in nanosised pores is more frequently used because of the low cost and the better energetic efficiency of process. The electrodeposition is a preparation method which allows the controlled deposition from solution of metallic materials. Generally, such a solution contains dissolved salts of metals which are going to be deposited. Passing of a current through the electrochemical cell (formed by three electrodes: the reference electrode, the counter electrode and the working electrode) allows the ions migration from the electrochemical bath to working electrode and their deposition in metallic state. There are a large number of metals which can be deposited by using this method from aqueous solutions such as: Ni, Fe, Co, Ga, B, Cu, Cr, Zn, Ru, Rh, Pd, Ag, Au, Pt etc.

In the case of materials prepared by the electrochemical method, besides the condition that can be easily used for the process development, the quality of the synthesized material can be better controlled by fine-tuning the electrolyte composition and electrolysis parameters control such as: the applied potential, the current density, electrical charge, temperature and the type of the electrolysis (potentiostatic or galvanostatic). The electrochemical method allows the preparation of nanowires with a high length/diameter ratio in polymeric membrane or anodised aluminium oxide membrane (AAO). The localised growth of straight and parallel nanowires on plane surfaces is a specific geometric feature that can be used to obtain nanosized interconnections for electronic and magnetic devices.

In the case of electrochemical cell used to prepare metallic nanowires, the anode is a platinium foil and the reference electrode is the saturated calomel electrode (ESC), silver electrode/silver chlorine (Ag/AgCl) or graphite electrode.

The working electrode is the electrode whose surface is used as support for the ions reduction from the solution. In the case of the electrochemical deposition of simple or multilayered nanowires the nanoporouse membrane, with pores which will be filled with metallic nanowires, is used as cathode in electrochemical cell. To be used for electrochemical deposition the membranes are prepared as follows:

- on one side of the nanoporouse membrane a 500 nm gold thin film is deposited by thermal evaporation in vacuum;
- the gold layer is physically isolated from the electrolyte by using a special insulator layer.

In this configuration, the deposited metallic layer is not in direct contact with the electrolyte, but only through the pores of the membrane, the electrochemical deposition being achieved only by pores.

The preparation of metallic nanowires by using nanoporous membrane involves a better knowledge of physical and electrochemical processes of deposition.

Among the most used membranes for the preparation of nanowires are the polycarbonate membranes and the alumina membranes. The polycarbonate membranes are obtained by the "track-etch" method. This method uses the bombardation with heavy atoms of a nonporous material to create holes. This step is followed by chemical treatment to transform the holes in nanopores. The nanoporouse membrane contains cylindrical pores of uniform diameters but which are randomly distributed on its surface. This type of membrane, commercially available (Nucleopore and Poretics companies), may contain pores with diameters between 10 nm and 800 nm with variable densities.

Alumina nanoporouse plane membranes (AAO) are obtained by anodization of aluminium foils in acids electrolytes containing bivalent or trivalent anions such as: oxalic acid $(COOH)_2$ (Li et al., 1999), sulphuric acid H_2SO_4 (Jessensky et al., 1998), or phosphoric acid H_3PO_4 (Li et al., 1998). One of the methods proposed in literature, which leads to the preparation of plane and good quality membrane, is the anodisation in two steps. This method was proposed for the first time by Masuda and Fukuda (Masuda & Fukuda, 1995). The characteristics of the prepared alumina membrane depend on the anodisation conditions (the concentration of the electrolyte used to modify the anodisaton conditions, the working temperature, the anodisation potential). Thus, by the modification of anodisation conditions alumina membrane with pores with diameters between 20 nm and 400 nm can be obtained. The alumina nanoporouse membrane can be also obtained by the combination of anodisation process and nanoindentation process. This technique consists in the creation of an array of defects on the aluminium surface which will serve as nucleation centre for pores in the next anodisation step. The nanoindentation technique allows the preparation of nanoporous membranes with ordered pores by one step anodisation process. In this case the distance between pores and the membrane porosity can be controlled as well. It is worth to be mentioned the fact that by using this technique, arrays of pores with different symmetries can be prepared (Masuda et al., 2001; Asoh et al., 2001; Vojkuvka et al., 2008).

The nanowires made up of noble metals and transition metals are the most important types of studied nanowires due to their versatility in applications such as biosensors or magnetic elements. In function of the application field the nanowires' properties can be studied either in the membrane or "free", after the dissolution of the membrane.

The magnetic nanowires represent a class of nanosized materials in the shape of nanowires intensively studied in the last years is. This family of nanowires is interesting because of their magnetical and transport properties (giant magnetoresistance, reversal magnetization in only one nanowire) being of significant interest due to their potential to work as sensing elements in chemical biological sensors or in optical and electronic devices. The special properties of nanowires can be used in various applications (spintronics, miniaturization of magnetic sensors, ultrahigh-density magnetic storage media, etc.).

The interesting physical properties of magnetic nanowires reside in their geometry and in their dimensionality. The studies presented in the literature on simple magnetic nanowires based on Fe, Co and Ni show that the magnetic properties of nanowires materials are different from the bulk material. This is especially related to the shape anisotropy (Nielsch et al., 2001; Sarkar et al., 2007; Nguyen et al., 2006). The research studies show at the same time that the magnetic properties of nanowires are function of the pH value of the preparation solution. For instance, depending on the pH value of the solution, the cobalt nanowires present two different crystallographic structures: hexagonal or cubic. Thus, the cobalt nanowires prepared at pH 3 have a cubic structure while the nanowires prepared at a pH ranging between 3,5 and 6 present a hexagonal structure (Li et al., 2004, Encinas et al., 2002; Ren et al., 2009, Sanchez-Barriga et al., 2007) which confer different magnetic properties to the nanowires synthesized from different pH solutions.

The giant magnetoresistance (GMR) studies of magnetic nanowire arrays started in the nineties (Piraux et al., 1994) and is continuing nowadays (Nasirpouri et al., 2007; Huang et al., 2009). The GMR effect is observed in magnetic multilayered nanowires when the ferromagnetic elements are layered with nonmagnetic elements (Figure 1).

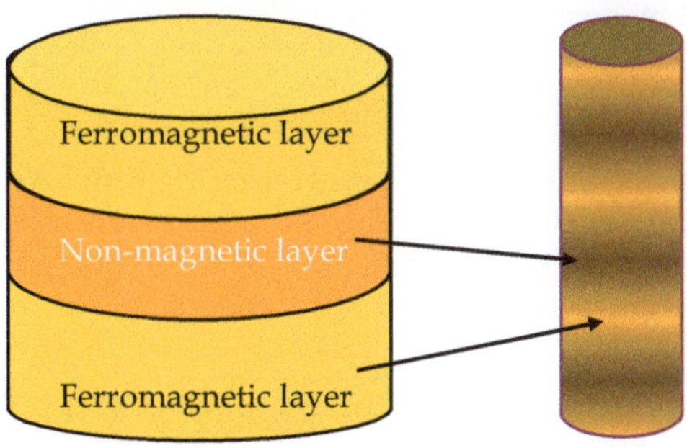

Fig. 1. Sketch of multilayered magnetic nanowires with GMR effect

The advantage of the use of multilayer nanowires (especially of NiFe/Cu magnetic nanowires) was intensively studied. The magnetic and magnetoresistance properties of this type of nanowires depend on the NiFe and Cu layers thickness (Chiriac et al., 2009).

The noble metal nanowires sequentially deposited (multilayered nanowires such as: Au/Pt, Au/Ag, or Ag/Pt) can be used as "bar-codes" in biological testing (Nicewarner-Peña et al.,

2001). Thus, sequences of different metals in a single nanowire adsorb different molecules which can be used to simultaneously detect different biological molecules.

Further on, a conventional method of synthesis of gold, silver and platinum simple nanowires and gold/platinum multilayer nanowires by electrochemical deposition will be described. In this work we used a VOLTALAB 10 PGZ 100 potentiostat in order to control the applied voltage during the electrodeposition. After electrodeposition is complete, the AAO template filled with noble metals were characterized by scaning electron microscopy (SEM) by using a JEOL microscope equipped with energy dispersive X-ray spectroscopy (EDS) analysis tool and by current-atomic force microscopy (I - AFM) using a Park microscope.

2. Experimental

The nanowires were growth inside an anodic aluminum oxide (AAO) template provided by Whatman. This template has a specific pore size of 200 nm and a thickness of 50 μm. For performing the electrochemical deposition, we used a three- electrode cell: as reference we used SCE for gold and platinum electrodeposition and, in order to avoid the precipitation of the silver chloride during the electrodeposition, we used a graphite electrode for silver electrodeposition. For all the experiments, as counter electrode we used Pt foil. Prior to electrodeposition, an adhesion layer of Au film was spread onto one side of the AAO template by thermal evaporation in order to cover the pores completely, and to serve as the working electrode during electrochemical deposition. All the experiments were performed at room temperature. The electrodeposition experiments were performed by pulsed electrodeposition (Inguanta, 2009). Platinum nanowires were growth in aqueous solution of H_2PtCl_6 5 mM/L and HCl 0.1M by applying a dc current of -0.2 V for 3 s and 0 V for 1 s. In the case of gold nanowires deposition we have used an aqueous solution of $HAuCl_4$ 5 mM/L and H_3BO_3 0.5 M. The electrodeposion was performed by applying a dc current of -1.3 V for 5 s and 0 V for 1 s The silver nanowires were deposited from an aqueous solution of $AgNO_3$ 30 g/L and H_3BO_3 45 g/L at -0.7 V for 5 s and 0 V for 1 s. The electrodeposition potential was determined by linear voltammetry.

3. Results and discussions

After electrodeposition was carried out, the cross section of the AAO template filled with noble metals were characterized by scanning electron microscopy (SEM) using a JEOL microscope. Figure 2 shows the SEM micrographs of the cross sections of the AAO: filled with platinum nanowires (Figure 2a), filled with gold nanowires (Figure 2b), filled with silver nanowires (Figure 2c).

The image analysis show that the membranes are homogeneus filled with nobles metals. The growth rate of metals nanowires is changing in function of the nature of the electrodeposited metals: for platinum deposition, the growth rate - 2 μm/h, for gold - 18 μm/h and for silver - 11 μm/h. After the deposition, the AAO template was dissolved by immersing it in a KOH 5M solution in order to liberate the noble metals nanowires. After the dissolution of the template, the nobles metal nanowires are rinsed several times with distilled water in order to remove the potassium hydroxide from the nanowires surface. In Figure 3 are presented the SEM images of the noble metals nanowires free of the alumina template.

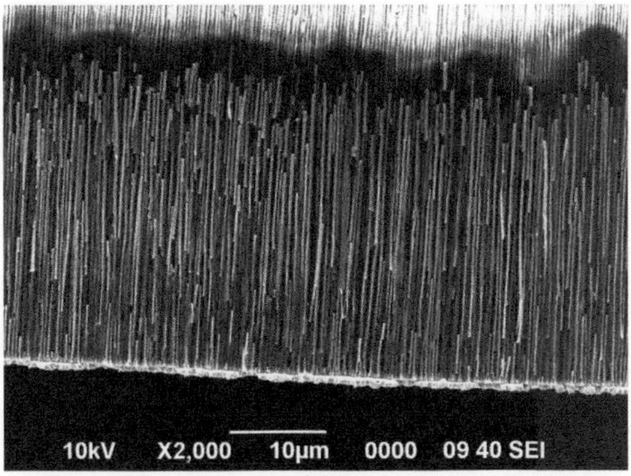

(a)

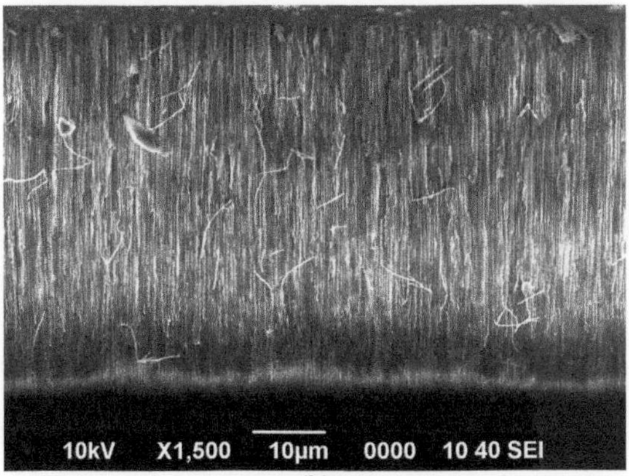

(b)

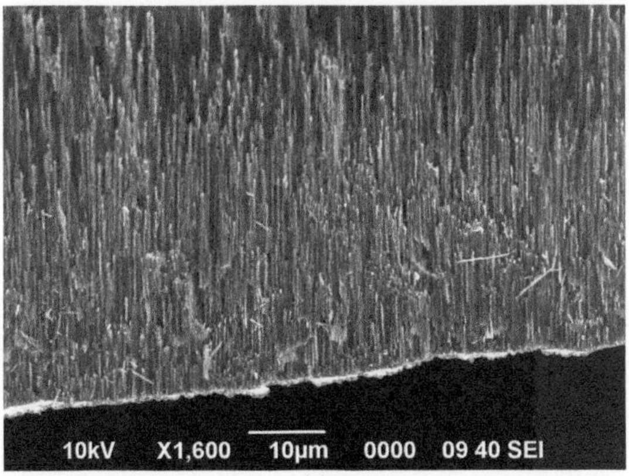

(c)

Fig. 2. Cross section SEM micrographs of a AAO template filled with platinum nanowires (a), gold nanowires (b), silver nanowires (c)

(a)

(b)

(c)

Fig. 3. SEM micrographs of noble metals nanowires liberate from the AAO template: platinum nanowires (a), gold nanowires (b), silver nanowires (c).

The freed nanowires were collected from the hydroxide solution via centrifugation and rinsed several times with distilled water. Thereafter, the nanowires were submitted to EDS analysis. The EDS spectra (Figure 4) show that the obtained nanowires do not contain impurities (the detected elements are platinum, gold, silver and titanium). The titanium tracks present in all the spectra showed in Figure 4 comes from the sample holder whereas the gold comes from the thin layer deposited by thermal evaporation in vacuum which ensured the electrical conductivity.

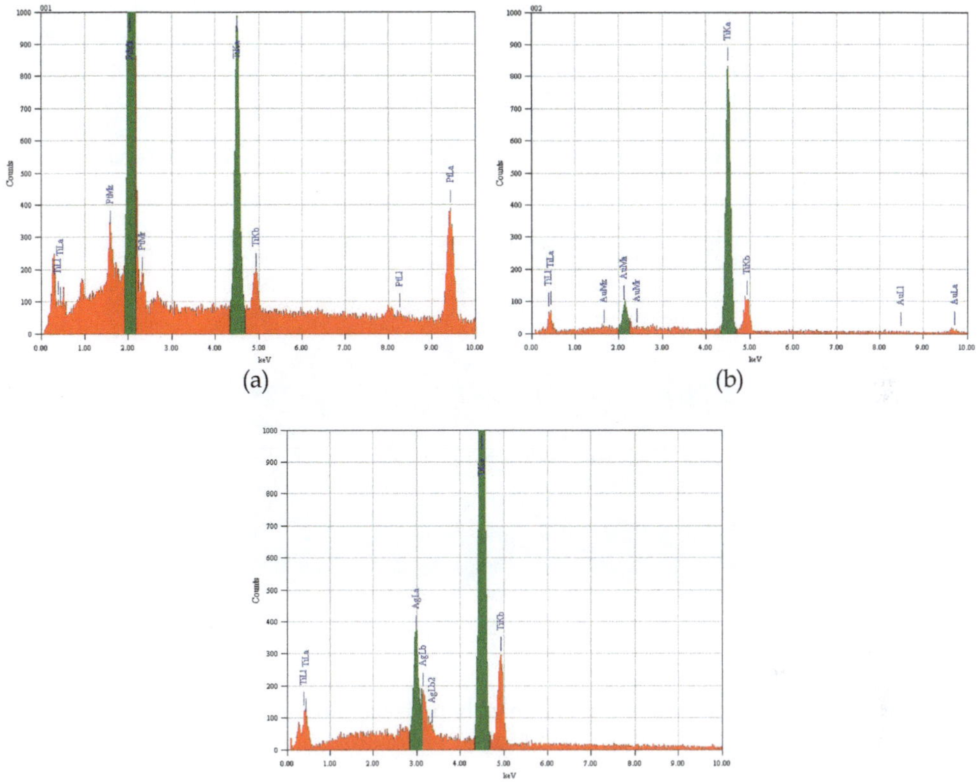

(a) (b) (c)

Fig. 4. EDS analysis of platinum nanowires (a), gold nanowires (b), silver nanowires (c)

The obtained noble metals nanowires are individually characterized by current atomic force microscopy (I - AFM). For performing an accurate analysis the surfaces must be very smooth. Therefore, after electrodeposition, the electrodeposited alumina samples are submitted to a mechanical polishing process by using diamond (particles size - 3 μm) and Syton (particles size - 20 nm). The role of this step is to bring the nanowires to the same length on the surface and to obtain very smooth surfaces. After each polishing step, the AAO surface is visualized with the SEM microscope. Figure 4 shows the top-view SEM micrograph of the mechanically polished alumina membrane filled with platinum (Figure 5a), gold (Figure 5b), and silver (Figure 5c).

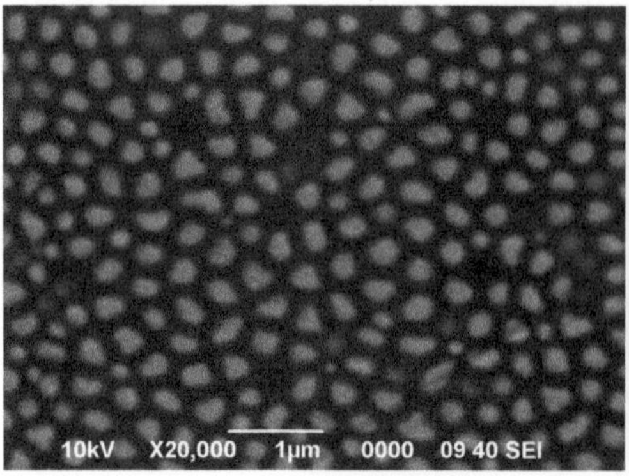

(a)

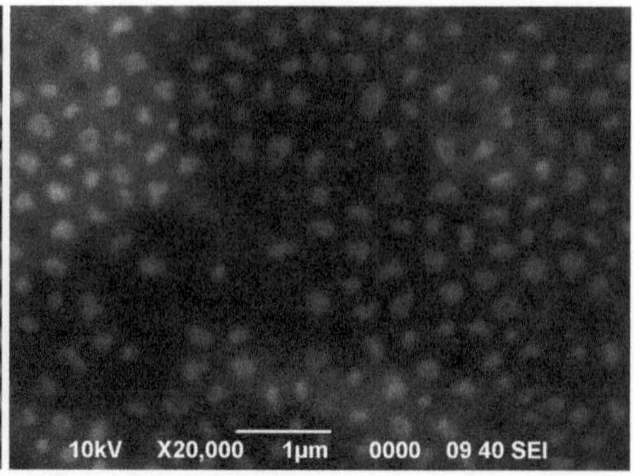

(b)

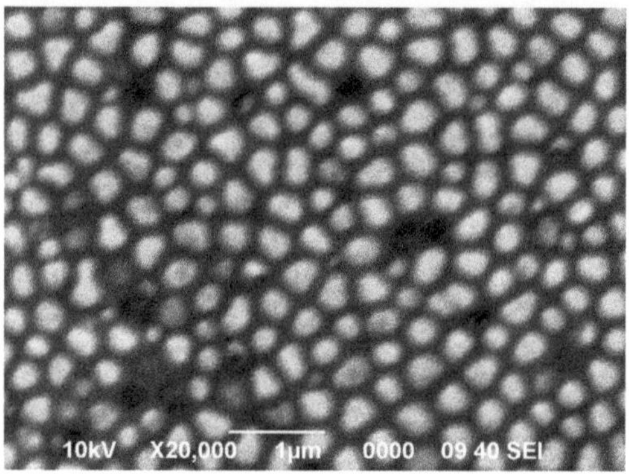

(c)

Fig. 5. Top-view SEM micrograph of mechanically polished alumina membrane filled with platinum nanowires (a), gold nanowires (b), silver nanowires (c).

This top-view SEM micrograph shows that noble metal nanowires are of the same diameter and shape as the alumina membrane pores. When the polishing process is finished, the surface of the AAO template filled with nanowires is examined by current-atomic force microscopy (I – AFM) by applying a +1 V dc bias current between the AFM tip and the sample's surface. In I-AFM mode, a conductive AFM tip scans the surface while it is in contact. This technique is able to image simultaneously both the topography and the conductivity of the surface. The current flowing between the tip and the sample gives us information about the surface conductivity of the sample. Contact topography image is generated by using a feedback loop to maintain the constant tip deflection whereas the I-AFM image is generated by measuring the current flow. In Figure 6 is presented the topographically and the electrically images of mechanically polished alumina membrane filled with silver nanowires.

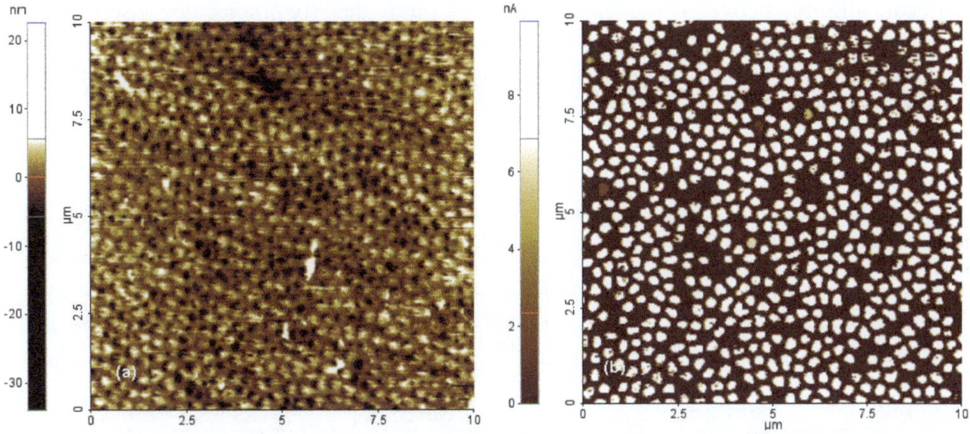

(a) (b)

Fig. 6. I - AFM images of mechanically polished alumina membrane filled with silver nanowires; the topography is shown in (a), and the simultaneously recorded (at +1 V bias voltage) surface conductivity in (b)

The topographical image of the nanowires correlates very well with the peaks on the current map. Close to 100% of nanowires were found to be conductive.

Nanowire functionalization

The functionalization of nanowires with (bio)molecules represents a chemical process in which a strong covalent bond is formed between the nanowire and the (bio)molecules (Mbindyo et al., 2001).

The functionalization of metallic nanowires with biomolecules represents one of the recent applications of nanomaterials. The unique physical properties of nanomaterials to recognise biomolecules in a selective mode can lead to the miniaturisation of biological sensors.

Although the use of nanowires in biosensors is of high interest, the electrochemical synthesis of simple and multilayered nanowires which contain noble metals is difficult, the influence of the electrochemical deposition parameters not being very well known.

There are a significant number of methods based on chemical approaches for surface functionalization. Well-documented collections of bioconjugation and functionalization techniques are available now. Bioconjugation involves the linking of two or more molecules to form a new complex having the combined properties of its individual components (Hermanson, 2008) As a straightforward example, we recommend for a detailed analysis the Hermanson's collection of methods (Hermanson, 2008) that can be used for a lot of functionalization processes.

In the case of metallic nanowires, the functionalization with different organic natural or synthetic molecules follows commonly the same way as for their bulk counterparts. However, high differences occur when the specific magnetic, optic, electric etc. properties are investigated and compared.

As a basic rule, the functionalization of the metallic nanowires have to take into account the chemical affinity between the metal surfaces and the (bio)molecules used. It was experimentally observed that the chemical properties of the surfaces play a crucial role in the binding process, the chemical groups imposing variations in reactivity for different metallic surface (table 1).

Ligand	Name	Surface for modification	Proposed linkage
R-SH R-S-S-R′	Thiols Disulfides	Au, Ag, Cu, Hg, Fe	R-S-Surface
R-CN	Cyanides	Pt, Pd	R-CN-Surface
R-(CO)-OH	Carboxilic Acids	Metal oxides	R-(CO)-Surface
R-(PO_2)-OH	Phosphonates	Metal oxides	R-(PO_2)-Surface
R_2-Si-O-R	Siloxanes	Metal oxides	R_2-Si-O-Surface
R-(CO)-NH-OH	Hydroxamic acids	Metal oxides	Surface-O-(CR)-(NH)-O-Surface

Table 1. The most used chemical groups for surface functionalization of different metals (Reich et al., 2011)

Here, we are focusing mainly on the functionalization of gold-based nanowires as representative ones for the potential applications in the biomedical field because gold nanomaterials have proven to be versatile biomedical tools due to their particular structural and physico-chemical properties (Ray et al., 2011).

For functionalization of gold nanowires, a preferred method is the self-assembling of (bio)molecule monolyers. For instance, a mercaptoundecanoic acid ($HS(CH_2)_{10}CO_2H$) was used to coat Au-based nanowires resulting in surfaces functionalized with carboxylate groups that allowed the use of carbodiimide chemistry to conjugate primary amine groups of a capture antibody to the carboxylate groups on the nanowires. (Tok et al., 2006; Hermanson, 2008).

Single-strand DNA can be also specifically modified in order to meet the affinity requirements of a metallic surface for coupling. For example, in the case of a gold nanowire, a thiol group was inserted in the 5′ position of a single-strand DNA whereas tetramethyl rhodamine was introduced in the 3′ position. The optical images showed the single-strand DNA reacted with gold nanowires through the thiol groups (Mbindyo et al., 2001).

At the surface level, multi-segment nanowires can make available a diversity of chemical properties that can be used to selectively functionalize the metallic segments. Because the metallic stripes present different chemical reactivity and, therefore, reacts differently towards the chemical groups of the biomolecules, a selective functionalization can be carried out in function of the metals making up the nanowire. For example, a Au-Pt-Au nanowire can be functionalized both with thiols and isocyanides. Due to affinity of thiols towards Au, a self-assembled-monolayer of 2-mercaptoethylamine can be formed on the gold surface whereas a butaneisonitrile monolayer attaches to the Pt segments (Kovtyukhova et al., 2002). The biomolecules can be modified with fluorescent markers in order to spatially discriminate the position of the biomolecules along the stripped nanowires.

Gold nanowires with additions of magnetic materials, such as nickel, can be also selectively functionalized. For example, Au surface can be functionalized with thiol-based hexa(ethylene glycol) groups whereas Ni surfaces were functionalized with palmitic acid. Fluorescently-marked proteins bound to hydrophobic palmitic acid lead to a bright fluorescence whereas thiol-based hexa(ethylene glycol) groups do not allow the protein to attach to it (Birenbaum et al., 2003). A secondary role of the nickel segments is to endow the nanowires with magnetic properties.

For certain conditions, gold nanostructures present unexpectedly some ferromagnetic or paramagnetic-like properties. Thus, it has been shown that gold nanostructures capped with alkanethiols present an important ferromagnetic behaviour as compared with their non-functionalized counterparts. For instance, the simultaneous presence of Au–Au and Au–S bonds, conjugated with the creation of an ordered self-assembled monolayer shell, is considered key parameters for the ferromagnetic-like behaviour revealed by the thiol-functionalized gold nanostructures. The magnetic properties become obvious and manifest when the capping organic molecules form self-assembled monolayers on gold substrates. It was experimentally established that the simultaneous presence of Au–Au and Au–S bonds is required to observe ferromagnetic behaviour in thiol-functionalized nanostructures. Polymeric-like phases (-Au–S–Au–S– bonds) do not show magnetization properties. (Guerrero et al., 2008).

Multi-component nanowires based on three Ni/Au/Ni segments can be also fabricated and functionalized. The preparation of multilayered nanowires allows the selective functionalization of different segments since the metallic segments present different chemical characteristics (Kovtyukhova & Mallouk, 2002). For example, a thiol-modified single strand DNA and a biotinylated peptide were tailored to selectively attach to the gold and nickel segments, respectively, by which a F1-ATPase motor can be bound only to the nickel segment of the nanowires by using the biotin-streptavidin linkage. Also, the gold segments of nanowires were functionalized by using fluorescent single strand DNA molecules. The process is based on the strongly binding between thiol groups on the single strand DNA and the gold surface. Also, in order to be optically detected the nickel segments of nanowires were functionalized by using fluorescent biotinylated peptide. (Ren et al., 2006).

Nanowire-based detection of disease-specific DNA

From the medical analysis standpoint, the most important biomolecules used for diagnosis are antibodies and DNA. The discovery of specific target DNA sequences of medical interest in the incipient phase of a disease such as tumor or viral pathology is correlated with an accurate assessment of patient's prognosis and with an appropriate way to monitor therapy. Usually, these specific DNA sequences are detected and quantified using molecular techniques such as Polimerase Chain Reaction (PCR), Restriction Fragment Length Polymorphism (RFLP), Real Time - Polimerase Chain Reaction (RT-PCR) along with electrophoretic migration in agarose gel. Regarding this issue, an alternative technique to the conventional ones could be the use of a bioassay method based on metallic nanowires that specifically detect and qualitatively identify the amplified target DNA sequences obtained by using specific modified primers.

Following the general tendency of nanowires' applications in biomedical domain, we tested the ability of the naked gold-based nanowires to be used in a biodetection assay for identification of a DNA sequence, specific for FLT3 gene mutation, responsible for acute mieloblastic leukemia.

Given in a synthetic hierarchy, the main steps of the procedure were performed as follows:

1. Separation and purification of specific genomic DNA from the blood of patients with acute mieloblastic leukemia, in order to detect mutation of FLT3 gene;
2. Amplification of target DNA sequence through PCR amplification by using specific primers modified at their 3′ ends with thiols, i.e. HS- chemical groups, simultaneously with other primers having their 5′ ends modified with a fluorophore, i.e. cy5;
3. Immobilization of the obtained PCR products on the surface of the metallic nanowires and further detection through a fluorescence-based analysis system;
4. For comparison, detection of the same PCR products was made by using a gel electrophoresis migration method.

In an additional point-to-point explanation, we should make clear some aspects related to the above synthetically presented steps of the procedure. First, the DNA separation from patients' blood followed a general and validated procedure (Miller et al., 1988; Beutler et al., 1990). Second, amplification of the target DNA sequence was made in the presence of commercial primers, i.e. short single stranded DNA, specific to the "diseased" DNA. Third, due to the thiol groups, which quickly and specifically bind to gold surfaces, the amplified DNA was immobilized on gold-platinum nanowires. The nanowire-DNA structures were investigated through a fluorescence-based analysis system by measuring the fluorescence generated by the fluorophore-tagged amplicons (i.e. products resulted from PCR amplification process) immobilized on nanowires (figure 7).

Finally, in order to basically validate the method, a fluorescence-based gel electrophoresis migration method was used as a comparison tool for detection of the same PCR products (figure 7).

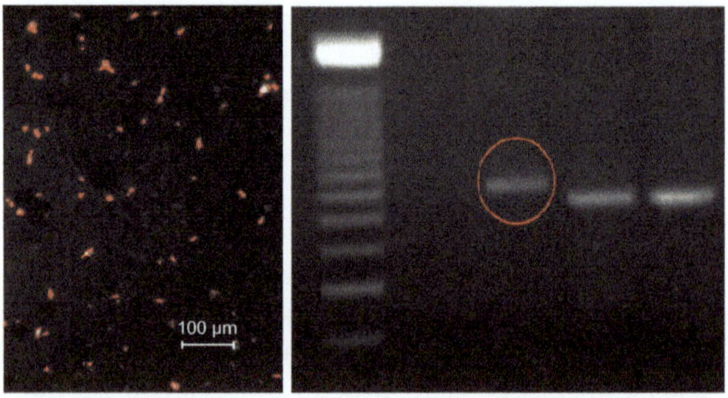

(a) (b)

Fig. 7. (a) Nanowires detection through the fluorescence-based analysis system performed by measuring the fluorescence generated by the fluorophore-tagged PCR-amplified products immobilized on nanowires; (b) fluorescence-based gel electrophoresis migration of PCR-amplified products showing FLT3 gene mutation (inside the red circle)

From figure 7(a), red spots of DNA-nanowires complexes are well observed. As obviously can be seen, generally, the nanowires are not individually spread out between the two laminas of the microscope, but in small groups. This behaviour is due to the typical physical forces governing interactions in liquids. However, the image shows the successful immobilization and qualitative detection of DNA, specific to FLT3 gene involved in acute mieloblastic leukemia.

However, in spite of the successful qualitative detection, based on a comparative study, that emphasized the usefulness of the nanowires-based bioassay method for specific biomedical issues, further analysis and tests are needed in order to certify the efficiency, sensitivity and specificity of this method.

4. Conclusion

In synthesis, by using a conventional electrodeposition process, the arrays of single Pt, Au and Ag nanowires have successfully been fabricated by pulsed electrodeposition. The obtained nanowires have been investigated by SEM and I - AFM. The results showed that the obtained nanowires have a diameter of about 200 nm and a length of several micrometers. All the samples has been mechanically polished and we have showed that the AAO membranes are fully filled with metallically compound. The I - AFM microscopy have showed that the as-obtained nanowires are continuous inside the membrane.

The nanowires were used to immobilize a disease-related DNA that was further detected by using a fluorescence-based analysis system. Also, the same disease-related DNA was detected through gel electrophoresis migration.

The comparative study showed the target amplified DNA was successfully detected by using metallic nanowires, the entire detection process being, in principle, simple.

The results also underlined the nanowires-based bioassay method could be used for specific biomedical assays in one condition: further analysis and tests in order to certify the efficiency, sensitivity and specificity of this method have to be carry out.

5. References

Asoh, H.; Nishio, K.; Nakao, M.; Tamamura, T. & Masuda, H. (2001). Conditions for Fabrication of Ideally Ordered Anodic Porous Alumina Using Pretextured Al, *Journal of the Electrochemical Society,* Vol.148, No.4, pp. B152-B156, Print ISSN 0013-4651, Online ISSN 1945-7111

Birenbaum, N.S.; Lai, B.T.; Chen, C.S.; Reich, D.H. & Meyer, G.J. (2003). *Selective Noncovalent Adsorption of Protein to Bifunctional Metallic Nanowire Surfaces,* Langmuir, Vol.19, No.23, pp. 9580 – 9582, Print ISSN 0743-7463, Online ISSN 1520-5827

Birenbaum, S.; Lai, T. B.; Reich, D. H.; Chen, C. H. & Meyer, G. J. (2003). Selective Noncovalent Adsorption of Protein to Bifunctional Metallic Nanowire Surfaces, *Langmuir,* vol. 19, No.23, pp. 9580–9582, ISSN 0743-7463

Björk, M. T.; Ohlsson, B. J.; Sass, T.; Persson, A. I.; Thelander, C.; Magnusson M. H.; Deppert, K.; Wallenberg, L. R. & Samuelson, L., (2002). *One-dimensional Steeplechase for Electrons Realized,* Nano Letters, Vol.2, No.2, pp.87 – 89, ISSN 1530-6984

Chiriac, H.; Dragos, O.G.; Grigoras M.; Ababei G. & Lupu N. (2009). *Magnetotransport Phenomena in [NiFe/Cu] Magnetic Multilayered Nanowires,* IEEE Transactions on magnetics, Vol.45, No.10, pp. 4077 – 4080, ISSN 0018-9464

Encinas, A.; Demand, M.; George J.-M. & Piraux L. (2002). Effect of the pH on the microstructure and magnetic properties of electrodeposited cobalt nanowires, *IEEE Transactions on Magnetics*, Vol.38, No.2574, ISSN 0018-9464

Guerrero, E.; Munoz-Marquez, M.A.; Fernandez-Pinel E.; Crespo P.; Hernando A. & Fernandez, A. (2008). Electronic structure, magnetic properties, and microstructural analysis of thiol-functionalized Au nanoparticles: role of chemical and structural parameters in the ferromagnetic behaviour, *Journal of Nanoparticle Research*, Vol.10, pp.179–192, No.1, Print ISSN 1388-0764 Online ISSN 1572-896X

Hermanson, G.T. Hermanson (2008). *Bioconjugate Techniques*, Elsevier, ISBN 978-0-12-370501-3, London, UK

Huang, X.; Tan L.; Cho H. & Stadler; B. J. H. (2009).; *Journal of Applied Physics*, Vol.105, 07D128, Print ISSN 0021-8979, Online ISSN 1089-7550

Huang, Y.; Duan, X.F.; Cui, Y. & Lieber, C.M. (2002). Gallium Nitride Nanowire Nanodevices, *Nano Lettres*, Vol. 2, No.2, pp. 101-104, ISSN 1530-6984

Inguanta, R.; Piazza, S. & Suenseri, C. (2009). Influence of the electrical parameters on the fabrication of copper nanowires into anodic alumina templates, *Applied Surface Science* Vol.255, pp. 8816–8823, ISSN 0169-4332

Jessensky, O.; Müller, F. & Gösele U. (1998). *Self-organized formation of hexagonal pore arrays,* Applied Physics Letter, Vol.72, pp. 1173-1175, Online ISSN 1077-3118

Kim, F.; Song, J.H. & Yang, P.D. (2002). *Photochemical synthesis of gold nanorods, Journal of American Chemical Society,* Vol.124, No.48, pp. 14316-14317, ISSN 0002-7863

Kovtyukhova, N. I. & Mallouk, T. E. (2002). Nanowires as Building Blocks for Self-Assembling Logic and Memory Circuits, *Chemistry - A European Journal*, Vol.8, No. 19, pp. 4354-4363, ISSN 1521 3765

Kovtyukhova, N.I. & Mallouk, T.E. (2002). Nanowires as Building Blocks for Self-Assembling Logic and Memory Circuits, *Chemistry – A European Journal,* Vol.8, No.19, pp. 4355-4363, Print ISSN 0947-6539, Online ISSN 1521-3765

Li, A.P.; Müller, F.; Birner, A.; Nielsch, K. & Gösele, U. (1999). Fabrication and Micro-Structuring of Hexagonally Ordered Two-Dimensional Nanopore Arrays in Anodic Alumina, *Advanced Materials*, Vol.11, No.6, pp. 483-487, Print ISSN 0935-9648, Online ISSN 1521-4095

Li, F., Zhang, L. & Metzger, R. M., (1998). *On the Growth of Highly Ordered Pores in Anodized Aluminum Oxide, Chemistry of Materials,* Vol.10, No.9, pp. 2470 – 2480, Print ISSN 0897-4756, Online ISSN 1520-5002

Li, F.; Wang, T.; Ren L. & Sun J. (2004). Fabrication and magnetic properties of Co nanowire arrays of different crystal structures, *Chinese Science Bulletin*, Vol.49, No.1532, Print ISSN 1001-6538, Online ISSN 1861-9541

Masuda, H., Fukuda, K., (1995). Ordered Metal Nanohole Arrays Made by a Two-Step Replication of Honeycomb Structures of Anodic Alumina, *Science*, Vol.268, No. 5216, 1466 – 1468, Print ISSN 0036-8075, Online ISSN 1095-9203

Masuda, H.; Asoh, H., Watanabe, M., Nishio, K.; Nakao, M. & Tamamura, T. (2001). Square and Triangular Nanohole Array Architectures in Anodic Alumina, *Advanced Materials*, Vol.13, No.189, pp. 189-192, Print ISSN 0935-9648, Online ISSN 1521-4095

Mbindyo, J. K.; Reiss, B. D.; Martin, B. R.; Keating, C. D.; Natan, M. J.; Mallouk, T. E. (2001), DNA-Directed Assembly of, Gold Nanowires on Complementary Surfaces,

Advanced Materials, Vol.13, No. 4, pp. 249-254, Print ISSN 0935-9648, Online ISSN 1521-4095

Mbindyo, J.K.N.; Reiss, B.D.; Martin, B.R.; Keating, C.D.; Natan, M.J.; Mallouk, T.E. (2001). *DNA-Directed Assembly of Gold Nanowires on Complementary Surfaces*, Advanced Materials, Vol.13, pp. 249–254, Print ISSN 0935-9648, Online ISSN 1521-4095

Miller, S.A., Dykens, D.D. & Polesky, H.F (1988). *A simple salting out procedure for extracting DNA for human nucleated cells*, Nucleic Acids Research, Vol.16, No.3, p. 1215, Print ISSN 0305-1048, Online ISSN 1362-4962

Nasirpouri, F.; Southern, P.; Ghorbani, M.; Irajizad, A. & Schwarzacher, W. (2007). GMR in multilayered nanowires electrodeposited in track-etched polyester and polycarbonate membranes, *Journal of Magnetism and Magnetic Materials*, Vol.308, No.1, pp. 35–39, ISSN 0304-8853

Nguyen, T. M.; Cottam, M. G.; Liu, H. Y.; Wang, Z. K.; Ng, S. C.; Kuok, M. H.; Lockwood, D. J.; Nielsch, K. & Gösele, U. (2006). Spin waves in permalloy nanowires: The importance of easy-plane anisotropy, *Physical Review B*, Vol.73, 140402R, ISSN 1098-0121

Nicewarner-Peña, S.R.; Freeman, R.G.; Reiss, B.D.; He, L.; Peña, D.J.; Walton, I.D.; Cromer, R.; Keating, C.D. & Natan, M.J. (2001). *Submicrometer Metallic Barcodes*, Science, Vol.294, No.5540, pp. 137 – 141, Print ISSN 0036-8075, Online ISSN 1095-9203

Nielsch, K.; Wehrspohn, R. B.; Barthel, J.; Kirschner, J.; Gösele, U.; Fischer, S. F. & Kronmüller, H., (2001). Hexagonally ordered 100 nm period nickel nanowire arrays; *Applied Physics Letter*, Vol.79, No.1360, Print ISSN 0003-6951, Online ISSN 1077-3118

Parkin, S.S.P. (1995). Giant Magnetoresistance in Magnetic Nanostructures, Annual Review of Materials Science, *Vol.25, pp. 357-388, ISSN 0084-6600*

Piraux, L.; George J. M., Despres, J. F.; Leroy C.; Ferain E.; Legras R.; Ounadjela, K. & Fert, A. (1994). Giant magnetoresistance in magnetic multilayered nanowires, *Applied Physics Letters*, Vol. 65, No.19, pp. 2484–2486, Print ISSN 0003-6951, Online ISSN 1077-3118

Ray, S.; Reddy, P. J. & Choudhary, S. (2011). Emerging nanoproteomics approaches for disease biomarker detection: A current perspective, *Journal of Proteomics*, in press, ISSN 1874-3919

Reich, D.; Meyer G.; Chien. C.-L.; Chen. C. & P.C. Searson (2006). Multifunctional magnetic nanowires, United State Patent, 7,132,275 B2. Available from http://www.patentstorm.us/patents/7132275/description.html

Ren, Q.; Zhao, Y.-P.; Yue, J.C. & Cui, Y.B. (2006). Biological application of multi-component nanowires in hybrid devices powered by F1-ATPase motors, *Biomedical Microdevices*, Vol. 8, No.3, pp. 201–208, Print ISSN 1387-2176, Online ISSN 1572-8781

Ren, Y.; Liu, Q.F.; Li, S.L.; Wang, J.B. & Han, X.H. (2009). *The effect of structure on magnetic properties of Co nanowire arrays*, Journal of Magnetism and Magnetic Materials, Vol.321, pp. 226 – 230, ISSN 0304-8853

Sanchez-Barriga, J.; Lucas, M.; Rivero, G.; Marin, P. & Hernando A., (2007). *Magnetoelectrolysis of Co nanowire arrays grown in a track-etched polycarbonate membrane*, Journal of Magnetism and Magnetic Materials, Vol.312, No.1, pp. 99–106, ISSN 0304-8853, ISSN 0304-8853

Sarkar, J.; Khan, G.G. & Bassumallick, A., (2007). *Nanowires: properties, applications and synthesis via porous anodic aluminium oxide template,* Bulletin of Materials Science, Vol.30, No.3, pp. 271 – 290, Print ISSN 0250-4707 Online ISSN 0973-7669

Tok, J.B.; Chuang, F.Y.; Kao, M.C.; Rose, K.A.; Pannu, S.S.; Sha, M.Y.; Chakarova, G., Penn, S.G. & Dougherty G.M. (2006). Metallic Striped Nanowires as Multiplexed Immunoassay Platforms for Pathogen Detection, *Angewandte Chemie International Edition*, Vol.45, pp. 6900 –6904, Print ISSN 1433-7851, Online ISSN 1521-3773

Vojkuvka, L.; Marsal, L. F., Ferre-Borrull, J.; Formentin, P. & Pallares, J. (2008). Self-ordered porous alumina membranes with large lattice constant fabricated by hard anodization, *Superlattices and Microstructures*, Vol.44, No.4 – 5, pp. 577-582, ISSN 0749-6036

Xu, D.; Sriram, V.; Ozolins, V.; Yang, J.-M.; Tu K.N.; Stafford G.R. & Beauchamp C. (2009). In situ measurements of stress evolution for nanotwin formation during pulse electrodeposition of copper, *Journal of Applied Physics*, Vol. 105, No.2. p. 023521 - 023521-6, Print ISSN 0021-8979, Online ISSN 1089-7550

Xu, J. & Wang, K. (2008). *Pulsed electrodeposition of monocrystalline Ni nanowire array and its magnetic properties*, *Applied Surface Science*, Vol.254, pp. 6623–6627, ISSN 0169-4332

Yu, Y.Y.; Chang, S.S.; Lee, C.L. & Wang, C.R.C. (1997). *Gold Nanorods: Electrochemical Synthesis and Optical Properties,* Journal of Physical Chemistry *B,* Vol.101, No.34, pp. 6661-6664, Print ISSN 1089-5647, Online ISSN 1520-5207

6

Nanolithography in the Evanescent Field of Noble Metals

Yong Yang[1], Guanxiao Cheng[2], Shaolin Zhou[3], Lixin Zhao[1] and Song Hu[1]
[1]Institute of Optics and Electronics, Chinese Academy of Sciences, Chengdu
[2]College of Electronic Science and Technology, Shenzhen University, Shenzhen
[3]School of Electronic and Information Engineering, South China University of Technology Guangzhou
China

1. Introduction

With the rapid progress of high integrity in integrity circuit(IC) field, the finer resolution of optical lithography became more and more urgent which spurs the scientists to put a premium eye on resolution, throughput and reliability. According to Abbe's theory, the spatial resolution can be improved by using either shorter wavelengths or higher numerical apertures. Although the semiconductor industry has made significant progress in increasing the lithography resolution in the past decades, further improvement of the resolution by accessing shorter wavelengths is facing critical challenges due to the availability of optical materials with suitable refractive index. The expansion of nanoscale science and engineering will require flexible, high spatial resolution, and low-cost nanolithographic techniques and systems other than those employed in the semiconductor industry, for reasons of both cost and limited flexibility. So far, optical projective lithography technique is of dominance in optical lithography, resolution can be extended to 32 *nm* or even less by using ArF light resource combined with immersion technique; however, resolution below 32 *nm* meets great difficulty till optical lithography being washed out. According to the International Technology Roadmap for Semiconductors(ITRS2009 Edition), next generation lithography techniques for feature size 32 *nm* and below are 193 *nm* immersion double patterning, 193 *nm* immersion with other fluids and lens materials, Maskless Lithography(ML2)and imprint etc., but the techniques of breaking the limitation of 32 *nm* resolution is far from mass production; they met the difficulties of complicated fabrication procedure, low efficiency and production. A number of near-field nanolithography techniques have been explored recently by some research groups.

In the early 20th century, scientists had observed the phenomena of Surface Plasmons Polaritons (SPPs) in reflective spectrum of metallic grating. In 1902, American scientist R. W. Wood found the totally different diffraction property of light when light was introduced to metallic grating and dielectric grating. Due to the lack of reasonable explanation for such strange phenomena, his report hadn't drawn much attention from other scientists. Till 1957, American scientist R. H. Ritchie reported that the free electrons of metal surface can be resonantly excited by light, which caused the SPPs propagated along the metallic surface. Since then, SPPs have drawn more concentrations from scientists to investigate its novel application.

With the development of microfabrication and nanotechnology, more and more attentions from scientists were paid to SPPs. T. W. Ebbesen reported the optical extraordinary transmision through subwavelength holes arrays. Leaec and his team found more details of diffraction property when light passed through subwavelength hole arrays, which will caused intensive transmission and small diffraction angle. British professor Barnes proved the reasonable existence of extraordinary transmission caused by SPPs. American scientist and his team made further experiments to prove the feasibility of super-resolution realized by SPPs metallic structure. So far, SPPs had shown its significant potential in several fields, such as optical super-resolution, bio-sensor, nano-photon devices etc..

J. Pendry predicted that a slab of negative refractive index material has the power to focus all Fourier components of a two-dimension image. According to Ebbesen's paper, SPPs are waves that propagate along the surface of a conductor, usually a metal. These are essentially light waves that are trapped on the surface because of their interaction with the free electrons of the conductor. The intensity of SPPs propagating along a smooth surface decreased due to the intrinsic absorption of metal. In order to investigate the influence of gain-assisted metamaterial in subwavelength optical lithography, we proposed a subwavelength optical lithography method based on SPPs. We computed and analyzed the distribution of optical field by the method of Finite Element Analysis (FEA). The results show that the images of object can be reconstructed by the structure.

In recent years, a flurry of activity in the fundamental research and development of surface plasmon based structures and devices were reported. The research on Nanolithography Based on Nobel Metals in this chapter is a step in the direction of providing affordable, highly flexible nanolithography. In this chapter, we present the case of nanolithography employed noble metal as objective lens to focus illumination light. The high-resolution surface plasmonic nanolithography has been investigated by using optical proximity exposure in the evanescent near field in nano-filmed noble metals. Sub-diffraction-limited feature size can be resolved by using I-line illumination exposure. And the simulations and experiments showed that sub-diffraction-limited feature size can be resolved by our given method, as illustrated in Fig. 1.

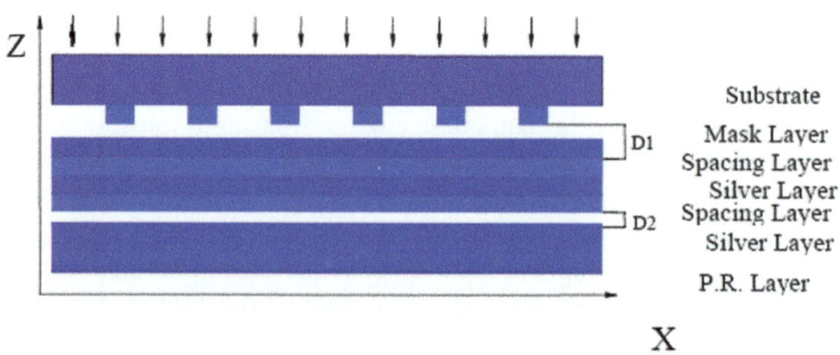

Fig. 1. Schematic of nanolithography by using noble metals. Multiple layers composing of noble metals and matched spacing imaged the mask on the photoresist layer. Patterns of arbitrary geometry on mask are reconstructed by the hetero-structure.

2. Surface plasmons plaritons

Fig. 2 shows the charges and the electromagnetic field of SPPs propagating along smooth surface. Assuming SPPs propagate along the x-axis with its propagating constant of k, the electromagnetic waves can be described by:

$$E = E_0^{\pm} \exp[+i(k_x x \pm k_z z - i\omega t)]\,, \tag{1}$$

Where, + for $z \geq 0$, - for $z \leq 0$, and with imaginary k_z, which causes the exponential decay of the field E_z. The wave vector k_x is parallel to the x direction, and

$$k_x = 2\pi / \lambda_p\,, \tag{2}$$

where, λ_p is the wavelength of the plasma oscillation.

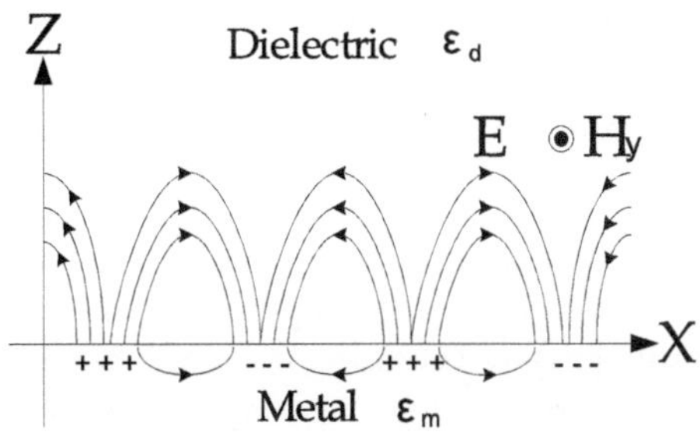

Fig. 2. Schematic of the charges and the electromagnetic field of SPPs propagating in the x-direction.

From Maxwell's equations and continuity at the boundary of metal and dielectric, the following dispersion relation of SPPs are derived:

$$\begin{aligned} k_x &= k_0 \sqrt{\varepsilon_m \varepsilon_d / (\varepsilon_m + \varepsilon_d)} \\ k_{zm} &= k_0 \varepsilon_m / \sqrt{\varepsilon_m + \varepsilon_d} \\ k_{zd} &= k_0 \varepsilon_d / \sqrt{\varepsilon_m + \varepsilon_d} \end{aligned}\,, \tag{3}$$

where $k_0 = 2\pi / \lambda_0$ is the wave vector of incident light in free space, k_{zm} and k_{zd} are the z component of wave vector for metal and dielectric respectively, and λ_0 is the wavelength of incident light. For any metal material, due to the metal's dispersion relation, there exists imaginary part of its permittivity which determines the skin depth of SPPs. However, the imaginary part of the dielectric can be ignored compared to the metal's imaginary part at the same incident light wavelength. So, we should shift our concentration to the imaginary part of material's permittivity.

As for the penetration of SPPs into either media, we can value it by its skin depth which was define by its field falls to $1/e$, that is

$$|z| = 1/|k_z|\,, \tag{4}$$

or in the metal medium:

$$|z_m| = \frac{2\pi}{\lambda}\sqrt{\frac{\varepsilon_{mR}+\varepsilon_d}{\varepsilon_{mR}^2}}, \tag{4a}$$

and in the dielectric medium:

$$|z_d| = \frac{2\pi}{\lambda}\sqrt{\frac{\varepsilon_{mR}+\varepsilon_d}{\varepsilon_d^2}}, \tag{4b}$$

where ε_{mR} is the real part of ε_m .

3. Nanolithography in the evanescent near field by using nano-filmed noble metal layer(s)

The sharpness of object can't be resolved by conventional lens due to the limitation by the wavelength of illumination light. J. B. Pendry had predicted that a slab of negative refractive index material has the power to focus all Fourier components of a 2D image. The super resolution of the negative index materials using silver layer, which was called 'superlens', can reconstruct the image of a pattern with line width of 40 nm (Fang, et al., 2005). They made mask, silver slab and photoresist integrity in Fang's experiment, which likes the traditional contact exposure of lithography. It is not practical in real application by Fang's method of nanolithography because each wafer needs its respective mask. The experiments of super resolution using silver slab was reported, and the line width with one fifth of illumination wavelength can be successfully resolved by the silver slab (Blaikie et al., 2006).

In order to investigate the influence of distance between mask and noble metal slab on imaging, we designed a separated 'superlens' with silver slab 100 nm away from mask. We analyzed the distribution of optical field by Finite Difference Time Domain (FDTD). The results show that the images of object can be reconstructed by the structure.

3.1 Nanolithography method

The exposure method of near-field nanolithography that was proposed by this paper is illustrated in Fig. 1, and the parameters of the materials are shown in Table 1. A UV transparent substrate with refractive index of n = 1.6 (@365 nm) is used for supporting the mask. The object layer with line width of 60 nm and pitch of 120 nm , which is made of Cr with refractive index of n=2.924 (@365 nm), acts as the function of mask in exposure. The air gap comes from the vacuum contact between mask and silver layer, which can be viewed as a kind of practical nanolithography technique.

In Fig. 1, the spacing layer with refractive index of n=1.517 (@365 nm) acts as the following two functions: 1) to match the surface plasmon polaritons resonating conditions; 2) to protect the surface of silver slab.

In order to explore the potential valid imaging distance between mask and photoresist, a repeatation of spacing layer and silver slab layer was followed after the first silver slab. The sample was exposed in i-line light that shone from the substrate side. For the convenience of description, we defined D1 as the distance between mask and silver slab, and D2 as the distance between silver slab and photoresist, as shown in Fig. 1.

Materials	Parameters	Thickness
Substrate	n=1.6@365nm	100nm
Mask	n=2.9@365nm	40nm
Spacing	n=1.52@365nm	40nm
Silver	ε=-2.4+i*0.2488@365nm	40nm
Photoresist	n=1.7@365nm	100nm

Table 1. Physical parameters of the materials.

3.2 When silver slab separated 40 nm from mask (i.e. D1=40 nm)

In order to save calculating time and PC resources, we computed the distribution of optical near-field intensity by the 2-D FDTD method. We chose cell size of $X \times Z = 2nm \times 1nm$, which is much smaller than both the exposure light wavelength and the mask's feature size. The timestep, according to Courant condition, should be:

$$TimeStep = 2\sqrt{2}N^{1/3} \tag{5}$$

N is the total cells of computing area. We chose 3500, by which the amplitude of electric field already became steady. The distance of 40 nm comes from the spacing layer. At this situation, it is a kind of ideal condition, because mask and spacing layer had a hard contact. The surface plasmons polaritons of two interfaces between silver slab and its surroundings can magnify the evanescent waves that carried the detailed information of object. When D1=40 nm, we calculated the following four conditions: D2=0 nm, 20 nm, 40 nm, 60 nm. Fig. 3 shows the distribution of electromagnetic (abbreviated to EM afterwards in the paper) field respectively.

It was found that the image of mask can be clear resolved by the method. Fig. 3(a) shows better result, however, Fig. 3(b)-(d) showed worse results due to the exponential decay of the evanescent waves came from the exit side of the interface between silver slab and photoresist. The strong contrast of EM field may come from the edge effect of the evanescent waves. We chose the 10 nm cross-section of photoresist layer to compare the imaging result of silver slab. When D1=40 nm and D2 varied from 0 nm to 60 nm, the distribution of optical field in the section was shown in Fig. 3. When D2 changed from 0 nm to 60 nm, the amplitude reduced to about a half under the same condition, but the high contrast of lines still can be clearly observed in Fig. 4. The amplitudes of lines were relatively uniform when D_2=20 nm and 60 nm compared with D_2=0 nm. If the parameters of photoresist were under better control, the lines of images will be more uniform. However, compare with the condition of D2=0 nm, the depth of lines in photoresist will be shallower at the same exposure condition.

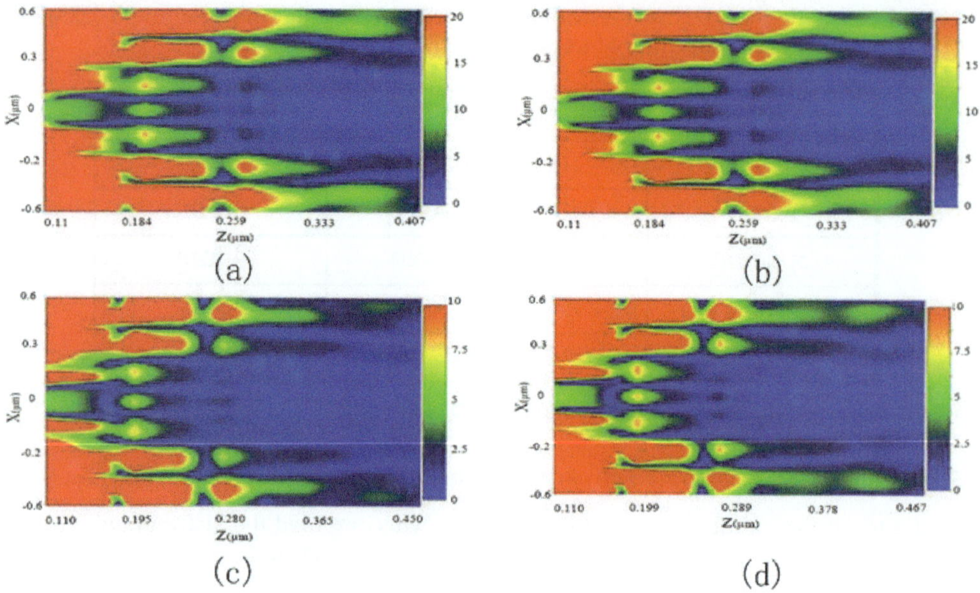

Fig. 3. The distribution of EM field in the model. Photoresist layer lies (a) between Z=0.31 μm and Z=0.41 μm, (b) between Z=0.33 μm and Z=0.43 μm, (c) between Z=0.35 μm and Z=0.45 μm, (d) between Z=0.7 μm and Z=0.47 μm.

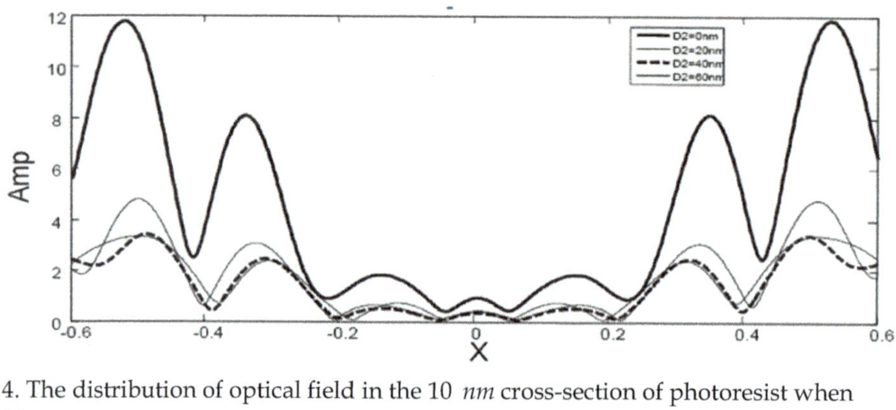

Fig. 4. The distribution of optical field in the 10 nm cross-section of photoresist when D1=40 nm.

3.3 When silver slab separated 60 nm and 80 nm from mask (i.e. D1=60 nm, 80 nm)

When D1=60 nm, 80 nm, we calculated four conditions respectively for each D1. We still chose the 10 nm cross-section of photoresist layer to investigate the optical field. In order to show the clear comparison results, we give the final comparison of the amplitude instead of EM distribution figures for each condition of different D2. Fig. 5 and Fig. 6 showed the result respectively when silver slab separated 60 nm and 80 nm from mask.

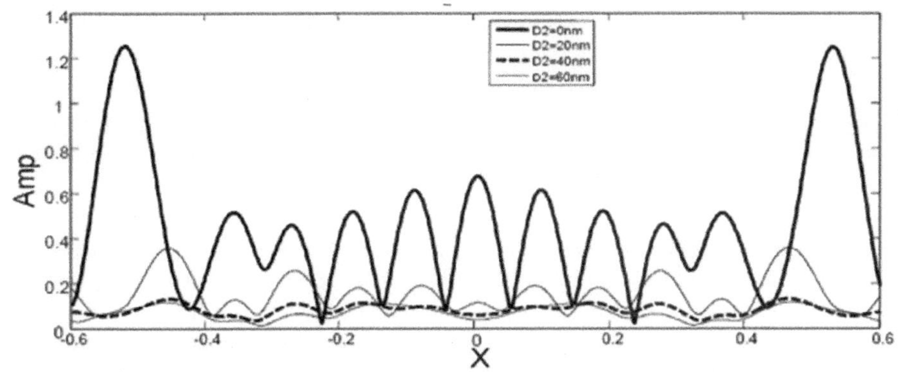

Fig. 5. The final comparison when D_1=60 nm while D_2=0 nm , 20 nm , 40 nm and 60 nm .

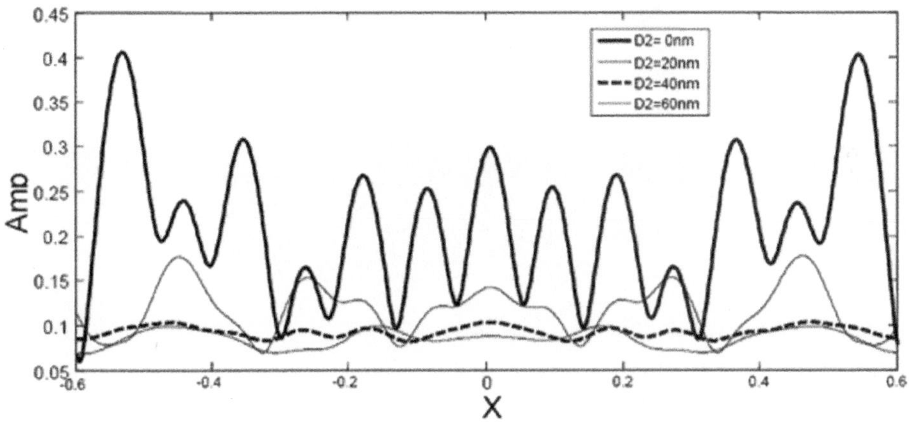

Fig. 6. The final comparison when D1=80 nm while D2=0 nm , 20 nm , 40 nm and 60 nm .

It was found that there came out extra fringes in Fig. 5 and Fig. 6. This kind of phenomena may be caused by the strong interference effects among evanescent waves. The image of mask still can be resolved in photoresist layer by proper choice of materials and exposure conditions. These results showed a bad conformity between mask and recorded image in photoresist layer. On the other hand, it gave us a hint to realize better resolution of optical lithography by reasonably using the interference effect.

3.4 When silver slab separated 100 nm from mask (i.e. D1=100 nm)

The evanescent waves cannot propagate to a long distance due to its exponential attenuation. The intensity of evanescent waves decays with a characteristic length Z_0 :

$$Z_0 = \frac{n}{k_t \sqrt{\sin^2 \theta_1 - n^2}} \tag{6}$$

where $n = n_2 / n_1$ is the relative refractive index of two surrounding media; θ_1 is the incidence angle of light from optically denser media to optically thinner media; where $k_t = 2n_2\pi / \lambda$, λ is the wavelength of incident light. Theoretically, Z_0 can be $100\,nm$ by calculation.

In order to explore the potential imaging property of silver slab, we increased D1 as much as possible. Considering the evanescent waves may diminish when $D_1=100\,nm$, so we calculated the condition of D2=0 nm only, the distribution of electromagnetic filed was shown in Fig. 7. It was found that the image of mask still can be resolved clearly in photoresist layer with good uniformity of imaged lines. We investigated the distribution of optical field in the $10\,nm$ cross-section of photoresist layer. The distribution of optical filed in the transverse section of photoresist was shown in Fig. 8. The image of mask can be resolved with high contrast. With proper choice of exposure condition and materials, the information of mask can be transferred to photoresist layer, and the image of mask can be reconstructed by the silver slab layer.

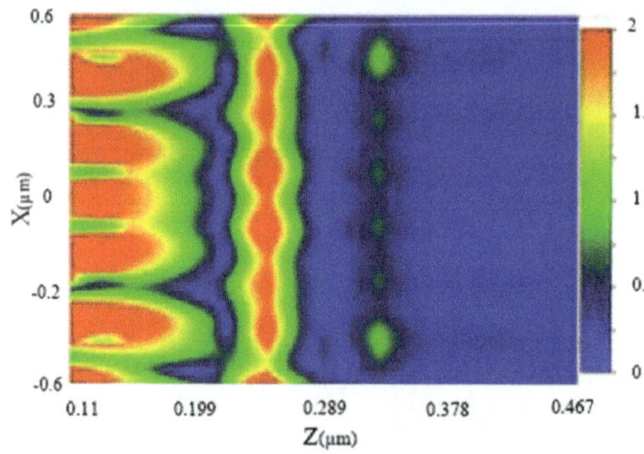

Fig. 7. The distribution of electromagnetic field in the model (Both X and Z is in unit of μm . Photoresist layer lies between Z=0.37 μm and Z=0.47 μm).

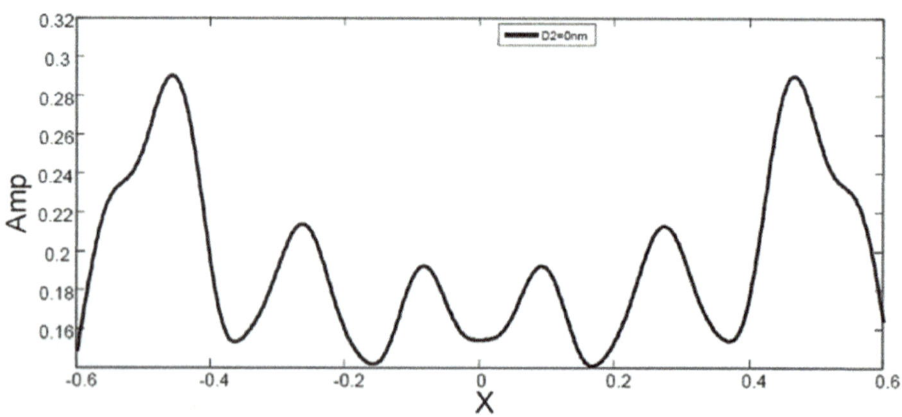

Fig. 8. The distribution of optical field in the $10\,nm$ cross-section of photoresist when silver slab separated $100\,nm$ from mask.

In brief, the image of mask can be transferred to the photoresist layer by the enhancement function of surface plasmons polaritons in silver slab. We calculated the 2D distribution of electromagnetic field in our model; the results showed that the image of mask with feature size of 60 nm can be resolved in photoresist layer when silver slab separated 100 nm from mask. By proper design and choice of material, nanolithography with better resolution can be realized by the very function of silver slab.

3.5 Nanolithography experiments

In order to further investigate the potential resolution of multiple layer of hetero-structure, we design a nanolithography composing of dual silver-dielectric configuration. Fig. 9 shows the distribution of electric filed in the evanescent field of nano-filmed noble metals under the illumination of i-line light. The parameters of the configuration are shown in Table 2.

Materials	Parameters	Thickness
Substrate	n=1.6@365nm	100nm
Mask	n=2.9@365nm	50nm
Spacing	n=1.52@365nm	20nm
Silver	ε=-2.4+i*0.2488@365nm	20nm
Photoresist	n=1.7@365nm	100nm

Table 2. Physical parameters of the materials.

For the pattern with feature size of 30 nm, we can find that the pattern of mask can be resolved in the photoresist. In fact, the metal have its intrinsic property of absorption which cause the loss of the light travelled along the surface, while the surface plasmons polaritons have its local enhancement. So, we can reasonably apply the both property to transfer the information of the mask to a very long distance. Theoretically, that mean we can apply a lot of metal-dielectric to play the function of lens.

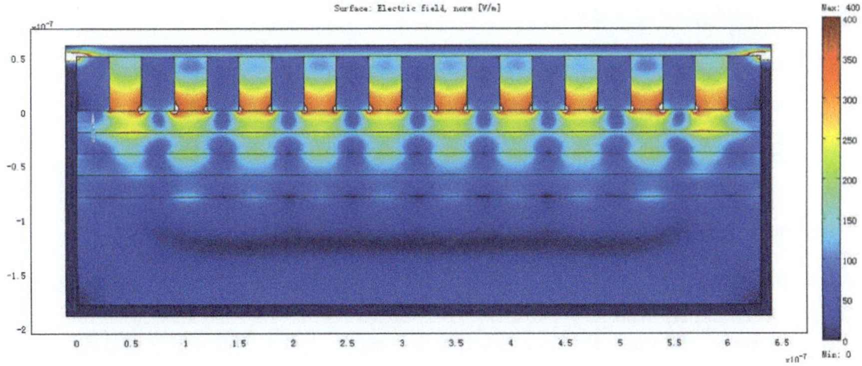

Fig. 9. Electric field distribution of dual-layered hetero-structure for CD=30 nm .

3.6 Nanolithography by dual metal-dielectric hetero-structure

We designed a nanolithography system to demonstrate the technique proposed by the paper, the parameters of our system and the process are shown in Table 2 which is slightly different from the condition of simulation.

Materials	Thickness
Substrate (Quartz)	100nm
Mask (Chromium)	40nm
Spacing (PMMA)	40nm
Silver	30nm
Photoresist (ARP-3170)	100nm
Light Source	i-line (365nm)
Optical Intensity	12.8mW/cm2
Exposure	15s
Development	30s

Table 3. Experiment parameters of our nanolithography system.

The process of our nanolithography should follow a different method from the traditional one. The detailed process is shown in Fig. 9. We should pay special attention to the thickness of each layer during the process. For the precision control of the thickness, it took us a lot of time to finish the process.

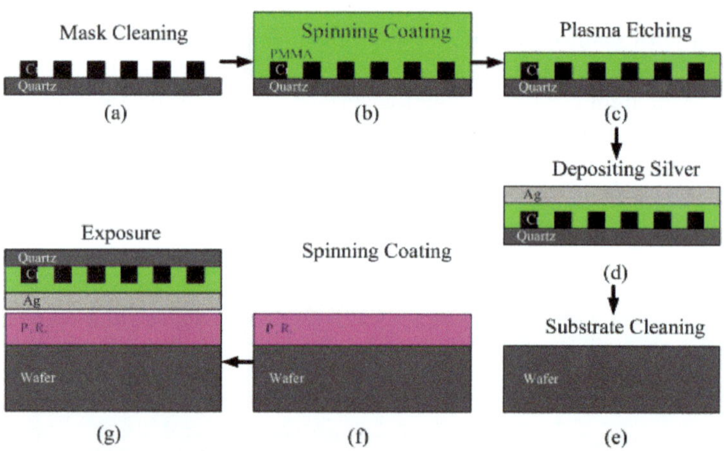

Fig. 10. The special process for the nanolithography.

Fig. 9 is the SEM image of our experiment. It was found that the pattern with feature size of 100 nm has better quality than the pattern with feature size of 50 nm has. As for the pattern with feature size of 50 nm, the worse quality may come from the mask fabrication procedure. Because the fabrication of mask with feature size of 50 nm is also a great challenge to FIB or EBL.

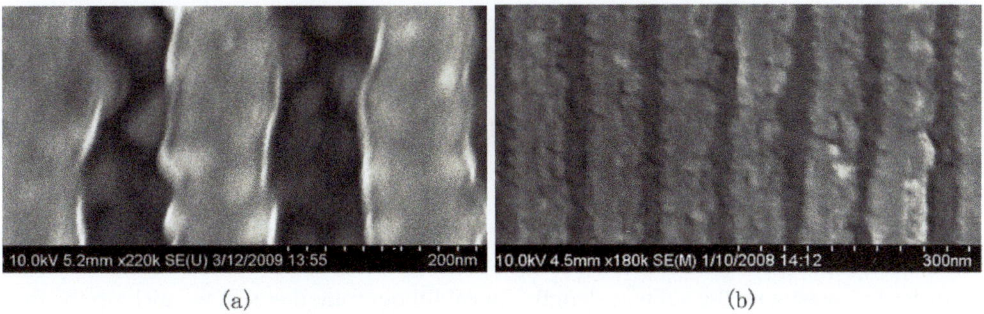

Fig. 11. The experimental results of our nanolithography method. (a) for the feature size of 100 nm and (b) for the feature size of 50 nm.

In brief, the experiment mentioned above demonstrates the feasibility of using noble metal as a special lens which can realize the subwavelength resolution. We believe in that Surface plasmons polaritons will be used as one of the promising tools in the nanometric device and special IC areas for the purpose of high resolution regardless of low yield.

4. Nanolithography by using gain-assisted metamaterials

Seen from equation(3), it seems that there is no other method to reduce the metallic losses to increase the SPPs propagation length in the metal-dielectric configuration.

For the configuration described in Fig. 2, metal got its intrinsic property of absorption; we cannot do anything to metal. However, there seems to be an alternative method to compensate for the losses in the metal. It was reported that if the configuration of metal-dielectric(without gain) was replaced by the configuration of metal-dielectric(with gain), the losses in the metal will be decreased and cause longer propagation length. Based on the report, we proposed a subwavelength optical lithography method which can be simplified in Fig. 9.

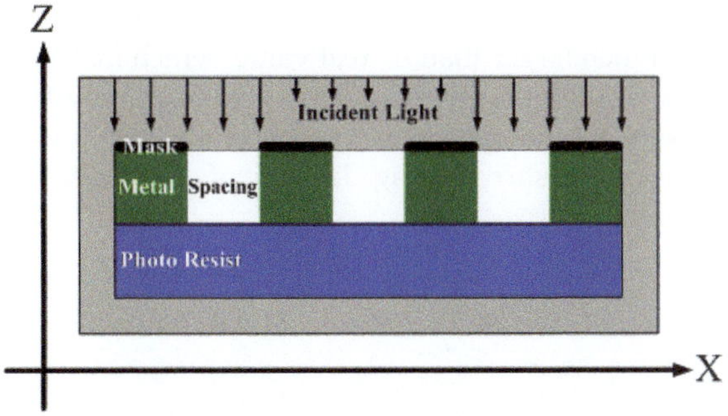

Fig. 12. Sketch of subwavelength optical lithography method.

In Fig. 9, the incident light was introduced to the system with perpendicular direction to the surface; mask acts as an object which blanks the incident light; the photo resist imaged the image of the mask. Due to the diffraction limitation of light, mask with feature size below half wavelength of incident light cannot be resolved by the conventional exposure system.

In our subwavelength optical lithography system, we designed a mask with feature size of $60\,nm$ to demonstrate the resolution of our system. The mask is designed with period of $120\,nm$.

According to the excitation condition of SPPs, once the materials got matched, SPPs will propagate along the interface between metal and dielectric, and we can record the image in photo resist to study the resolving capacity of our subwavelength optical exposure system.

4.1 Computation and analysis

In order to investigate the subwavelength optical lithography method, we pick up the finite element method(FEM) to study the distribution of optical field in photo resist.

The ultraviolet light with wavelength of $365\,nm$ is perpendicular to the surface; the physical parameters of material are shown in Table 3. All the parameters are taken from [Edward D. Palik, Handbook of Optical Constants of Solids. London: Academic Press. 1985] which is highly accepted in optics field.

Materials	Parameters
Mask	ε=-8.8203+i*9.1858
Spacing	ε=2.3013+i*0.0014
Silver	ε=-2.5619+i*0.6015
Photoresist	ε=2.5918+i*0.0097

Table 4. Physical parameters of the materials.

We selected an area of $w \times h = 0.72\mu m \times 0.205\mu m$, as shown in Fig.3, to compute the distribution of optical field when the light with TM mode is introduced into the system. The computation area is surrounded by Perfect Matched Layer(PML) to simplify the simulation process. In order to merely investigate the propagation length of SPPs, we simplified mask with its imaginary part much larger than its real value, which makes mask absorbed most of the incident light to blank the passage. In order to get better results, we picked up higher order of element to carry the computation. In our model, SPPs will propagate along the interface between metal and its surrounding. The size of metal is $w \times h = 0.06\mu m \times 0.06\mu m$.

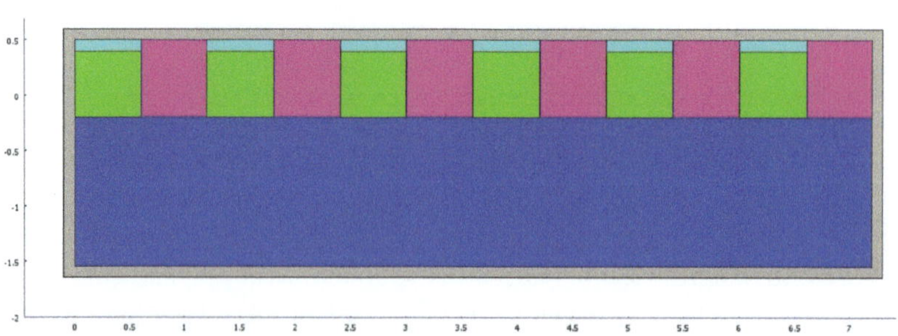

Fig. 13. FEM computation area of model.Fig.3

We chose the electric field to analysis the interaction between photoresist and light which is widely applied in the simulation of optical lithography. Fig. 11(a) and Fig. 11(b) show the distribution of electric filed in our system under ideal condition and real condition, respectively. Both of them show that the SPPs can reach to the exit side of the light passage. The intensity difference caused by the following conditions: in ideal condition, we set the permittivity of metal without no imaginary part, and we made both material perfect matched for the excitation condition of SPPs; in real condition, we set metal with imaginary part, and we made little bit difference between the metal's $|\varepsilon_{mR}|$ and its surrounding's $|\varepsilon_{dR}|$ which will cause the excitation condition little mismatched.

In order to investigate the propagation length, we extended the size of metal to $w \times h = 0.06\mu m \times 0.18\mu m$ while other conditions keep fixed as same. Fig. 12(a) and Fig. 12 (b) show the distribution of electric filed in our system under ideal condition and real condition, respectively. Both of them show that the SPPs can reach to the exit side of the passage. For the ideal condition, the enhanced effect in between the passage might be caused by the coupling of SPPs on both surfaces of the passage. Because there is no imaginary part in metal's permittivity, more electromagnetic power can be transferred to the exit surface. However, in the real condition, the uniform distribution was exhibited.

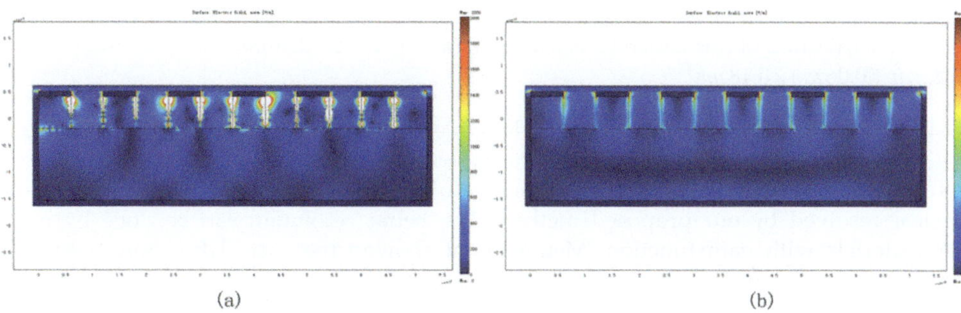

Fig. 14. Distribution of electric filed. (a) for ideal model and (b) for real model.

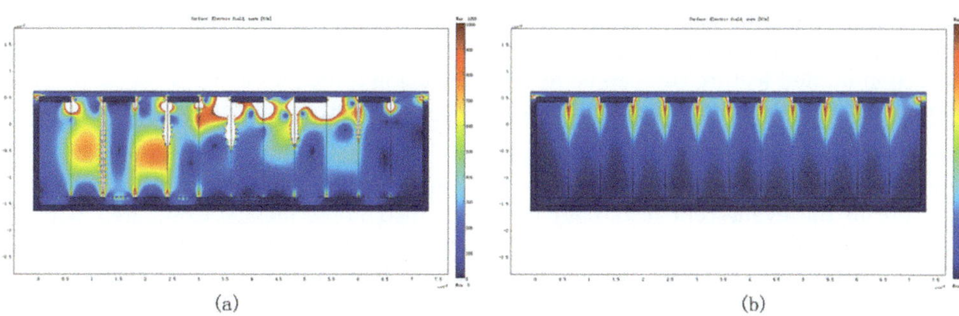

Fig. 15. Distribution of electric filed. (a) for ideal model and (b) for real model.

We pick up the exit surface of the passage to compare the optical lithography result, as shown in Fig. 13(a) and Fig. 13(b). In the both figures, the black solid line and dashed dot line signed for ideal condition and real condition, respectively. Fig. 13(a) and Fig. 13(b) showed the distribution of the electric field with metal size of $w \times h = 0.06\mu m \times 0.06\mu m$ and $w \times h = 0.06\mu m \times 0.18\mu m$ respectively. We found that the uniformity of real condition is much better than that of ideal condition, which might be caused by the interference among the SPPs on the exit surface.

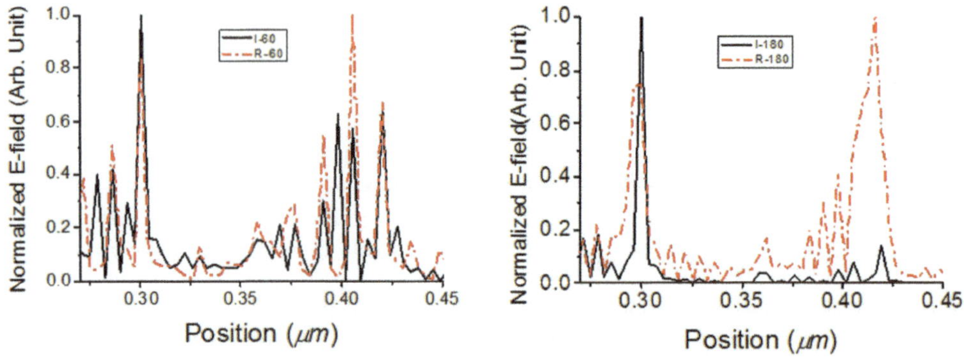

(a) (b)

Fig. 16. Distribution of normalized electric filed. (a) for $w \times h = 0.06\mu m \times 0.06\mu m$ and (b) for $w \times h = 0.06\mu m \times 0.18\mu m$.

We investigated the potential propagation length of SPPs on metal-dielectric-metal structure by FEM method. And we demonstrated the subwavelength optical lithography realized by gain-assisted metamaterial. The computation results showed that feature size of $60\,nm$ can be clear resolved by our proposed method. The better resolution can be done by using metamaterials with gain function. Metamaterial showed the special function which can compensate for the intrinsic loses caused by metal.

5. Conclusion

Two types of nanolithography in the evanescent field of noble metal are discussed in this chapter. Both of them focus on the evanescent near field in nano-filmed noble metals. Sub-diffraction-limited feature size can be resolved by using i-line illumination exposure. Better resolution up to $50\,nm$ was successfully simulated and tested in this chapter.

The first applies single or multiple layers of metal-and-dielectric to realize the nanolithography. Nanolithography has been investigated by using optical proximity exposure in the evanescent near field in nano-filmed noble metals. Compared with the model of original superlens, we separated the superlens $100\,nm$ away from the mask, under the illumination of i-line light, the initial simulation shows that the sub-diffraction-limited feature as small as $60\,nm$ line width can be clearly resolved without hard contact between mask and nano-filmed noble metal. By proper design of the materials and the parameters of nano-filmed layers, better resolution can be realized. The experiment showed that pattern with feature size of $50\,nm$ can be reconstructed in the photoresist by our nanolithography method. The experimental results showed the good consistency between the simulation and the experiments. As for the pattern with feature size of $50\,nm$, the worse uniformity of lines may come from the following aspects: the uniformity of the light source, the worse quality of the mask, the bad precision for the thickness of different layer and the process itself. Anyway, we do believe that we can get better results once we optimize the whole process of the nanolithography.

The second mainly focuses on the gain-assisted metamaterials applied in nanolithography. Surface Plasmon polaritons are electromagnetic waves that propagate along the surface of a conductor, usually a metal. It is shown that the gain-assisted metamaterial can compensate for the intrinsic absorption loss in metal. In this chapter, the propagation of surface plasmon polaritons on gain-assist metamaterial system is investigated. As an example, nanolithography has been considered by using optical proximity exposure in the evanescent near field of gain-assisted metamaterial layer. The evanescent waves carried the detailed information of the object which was defined by the high space frequency of the mask. With the enhancement of surface plasmon polaritons and gain-assisted metamaterials system, the evanescent waves can be propagated to a relatively far distance. Numerical computations by finite element analysis shows that better optimization of the gain-assisted metamaterials system can further improve the resolution. The computation result shows it will be an alternative nanolithography technique for the next generation lithography.

In brief, a plasmonic structure for imaging and super focusing is a new approach besides the concept of negative refractive index. It is possible to realize imaging resolution beyond diffraction limit with a certain working distance within several wavelengths range. To realize this target, one of technical challenges is that how to transfer the high spatial frequency near-field signals from evanescent wave to propagation wave. The other challenge is that how to amplify the near-field evanescent wave from conventional $\sim 180nm$ to be $\sim 1\mu m$ or even several wavelengths in free space.

6. Acknowledgement

The project is financially supported by the National Natural Science Foundation of China (NSFC) under project No. 60906049 and Guangdong Natural Science Foundation (Grant No. S2011040000711). The authors highly appreciate the valuable discussions and suggestions from their colleagues.

7. References

Barnes, W. L., Dereux, A. & Ebbesen, T. W. Surface plasmon subwavelength optics. Nature 424, 824-830 (2003).

D. O. S. Melville, R. J. Blaikie, “Submicro imaging with a planar silver lens”, Appl. Phys. Lett. 84(2004)4403.

D. O. S. Melville, R. J. Blaikie, “Super-resolution imaging through a planar silver layer”, Optics Express 13(2005)2127.

Edward D. Palik, Handbook of Optical Constants of Solids. London: Academic Press. 1985.

Guanxiao Cheng, Chao Hu, Ping Xu, and Tingwen Xing, "Zernike apodized photon sieves for high-resolution phase-contrast x-ray microscopy," *Opt. Lett.* 35, 3610-3612 (2010).

Guanxiao Cheng, Tingwen Xing, Wumei Lin, Jinmei Zhou, Chuankai Qiu, Zhijie Liao, Yong Yang, Lei Hong, Jianling Ma, “Photon sieve array x-ray maskless nanolithography,” *Proc. SPIE* 6517, 651736 (2007).

Guanxiao Cheng, Tingwen Xing, Yong Yang, Jianling Ma, "Experimental characterization of optical properties of photon sieve," *Proc. SPIE* 6724, D7240 (2007).

Guanxiao Cheng, Tingwen Xing, Yong Yang, Jianling Ma, "Resolution enhancement of photon sieve based on apodization," *Proc. SPIE* 6832, 83229 (2008).

Heinz Raether. Surface Plasmons on Smooth and Rough Surfaces and on Gratings. New York: Springer-Verlag.

J.B. Pendry, "Negative refraction makes a perfect lens", Physical Review Letters, Vol.85, No.18 (2000).

J.G. Goodberlet, H. Kavak, "Patterning sub-50nm features with near-field embedded-amplitude masks", Appl. Phys. Lett. 81 (2002) 1315.

Lezec, H. J. et al. Beaming light from a subwavelength aperture. Science, 2002(107): 1895.

M.M. Alkaisi, R.J. Blaikie, S.J. McNab, R. Cheung, D.R.S. Cumming, "Sub-fiffracion-limited patterning using evanescent near-field optical lithography", Applied Physics Letters, Vol.75, No.22(1999).

Maziar P. Nezhad, Kevin Tetz, Yeshaiah Fainman. Gain assisted propagation of surface Plasmon polaritons on planar metallic waveguides. Optics Express, Vol.12 No. 17(2005): 4072-4079

Nicholas Fang, Hyesog Lee, Cheng Sun, Xiang Zhang, "Sub-diffraction-limited optical imaging with a silver superlens", Science, Vol. 308(2005).

Ritchie, R. H. Plasma losses by fast electrons in thin films. Phys. Rev., 1957(106):874-881

Shaolin zhou, Feng Xu, Song Hu, Xiaoping Tang, "Positioning scheme based on Grating modulation and phase imaging in lithography", Proc. SPIE, the 5th International Symposium on Advanced Optical Manufacturing.

Shaolin Zhou, Yong Yang, Lixin Zhao, and Song Hu, "Tilt-modulated spatial phase imaging method for wafer-mask leveling in proximity lithography" Opt. Lett. 35, 3132-3134(2010).

Shaolin Zhou, Yongqi Fu, Xiaoping Tang, S. Hu, Wangfu Chen, and Yong Yang, "Fourier-based analysis of moiré fringe patterns of superposed gratings in alignment of nanolithography," Opt. Express 16, 7869-7880 (2008).

T. W. Ebbesen, H. J. Lezec, H. F. Ghaemi, T. Thio, and P. A. Wolff, Extraordinary optical transmission through sub-wavelength hole array, Nature 391, 667-669 (1998).

W. L. Barnes, W. A. Murray, J. Dintinger, E. Devaux, T. W. Ebbesen. Surface Plasmon polaritons and their role in the enhanced transmission of light through periodic arrays of subwavelength holes in a metal film. Phys. Rev. Lett., 2004(92): 107401

Xiangang Luo, Teruya Ishihara, "Surface Plasmon resonant interference nanolithography technique", Applied Physics Letters, Vol. 84, No.23 (2004).

Yong Yang, Hanmin Yao, Song Hu, Guanxiao Cheng, "Nanolithography in the evanescent field by using silver layer", Proc. SPIE 6724, 77241A(2007).

Yong Yang, Song Hu, Hanmin Yao, Guanxiao Cheng, Chunmei Zhang, Wei Yan, "Nanolithography in the Evanescent Near Field by Using Nano-filmed Noble Metal Layers," *Proc. SPIE* 6724, 77241A (2007).

Yong Yang, Yongqi Fu, Hanmin Yao, Song Hu, Shaolin Zhou, Wei Yan, Wangfu Chen, Guanxiao Cheng, Zhan Li, "Beam Splitter Achieved by Using Metallic Structure with Nanoslits," *J. Comput. Theor. Nanosci.* 6(5),1030-1033,(2009).

7

Organic–Inorganic Mesoporous Silica Nanotube Hybrid Anodic Alumina Membranes for Ultrafine Filtration of Noble Metal Nanoparticles

Sherif A. El-Safty[1,2] and Nguyen Duc Hoa[1]
[1]Materials Research Laboratory for Environmental and Energy National Institute for Materials Science (NIMS), Tsukuba-shi Ibaraki
[2]Graduate School for Advanced Science and Engineering Waseda University, Shinjuku-ku, Tokyo
Japan

1. Introduction

Novel metal nanoparticles (NPs), such as gold (Au), platinum (Pt), palladium (Pd), and silver (Ag) have gained increased attention due to their distinctive physical and chemical properties as well as effective applications in catalysis, imaging, gene expression, chemical sensing, biolabeling, disease diagnostics, disease therapeutics, nanoscale electronics, and photonics. The physical and chemical properties as well as potential applications of these NPs are strongly dependent on their shapes, sizes, and structural geometries. However, conventional synthesis methods produce novel metal NPs with broad size distributions and uncontrollable precise morphologies. Therefore, it is necessary to develop a simple, effective, and fast method for separating NPs with different sizes and shapes.

Engineered control over one-dimensional (1D) mesoporous materials inside an anodic alumina membrane (AAM) has led to widespread advances in the separation of biomolecules or NPs. However, several challenges remain, particularly in applying AAM as robust nanofilters for the separation of high concentration NPs. With the aim of controlling the design functionality of nanofilters for size-exclusion cut-off separation, this book chapter focuses on the successful, up-to-date development of mesofilter designs as promising filter candidates for size-exclusive separation of NPs, in which multifunctional surface coating of the pore channels of the AAM with organic coupling agents facilitated the production of extremely robust constructed membrane sequences without the formation of air gaps among NTs. This practical nanofilter design shows evidence of controlled assessment processes that involve the evaluation of intrinsic filter properties (e.g., long-term stability, separation efficiency, and reusability, among others).

This chapter is designed to include night parts. After the introduction, an overview of the recent synthesis and characteristics of noble metal NPs (i.e., Ag, Au, Pt, and Pd) are provided. The advantages and disadvantages of conventional synthesis methods are also presented. In the third part, the basic concept of membrane separation, including filtration and nanofiltration technology, is briefly introduced. Different nanofiltration methods and their applications for small molecules, biomolecules, and NPs are discussed. In the fourth part, the design of the high-ordered mesoporous (HOM) silica NT membrane for nanofiltration is introduced, followed by a discussion of the advantages and disadvantages of this nanofiltration process compared with other traditional methods. In the fifth part, more details about the synthesis of the high-ordered mesoporous silica membrane fabricated inside the pore channels of AAM using different surfactants are presented. We targeted the important development of mesofilter membranes composed of mesoporous silica NTs (i.e., those that are perpendicular to the longitudinal axis of the nanochannels) and uniform multidirectional pores in nanoscale sizes that effectively enhance the size-based separation of noble metal NPs within seconds. In the sixth part, the characteristics of the fabricated HOM silica NT membrane are reported. In the seventh part, the application of synthesized silica NT membranes for noble metal NPs separation is introduced. Some segments of parts of seven and eight introduce the application of high-ordered mesoporous silica NT membrane for separating novel metal NPs and their performances. Finally, the conclusion and outlook are discussed in the last part.

2. Noble metal NPs

2.1 Gold nanoparticles

Au NPs are among the most studied noble metallic NPs due to their unique physical, chemical, and bio-compatible properties as well as their potential applications (Grzelczak et al., 2008). Au NPs have been applied in various fields, included of catalysis, imaging, gene expression, chemical sensing, biolabeling, disease diagnostics, disease therapeutics, nanoscale electronics, and photonics (Nath & Chilkoti, 2002). For instance, Au NPs can be used for molecular sensing based on either surface plasmon resonance or surface-enhanced Raman spectroscopy, in which the surface plasmon resonance wavelength of Au NPs depends on their shape, size, and the dielectric constant of the surrounding medium. Therefore, any change in the local environment of the NPs leads to variations in the surface plasmon resonance wavelength and absorption band, resulting in significant detection of molecular (Enders et al., 2011). Au NPs or colloids have been investigated since the 17th century. However, the first scientific document which reported the wet chemical synthesis of Au NPs was in 1857, when Michael Faraday prepared the Au colloids by reducing gold chloride with phosphorus in water and then investigated their optical properties (Faraday, 1857). He realized that the color of Au colloid solution was red but not the yellow of traditional bulk Au. Since then, the synthesis of Au nanostructures of different sizes and shapes has become an extensive topic. Various methods have been developed to synthesize Au NPs, including wet chemical reduction by sodium citrate, ascorbic acid, sodium boron hydride, or block-copolymers in the presence of stabilizer polymers, photoreduction, microwave-assisted method, and hydrothermal method. For instance, Polte et al. investigated the mechanism of Au NP formation in the classical citrate synthesis method (Polte et al., 2010). They prepared Au NPs by reducing tetrachloroauric acid using trisodium citrate at different temperatures and reactant concentrations. The mechanistic growth of Au NPs via classical citrate synthesis method included nucleation and aggregation as well as slow and fast growth, resulting in particle sizes of 2, 3, 5.5 and 7.7 nm, respectively. Liu et al. reported the hydrothermal synthesis of spherical single crystalline Au NPs of particle size ranging from 5 to 25 nm under a basic condition (Lui et al., 2010). Hieda et al. reported the synthesis of Au NPs using plasma in aqueous solution; they used the pulsed power supply to generate discharges for the synthesis of Au NPs with exotic shapes, such as spheres, triangles, pentagons, and hexagons

(Hieda et al., 2008). Among the developed synthetic methods, wet chemical reduction of Au salts in the presence of stabilizers and reducing agents is more general and effective for the synthesis and control of Au NPs. The well-known and effective solution pathway consist of the polyol processes that include the use of poly alcohols (i.e., those containing multiple hydroxyl groups, -OH), such as ethylene glycol, propyleneglycol, diethyleneglycol, trimethyleneglycol, and butyleneglycol as a solvent and mild reducing agent (Chen et al., 2005, Carroll et al., 2011). The polyol reduction processes can be also assisted using some bases as reducing agents, including sodium borohydride or sodium bicarbonate. In the polyol processes, the polymer stabilizers, surfactants, or capping agents like poly(vinylpyrrolidone) (PVP) are used to enhance the homogenous dispersion of NPs during synthesis. In a typical synthesis, a solid inorganic precursor of Au is suspended or dissolved in liquid polyol. The solution is then stirred and heated to a given temperature, which can reach the boiling point of the polyol, in order to reduce Au metal ions. The stabilizer, such as polymer PVP, is also used to prevent the NPs from aggregation. For instance, Kim et al. reported the synthesis of Au NPs with different morphologies using a modified polyol process with the presence of the surface-regulating polymer (PVP) (Kim F. et al., 2004). Green and O'Brien reported the one phase reduction of Au (IV) chloride by sodium borohydride in hot tri-n-octylphosphine oxide as both a reaction medium and passivating ligand. The obtained Au NPs had particles of different shapes (i.e., square, triangular, and spherical) and sizes ranging from 10 to 100 nm (Green & O'Brien, 2000). Other studies have focused on the size-controlled synthesis of gold NPs, resulting in Au nanocrystals with various sizes and shapes (Xiao et al., 2011, Walker et al., 2001). Different morphologies of Au NPs have been synthesized by different pathways, as shown in Fig. 1. However, even the produced NPs still have different sizes in the optimal synthetic conditions. Thus, controlling precise shapes and/or sizes of NPs is still a challenge. A solution may be the development of general pathways, such as membrane filtration method, for size-selective separation.

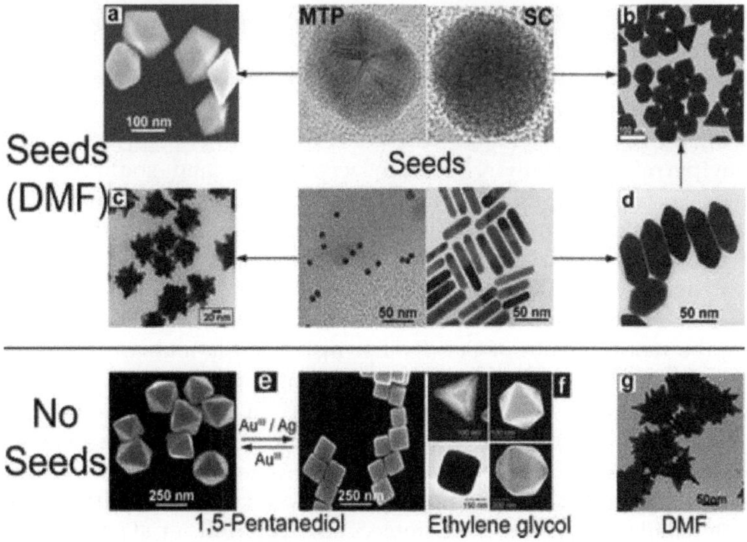

Fig. 1. TEM images of Au NPs with different shapes and sizes synthesized by different pathways (reprinted from Grzelczak et al., 2008).

2.2 Platinum nanoparticles

Recent interest on Pt materials focused on its nanostructures included of NPs, among others in order to improve their cost effectiveness and potential applications. Different morphologies of nanostructured Pt, such as nanocages, nanorods and NPs, have also been successfully fabricated (Onizawa et al., 2009, Petroski et al., 1998). The properties of Pt NPs are dependent on their shapes, sizes, and crystal structures, especially the crystallographic planes exposed on the surfaces of the NPs, thus the development of the facile method for the large-scale synthesis of Pt NPs with the ability to control the size, shape, or morphology is a crucial issue (Tsung et al., 2009). Recent research about the design of novel Pt and Pt-based nanomaterials with unique properties have been greatly intensified, because of their potential for new applications and improvement of current ones. The synthesis and control of the morphologies of Pt NPs have been extensively investigated for various applications, including catalytic applications in automobile exhaust purification, fuel-cell technology, and those for the petrochemical and fine chemical industries. To date, various methods have been developed for the synthesis of Pt NPs, including the sol−gel method, electrodeposition, electroless deposition and physical approaches, as well as the hydrothermal and solvothermal techniques (Chen A. et al., 2011). Among these, the polyol method is among the most popular pathways for synthesizing Pt NPs. This method has also been applied to synthesize metallic powders of Ru, Rh, Sn, Re, W, Pt, Au, Fe-Cu, Co-Cu, Ni-Cu, Fe, Co, Ni, Cu, Pd, and Ag (Kurihara et al., 1995, Siekkinen et al., 2006). For instance, the Xia group has introduced the process for the reduction of a Pt(II) or Pt(IV) precursor in the presence of a stabilizing polymer; the group achieved this by reducing agents, such as alcohols, sodium borohydride, or hydrogen gas (Herricks et al., 2004). In the polyol synthesis of Pt NPs, the type and amount of the surface capping-agents strongly affect the size and shape of the synthesized NPs. Polyvinylpyrrolidone is the most widely used capping agent in the synthesis of noble metal NPs. El-Sayed and coworkers controlled the shapes and sizes of the synthesis of Pt NPs by changing the concentration ratios of the capping polymer material and the platinum cations used in the reductive synthesis of colloidal particles in the solution (Ahmadi et al., 1996). The obtained products involved an abundance of different shapes, such as tetrahedal, polyhedal, irregular prismatic and cubic, and a wide particle size distribution ranging from 4 to 14 nm. Meanwhile, Miyazaki et al. synthesized the Pt NPs using polyvinylpyrrolidone, poly(N-isopropylacrylamide), and sodium polyacrylate as capping agents (Miyazaki et al., 2003). The NPs they produced had different shapes, such as hexagonal, square and triangular, depending on the capping agents used. Other methods, such as soft template (Attard et al., 1997), hard templates (Fukuoka et al., 2002) and electrochemical deposition (Tian et al., 2007), have been applied in order to control the morphologies of Pt nanocrystals. The catalytic-assisted synthesis is also an effective method for controlling the morphologies of Pt NPs. For instance, Xia and coworkers reported the synthesis of quasioctahedral shapes by simply reducing the H_2PtCl_6 precursor with PVP in aqueous solutions containing a trace amount of $FeCl_3$ (Lim et al., 2008). Another method for the synthesis of Pt NPs in an organic medium involves the Schiffrin method, in which aqueous $PtCl_6^{2-}$ ions are transferred to non-polar organic solvents by phase-transfer molecules, such as tetra-alkyl-ammonium salts reduced by borohydride treatment and capped with stabilizing alkyl isocyanide molecules (Horswell et al., 1999). The prepared particles have a well-defined crystalline structure and diameters ranging from 1 to 3 nm. Others methods, such as radiofrequency, sputtering, reverse micelles, electron beam lithography and chemical vapor deposition techniques, have also been developed for the synthesis of Pt NPs (Welch & Compton, 2006). In general, Pt NPs can be controlled using various synthesis methods with different conditions despite the shapes and sizes; thus, the final products still produced NPs of different shapes or sizes (Fig. 2).

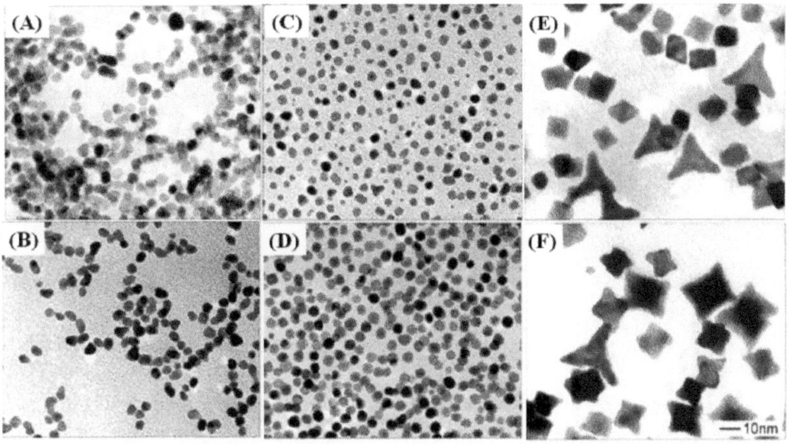

Fig. 2. TEM images of Pt NPs of different shapes and sizes (reprinted from Herricks et al., 2004)

2.3 Palladium nanoparticles

Pd is an important novel metal because of its unique physical and chemical properties and great potential applications in various fields; it can be used as a catalyst, for biological labeling, controlled drug delivery as well as electronic and optical devices (Nieto et al., 2008, Bianchini et al., 2009). However, their optical and catalytic properties and potential applications are dependent on their sizes, shapes, and morphologies. Therefore, the development of facile synthetic methods for the formation of Pd NPs with well-controlled size and shape is a very important topic that has been gaining attention in recent years (Horinouchi et al., 2006). In particular, the fabrications of noble metal NPs (including Pd) are generally conducted using wet chemical methods, such as the polyol process. A variety of Pd NPs with different shapes and sizes have been synthesized using a modified polyol reduction process in the presence of ethylene glycol and stabilizer (Xiong et al., 2005). For instance, Nemamcha et al. reported the synthesis of Pd NPs through the sonochemical reduction of Pd (II) nitrate in aqueous solution (Nemamcha et al., 2006). They prepared the starting solutions by adding different concentrations of Pd(II) nitrate in ethylene glycol and PPV. Afterwards, the resulting mixtures were irradiated with ultrasonic 50 kHz waves in a glass vessel for hours to obtain Pd NPs with particle sizes ranging from 3 to 6 nm, depending on the synthesis conditions. Roy et al. reported the characteristics of Pd NPs obtained by a one-pot reduction of Pd chloride in aqueous solution using citric acid in the presence of steric stabilizer polyvinylalcohol (Roy et al., 2006). The diameters of the Pd NPs can be controlled in a range of ~8–53 nm, using different synthesis conditions, such as reflux time, concentrations of metal ion, reducing agent, and capping polymer. The solution and seed-mediated growth methods have also been reported on the size and shape evolution of metal nanostructures. For instance, Kim et al. reported on the synthesis of monodisperse Pd

NPs by thermal decomposition of a Pd−surfactant complex (Kim SW. et al., 2003). They controlled the particle size of Pd NPs within 3.5–7 nm by varying the concentration of the stabilizing surfactant. Niu et al. introduced the shape-controlled synthesis of Pd nanocrystals via the seed-mediated method with CTAB, potassium iodide, and ascorbic acid as the surfactant additive and reductant, respectively (Niu et al., 2010). Several types of Pd nanocrystals, such as rhombic dodecahedral, cubic and octahedral have been prepared by manipulating potassium iodide concentration and the reaction temperature. However, synthesizing Pd NPs with fine and narrow particle size distribution is still a challenge. Pd NPs of different shapes and sizes fabricated by wet chemical route are shown in Fig. 3.

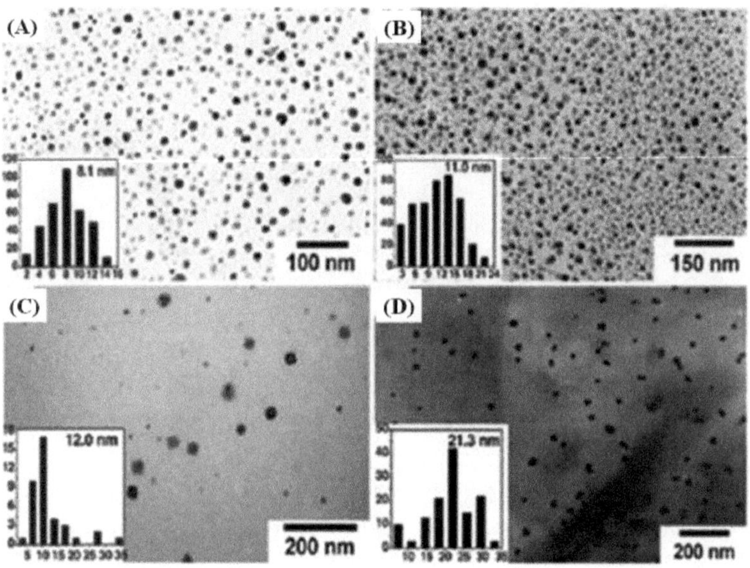

Fig. 3. TEM images and size distributions of Pd NPs of different morphologies (reprinted from Xia et al., 2009)

2.4 Silver nanoparticles

Ag NPs and their related nanostructures have been investigated extensively, because of their unique properties and great potential applications in oplasmonics, anti-bacterial materials, sensing, and spectroscopy. For instance, the anti-bacterial activity and good conductivity of Ag have been known since the ancient times and are now widely investigated for real applications (Kosmala et al., 1011). In addition, Ag NPs exhibit the surface plasmon resonance effect and strong bacterial resistance to antibiotics, making them ideal for biotechnological applications (Shrivastava et al., 2009). Ag NPs can be synthesized using various methods, such as chemical reduction (Pal et al., 2007), electrochemical (Santos et al., 2002), γ-radiation (Choi et al., 2005), laser ablation (Amendola et al., 2007), solvothermal (Yang et al., 2007), hydrothermal (Shen et al., 2011), photochemical (Henglein, 1998), and sonochemical (Salkar et al., 1999). Generally, different shapes and sizes of Ag NPs can be fabricated by wet chemical routes just simply varying the synthetic conditions. For instance, Dong et al. reported the synthesis of triangular Ag nanoprisms via stepwise reduction of Ag nitrate with sodium borohydride and trisodium citrate (Dong et al., 2010). They realized that the formation of the triangular nanoprisms is dependent on the molar ratios of sodium borohydride and trisodium citrate used in the reactions, in which a balance between the precursor contributed to the formation of the small spherical particles. The transformation of the spherical NPs is critical for the synthesis of the triangular nanoprisms. The TEM images of obtained silver NPs are shown in Fig. 4, and as can be seen, the Ag NPs have different shapes and sizes depending on the reduction time.

In summary, almost all aforementioned methods used the reducing agents to reduce the Ag salts to form Ag NPs with different shapes and sizes by carefully controlling the reducing environment and stabilizer concentration during synthesis. However, none of the above methods appear suitable for general utility. Most methods offer an acceptably narrow size distribution; however, they do not offer much variability in the size that can be produced. The preparation of Ag NPs with homogenous particle size is highly important in increasing their potential applications.

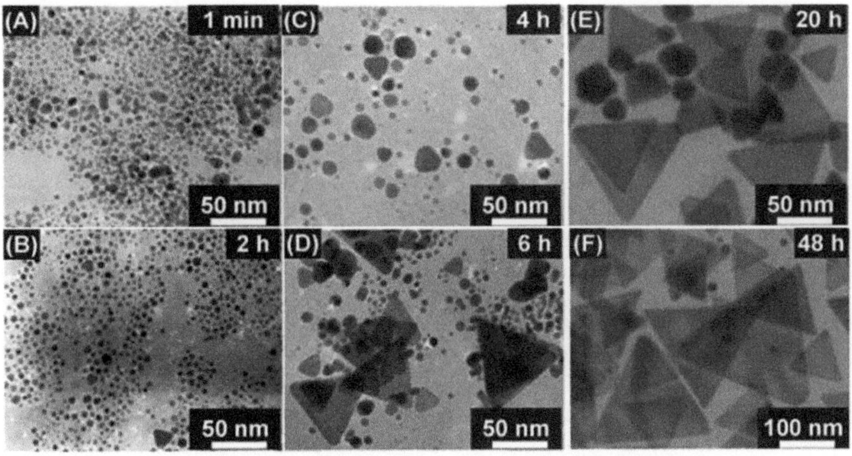

Fig. 4. TEM images of Ag NPs obtained through different time duration reductions of silver nitrate by sodium borohydride (reprinted from Dong et al., 2010)

2.5 Optical properties of novel metal NPs

Noble metal NPs exhibit diverse and unique characteristics, including catalytic, electronic, and optical properties. Among other, their optical properties are most interesting because of the interactions of electromagnetic field from the light and NPs, resulting different colors of NPs. For instant, the color of Au colloidal solution can be changed from red to light blue by varying the size and shapes of NPs (Jones et al., 2011). The abnormal behaviors have been attributed to the interaction between the electromagnetic radiation and the conduction band electrons (free electron cloud) of NPs (Fig.5).

When the oscillating electric field from incident light interacts with the delocalized electrons of a metal, the free electron cloud is perturbed and displaced from the metal framework, thus generating a dipole in the NP. However, the Coulombic attraction from the positively charged nuclei of the metal pulls the electron cloud back to its initial position. Due to the large difference in mass, the heavy nuclei remain in a fixed position, whereas the lighter electrons experience motion. With the electric field component of incident light acting as a sinusoidal driving force and Coulombic attraction acting as a restoring force, all components of a harmonic oscillator are present. Consequently, resonant conditions can be achieved when light is coupled in phase to the natural frequency of the plasmon oscillation. At these resonant conditions, the metal structures absorb the maximum amount of incident electromagnetic radiation, thereby causing the greatest amount of charge displacement. The oscillation wavelength depends strongly on the surrounding medium, size and shape of NP, which lead to different surface plasmon bands of NP solutions and different colors, as shown in Fig. 6.

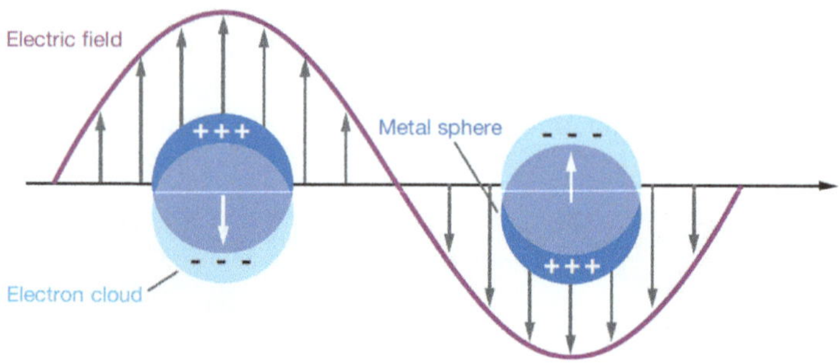

Fig. 5. Schematic diagram illustrating a localized surface plasmon (reprinted from Willets & Duyne, 2007)

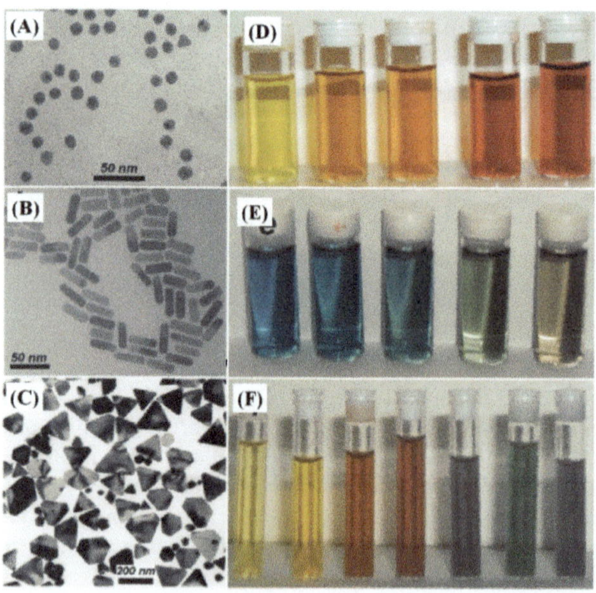

Fig. 6. Left: Transmission electron micrographs of Au nanospheres and nanorods (a,b) and Ag nanoprisms (c, mostly truncated triangles). Right: Photographs of colloidal dispersions of Au-Ag alloy NPs with increasing Au concentration (d), Au nanorods of increasing aspect ratio (e), and Ag nanoprisms with increasing lateral size (f) (reprinted from(Liz-Marzán, 2004))

3. Basic concepts of membrane separation

3.1 Overview of membrane separation

The separation of different molecules, ions, and particles from their mixture is a critical issue in the fields of nanoscience, nanotechnology, and especially in pharmacy and biology (Han et al., 2008). In recent years, several separation methods have been developed, including membrane filtration, size-exclusion chromatography, and electrophoresis. Membrane-based separations are becoming increasingly relevant for a number of applications due to their low energy requirements, potentially low fabrication cost, and steady-state operation (Jang et al., 2011). The membrane filtrations have been applied for the separation of gas phases, liquid–liquid phases, submicron particles from water, organic mixture, and hydrocabon mixture (Gin et al., 1011). They are very effective in separating molecules of different sizes, solids from liquids, and/or large size particles from small sizes by imperatively imposing the mixture solution through the proper porous membranes (Soria, 1995, Asatekin et al., 2011). Synthetic membranes have been developed and used in many separation processes, from industrial-scale ones (e.g., separating atmospheric gases for medical and industrial use and removing salt from seawater) to smaller-scale processes in chemical synthesis and purification, in which the membranes work by forming a barrier between two phases (e.g., salt water and fresh water) that restrict the movement of some molecules while allowing others to pass through. Diverse types of membrane filtration have been investigated and manufactured for the separation of gases, molecules, and particles of different sizes, permeabilities, and hydrophobicities. The membranes are classified as either dense (nonporous) or porous, depending on how the molecules (filtrates) move across the barrier. The porous membranes are further divided into asymmetric and symmetric. Those membranes have been classified based on their porous characteristics (Table 1).

Membrane barrier structure	Trans-membrane gradient		
	Concentration	Pressure	Electrical field
Non-porous Microporous dp≤2 nm Mesoporous dp=2–50 nm Macroporous dp=50–500 nm	P evaporation (PV) Dialysis (D) Dialysis	Gas separation (GS) Nanofiltration (NF) Ultrafiltration (UF) Microfiltration (MF)	Electrodialysis (ED) Electrodialysis

Table 1. Classification of membranes and membrane process for separation via passive transport (Ulbricht, 2006)

The membranes are classified as non-porous (dense), microporous ($d_p \leq 2$ nm), mesoporous (d_p=2–50 nm), and macroporous (d_p=50–500 nm) based on the barrier structure. These membranes are also classified if they are intended for gas separation (GS), reverse osmosis (RO), nano-filtration (NF), ultra-filtration (UF), and micro-filtration (MF) based on the trans-membrane gradient. The separation based on the gradient of concentration is classified into pressure evaporation and dialysis. However, the classifications are relative and associated. For instance, the pore sizes of RO, NF, UF, and MF fall within the ranges of

10^{-4}–10^{-3}, 10^{-3}–10^{-2}, 10^{-2}–10^{-1} and 10^{-1}–10^{1} μm, respectively (Wagner, 2001). The GS is used for air separation or natural gas purification, such as H_2, He, O_2, N_2, CO_2, C_2H_6, C_3H_8, and CH_4. The RO membrane is applied in liquid/liquid separation, where the water is the only material passing through the membrane. RO is effectively used to purify water, especially in successful applications for desalination of sea and brackish water, waste treatment, and various separations in chemical, food, pharmaceutical, and other industries; all dissolved and suspended material are rejected through the membrane. The NF rejects only ions with more than one negative charge, such as sulfate or phosphate, while passing single charged ions. NF also rejects uncharged, dissolved materials, and positively charged ions according to the size and shape of the molecule in question. UF is a process, in which the high molecular weight component (e.g., protein and suspended solids) are rejected, whereas all low molecular weight components pass through the membrane freely. Mono- and di-saccharides, salts, amino acids, organics, inorganicacids, or sodium hydroxide are not rejected by UF. The MF is a process where in only suspended solids are rejected, whereas proteins pass the membrane freely.

The membrane filtration can be classified as organic (e.g., polyethylene terephthalate, cellulose acetate, polyvinylidenedifluoride) and inorganic (e.g., carbon, alumina, silica) types based on the materials. In addition, they can be classified as having flat and hollow fiber forms based on the geometries. For instance, the commercial ordered porous organic polymeric membrances are track-etched polycarbonate and polyethylene terephthalate, whereas the commercial available inorganic membranes are glass, metal, zirconia, zeolite, and alumina (Robeson, 2008, Choi et al., 2009). Figure 7 shows the SEM images of hollow fiber poly(vinylidenefluoride), (a) hollow fiber carbon, (b) photos of commercial track etched polycarbonate, and the (c) high-ordered porous anodic alumina membranes (d).

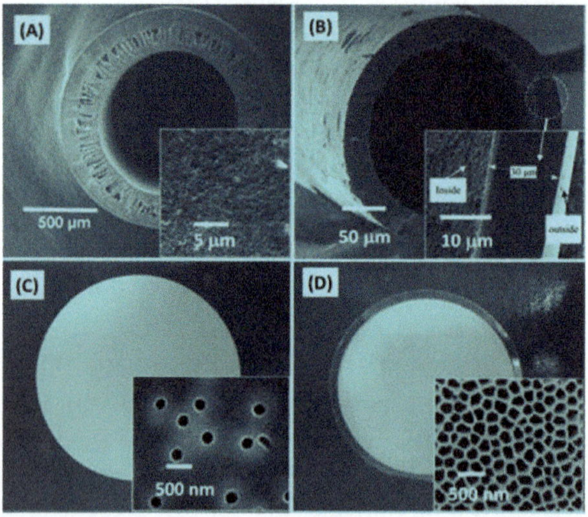

Fig. 7. The hollow fiber and flat membranes: (a) SEM images of hollow fiber poly(vinylidene fluoride) membranes (reprinted from Hashim et al., 2011); (b) Hollow fiber carbon membranes (reprinted from He at al., 2011) and photos of 47 mm diameter; (c) track-etched polycarbonate;(d) and anodic alumina membranes (insets are respective SEM images).

The membrane can be designed with the most appropriate geometry, size, and characteristics, depending on the purpose of application. Polymeric membrane filtrations are more popular than their inorganic counterparts, because they have the advantage of easy handling and better functionality (high selectivity). They are also suited for large-scale, low-cost fabrication processes, and can be designed with different structures (e.g., hollow, porous fibers or porous films). However, they suffer from several limitations such as low permeance, high temperature instability, as well as swelling and decomposition of organic solvents. On the other hand, inorganic membranes, such as porous anodic alumina, zeolite and porous silica possess high permeability, tunable selectivity, and high thermal and chemical resistance (Hsieh, 1991). The potential application of inorganic membranes is widely recognized since the production of high-quality porous ceramic membranes for large-scale industrial usage have been realized. Some inorganic membranes, such as mica, glass and alumina, are commercially available nowadays. Despite the extensive development and investigation of inorganic membranes for the separation of liquids and gases, reports on porous membranes for NP separation are rare, especially those for separating noble metals that are less than 10 nm in particle size.

3.2 Ultrafiltration and nanofiltration membranes

For the separation of NPs that are less than 10 nm in size, membranes should have proper porous structures, with a homogenous pore size distribution. Ultrafiltration and nanofiltration membranes are actually ideal candidates for separating NPs and large molecules, such as proteins, because they have pore sizes in the range of 1–100 nm. Ultrafiltration and nanofiltration membranes have been extensively used in industry and research for purifying and concentrating macromolecular solutions (Iglesias et al., 2009). They are effective in separating proteins and NPs due to the simplicity of the process and their tunable proper pore sizes. Ultrafiltration and nanofiltration membranes also enable the continuous separation of a large volume of feed, in which the feed stream is applied using a relative high pressure that forces the mixture against the porous membrane. The solutes with high molecular weight and large particles are retained, while those with low molecular weight and small particles pass through the membrane. Using ultrafiltration membranes, it is easy to separate the large and small particles as well as the large and small molecules through the size cut-off sieving processes. Polymer ultrafiltration membranes used for separation by a variety of food, biological, and pharmaceutical systems, as well as by the water purification and treatment industries, have been well-developed and are commercially available. However, polymeric membranes are not effective in separating NPs, because of their low porous density. In addition, they exhibit lower selectivity, poor thermal and chemical stability than inorganic counterparts. Inorganic membranes have high thermal chemical stabilities, and some inorganic membranes, such as glass, metal, alumina, zirconia, zeolite, and carbon membranes are commercially available. Despite that, these membranes vary greatly in pore size, support material, and configuration; in addition, they are not efficient in separating noble metal NPs. Furthermore, the design of a proper anisotropic porous inorganic membrane with higher flux, better selectivity, thermal stability, and chemical stability for the separation of large molecules and NPs remains a challenge. The fabrication design of the hierarchal mesoporous filter membranes with nano-space pores and of chemically robust and mechanically stable NTs can solve those problems.

3.3 Filtration flow methods

During membrane filtration, flow directions or filtration methods have a great impact on separation performances. Fig.8 illustrates two types of flow filtration, namely, tangential (cross) flow (A) and dead-end flow (B). In the tangential flow filtration shown in Fig. 8(A), the feed stream is applied tangential to the surface of the porous membrane (i.e., crossing the pore channels of the membrane). This feeding flow method minimizes membrane fouling, maintains a high filtration rate and ensures higher product recovery, because the sample remains safely in solution separation and the retentate stays on the surface of the membrane. To be usable for tangential flow filtration, the membranes are designed as tubular or hollow fiber membranes, as shown in Figs. 7(A)-(B). In the dead-end filtration shown in Fig. 8(B), the feed is applied perpendicularly to the surface of the membrane (i.e., feed stream flows through the pore channels of the membrane). This flow method enables rapid separation, because the feed is forced against the membrane surface. To be usable for dead-end flow filtration, the membranes are designed as flat sheets, as shown in Figs. 7(C)-(D). However, flat sheet membranes require the product stream to spread across the entire surface of the individual sheets prior to recirculation. This non-uniform flow path causes build-up and product loss in the "membrane corners" that see slower flow rates. In addition, dead-end filtration results in a build-up of product on the membrane surface; this may damage the product and cause lower recovery, thereby "fouling" the membrane.

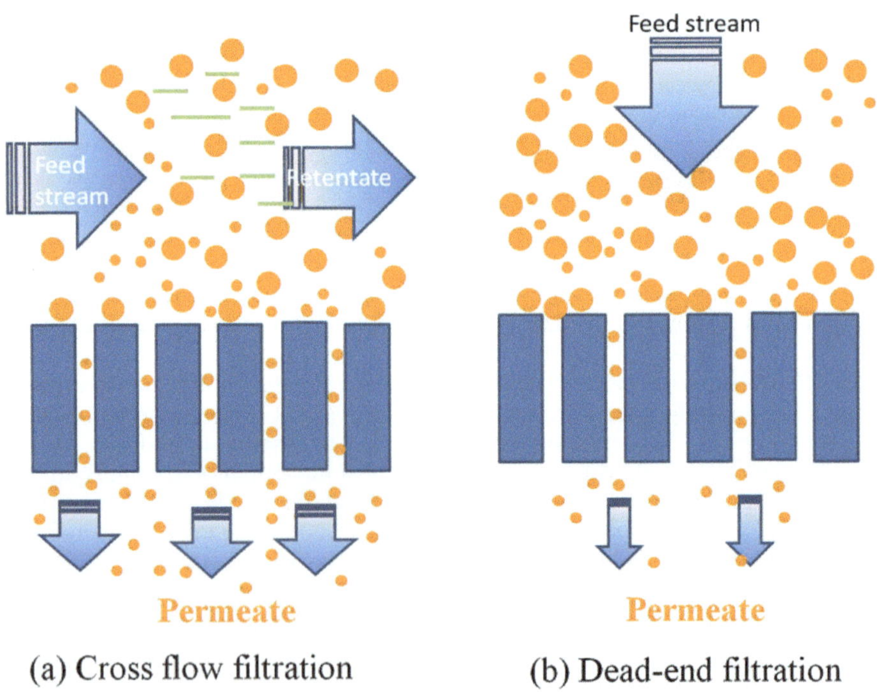

Fig. 8. Schematic diagram of (a) tangential (cross) flow filtration and (b) dead-end filtration

4. Design of HOM silica NTs membranes for nanofiltration

Silica NTs have been investigated extensively since their discovery in 1995 (Nakamura &. Matsui, 1995). Thereafter, enormous efforts have been devoted to identify their synthesis, fundamental properties, and applications. To date, researchers have been able to fabricate nanofibers, NTs, nanowires, and nanorods of high alignment and with the ability to control the porous structures by combining the use of soft and hard templates. Substantial efforts have particularly been devoted to the fabrication of 1D or two-dimensional (2D) hexagonally ordered or stacked doughnut (circular) silica mesostructures inside the columnar pores of the AAM nanochannels using template-guided syntheses. The synthesis pathway is the so-called dual template method has been widely used to fabricate a range of 1D and 2D mesostructures of silica, carbon, polymer, and metals. However, controlling the mesopore geometry and dimension into the AAM nanochannels has generally required labor-intensive or complicated design strategies that involve aging time, temperature, pH, and humidity. Table 2 provides a brief summary of the fabrication of hierarchical mesoporous structures inside the 1D nanochannels synthesized in AAM templates. A feasible control over the engineering of 2D hexagonal and 3D mesopore cubic *Im3m* cage structures in the entire silica NTs that are well-aligned with AAM nanochannels has also been recently reported (see Table 2).

In general, dual template pathways have been applied in synthesizing silica NTs of porous structures (see table 2). This synthetic pathway uses the AAM pore channels as the hard template to maintain the tubular structure of the silica, whereas the self-assembly of the liquid crystal or surfactant is used as the soft template to control the porous structure of the silica. This general synthetic pathway can be extended by varying synthetic conditions, including the composition and type of surfactants for synthesizing various mesoporous structures of silica NTs. The use of dual template for the synthesis of highly ordered mesoporous silica NTs has some advantages, including the ability to control the porous structure of silica NTs, homogenous pore size distribution, and the enhancement of the mechanical stability of membrane. However, to improve the mesoporous membranes synthesized using AAM as hard templates, the following chemical and mechanical stability problems must be addressed: (1) synthesis of robust, tubular structures into the entire AAM nanochannels; the (2) formation of unconnected pores and irregular shapes due to the detachment of mesoporous silica NTs from the AAM walls; (3) creation of air gaps within the shrinking mesoporous silica upon high-temperature calcination of soft templates; and (4) low-term chemical stability of the membrane within the separation reuse-cycles (El-Safty et al., 2010, 2011). Normally, the manufacturing defects in mesoporous silica NT hybrid AAM limit the long-term storage stability of NTs inside AAM (even for a month) and reduce the potential of NT filtration systems to control the filtration rates and diffusivity of NPs. Therefore, the fabrication design of chemically robust and mechanically stable NTs as well as hierarchal mesoporous filter membranes with nano-space pores remains a key challenge in the field of materials science.

The development of a new strategy leading to synthetically constructed mesostructures with tunable pore sizes throughout the 1D nanochannels is highly desirable for the size-exclusion separation of macromolecules and NPs. As a more detailed example, a method for the fabrication of hierarchal mesofilters based on the nanolinker approach templated inside the nanochannels of AAM has been successfully developed in recent years. In this development method, impermeable layers coated with silica NT hybrid AAM have been used to create chemically robust and mechanically stable designs, as shown in Fig. 9.

Structure types	Synthesis techniques and surfactants [a]	Properties	Application	Ref.
Mesoporous silica NFs	Vapor phase synthesis (PVS), using and F127, CTAB, DTAB, and OTAB templates	Pore size= 9 and 7 nm, S_{BET}=651 and 736 m^2g^{-1}	Catalytic Knoevenagel condensation	(Lee et al., 2009)
Mesoporous silica NFs	Sol-gel process, using P123 template	Pore size=7 nm; S_{BET} = 73.0 and 28.7m^2 g^{-1}	---[b]	(Gong et al., 2009)
Mesoporous silica NFs	Vapor-deposition hydrolysis, using OTAB, DeTAB, DTAB, TTAB, and CTAB surfactants	S_{BET} = 554, 610, and 780 m^2 g^{-1}	---[b]	(Lee et al., 2008)
Mesoporous carbon NFs	Sol-gel process, using P123 template	---	---[b]	(Cott et al., 2006)
Mesoporous silica NRs	Electrophoretic deposition, using P123 template	Pore size 10 nm	---[b]	(Hill et al., 2009)
Mesoporous silica NRs	Dip-coating method, using Brij 56 template	Pore size 4-5 nm	Gas separation	(Yoo et al., 2006)
Mesoporous silica NRs	sol-gel dip-coating method, P123 templates	---	Templates for synthesis of Ag, Ni, Cu_2O NWs	(Wu et al., 2004)
Mesoporous silica NRs	EISA method, CTAB, P123 and Brij 56	Pore size 8-9 nm; S_{BET}= 43- 55 m^2g^{-1}	---	(Platschek et al., 2008)
Mesoporous silica NRs	Sol-gel casting method, P123 template	Pore size 3-30 nm	Templates for synthesis of Cu, Ag, Te NWs	(Moses et al., 2010)
2D mesoporous silica NSs	Microemulsion liquid crystal phases of CTAB, DDAB, Brij. 30, 35, 56, 58, 76, 78, 97, 98 surfactants	S_{BET} =28.6-100 m^2g^{-1} Pore size= 3.7 - 5.1 nm	Separation of NPs	(El-Safty et al., 2010, 2011)
3D mesocage silica NTs	Microemulsion liquid crystal phases of F108, F127 surfactants	S_{BET} =100- 310 m^2/g	Separation of biomolecules	(El-Safty et al., 2010, 2011)

[a]Pluronic P123 ($EO_{20}PO_{70}EO_{20}$); Pluronic F127 ($PEO_{100}PPO_{70}PEO_{100}$), Pluronic F108 ($EO_{141}PO_{44}EO_{141}$), hexadecyltrimethylammo-niumbromide (CTAB), dilauryldimethylammonium bromide (DDAB), dodecyltrimethly-ammoniumbromide(DTAB),octyltrimethyl-ammoniumbromide(OTAB),decyltrimethyl-ammoniumbromide(DeTAB),tetradecyltrimethyl-ammoniumbromide (TTAB). [b] Not used for any applications.

Table 2. Fabrication, properties, and feasible applications of mesoporous structures inside nanowires (NWs), Nanofibers (NFs), nanorods (NRs), nanostrands (NSs), and nanotubes (NTs) hybrid AAM channels

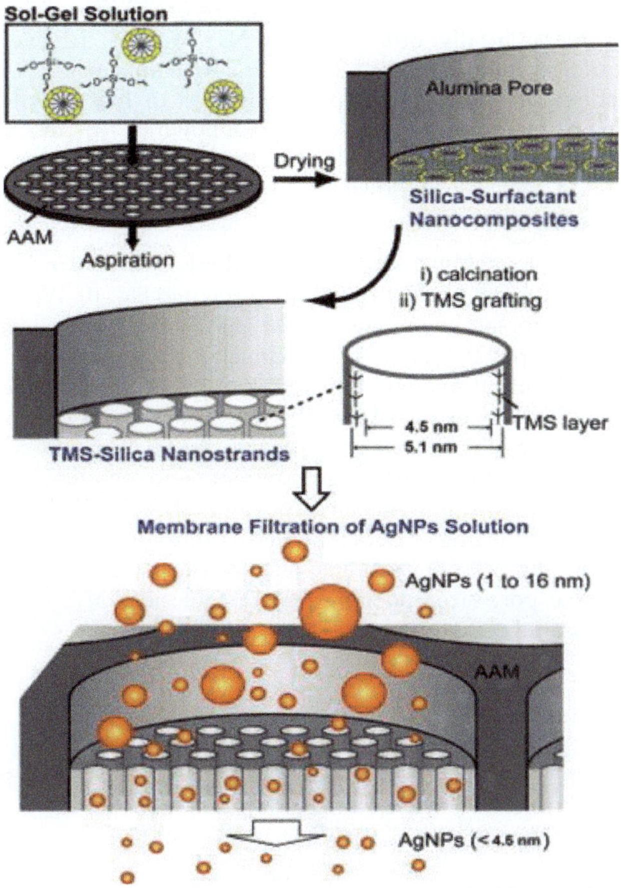

Fig. 9. Schematic design for the fabrication of mesoporous silica hybrid AAM and the size-exclusive separation system (reprinted from El-Safty et al., 2011)

The synthesis of densely-engineered mesofilters has been achieved using the following steps. First, the 2D hexagonal and 3D cubic *Im3m* mesoporous silica-NTs were synthesized inside the AAM or TMAC-AAM linkers using a direct templating method with cationic or nonionic surfactants. A key advantage of this approach is the capability to control the building of mesopores throughout the AAM nanochannels. Notably, the use of the direct templating method can lead to a more thermodynamically stable phase with hexagonal and 3D cubic *Im3m* structures inside the AAM nanolinkers. In addition, layered surface chemistry has been achieved by coating the inner mesopores inside the silica NTs with hydrophobic trimethylsilyl (TMS) groups. The hydrophobic TMS coating in mesoporous NTs can lead to the treatment of the tortuous-pore openings that form during the removal of the soft templates. In this approach, the organic monolayers (TMS) are densely packed and are highly stable under thermal conditions. This mesofilter membrane design is the key to broadening the nanofiltration applications for bio-molecule and NP separation.

5. Fabrication of HOM silica NTs using direct–templating method

The formation of hierarchically ordered silica NTs can be achieved by templating microemulsion liquid crystal phases with different surfactants inside AAM to control the mesoporous structures of the NTs. A typical synthesis of highly ordered mesoporous silica NTs on AAM templates includes the following: (i) the fabrication of a microemulsion precursor, and (ii) penetration of the microemulsion precursor in the pore channels of the AAM template, followed by (iii) calcinations at high temperature or extraction to remove the surfactants.

The schematic diagram of a system used for the synthesis of highly ordered silica NTs and the nanofiltration separation of noble metal NPs is shown in Fig. 10. When used to fabricate the HOM silica NT membrane, the pristine AAM is settled in the system, after which the precursor solution is dropped to fill in its pore channels with the assistance of slight vacuum pumping. After drying and calcinations, the HOM silica NTs were obtained inside the pore channels of AAM. However, when used for NPs separation, the HOM silica NT membrane is settled in a filtering system instead of a pristine AAM; then, the solution of NPs is dropped into the filtering funnel. Using vacuum pumping, the solution is separated through the membrane. In this regard, the NPs may go through the membrane with a smaller particle size than the pore diameter of NTs, whereas those with a larger particle size are retained in the filtering tunnel. The separated size of NPs is controlled by the pore size of silica NTs; in turn, the pore size and pore structure of silica NTs can be controlled by varying the surfactants and compositions.

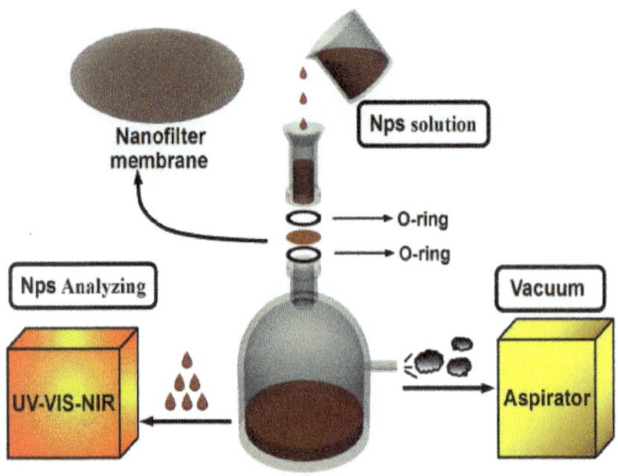

Fig. 10. Schematic diagram of a nanofiltration system used for the synthesis of highly ordered silica NT membrane and the NP filtration (El-Safty et al., 2010, 2011)

5.1 Synthesis of HOM silicas NTs using Brij–type (C_xEO_y) templates

Typically, Brij-type surfactants were dissolved in a mixture of ethanol and acidic aqueous solution (H_2O/HCl). The resultant mixture was stirred for 30 min under refluxing conditions. Then, tetraethyl orthosilicate (TEOS) was added dropwise to the mixture, with continuous stirring at 30 °C for 2.5 h. The precursor solution was added dropwise into the AAM. Moderate aspiration was applied using an aspirator (Ulvac, Model MD A-15), leading to the penetration of the precursor solution into the columnar alumina pores. The resultant surfactant-silica hybrid membrane was then dried at room temperature for 1 h. The surfactant was removed by calcination at 600 °C for 2 h under air, using a Yamato FO 310 muffle furnace with a temperature rate of 2 °C/min (El-Safty et al., 2010).

5.2 Synthesis of HOM silica NTs using CTAB templates

Briefly, the microemulsion phase domains containing TMOS, cationic surfactants (DDAB, CTAB), decane, HCl-H_2O, and EtOH at their molar compositions of 1:0.165:2.08:8.284 and 1:0.209:2.08:8.284 for DDAB and CTAB systems, respectively, were obtained by mixing thoroughly for 30 min. The penetration of the microemulsion domains into the AAM was done by means of a gentle vacuuming pressure of ≤0.04 MPa. The silica NTs were allowed to stand in a sealed container at 45 °C for 10 h to complete the drying process. The AAM embedded microemulsion phases were calcined at 600 °C in air for 3 h to remove the organic compounds and form the mesoporous silica NTs (El-Safty et al., 2010).

5.3 Synthesis of HOM silicas NTs using F108 templates

Quaternary microemulsion liquid crystalline phases of triblock copolymer F108/C_{12}-alkane/TMOS/H_2O-HCl were used to facilitate the template-guided synthesis of 3D mesocage surfactant/silica phases inside AAM. In this direct synthesis, the precursor solution containing TMOS, F108, dodecane, H_2O-HCl, and ethanol at a respective molar composition ratio of 1:0.00732:0.313:2.59:16.56 was applied dropwise into the AAM nanochannels. Penetration of the precursor solution into the membrane was achieved by means of vacuum at a starting pressure of ≤0.04 MPa, thus allowing control of the silica-NTs inside the membrane pores. The addition of C12-alkane led to the formation of 3D cubic *Im3m* structures with large cage pores and uniform open entrances. The hybrid organic-inorganic membrane was allowed to stand in a sealed container at 45 °C for 10 h to complete the drying process. The organic moieties were removed by calcination at 600 °C under air (El-Safty et al., 2010, 2011).

5.4 Synthesis of HOM silicas NTs using F127 templates

The quaternary microemulsion liquid crystalline phases of triblock copolymer F127/C_{12}-alkane/TMOS/H_2O-HCl (1.4 g of F127, 0.7 g of dodecane, 2 g of TMOS, 1.25 g of H_2O-HCl, and 10 g of ethanol) were used to facilitate template-guided synthesis of 3D mesocage surfactant/silica phases. The precursor solution was applied dropwise into the TMAC-AAM nanochannels (TMCS=trimethylcholrosilane)). Penetration of the precursor solution into the membrane was achieved by means of vacuum at a pressure of ≤ 0.04 MPa. It is important to note that the use of the quaternary microemulsion liquid crystalline phase could lead to a more thermodynamically stable phase with a high curvature of cubic *Im3m* geometry. Furthermore, the addition of C_{12}-alkane led to the formation of 3D cubic *Im3m* structures with large cage pores and uniform open entrances. The hybrid organic-inorganic membrane was allowed to stand in a sealed container at 45 °C for 10 h to complete the drying process. Next, the hybrid organic-inorganic membrane was washed thoroughly with ethanol/acidified H_2O for several cycles in a Soxhlet apparatus to remove the F127 surfactant template; this was then dried at 60 °C.

In general, the quaternary microemulsion liquid crystalline phases of triblock copolymers were used as a template-guided synthesis of cubic *Im3m* mesocage pores inside the NTs. The 3D mesopore cubic *Im3m* cage array with uniform entrances (~5 nm) and large cavities (~17.3 nm) inside AAM allowed the development of size-exclusion nano-filter membranes (El-Safty et al., 2010, 2011).

6. Characteristics of HOM silica NTs membranes

The study of vertically-aligned nanochannels and the coexistence of circular or columnar mesopores and mesostructured geometries inside the AAM channels has gained considerable research attention in recent years. Such architectures inside the AAM channels are very effective for the size-exclusion separation of macromolecules and NPs. In this section, the characteristics of fabricated HOM silica NTs inside nanopores of AAM are investigated using various tools of analyses, such as nitrogen adsorption/desorption isotherm, small-angle X-ray scattering (SAXS), field emission scanning electron microscopy (FE-SEM), and high-resolution transmission electron microscopy (HRTEM). Details about the characterization method can be found in references (Hoa & El-Safty, 2011).

The advanced SAXS technique is one of the most powerful methods in investigating the mesoporous structure of HOM silica NTs. As reported in recent publications, silica NTs prepared using different surfactants and compositions have different mesophase structures, such as hexagonal and cubic *Im3m* or disordered mesostructures (El-Safty et al., 2010, 2011). As an example, Fig. 11 (A) shows the SAXS patterns of porous silica NTs fabricated inside the AAM pore channels using triblock copolymers of Pluronic F127 as a template. The silica NTs exhibit the cubic *Im3m* mesostructure. The inset of Fig. 11(A) is the structural parameters of fabricated HOM silica NTs characterized by SAXS.

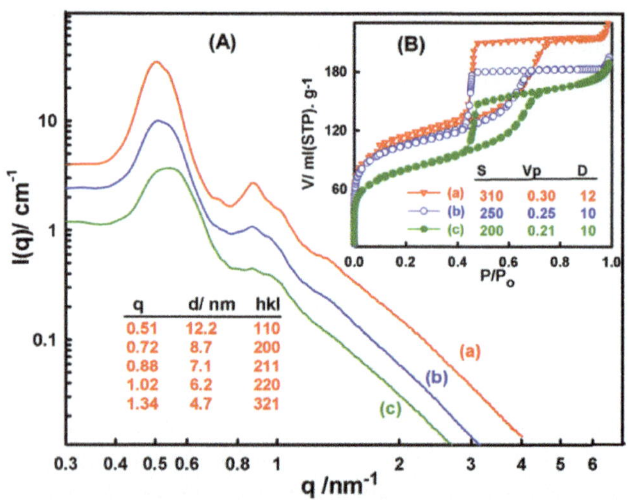

Fig. 11. SAXS (A) and N_2 isotherms (B) of mesofilter membranes based on mesoporous cubic *Im3m* cage silica NT hybrid AAM before (a) and after (b) modification with TMS groups and after removal of the blocked proteins from the AAM surfaces and pores during the second use of the mesofilters (c) (reprinted from El-Safty, 2011).

Nitrogen adsorption/desorption isotherm was used for characterizing the mesoporous materials, including porous silica NTs fabricated inside the AAM pore channels. From the N_2 adsorption isotherm data, various parameters, such as the porous structure, pore size, pore volume and specific surface area, can be estimated. For instance, the mesoporous silica NT membranes have a large surface area (S_{BET}~ 100–310 m^2/g), as shown in Fig. 11(B). The average pore size of HOM silica NTs synthesized using F127 surfactant is about 12 nm. The NTs are stable and retain their mesoporous structures after removal of surfactant, functionalization with the TMS group and use for separation.

FE-SEM images shown in Figs. 12(A)–(B) were recorded with partial etching of the alumina matrix with 8% H_3PO_4 of the silica NT hybrid AAM membrane. SEM micrographs show evidence of the formation of silica NTs inside the 200 nm diameter nanochannels of AAM. After complete etching of the alumina matrix, SEM micrographs reveal that the silica NTs fabricated are well aligned within the AAM pores (Figs. 12 (C)–(D)). Silica NTs with regular and continuous alignment along the perpendicular axis are also observed (Fig. 12(E)) even after the TMS coating inside the pores (Fig. 12(F)). The results, in general, also indicate that the open pore silica NTs fabricated using structure directing surfactants fill the diameter (~200 nm) and most of the length (~60 μm) of the AAM pores (Fig. 12(E)). The NT alignment observed in Fig. 12(F) may have been attained within the inclusive TMS immobilization process inside AAM.

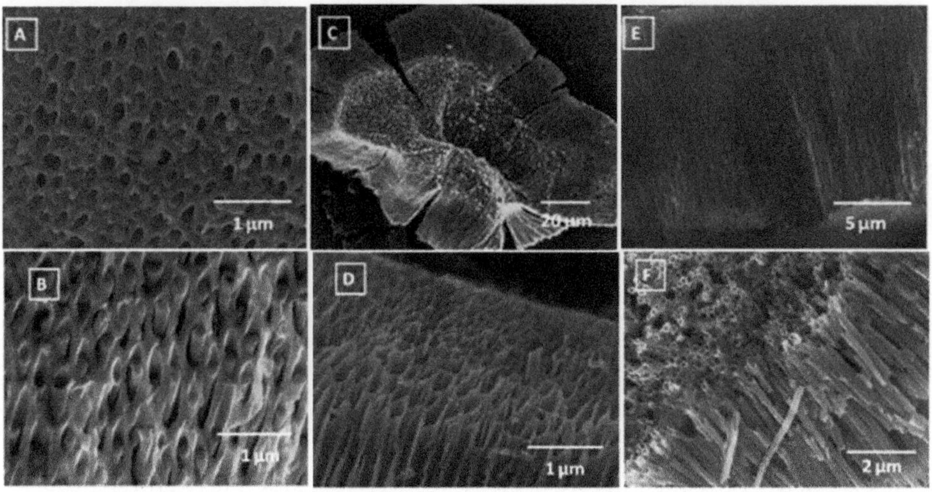

Fig. 12. FE-SEM cross section [(A), (B)] silica NT hybrid AAM membrane after partial etching of the alumina matrix. FE-SEM cross section [(C), (D)] and plan-view (E) images of vertically aligned silica NTs. FE-SEM plan-view image (F) recorded with TMS-silica NTs. (reprinted from El-Safty et al., 2010).

Meanwhile, the HR-TEM images of silica and TMS-silica NTs after the removal of the alumina matrix are shown in Fig. 13. The HR-TEM images clearly reveal large domain sizes of ordered hexagonal mesoporous networks running along the 1D AAM nanochannels. However, the HR-TEM profiles show evidence of the successful formation of hollow mesostructures with a predominantly columnar orientation. The most prominent feature in the HR-TEM images is the uniform mesoporosity and continuously ordered channels present along all axes of the hexagonal mesoporous silica and TMS-silica NTs that are not significantly distorted under the synthetic conditions. Moreover, the HR-TEM images show evidence that the synthetic route allowed for the controlled formation of free-standing, aligned strands, which permitted a high flux and transport of spherical NPs through the TMS-silica mesopores inside the AAM channels without substantial hindrance. The top-view TEM image shown in Fig. 13(D) indicates variability in the columnar or circular ordering in the AAM channels during the fabrication of silica NTs using this approach.

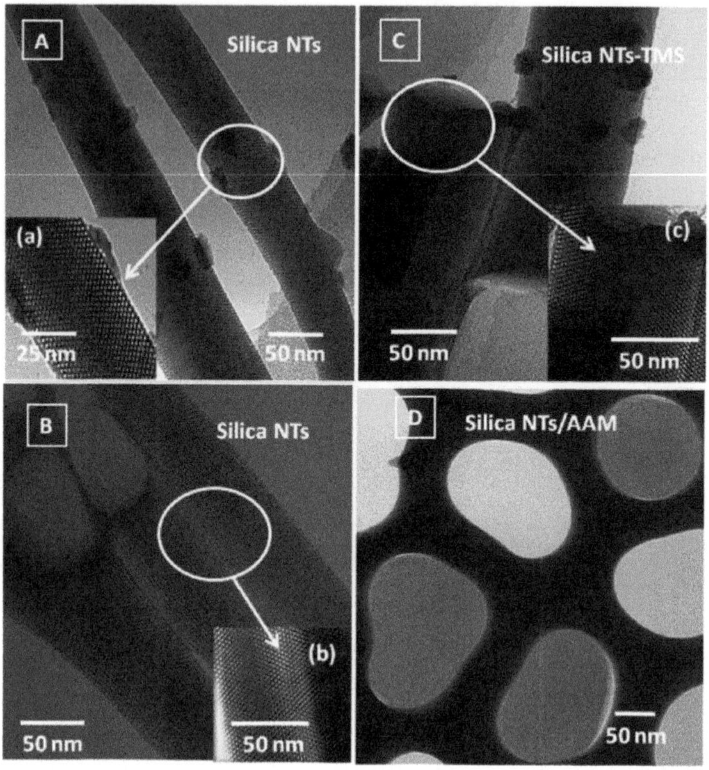

Fig. 13. HR-TEM images of silica [(A), (B)] and TMS-silica (C) NTs after etching of the alumina matrix. (D) TEM image recorded on the top-view of silica NTs hybrid AAM (reprinted from El-Safty et al., 2010)

7. Separation of NPs using HOM silica NTs membranes

The performance of the mesofilter membranes synthesized using a template-guided synthesis for the separation of noble metal NPs, such as Ag, Au, and Pt with different shapes and sizes was evaluated. To synthesize these NPs, two-phase redox reactions were used to control the fabrication of the organic solution-phase Ag NPs with a size distribution ranging from 1 to 16 nm. Initially, 0.081 g of sodium laureate (SL) was dissolved in 5 ml of water. This SL solution was added dropwise to the $AgNO_3$ aqueous solution

($AgNO_3$:0.091 g in 5 ml of water), yielding white precipitates of Ag-SL salt. Collected Ag-SL salt was rinsed with water 10 times and then washed with a small amount of methanol thrice. An organic Ag-SL solution was formed by adding 20 ml of toluene to the methanol (5 ml) solution of Ag-SL salt (0.09 g). Sodium tetrahydroborate solution (Aldrich, 0.0063 g in 10 ml of water) was then added dropwise into the organic Ag-SL solution with vigorous stirring. The toluene phase immediately became yellow. After 10 min of stirring, a yellow toluene phase was collected and stored at 4 °C (Mekawy et al., 2011).

The NPs of gold and platinum were synthesized as well, and the details are presented below. A mixture of 40 mg AuCl, and 1 g oleylamine was dissolved in 10 ml chloroform and stirred for 2h at 60 °C. To this mixture composition, 20 ml acetone was added to precipitate the Au NPs. The precipitated Au NPs were then collected and thoroughly washed with deionized water and ethanol by means of a centrifuge. The Au NPs were dried in air at 45 °C. The synthesis of the PtNPs with octadecylamine followed the same procedures; however, the Pt compound dipotassium hexachloroplatinum (K_2PtCl_6,) was used as the Pt source. The Pt or AuNPs were dispersed in 2 ml of chloroform for subsequent separation.

The schematic design for the size-exclusive separation system of wide distribution sizes of NPs based on TMS-immobilized silica hybrid AAM for producing ultrafine, uniformly-spherical NPs is shown in Fig. 14(A). Typically, the rigid mesoporous silica NTs inside AAM channels having a pore diameter of ~4 nm were used as ultrafine filtration systems (nanofiltration) for separating noble metal NPs fabricated in a wide-range of sizes and spherical/pyramidal morphologies. The grafting of hydrophobic trimethylsilyl (TMS) groups onto the inner pores of the silica NTs walls is very important in creating size- and

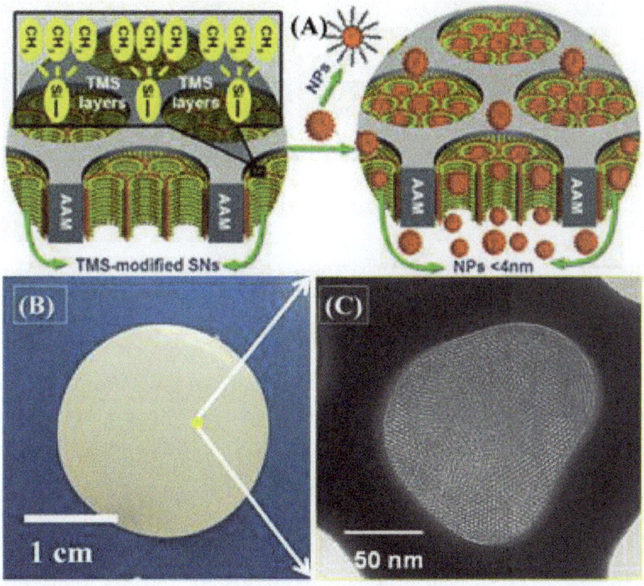

Fig. 14. (A) Schematic design for the size-exclusive separation system, (B) photo of a NT membrane used for NPs separation, and (C) a top-view HRTEM image of a highly ordered porous silica NT.

shape-exclusive separation systems, because the presence of the TMS groups allows the silica NTs to retain their inner pore structure without distortion during the pressurized separation assays that could cause tortuous-pore membranes. Fig.14 shows the size-exclusive porous silica NT membranes, a photo of an NT membrane after calcination (B) and a top-view HRTEM image (C) of a highly ordered porous silica NT used for NP separation.

8. Separation performance of NPs using HOM silica NTs membranes

The separation performance of NPs can be evaluated through quantitative analysis, thus validating the efficiency of the mesoporous TMS-silica hybrid membrane as a size-exclusive separator of NPs. Figs. 15 (A)-(C) show the adsorption spectrum and color change of NPs before and after separation, respectively. As revealed in Fig. 15(A), 20% of the original amount of Ag NPs can be separated in monodispersed and uniform Ag NPs within minutes. However, large NPs with size >4.5 nm are caught in the pores of NTs or on the surface of the hybrid AAM. The color of the membrane changed after the separation process due to the blockage of large NPs, as shown in Fig. 15 (A), inset. The membrane also exhibited effective separation of other NPs, such as gold (Fig. 15(B)) and CdS (Fig. 15(C)). Generally, the amount of blocked NPs depends on the shapes and sizes of NPs in the feed solution. Spherical ones with particles smaller than the pore size of the NT membrane are filtered through the membrane, whereas non-spherical ones with large particles are caught in the pores of NTs or on the surface of the membrane.

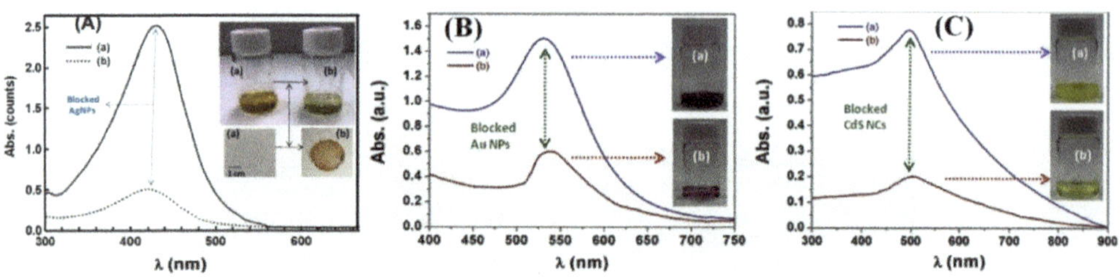

Fig. 15. Adsorption spectrum and color change of NPs before and after separation: (A) Silver NPs; (B) Au NPs; (C) CdS NPs (reprinted from El-Safty et al., 2010, 2011)

Further evaluation of the separation performances of NPs using HRTEM images is shown in Fig. 16. The HRTEM image of Ag NPs before separation (Fig. 16(A)) shows that the NPs are spherical and irregularly-shaped particles with a wide particle size distribution (Fig. 16(D)). After the simple separation process, the TEM analysis shows uniform spherical and regularly shaped particles (Figs. 16(B) and (C)) less than 4.5 nm in diameter, indicating the ultra-fine separation of Ag NPs by size and shape (see the histograms, Figs. 16 (E) and (F)).

The silica NT membrane is not only effective in the separation of Ag NPs, it also performs excellently in the separation of other NPs, including of Au, Pt, CdS, and ZnS. The HR-TEM images of the NPs before filtration show irregularly-shaped NPs with a wide distribution of sizes, as shown in Figs. 17(A), (C), and (E). In contrast, the HR-TEM images taken after filtration, which occurred on a second timescale, show uniformly spherical and regularly-shaped NPs less than 4.5 nm in diameter, indicating the ultra-fine filtration of NPs. Particles size distributions of NPs before and after separation are shown in the insets, respectively.

The size-exclusion cut-off (<4.5 nm) of the membrane filters agrees well with the NLDFT pore-size distribution of the TMS-silica NT. During the filtration process, small-sized particles easily go through the pore channels, whereas the large particles are retained on the surface of membrane. Because the NTs have nanopores with uniform sizes, thus HOM silica NT membranes are effective for the NPs separation. Despite that, after using several times, the particles would stack and block the pore channels, leading to insufficient separation. Therefore, the cleaning processes are required for the reuse of the HOM silica NT membranes. For the cleaning processes, the stacked membranes were settled upside down in the nanofiltration system (Fig. 10). Thereafter, the distilled water and ethanol were filtrated through membrane with the assistance of slight vacuum pumping.

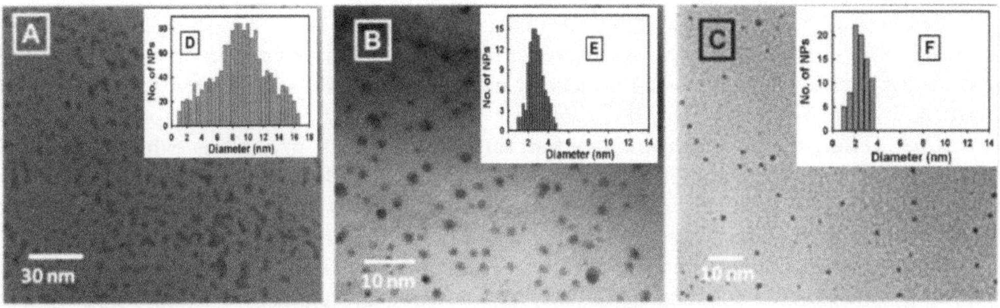

Fig. 16. TEM images of Ag NPs: (A) as-made, (B), (C) after separation through the NT membranes. Insets are correlative particle size distribution (reprinted from Mekawy et al., 2011).

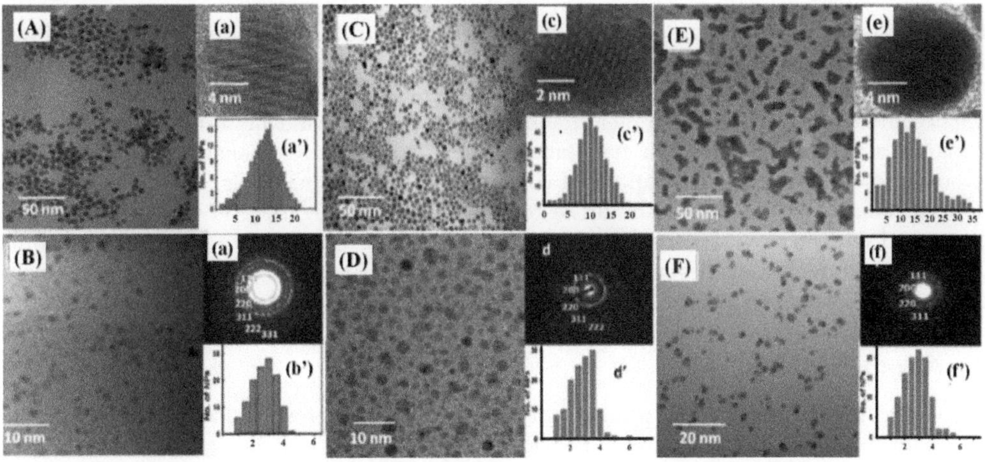

Fig. 17. HR-TEM micrographs of spherical Ag, Au, and Pt NPs before [(A), (C), (E)] and after [(B), (D), (F)] separation using HOM silica NTs membranes. Insets [(A), (C), (E)] are high-magnification TEMs of one NP [(a), (c), (e)] and histogram of NP patterns [(a'), (c'), (e')]. Insets [(B), (D), (F)] are the ED [(b), (d), (f)]and histograms [(b'), (d'), (f')] of Ag, Au, and Pt NPs with fcc phases of cubic *Fm3m* symmetry, respectively (reprinted from El-Safty et al., 2010, 2011)

9. Conclusion and outlook

The development design of hierarchically ordered TMS-silica NTs inside 1D columnar channels has enhanced the potential ability and functionality of the application of membranes in size-exclusive separation systems of particles and small molecules. To design a robust nanofilter, cationic surfactants are used as soft templates for predicting ordered assemblies of surfactant/silica composites through strong interactions within AAM pockets. This approach forms densely-packed NTs in the entirety of the AAM channels. These NTs are coated with layers of organic moieties to create a powerful technique for ultrafine filtration. The nanofilter membranes can then be used effectively to separate several NPs.

The developed membranes are effective in separating novel NPs with particle sizes of less than 4.5 nm. This precise size-cutoff separation is very important in investigating and understanding the quantum size effects in nanomaterials. Physicists have predicted and confirmed that metal NPs in the diameter range of 1-10 nm can display electronic structures, reflecting the electronic band structure of the NPs, as well as exhibit abnormal optical properties. The smaller spherical NPs exhibit a single and sharp localized surface plasmon resonance absorption band in the visible region, while larger spheres exhibit increased scattering and have peaks that broaden significantly and shift towards longer wavelengths. Additionally, the membranes exhibit excellent size-cutoff separation of NPs and help enhance the dispersal of the NPs, with the aggregated NPs becoming mono-dispersed NPs when forced to go through the membrane. The enhancement of dispersal and precise size-cutoff separation of NPs can increase the potential applications of noble metal NPs in biological, optoelectronic and sensing areas, where their optical properties are strongly influenced by interparticle interactions, shapes, and sizes.

Furthermore, multiple strategies are being pursued for the development of nanofilter membranes that fulfill all requirements of high-performance molecular transport and the size-exclusion selectivity of biomolecules and small particles. However, research on the science and engineering of such membranes is needed to combine all of these attributes into practical systems. Advances in nanofabrication methods have opened up exciting new avenues to overcome the limitations of conventional porous materials; however, practical, scalable fabrication processes should also be created. Indeed, there is still a demand for controlled fabrication of mesoscale pores inside metals or metal oxide NTs or NW hybrid AAM. The formation of mesoporous structures inside NTs or NWs can open exciting new applications, including electronics, optics, gas and bio-sensors, and nano-fluidic devices.

10. Acknowledgements

We would like to express our gratitude to the Japan Society for the Promotion of Science (JSPS) for financial support (Grant No. P09606).

11. References

Ahmadi, T.S.; Wang, Z.L.; Green, T.C.; Henglein, A. and El-Sayed, M.A. (1996). Shape-Controlled Synthesis of Colloidal Platinum Nanoparticles, *Science*, Vol. 272, pp. 1924-1925

Asatekin, A. ad Gleason, K.K. (2011). Polymeric nanopore membranes for hydrophobicity based separations by conformal initiated chemical vapor deposition, *Nano Lett.* Vol. 11, pp. 677–686.

Attard, G. S.; Bartlett, P. N.; Coleman, N. R. B.; Elliott, J. M.; Owen, J. R. and Wang, J. H. (1997). Mesoporous Platinum Films from Lyotropic Liquid Crystalline Phases, *Science*, Vol. 278, pp. 838–840.

Bianchini, C. and Shen, P. K. (2009). Palladium-Based Electrocatalysts for Alcohol Oxidation in Half Cells and in Direct Alcohol Fuel Cells, *Chem. Rev.*, Vol. 109, pp. 4183–4206.

Carroll, K. J.; Reveles, J. U.; Shultz, M. D.; Khanna, S. N. and Carpenter, E. E. (2011). Preparation of Elemental Cu and Ni Nanoparticles by the Polyol Method: An Experimental and Theoretical Approach, *J. Phys. Chem. C*, Vol. 115, pp. 2656–2664.

Chen, A. and Holt-Hindle, P. (2010). Platinum-Based Nanostructured Materials: Synthesis, Properties, and Applications, *Chem. Rev.*, Vol. 110, pp. 3767–3804.

Chen, Y.; Gu, X.; Nie, C.G.; Jiang, Z.Y.; Xie, Z.X. and Lin, C. J. (2005). Shape controlled growth of gold nanoparticles by a solution synthesis, *Chem. Commun.*, pp. 4181–4183.

Choi, J.; Jeong, H.K.; Snyder, M.A.; Stoeger, J.A. Masel R.I. and Tsapatsis, M. (2009). Grain boundary defect elimination in a zeolite membrane by rapid thermal processing, *Science*, Vol. 325, pp. 590–593.

Choi, S.H.; Zhang, Y.P.; Gopalan, A.; Lee, K.P. and Kang, H.D. (2005). Preparation of catalytically efficient precious metallic colloids by γ-irradiation and characterization, *Colloids Surf. A*, Vol. 256, pp. 165–170.

Cott, D. J.; Petkov, N.; Morris, M. A.; Platschek, B.; Bein, T. and Holmes, J. D. (2006). Preparation of oriented mesoporous carbon nano-filaments within the pores of anodic alumina membranes, *J. Am. Chem. Soc.*, Vol. 128, pp. 3920–3921.

Dong, X.; Ji, X.; Jing, J.; Li, M.; Li, J. and Yang, W. (2010). Synthesis of Triangular Silver Nanoprisms by Stepwise Reduction of Sodium Borohydride and Trisodium Citrate, *J. Phys. Chem. C*, Vol. 114, pp. 2070–2074.

El-Safty, S.A.; Mekawy, M.; Yamaguchi, A.; Shahat, A.; Ogawa, K. and Teramae, N. (2010). Organic–inorganic mesoporous silica nanostrands for ultrafine filtration of spherical nanoparticles, *Chem. Commun.*, Vol. 46, pp. 3917–3919.

El-Safty, S.A.; Shahat, A.; Mekawy, M. Hoa, N. Warkocki, W. and Ohnuma, M. (2010). Mesoporous silica nanotubes hybrid membranes for functional nanofiltration, *Nanotechnology*, Vol. 21, 375603.

El-Safty, S.A.; Shahat, A.; Awual, Md. R. and Mekawy, M. (2011). Large three-dimensional mesocage pores tailoring silica nanotubes as membrane filters: nanofiltration and permeation flux of proteins, *J. Mater. Chem.*, Vol. 21, pp. 5593–5603.

El-Safty, S.A.; Shahat, A.; Warkocki, W. and Ohnuma, M. (2011). Building-Block-Based Mosaic Cage Silica Nanotubes for Molecular Transport and Separation, *Small*, Vol. 7, pp. 62–65.

El–Safty, S.A. (2011). Designs for Size–Exclusion Separation of Macromolecules by Densely Engineered Mesofilters, *Trends Anal. Chem.*, Vol. 30, 447–458.

El–Safty, S.A. and Shenashen, M.A. (2011). Size–selective separations of biological macromolecules on mesocylinder silica arrays, *Anal. Chim. Act.*, pp.151–161.

Enders, D.; Nagao, T.; Pucci, A.; Nakayama, T. and Aonoa, M. (2011). Surface-enhanced ATR-IR spectroscopy with interface-grown plasmonic gold-island films near the percolation threshold, *Phys. Chem. Chem. Phys.*,Vol. 13, pp. 4935–4941.

Faraday, M. (1857). The Bakerian lecture: experimental relations of gold (and other metals) to light, *Philos.Trans.R.Soc.London*, Vol. 147, 145.

Fukuoka, A.; Araki, H.; Sakamoto, Y.; Sugimoto, N.; Tsukada, H.; Kumai, Y.; Akimoto, Y. and Ichikawa, M. (2002). Template Synthesis of Nanoparticle Arrays of Gold and Platinum in Mesoporous Silica Films, *Nano Letters*, Vol. 2, pp. 793–795.

Gin D.L. and Noble, R.D. (2011). Designing the next generation of chemical separation membranes, *Science*, Vol. 332, pp. 674–676.

Gong, Z.; Ji, G.; Zheng, M.; Chang, X.; Dai, W.; Pan, L.; Shi, Y. and Zheng, Y. (2009). Structural Characterization of Mesoporous Silica Nanofibers Synthesized Within Porous Alumina Membranes, *Nanoscale Res. Lett.*, 2009, 4, 1257–1262.

Green, M. and O'Brien, P. (2000). A simple one phase preparation of organically capped gold nanocrystals, *Chem. Commun.*, pp. 183–184.

Grzelczak, M.; Pérez–Juste, J.; Mulvaney P. and Liz–Marzán, L. M. (2008). Shape control in gold nanoparticle synthesis, *Chem. Soc. Rev.*, Vol. 37, pp. 1783–1791.

Han, J.Y.; Fu, J. P. and Schoch, R. B. (2008). Molecular sieving using nanofilters: Past, present and future. *Lab. Chip*, Vol. 8, pp. 23–33.

Hashim, N. A.; Liu, Y. and Li, K. (2011). Preparation of PVDF hollow fiber membranes using SiO_2 particles: The effect of acid and alkali treatment on the membrane performances, *Ind. Eng. Chem. Res.*, Vol. 50, pp. 3035–3040.

He, X.; Lie, J.A.; Sheridan, E. and Hagg, M.B. (2011). Preparation and Characterization of Hollow Fiber Carbon Membranes from Cellulose Acetate Precursors, *Ind. Eng. Chem. Res.* Vol. 50, pp. 2080–2087.

Henglein, A. (1998). Colloidal Silver Nanoparticles: Photochemical Preparation and Interaction with O_2, CCl_4, and Some Metal Ions, *Chem. Mater.*, Vol. 10, pp. 444–450.

Herricks, T.; Chen, J. and Xia, Y. (2004). Polyol synthesis of platinum nanoparticles: control of morphology with sodium nitrate, *Nano Lett.*, Vol. 4, pp. 2367–2371.

Hoa, N.D and El–Safty, S.A. (2011). Highly sensitive and selective volatile organic compound gas sensors based on mesoporous nanocomposite monoliths, *Anal. Methods*, Vol. 3, pp. 1948–1956.

Hoa, N.D and El–Safty, S.A. (2011). Gas nanosensor design packages based on tungsten oxide: mesocages, hollow spheres, and nanowires, *Nanotechnology* Vol. 22, pp. 485503.

Hoa, N.D and El–Safty, S.A. (2011), Meso- and Macroporous Co_3O_4 Nanorods for Effective VOC Gas Sensors, *J. Phys. Chem. C*, Vol.115, pp 8466–8474.

Hoa, N.D and El–Safty, S.A. (2011). Synthesis of Mesoporous NiO Nanosheets for the Detection of Toxic NO_2 Gas, *Chem. Eur. J.* Vol. 17, pp. 12896 - 12901.

Horswell, S.L.; Kiely, C.J.; O'Neil, I.A. and Schiffrin, D.J. (1999). Alkyl Isocyanide-Derivatized Platinum Nanoparticles, *J. Am. Chem. Soc.*, Vol. 121, p. 5573

Horinouchi, S.; Yamanoi, Y.; Yonezawa, T.; Mouri, T. and Nishihara, H. (2006). Hydrogen Storage Properties of Isocyanide-Stabilized Palladium Nanoparticles, *Langmuir*, Vol. 22 (4), pp. 1880–1884.

Hill, J. J.; Cotton, S.P. and Ziegler, K.J. (2009). Alignment and Morphology Control of Ordered Mesoporous Silicas in Anodic Aluminum Oxide Channels by Electrophoretic Deposition, *Chem. Mater.*, Vol. 21, pp. 1841-1846.

Hieda, J.; Saito, N.; Takai, O. (2008). Exotic shapes of gold nanoparticles synthesized using plasma in aqueous solution, *J. Vac. Sci. Technol. A*, Vol. 26, pp. 854–856

Hsieh, H.P. (1991). Inorganic membrane reactors, *Catal. Rev. Sci. Eng.* Vol. 33, pp. 1–70.

Iglesias, R.A.; Tsow, F.; Wang, R.; Forzani, E. S. and Tao, N. (2009). Hybrid separation and detection device for analysis of benzene, toluene, ethylbenzene, and xylenes in complex samples, *Anal. Chem.*, Vol. 81, pp. 8930–8935.

Jang, K.S.; Kim, H.J.; Johnson, J.R.; Kim, W.; Koros, W.J.; Jones, C.W. and Nair, S. (2011). Modified mesoporous dilica gas separation membraneson polymeric hollow fibers, *Chem. Mater.* Vol. 23, pp. 3025–3028.

Jones, M. R.; Osberg, K. D.; Macfarlane, R. J.; Langille, M. R. and Mirkin, C. A. (2011). Templated Techniques for the Synthesis and Assembly of Plasmonic Nanostructures, *Chem. Rev.* Vol. 111, pp. 3736–3827.

Kim, F.; Connor, S.; Song, H.; Kuykendall, T. and Yang, P. (2004). Platonic Gold Nanocrystals, *Angew. Chem., Int. Ed.*, Vol. 43, pp. 3677–3682.

Kim, S.W.; Park, J.; Jang, Y.; Chung, Y.; Hwang, S.; Hyeon, T. and Kim, Y.W. (2003). Synthesis of Monodisperse Palladium Nanoparticles, *Nano Letters,* Vol. 3, pp. 1289–1291.

Kosmala, A.; Wright, R.; Zhang, Q. and Kirby, P. (2011). Synthesis of silver nano particles and fabrication of aqueous Ag inks for inkjet printing, *Mat. Chem. Phys.* Vol. 129, pp. 1075–1080.

Kurihara, L. K.; Chow, G. M. and Schoen, P. E. (1995). Nanocrystalline metallic powders and films produced by the polyol method, *Nano. Mat.*, Vol. 5, pp. 607–613.

Lee, K.J.; Min, S. H. and Jang, J. (2008), (2009). Vapor–Phase Synthesis of Mesostructured Silica Nanofibers Inside Porous Alumina Membranes, *Small,* Vol. 4, pp. 1945–1949. (b) Mesoporous Nanofibers from Dual Structure–Directing Agents in AAO: Mesostructural Control and their Catalytic Applications, *Chem. Eur. J.* Vol. 15, 2491–2495.

Lim, B.; Lu, X.; Jiang, M.; Camargo, P.H.C.; Cho, E.C.; Lee, E.P. and Xia, Y. (2008). Facile synthesis of highly faceted multioctahedral Pt nanocrystals through controlled overgrowth, *Nano Lett.*, Vol. 8, pp. 4043–4047.

Liu, Z.; Zu, Y.; Fu, Y.; Meng, R.; Guo, S.; Xing, Z. and Tan, S. (2010. Hydrothermal synthesis of histidine–functionalized single–crystalline gold nanoparticles and their pH–dependent UV absorption characteristic, *Coll. Surfaces B.*, Vol. 76, pp. 311–316.

Liz–Marzán, L.M. (2004). Nanometals: formation and color, *Mater. Today*, pp. 26–31.

Mekawy, M.; Yamaguchi, A.; El-Safty, S.A.; Itoh, T. and Teramae N. (2011). Mesoporous Silica Hybrid Membranes for Precise Size-Exclusive Separation of Silver Nanoparticles" *J. Colloid Interface Sci.* Vol. 355, pp. 348-358.

Miyazaki, A.; Balint, I. and Nakano, Y. (2003). Morphology Control of Platinum Nanoparticles and their Catalytic Properties, *J. Nano. Res.* Vol. 5, p. 69

Moses, J.; Khn, R.; Dblinger, M. and Bein, T. (2010). Electrodeposition of Copper and Silver Nanowires in Hierarchical Mesoporous Silica/Anodic Alumina Nanostructures, *Chem. Mater.*, Vol. 22, pp. 5430-5436.

Nakamura H. and Matsui, Y. (1995). Silica Gel Nanotubes Obtained by the Sol-Gel *J. Am. Chem. Soc.*, Vol. 117, pp. 2651- 2652.

Nath, N. and Chilkoti, A. (2002). A Colorimetric Gold Nanoparticle Sensor To Interrogate Biomolecular Interactions in Real Time on a Surface, *Anal. Chem.* Vol. 74, pp. 504-509.

Nemamcha, A.; Rehspringer, J.L. and Khatmi, D. (2006). Synthesis of Palladium Nanoparticles by Sonochemical Reduction of Palladium(II) Nitrate in Aqueous Solution, *J. Phys. Chem. B*, Vol. 110, pp. 383-387.

Nieto, J.T.; Arvalo, A. and Garca, J.J. (2008). Catalytic Desulfurization of Dibenzothiophene with Palladium Nanoparticles, *Inorg. Chem.*, Vol. 47, pp. 11429-11434.

Niu, W.; Zhang, L. and Xu, G. (2010). Shape-Controlled Synthesis of Single-Crystalline Palladium Nanocrystals, *ACS Nano*, Vol. 4, pp. 1987-1996.

Onizawa, S.; Aoshiba, K.; Kajita, M.; Miyamoto, Y. and Nagai, A. (2009). Platinum nanoparticle antioxidants inhibit pulmonary inflammation in mice exposed to cigarette smoke, *Pulm. Pharmacol. Ther.* Vol. 22(4), pp. 340-349.

Pal, A.; Shah, S. and Devi, S. (2007). Synthesis of Au, Ag and Au-Ag alloy nanoparticles in aqueous polymer solution, *Colloids Surf. A*, Vol. 302, pp. 51-57.

Petroski, J.M.; Wang, Z.L.; Green, T.C. and El-Sayed, M.A. (1998). Kinetically controlled growth and shape formation mechanism of platinum nanoparticles, *J. Phys. Chem. B*, Vol. 102, pp. 3316-3320.

Platschek, B.; Petkov, N.; Himsl, D.; Zimdars, S.; Li, Z.; Khn, R. and Bein, T. (2008). Vertical Columnar Block-Copolymer-Templated Mesoporous Silica via Confined Phase Transformation, *J. Am. Chem. Soc.*, Vol. 130, pp. 17362-17371.

Polte, J.; Ahner, T.T.; Delissen, F.; Sokolov, S.; Emmerling, F.; Thnemann, A. F. and Kraehnert, R. (2010). Mechanism of Gold Nanoparticle Formation in the Classical Citrate Synthesis Method Derived from Coupled In Situ XANES and SAXS Evaluation, *J. Am. Chem. Soc.*, Vol. 132 (4), pp. 1296-1301.

Robeson, L. M. (2008). The upper bound revisited, *J. Membr. Sci.* Vol., 320, 390-400.

Roy, P.S.; Bagchi, J. and Bhattacharya, S. K. (2009). Size-controlled synthesis and characterization of polyvinyl alcohol coated palladium nanoparticles, *Transition Met. Chem.*, Vol. 34, pp. 447-453.

Salkar, R. A.; Jeevanandam, P.; Aruna, S. T.; Koltypin, Y. and Gedanken, A. (1999).The sonochemical preparation of amorphous silver nanoparticles *J. Mater. Chem.*, Vol. 9, pp. 1333-1335.

Santos, I. P. and Liz-Marzán, L.M. (2002). Synthesis of Silver Nanoprisms in DMF, *Nano Lett.*, Vol. 2, pp. 903–905.

Shen, J.; Shi, M.; Yan, B.; Ma, H.; Li, N. and Ye, M. (2011). One-pot hydrothermal synthesis of Ag-reduced graphene oxide composite with ionic liquid, *J. Mater. Chem.*, Vol. 21, pp.7795–7801.

Shrivastava, S.; Bera, T.; Singh, S. K.; Singh, G.; Ramachandrarao, P. and Dash, D. (2009). Characterization of Antiplatelet Properties of Silver Nanoparticles, *ACS Nano*, Vol. 3, pp.1357–1364

Siekkinen, A. R.; McLellan, J. M.; Chen, J. and Xia, Y. (2006). Rapid synthesis of small silver nanocubes by mediating polyol reduction with a trace amount of sodium sulfide or sodium hydrosulfide, *Chem Phys Lett.* Vol. 432, pp. 491–496.

Soria, R. (1995). Overview on industrial membranes, *Catal. Today*, Vol. 25, pp. 285–290.

Tian, N.; Zhou, Z.Y.; Sun, S.G.; Ding, Y. and Wang, Z.L. (2007). Synthesis of Tetrahexahedral Platinum Nanocrystals with High-Index Facets and High Electro-Oxidation Activity, *Science*, Vol. 316, pp. 732–735.

Tsung, C.K.; Kuhn, J.N.; Huang, W.; Aliaga, C.; Hung, L.; Somorjai, G. A. and Yang, P. (2009). Sub-10 nm Platinum Nanocrystals with Size and Shape Control: Catalytic Study for Ethylene and Pyrrole Hydrogenation, *J. Am. Chem. Soc.*, Vol. 131, pp. 5816–5822.

Ulbricht, M. (2006). Advanced functional polymer membranes, *Polymer*, Vol. 47, pp. 2217–2262.

V. Amendola, S. Polizzi, and M. Meneghetti, Free Silver Nanoparticles Synthesized by Laser Ablation in Organic Solvents and Their Easy Functionalization, *Langmuir*, 2007, 23, 6766–6770

Wagner, J. (2001). Practical tips and hints, in: Membrane Filtration Handbook, Chemical Engineering, 2nd Edn, 2001

Walker, C. H.; John, J.V.St. and Neilson, P.W. (2001). Synthesis and Size Control of Gold Nanoparticles Stabilized by Poly(methylphenylphosphazene), *J. Am. Chem. Soc.*, Vol. 123, pp. 3846–3847.

Welch, C. M. and Compton, R.G. (2006). The use of nanoparticles in electroanalysis: a review, *Anal. Bioanal. Chem.* , Vol. 384, pp. 601–619.

Willets, K. A. and Duyne, R. P. V. (2007). Localized Surface Plasmon Resonance Spectroscopy and Sensing, *Annu. Rev. Phys. Chem.* Vol. 58, pp. 267–297.

Wu, Y. ; Livneh, T.; Zhang, Y. X.; Cheng, G.; Wang, J.; Tang, J.; Moskovits, M. and Stucky G.D. (2004). Templated Synthesis of Highly Ordered Mesostructured Nanowires and Nanowire Arrays, *Nano Lett.*, Vol. 4, pp. 2337–2342.

Xiao, J. and Qi, L. (2011). Surfactant-assisted, shape-controlled synthesis of gold nanocrystals, *Nanoscale*, Vol. 3, pp. 1383–1396.

Xia, Y.; Xiong, Y.; Lim, B. and Skrabalak, S.E. (2009). Shape-Controlled Synthesis of Metal Nanocrystals: Simple Chemistry Meets Complex Physics?, *Angew. Chem. Int. Ed.*, Vol. 48, pp. 60–103.

Yang, Y.; Matsubara, S.; Xiong, L.; Hayakawa, T. and Nogami, M. (2007). Solvothermal Synthesis of Multiple Shapes of Silver Nanoparticles and Their SERS Properties, *J. Phys. Chem. C*, Vol. 111, pp. 9095–9104.

Xiong, Y.; Chen, J.; Wiley, B.; Xia, Y.; Aloni, S. and Yin, Y. (2005). Understanding the Role of Oxidative Etching in the Polyol Synthesis of Pd Nanoparticles with Uniform Shape and Size, *J. Am. Chem. Soc.*, Vol. 127, pp. 7332–7333.

Yoo, S.; Ford, D. M. and Shantz, D. F. (2006). Synthesis and characterization of uniform alumina–mesoporous silica hybrid membranes, *Langmuir, Vol. 22, pp.* 1839–1845.

8

New Technology for the Synthesis of New Materials Based on Cellulose and Sorption of Noble Metals

Vaso Bojanić[1] and Miomir Pavlović[2]
[1]*University of Banja Luka, Faculty of Agriculture*
[2]*University of Eastern Sarajevo, Faculty of Technology Zvornik*
Republic of Srpska, B&H

1. Introduction

Cellulose is continuously updated biopolymer through photosynthesis and, hence, it is inexhaustible raw material for new materials and new technology (Bojanic et al., 1988a; Granja et al., 2006; Hubbe et al., 2008, 2011; Jovanovic et al., 2002; Kamel, 2007; Puoci et al., 2008; Schwanninger et al., 2004; Vainio, 2007; Wu et al., 2007; Zhang et al., 2010). Modification of biopolymers has been given scientific and practical importance. Grafting is one of the best methods making the synthesis of new materials and their applications virtually unlimited (Achilleos & Vamvakaki, 2010; Bhattacharya & Mirsa, 2004; Cohen Stuart et al., 2010; Crini, 2005; Gandini, 2008; Li et al., 2011; Lu et al., 2008; Roy et al., 2009; Petrovic et al., 2010; Sharma et al., 2010; Xin et al., 2011). Cellulose is, due to its chemical and sub-molecular structure, mechanically resistant and chemically stable. Such properties are of great importance for the chemical and electrochemical modification and represent a subject to numerous studies aimed at obtaining new materials with special properties for specific applications (Anderson, 2000; Bonne, 2008; Chmielewska et al., 2010; Cao et al., 2007; Hu et al., 2009; Kim et al., 2010; Pinto & Maaroufi, 2005; Spiridon et al., 2011; Vitz et al., 2010; Wang, 2008; Zhou et al., 2011). Grafting reactions represent possible solutions for changing chemical, physical and mechanical properties of the cellulose molecules in desired direction. Modification of cellulose leads to formation of the new ionic polymers (Heinze, 1998) and grafting of N-vinyl pyrrolidone (Gupta & Sahoo, 2001; Chauhan et al., 2005), styrene, methyl methacrylate, methyl acrylamide (Coshun & Temuz, 2005; Sharma & Chauhan, 2009) acrylamide and acrylic acid (Chauhan & Lal, 2003), grafting of acrylamide (Chauhan et al., 2003) and 4-vinylpyridine (Chauhan et al., 2000; Kaur & Dhiman, 2011), and ethyl acrylate (Kalia et al., 2011) on cellulose have been studied. New materials based on biopolymers and, especially, modification of cellulose and lignin, have been used as semipermeable membranes, ion-exchangers and matrices for medicaments (Bilba, 1998; Hubicki et al., 2008; Maliyekkal et al., 2010; Nada et al., 2007; Ozdemir et al., 2006; Parajuli, 2006; Rodriguez et al., 2009; Saarinen, 2008; Vlasankova & Sommer, 1999; Wang, 2005; Xu, 2005). In order to obtain grafted cellulose copolymers with 4-vinylpyridine, vinylimidazole, 1-vinyl-2-pyrrolidone, 9-vinylkarbazole and other vinyl monomers containing double bonds capable for copolymerization with vinyl monomers have been introducted in cellulose. For the induction of double bonds in cellulose, the reaction with acyloyl chloride from which cellulose acrylate has been produced under equal concentrations of reactants and an equal reaction time under procedure described by (Akelah & Sherrington, 1981) is used. They have synthesized acrylate cellulose with degree of substitution, *DS* = 0,81 where on C_6 - polysaccharide unit in average came 0.81 of acrylic residue with grafted styrene on it, but without consideration of the optimization of cellulose acrylate synthesis. The procedure for the synthesis of cellulose acrylate has changed itself. Its optimization was done and a new molar ratio of reactants was given in reaction mixture with different reaction times. Grafting reaction was conducted by radical polymerization of 4-vinylpyridine, 1-vinylimidazole, 1-vinyl-2-pirrolidone and 9-vinylcarbazole. Grafed copolymers were synthesized (new materials): cellulose-poly-4-vinylpyridine, cellulose-poly-1-vinylimidazole, cellulose-poly-1-vinyl-2-pyrrolidine and cellulose-poly-9-vinylcarbazole. Quaternization of pyridine ring was performed on grafted cellulose copolymers: cellulose-poly-4-vinylpyridine and cellulose-poly-1-vinylimidazole with methyl iodide. New materials based on cellulose in ionic form at constant potential were synthesized by electrolysis. Electrochemical modification was performed on new cellulose materials with different ions originating from the used basic electrolytes. Thermal and ion-exchanger properties were determined on synthesized new cellulose based materials and they were used as selective ion-exchaners for extraction of noble metals from aqueous solutions. Optimization of the cellulose acrylate synthesis process was used as a model for the synthesis of grafted copolymers of lignin and tannin.

2. Optimization of cellulose acrylate synthesis process and grafting of 4-vinylpyridine, 1-vinylimidazole, 1-vinyl-2-pyrrolidone, and 9-vinylcorbazole

This experiment was performed with a sample of powdered microcrystalline cellulose form Aldrich company. Before using the cellulose, powder was subsequently rinsed with water, ethanol, methanol and acetone, and then dried at 50°C. Parameters important for optimization of the synthesis of cellulose acrylate and that were considered were: molar ratio of reactants in reaction mixture of cellulose/potassium-t-butoxide/acryloyl chloride = 1:3:10, and different synthesis reaction times of cellulose acrylate of 1,3,5,8 and 10 hours. Cellulose acrylate and grafted cellulose copolymers with 4-vinylpyridine, 1-vinylimidazole, 1-vinyl-2-pyrrolidone and 9-vinylkarbazole were synthesized according the methods and procedures described by (Bojanic, 1994, 2010). Cellulose acrylate (Cell-acrylate) was synthesized by making 3.24 g-0,02 mol of cellulose swell in 50 ml of acetonitrile for one hour at room temperature. Potassium-t-butoxide 6.72 g-0.06 mol was dissolved with mild heating in 50 ml of acetonitrile and the reaction with the cellulose solution at room temperature, with mixing, during 4 hours took place. Formed potassium-cellulosate was not isolated, and overplus of acryloyl chloride was added in it, namely 18g-0.2 mol in 50 ml of acetonitrile at 25^0C. Reaction mixture was mixed and refluxed during 10 hours. Cellulose acrylate was then washed with water, ethanol, methanol and acetone and dried at 50^0C. 4.65 g of product was obtained which has a band at 1720 cm^{-1} from ester carbonyls in IR-spectrum in KBr. Elemental analysis of samples taken after 1, 3, 5, 8, 10 hours provides optimum time of 10 hours. During this period, the reaction gave a product with a degree of substitution (DS) of 2.4. The same procedure was repeated for lignin and tannin, with the difference that instead of cellulose the same weight amount of lignin and tannin was taken. The only difference is that the tannin acrylate was not elutriated with water because it is being dissolved in it. The resulting tannine acrylate and lignin acrylate have characteristic band at 1720 cm^{-1} in the IR-spectrum. Synthesis of cellulose poly-4-vinylpyridine (PVP-Cell) was performed as follows: Grafting of 4-vinylpyridine on cellulose acrylate was done by adding 1 g-0.0046 mol of acrylate cellulose, 5ml-4,88g-0,046 mol of 4-vinylpyridine and 0.05 g azobisisobutyronitrile (AIBN) in 50 ml of acetonitrile. Reaction mixture was mixed and refluxed for 5 hours in a stream of nitrogen. 1,1g of Cell-PVP was obtained, and it was characterized by IR spectroscopy with characteristic IR-spectrum bands at 1620, 990 and 820 cm^{-1}. Sorption of gold after 1 and 24 hours was 98.12 and 99.63 wt.%. The capacity was 0.2 g Au/g ion-exchanger. In comparison, the capacity for Pd and Pt is about 0.03 g/g ion-exchanger.

Synthesis of lignin poly-4-vinylpyridine (Lig-PVP) was performed as follows: the procedure was the same as for the synthesis of Cell-PVP, but instead of cellulose, 1g of lignin was taken. 1,1g Lig-PVP was obtained, and it was characterized by IR spectroscopy with characteristic IR-spectrum bands at 1620, 990 and 820 cm^{-1}. Obtained Lig-PVP does not swell in water. However, in a solution containing Au, Pd and Pt ions, it swelled 100 wt%. Sorption of Au was fast and complete and the capacity was 0.211 g Au/g ion-exchanger. This capacity was confirmed in the strong acidic electrolytes in which both Pt and Pd are present. Regeneration with HCl (1:1) was complete. By checking the capacity with 0.1 M solution of Au it was noticed that the yield was 0.4 g Au/g ion-exchanger. The sorption of platinum metals was studied as well, and it was very good.

Synthesis of tannine poly-4-vinylpyridine (Tan-PVP) was performed as follows: the procedure was the same as for the synthesis of Cell-PVP, but instead of cellulose, 1g of tannine was taken. 1.1g of Tan-PVP was obtained, and it was characterized by IR spectroscopy with characteristic IR-spectrum bands at 1620, 990 and 820 cm^{-1}. Sorption of Au for Tan-PVP has not been tested because it is soluble in water.

Synthesis of cellulose poly-1-vinylimidazole (Cell-PVIm) was performed as follows: The procedure for the synthesis of Cell-PVP was repeated and grafting of 1-vinylimidazole 4 ml-0.044 mol on cellulose was performed. Cell-PVIm, 1.1g was obtained. Cell-PVIm was characterized in the IR-spectrum with characteristic bands at 1620, 1480, 910, 820 and 740 cm^{-1}. It did not swell in water and the sorption of gold was 0.23 g Au/g ion-exchanger. Synthesis of cellulose-poly-1-vinyl-2-pyrrolidone (Cell-P1V2P) followed the procedure for the synthesis of Cell-PVP and grafting of 1-vinyl-2-pyrrolidone, 5 ml - 0.047 mol, was performed afterwards. Cell-P1V2P, 1.1g was obtained with characteristic IR-spectrum bands at 1650, 1310 and 890 cm^{-1}. Copolymer blured by swelling and it changed color into brown-red. Sorption of Au after 1 and 24 hours was 25.50 and 54.08 wt.%. The capacity of Au was 0.20 g Au/g ion-exchanger. Synthesis of cellulose-poly-9-vinylcarbazole (Cell-P9VK) followed the procedure for the synthesis of Cell-PVP and grafting of 9-vinylcarbazole, 5g-0,047 mol, on cellulose was performed afterwards. 1.2 g of copolymer was obtained with characteristic IR-spectrum bands at 1580, 1310, 1210, 740 and 720 cm^{-1}. Sorption capacity of Au has not been tested. Synthesis of cellulose 1-methyl-poly-4-vinylpyridine iodide (Cell-1-Me-4-PVPJ) was derived as follows: Cell-PVP, 2g-0.0043 mol was mixed with 5ml-11.4 g-0.08 mol of methyl-iodide in 50 ml of dimethylformamide (DMF) and it was refluxed with constant mixing for 5 hours. The reaction produced 2.1 g of Cell-1-Me-PVPJ which was filtered and rinsed with water, acetone, ethanol and methanol, and finally dried at $50^{0}C$. Elemental analysis of Cell-1-Me-PVPJ was in good accordance with the theoretically calculated values. Sorption of Au after 1 and 24 hours was 99.03 wt.% and 99.89 wt.%. The capacity of Au was 0.24 g Au/g ion-exchanger. In contact with the Au solution, Cell-1-Me-PVPJ changed color into brown-red, which concludes the creation of a gold nanoparticles (Srivastava et al., 2008). Synthesis of lignin 1-metylpoly-4-vinylpyridine iodide (Lig-1-Me-PVPJ) and tannin 1-metylpoli-4-vinylpyridine iodide (Tan-1-Me-PVPJ) was performed in the same manner as the synthesis of Cell-1-Me-PVPJ with the difference that Lig-PVP and Tan-PVP with methyl iodide was used. Sorption of Au of the new obtained materials has not been tested. Synthesis of Cellulose 3-metylpoly-1-vinylimidazole iodide (Cell-3-Me-PVImJ) was performed the same as synthesis of Cell-1-Me-PVPJ, with difference that Cell-PVIm with methyl iodide was used. Cell-1-Me-3-PVImJ blured by swelling in contact with Au solution and it changed color into brown-red. The capacity of 0.20 g Au/g ion-exchanger was reached. The degree of polymerization of microcrystalline cellulose was determined by viscometer in cupriethylenediamine as solvent at 25°C, and elemental analysis was performed with the Perkin-Elmer instrument, model 240. IR-spectra of the initial cellulose and obtained copolymers were determined by Perkin Elmer spectrophotometer, model IRDMI-FTIR1724X, using KBr technique. Optimization of the synthesis process of cellulose acrylate was carried out by a series of experiments in which the impact of the relation between reactants in reaction mixture and reaction time on the cellulose acrylate has been examined, i.e. the degree of substitution (Bojanić, 2010). In the IR-spectrum, very pronounced band at 1720 cm^{-1} was observed, which corresponds to C = O group from esters.

Table 1. shows the dependence of content (C,%; H,%; O,%) from the reaction time, compared to the components in the reaction mixture (cellulose/potassium-t-butoxide/acryloyl chloride = 1:3:10). Samples for elementary analysis were taken after 1, 3, 5, 8 and 10 hours of reaction. It is evident from Table 1. that the content of C (%) in cellulose acrylate increases with increase in reaction time.

Microanalysis	% C	% H	% O
$C_9H_{12}O_6$	50.00	5.60	44.40
$C_9H_{12}O_6$ -1h	44.00	6.18	49.82
$C_9H_{12}O_6$ -3h	45.02	6.16	48.82
$C_9H_{12}O_6$- 5h	45.65	6.08	48.27
$C_9H_{12}O_6$- 8 h	46.50	6.03	47.47
$C_9H_{12}O_6$-10h	47.02	5.97	47.01

Table 1. Elemental analysis: C, H and O content in molar ratio cellulose/potasium-t-butoxide/acryloy chloride of 1:3:10 after various reaction times.

The percentage of subtitution of glucose units in cellulose (Y) is calculated according to equation (1) based on the results of elemental analysis:

$$Y = (A-B)/(C-D) \times 100 \quad (1)$$

a. % C specified in the grafted cellulose copolymer,
b. % C determined in the initial cellulose,
c. % C calculated at 100% derivatisation,
d. % C from the initial cellulose.

Extension of the reaction time increases the content of %C in acrylate as well as %Y, which, for the molar ratio of cellulose/potassium-t-butoxide/acryloyl chloride of 1:3:10 and reaction time of 10 hours, is Y=80.7%. The degree of substitution of cellulose acrylate (DS) is calculated from IR spectra using the relations of characteristic bands in equation (2):

$$DS = D_{1720}/D_{1410} \quad (2)$$

DS for cellulose acrylate obtained under optimal synthesis conditions is 2.4 and it is about three times higher than stated by other authors, where DS=0.81 and Y=27% (Akelah & Sherrington, 1981). The optimum conditions for cellulose acrylate synthesis are: molar ratio of cellulose/potassium- t-butoxide/acryloyl chloride = 1:3:10, with the reaction time of 10 hours, and they give DS=2.4 and Y = 80,7%. Fig.1 shows the synthesis of cellulose acrylate and grafting with vinyl monomers. Cell-acrylate (3) is copolymerised with 4-vinylpyridine (4), 1-vinylimidazole (5), 9-vinylcarbazole (6) and 1-vinyl-2-pyrrolidone (7), using AIBN as the initiator of the copolymerization. In the reaction between cellulose acrylate and 4-vynilpyridine Cell-PVP (8) grafed copolymer is formed and its structure was confirmed by IR-spectrum, with characteristic bands at 1620 and 820 cm^{-1}, characteristical for the pyridine molecule.

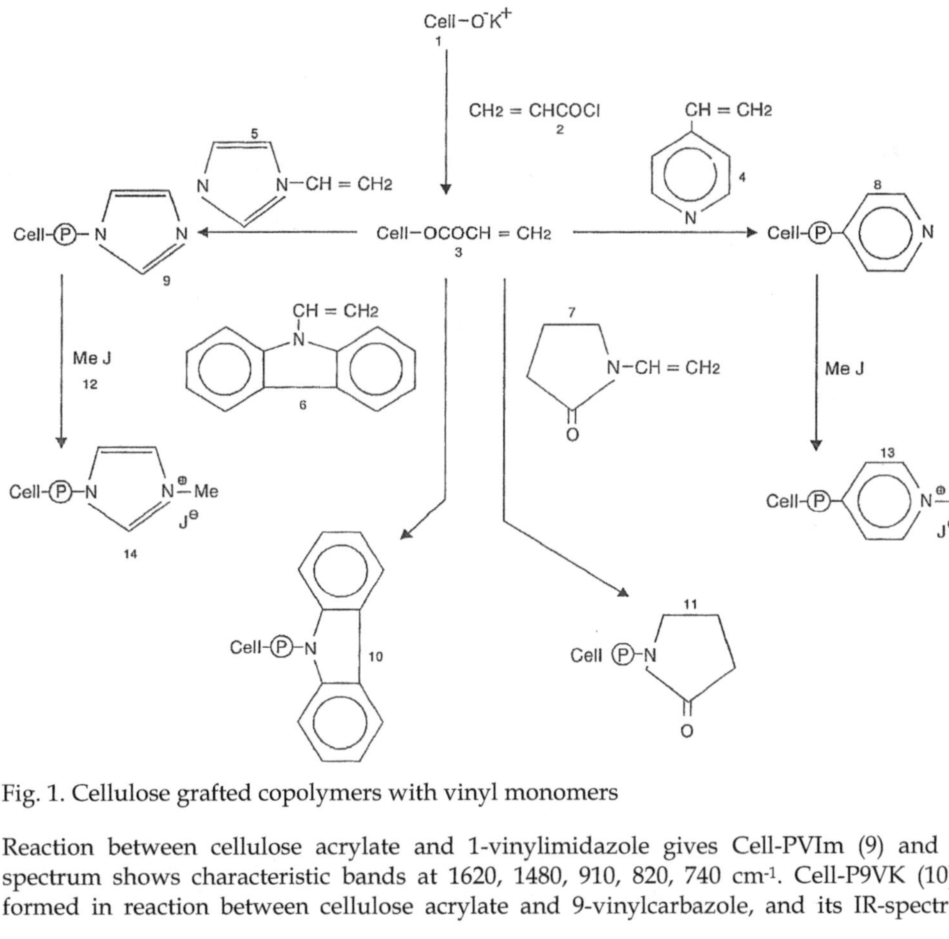

Fig. 1. Cellulose grafted copolymers with vinyl monomers

Reaction between cellulose acrylate and 1-vinylimidazole gives Cell-PVIm (9) and IR-spectrum shows characteristic bands at 1620, 1480, 910, 820, 740 cm^{-1}. Cell-P9VK (10) is formed in reaction between cellulose acrylate and 9-vinylcarbazole, and its IR-spectrum shows characteristic bands at 1580, 1310,1210, 740, 720 cm^{-1}. Reaction between cellulose acrylate and 1-vinyl-2-pyrrolidone gives Cell-P1V2P (11) and its IR-spectrum shows characteristic bands at 1650, 1310, 890 cm^{-1}. The reaction between Cell-PVP and methyl iodide produces Cell-1-Me-PVPJ (13), while Cell-PVPIm and methyl iodide produce Cell-3-Me-PVIm (14).

The results of elemental analysis of synthesized grafted cellulose copolymers are shown in Table2. In addition to the experimentally determined values, theoretical values for the content of carbon, hydrogen, oxygen, nitrogen and iodine in determined materials calculated from the gross formula of basic structural elements, assuming that all three hydroxyl groups in cellulose molecules have been substituted are shown in Table 2.

Materials	Segments units molar mass	Microanalysis				
		C(%)	H(%)	O(%)	N(%)	J(%)
Cellulose	$C_6H_{10}O_5$ 162	44.4 42.53	6.2 6.4	49.5 51.07	- -	-(Teor.) -(Exp.)
Cell-akrylate	$C_9H_{12}O_6$ 216	50.00 47.02	5.6 5.97	44.4 47.01	- -	- -
Cell-P1V2P	$C_{15}H_{12}O_7N$ 327	55.00 49.93	6.42 6.28	34.25 41.54	4.28 2.25	- -
Cell-P9VK	$C_{23}H_{23}O_6N$ 409	67.50 77.74	5.60 5.20	23.50 10.42	3.40 6.64	- -
Cell-PVP	$C_{16}H_{19}O_6N$ 321	59.80 52.43	5.90 6.05	29.90 36.75	4.36 4.77	- -
Cell-PVIm	$C_{14}H_{18}O_6N_2$ 310	54.19 47.90	5.80 5.91	30.97 40.43	9.00 5,76	- -
Cell-1Me-PVPJ	$C_{17}H_{22}O_6N_2$ 463	44.06 45.11	4.75 5.31	20.79 26.53	3.00 2.93	27.4 20.12
Cell-3Me-PVImJ	$C_{15}H_{21}O_6N_2J$ 452	39.82 42.63	4.64 5.39	21.35 32.61	6.19 5.25	28.00 14.12

Table 2. Elemental analysis results of synthetized cellulose copolymers

Based on elemental analysis, specific characteristics of a grafted copolymers of cellulose (mass fraction of grafted copolymer (X); relation of derivatizated parts/cellulose vinyl group (Z), the degree of grafting, wt%) are calculated and shown in Table 3.

Materials	Weight fraction of copolymers (X)	Ratio derivatives units/Cell,vinyl groups(Z)	Degree of grafting wt.%
Cell-P1V2P	0.16	0.18	17.80
Cell-P9VK	0.76	1.75	91.53
Cell-PVP	0.16	0.20	35.70
Cell-PVIm	0.05	0.06	19.34

Table 3. Some characteristic of grafted cellulose copolymers

Mass fraction of grafted copolymers is calculated from the following equations:

$$E=(1-X)A+64.8X \text{ (Cell-P1V2P)} \quad (3)$$

$$E=(1-X)A+87.0X \text{ (Cell-P9VK)} \quad (4)$$

$$E=(1-X)A+80.0X \text{ (Cell-PVP)} \quad (5)$$

$$E=(1-X)A+63.8X \text{ (Cell-PVIm)} \quad (6)$$

From the equations: E=%C found in grafted copolymer, 64.8=%C calculated on 1-vinyl-2-pyrrolidone segment, 87.0=%C calculated on 9-vinylcarbazole segment, 80.0=%C calculated on 4-vinylpyridine segment and 63.8=%C calculated on 1-vinylimidazole segment.

Relation of deriving parts/cellulose vinyl group, Z, is calculated from equation:

$$Z=[(X/Mx100)]/\ [(1-X)/F)xY] \quad (7)$$

where X and Y are calculated as stated before and F is the average molar mass of cellulose segment adjusted for the degree of subtitution.

$$F=(YxG)/100+[(1-Y)/162]/100 \quad (8)$$

From equation (8): G is theoretical molar mass of glucose derivates, 162 is theoretical molar mass of glucose residue in the cellulose molecular chain, M is theoretical molar mass of grafted monomers.

The degree of grafting of synthesized copolymers was determined according to equation:

$$\text{degree of grafting} = \%N/ATxM \quad (9)$$

where %N is mass percentage of nitrogen in grafted copolymer determined by elemental analysis and AT is atomic mass of nitrogen.

Vinyl monomers, 4-vinylpyridine and 1-vinylimidazole, due to present nucleophilic nitrogen are subject to reactions of nucleophilic substitutions. A typical reaction for the 4-vinylpyridine and 1-vinylimidazole is their quaternization with nucleophilic attack on alkyl halides or by protonation. This type of reaction was used for the synthesis of quaternary cellulose polypyridinium and cellulose polyimidazole copolymers. The reaction of Cell-PVP with methyl iodide yielded Cell-1Me-PVPJ and IR-spectrum showed a characteristic band at 1642 cm^{-1}, which corresponds to $C=N^{+}$- quaternized nitrogen in the pyridine ring. In reaction between Cell-PVIm and methyl iodide Cell-3Me-PVImJ was formed, whose structure was confirmed by IR-spectra with characteristic band at 1642 cm^{-1} for $C=N^{+}$, quaternized pyridine. To determine the iodine content in grafted copolymers Cell-1Me-PVPJ and Cell-3Me-PVImJ synthetized from Cell-PVP and Cell-PVIm by quaternization of nitrile atoms with methyl iodide coulometric method was applied.

Coulometric measurement at constant potential of second wave (E = 0.7 V) of cyclic voltammogram in a cell with a diaphragm and Pt-net anode (2 x 3 cm) after mixing (60 min) Cell-1Me-PVPJ grafted copolymer in 0.1 M solution of acetonitrile tetramethylammonium perchlorate melt (CH_3CN-Et_4NClO_4) was performed and current/time curve was recorded. The amount of electricity needed for complete oxidation of iodide ions into iodide was measured by electrical integrator and content of iodide in the synthesized Cell-1-Me-PVPJ sample was calculated to be 20.12 wt.%.

3. Electrochemical properties of grafted copolymers with 4-vinylpyridine and 1-vinylimidazole

Cellulose acrylate and grafted cellulose copolymers with 4-vinylpyridine and 1-vinylimidazole and their electrochemical transformations in various basic electrolytes were performed by the methods and procedures described in works by (Bojanic et al., 1996). Voltammetric measurements were performed in standard electrochemical cell, which consisted of saturated calomel electrode (SCE), platinum auxiliary electrode and polished platinum disc electrode as working electrode. Electrochemical modification of new cellulose materials was performed in electrochemical cell with a diaphram. Platinum net (4×5) cm was used as working electrode, cathode was from nickel and SCE was used as reference electrode. The cell was connected to the instrument Controvit-Tacussel with potentiostat and function generator type HI-TDK DT 2101 and x-z-y Guold recorder 3054 printer. The results of a simple electrochemical synthesis of new grafted cellulose copolymers in ionic form are shown, using electrolysis at constant potential. Electrochemical transformations of Cell-1Me-PVPJ and Cell-3Me-PVImJ into new grafted cellulose copolymers with different ions originating from the used basic electrolytes was carried out. Electrochemical synthesis of Cellulose 1-metylpoli-4-vinylpyridine perchlorate (Cell-1-Me-PVPClO4) was performed in the following manner: 200 mg of Cell-1-Me-PVPJ was oxidized in the anodic area of electrolytic cell with diaphragm using Pt-net anode (4×5) cm, Ni cathode and SCE in which was 0,1 M solution of $CH_3CNEt_4NClO_4$ at E= 0,7V until starting current drops to the residual current value. The resulting Cell-1-Me-PVPClO_4 is filtered, washed with water, ethanol, methanol and acetone and dried at 50°C. Electrochemical transformation of the anions on Cell-1-Me-PVPJ was preformed in the same manner as electrochemical synthesis of Cell-1-Me-PVPClO_4 and following new cellulose materials are obtained: Cellulose 1-methyl poly-4-vinylpyridine chloride (Cell-1-Me-PVPCl), Cellulose 1-methyl poly-4-vinylpyridine trifluoroacetate (Cell-1-Me-PVPCF_3COO), Cellulose 1-methyl poly-4-vinylpyridine nitrate (Cell-1-Me-PVPNO_3), Cellulose 1-methyl poly-4-vinylpyridine p-tosylate (Cell-1-Me-PVPp-TsO), Cellulose 1-methyl poly-4-vinylpyridine tetrafluoroborate (Cell-1-Me-PVPBF_4) and Cellulose 1-methyl poly-4-vinylpyridine hexafluorophosphate (Cell-1-Me-PVPPF_6). Electrochemical transformations on Cell-3-Me-PVImJ were performed in the same manner and under the same conditions as on Cell-1-Me-PVPJ and following new cellulose materials are obtained: Cellulose 3-methyl poly-1-vinylimidazole chloride (Cell-3-Me-PVImCl), Cellulose 3-methyl poly-1-vinylimidazole perchlorate (Cell-3-Me-PVImClO_4), Cellulose 3-methyl poly-1-vinylimidazole trifluoroacetate (Cell-3-Me-PVImCF_3COO), Cellulose 3-methyl poly-1-vinylimidazole nitrate (Cell-3-Me-PVImNO_3), Cellulose 3-methyl poly-1-vinylimidazole p-tosylate (Cell-3-Me-PVIm p-TsO), Cellulose 3-methyl poly-1-vinylimidazole tetrafluoroborate (Cell-3-Me-PVImBF_4) and Cellulose 3-methyl poly-1-vinylimidazole hexafluorophosphate (Cell-3-Me-PVImPF_6). All synthesized cellulose-based new materials show good sorption qualities towards noble metals. Sorption of gold was thoroughly examined only on Cell-1-Me-PVPCF_3. Sorption of Au was rapid and complete on it, and after 1 hour it was 99.53 wt.%, and after 24 hours was 99.89 wt.%. Capacity for Au was 0.22 g Au/g ion-exchanger and for palladium was 0.05 g Pd/g ion-exchanger. Electrochemical transformations of the anion on Lig-1-Me-PVPJ were performed under the same electrosynthesis conditions as on Cell-1-Me- PVPJ and following new materials based on lignin were obtained: Lignin 1-methyl poly-4-

vinylpyridine chloride (Lig-1-Me-PVPCl), Lignin 1-methyl poly-4-vinylpyridine perchlorate (Lig-1-Me-PVPClO$_4$), Lignin 1-methyl poly-4-vinylpyridine trifluoroacetate (Lig-1-Me-PVPCF$_3$COO), Lignin 1-methyl poly-4-vinylpyridine nitrate (Lig-1-Me-PVPNO$_3$), Lignin 1-methyl poly-4-vinylpyridine p-tosylate (Lig-1-Me-PVPp-TsO), Lignin 1-methyl poly-4-vinylpyridine tetrafluoroborate (Lig-1-Me-PVPBF$_4$) and Lignin 1-methyl poly-4-vinylpyridine hexafluorophosphate (Lig-1-Me-PVPPF$_6$). All synthesized new materials based on lignin showed good sorption qualities to noble metals but the capacities were not determend. Electrochemical transformations of the anion on Tan-1-Me-PVPJ were performed under the same electrosynthesis conditions as on Cell-1-Me-PVPJ and following new materials based on tannin were obtained: Tannin 1-methyl poly-4-vinylpyridine chloride (Tan-1-Me-PVPCl), Tannin 1-methyl poly-4-vinylpyridine perchlorate (Tan-1-Me-PVPClO$_4$), Tannin 1-methyl poly-4-vinylpyridine trifluoroacetate (Tan-1-Me-PVPCF$_3$COO), Tannin 1-methyl poly-4-vinylpyridine nitrate (Tan-1-Me-PVPNO$_3$), Tannin 1-methyl poly-4-vinylpyridine p-tosylate (Tan-1-Me-PVPp-TsO), Tannin 1-methyl poly-4-vinylpyridine tetrafluoroborate (Tan-1-Me-PVPBF$_4$) and Tannin 1-methyl poly-4-vinylpyridine hexafluorophosphate (Tan-1-Me-PVPPF$_6$). Sorption of noble metals of new tannin-based materials has not been tested. Table 4. shows numerous values of anode potential under which electrochemical transformation of grafted cellulose copolymers with 4-vinylpyrinine and 1-vinylimidazole has been carried out, as well as typical values for the anion absorption inducted into copolymer. Electrochemical cell with platinum disc electrode as working electrode was used for the study of electrochemical behavior of synthesized Cell-1-Me-PVPJ and Cell-3-Me-PVImJ in acetonitrile and tetraethylammonium perchlorate as conductive electrolyte.

Anion exchange	Solvent/Support Electrolyte(0,1M)	Applied Potential (V vs. SCE)	IR Bands (KBr)cm^{-1}
Cl$^-$	CH_3CN/Et_4NCl	0.7	
ClO_{4-}	CH_3CN/Et_4NClO_4	0.7	1095,650
CF_3COO^-	CH_3CN/CF_3COONa	0,7	1683,1204,834
NO_{3-}	$CH_3CN/H_2O_{(9:1)}$ NH_4NO_3	1.0	1380
p-TsO$^-$	CH_3CN/Et_4Np-TsO	0.7	1191,811
BF4$^-$	CH_3CN/Bu_4NBF_4	0.7	1120,1080
PF6$^-$	$CH_3CN/BuNPF_6$	0.7	841,558

Table 4. Electrochemical transformation of Cell-1-Me-PVPJ and Cell-3-Me-PVImJ

Cyclic voltamograms were recorded for Cell-1Me-PVPJ for a given solution composition at different mixing periods at 500 o/min. and they are shown in Fig.2. All recorded cyclic voltammograms of grafted cellulose copolymers in the ionic form show two anodic waves in the potential interval 0,35-0.65 V. Iodide ion shows two anodic waves in acetonitrile-tetraethylammonium perchlorate solution on the platinum electrode in the tested interval. The first anodic current maximum corresponds to potential of 0.35V and the second to potential of 0.65 V in relaton to SCE. The increase of the anodic waves maximum current is a result of ionic exchange reactions. Iodide is generated in reaction, as a function of time. With the extension of reaction time during the ionic exchange reaction, the amount of relased iodide increases, which is proportional to the current maximum of the first anodic wave on cyclic voltammogram in Fig. 2.

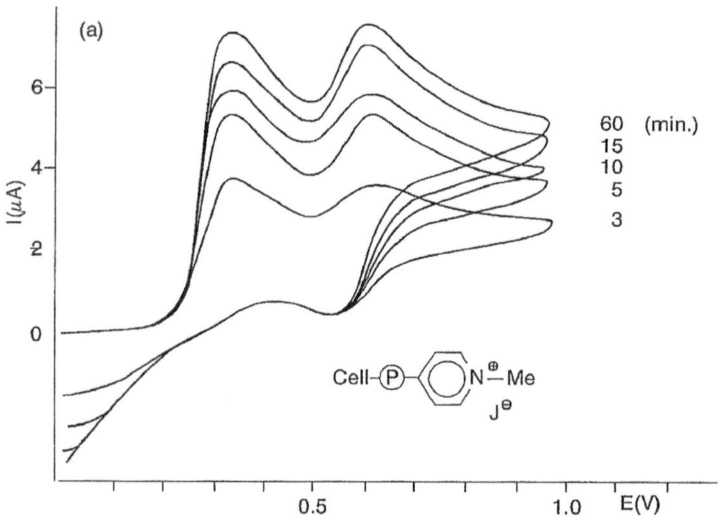

Fig. 2. Cyclic voltammograms at Pt-electrode (2r=1 mm) with scan rate of 0,1 Vs^{-1} in CH_3CN (25 ml)-Et_4NClO_4 (0.1 M):Cell-1Me-PVPJ (14.5 mg) at different time intervals.

Obtained cyclic voltammograms of disolved synthesized Cell-1-Me-PVPJ and Cell-3Me-PVImJ in acetonitrile in the presence of different electrolytes at different reaction times of ionic exchange are shown in Fig. 3.

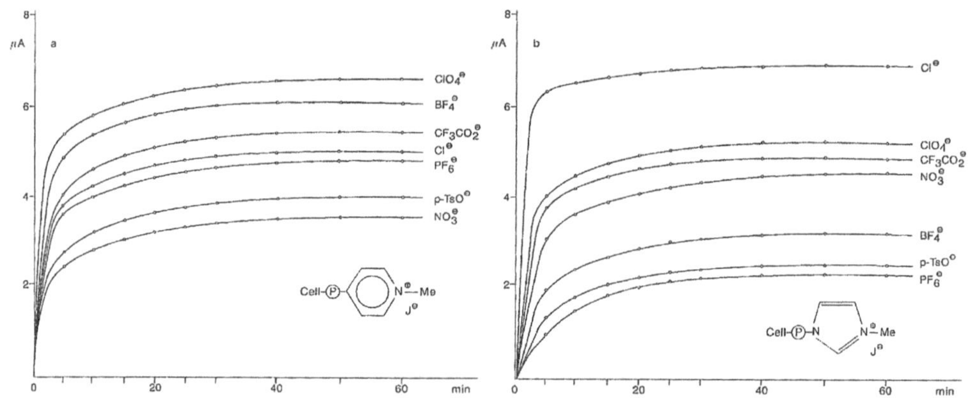

Fig. 3. Effect of time on anion-exchange reaction with the corresponding support electrolyte (0,1M) except for NH_4NO_3 (c= 0,025 M): 14,5 mg Cell-1-Me-PVPJ (a) and 14,5mg Cell-3-Me-PVImJ (b).

Curves shown in Fig. 3. indicate that the equilibrium state of iodide ionic exchange reaction with various anions is being reached after different reaction times, and that it depends on the type of anion. The amount of exchanged iodide from Cell-1Me-PVPJ and Cell-3Me-PVImJ and the concentration of basic electrolyte that displaces iodide are determined from the concentration calibration curve obtained with 1-methyl-gamma-picoline iodide and calculated equilibrium constants for the reaction of ionic exchange. Some other factors such as rate of anion diffusion in polymer matrix and ion-pair effect affect the speed of reaction of ionic exchanges. It has been shown that this reaction is sensitive to the nature of the solvents when the reaction is carried out in the presence of other anions (Bojanic et.al., 1996; Gunic & Tabakovic,I., 1988; Tabakovic,R., et al., 1992; Tabakovic,R., & Tabakovic,I., 1999). The equillibrium can be shifted in the direction of creating products through electrochemical oxidation of exchanged iodide. Electrochemical modification of Cell-1Me-PVPJ and Cell-3Me-PVImJ into new grafted cellulose copolymers in ionic form was performed as follows: Copolymer was mixed for one hour in a given solvent-basic electrolyte system and then it was subjected to preparative anodic oxidation under corresponding controlled potential on the plateau of the second wave of the current-potential curve which was determined by a rotating disk electrode. Preparative anodic oxidation of Cell-1Me-PVPJ and Cell-3Me-PVImJ was performed at the platinum net anode (4x5 cm) as described by (Bojanic et.al, 1996). Electrolysis was stopped when current slightly fell to the residual current value and the resulting grafted cellulose copolymers in ionic form were filtered and rinsed in water, ethanol, methanol, acetone and dried at 50^{0}C. None of the cyclic voltammograms of the formed copolymers recorded after completion of anodic oxidation in different solutions and acetonitrile showed the presence of anodic waves of iodide oxidation, not even in traces. This means that iodide was completely replaced with other anions and grafted cellulose copolymers as ions have been formed. Typical example of such analysis is shown in Fig. 4.

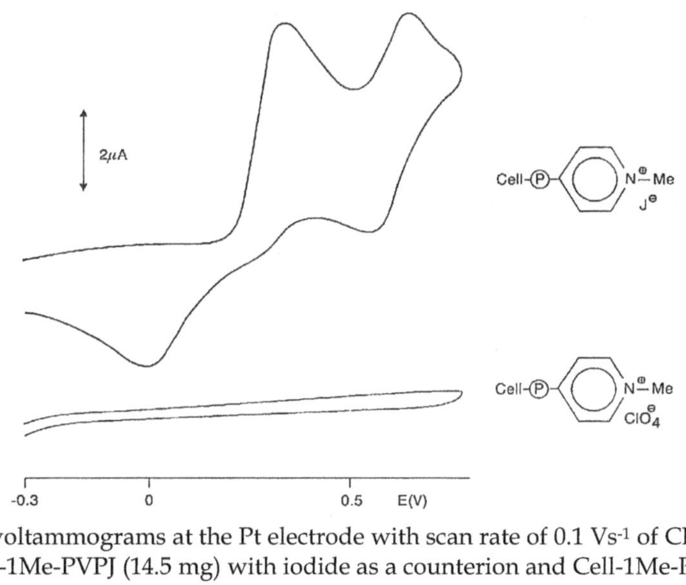

Fig. 4. Cyclic voltammograms at the Pt electrode with scan rate of 0.1 Vs^{-1} of CH_3CN-0,1M-Et_4NClO_4 Cell-1Me-PVPJ (14.5 mg) with iodide as a counterion and Cell-1Me-PVPClO$_4$ with perchlorate as a counterion

All synthesized grafted ionic cellulose copolymers were confirmed by IR spectroscopy and characteristic bands for the anions are shown in Table 4. The results on cellulose have shown that by simple electrochemical synthesis various new cellulose materials with 4-vinylpyridine and 1-vinylimidazole in ionic form can be obtained. New and simple methods for electrosynthesis of new materials based on cellulose, lignin and tannin have been developed by exchanging iodide anion with the one from the basic electrolyte. New technology for the synthesis of new cellulose based materials and sorption of noble metals has been successfully applied as a model for the synthesis of new materials based on lignin (Bojanic et al., 1998b) and tannin (Bojanic & Jovanovic, 2000). Following new cellulose based materials are synthesized with these new technologies: Cellulose acrylate, Cellulose-poly-1-vinyl-2-pyrrolidone, Cellulose-poly-9-vinylcarbazole, Cellulose-poly-4-vinylpyridine, Cellulose-poly-vinylimidazole, Cellulose-1-methylpoly-4-vinylpyridine iodide, Cellulose-3-methylpoly-1-vinylimidazole iodide, Cellulose-1-methylpoly-4-vinylpyridine perchlorate, Cellulose-1-methylpoly-4-vinylpyridine trifluoroacetate, Cellulose-1-methylpoly-4-vinylpyridine tetrafluoroborate, Cellulose-1-methylpoly-4-vinylpyridine chloride, Cellulose-1-methylpoly-4-vinylpyridine hexafluorophosphate, Cellulose-1-methylpoly-4-vinylpyridine p-tosylate, Cellulose-1-methylpoly-4-vinylpyridine nitrate, Cellulose-3-methyl poly-1-vinylimidazole perchlorate, Cellulose-3-methylpoly-1-vinylimidazole tetrafluoroborate, Cellulose-3-methylpoly-1-vinylimidazole trifluoroacetate, Cellulose-3-methylpoly-1-vinylimidazole chloride, Cellulose-3-methylpoly-1-vinylimidazole hexafluorophosphate, Cellulose-3-methylpoly-1-vinylimidazole p-tosylate, Cellulose-3-methylpoly-1-vinylimidazole nitrate. Also, with this new technology following new materials based on lignin have been synthesized: Lignin acrylate, Lignin-poly-4-vinylpyridine, Lignin-1-methylpoly-4-vinylpyridine iodide, Lignin-1-methylpoly-4-vinylpyridine trifluoroacetate, Lignin-1-methylpoly-4-vinylpyridine tetrafluoroborate, Lignin-1-methylpoly-4-vinylpyridine perchlorate, Lignin-1-methylpoly-4-vinylpyridine tosylate, Lignin-1-methylpoly-4-vinylpyridine hexafluorophosphate, Lignin-1-methylpoly-4-vinylpyridine chloride, Lignin-1-methylpoly-4-vinylpyridine nitrate; and new materials based on tannin: Tannin acrylate, Tannin- poly-4-vinylpyridine, Tannin-1-methylpoly-4-vinylpyridine iodide, Tannin-1-methylpoly-4-vinylpyridine perchlorate, Tannin-1-methylpoly-4-vinylpyridine tetrafluoroborate, Tannin-1-methylpoly-4-vinylpyridine trifluoroacetate, Tannin-1-methylpoly-4-vinylpyridine chloride, Tannin-1-methylpoly-4-vinylpyridine hexafluorophosphate, Tannin-1-methylpoly-4-vinylpyridine p-tosylate, Tannin-1-methylpoly-4-vinylpyridine nitrate.

4. Thermogravimetric and ion-exchanging properties of new cellulose based materials

4.1 Thermogravimetric analysis of cellulose copolymers grafted with 4-vinyl pyridine in ionic form

Thermal degradation of lignocellulosic material is not simple, but rather very complex process because it takes place through series of complex chemical reactions (Simkovic, 2007). This reaction is largely affected by nature of bonds and period of heating, the atmosphere in which it occurs, inorganic impurities and non-cellulose components (Ibrahim et al., 2011). Grafted cellulose copolymers, which have ion-exchanging qualities, represent materials that are showing different behavior during exposure to heating, depending on the nature of ion species within. Thermal degradation of cellulose, cellulose acrylate and some of its grafted copolymers in ionic form has been studied: Cell-PVP, Cell-1Me-PVPJ,Cell-1-Me-PVP·ClO_4, Cell-1-Me-PVPPF_6, Cell-1-Me-PVPBF_4, Cell-1-Me-PVPCF3COO, Cell-1-Me-PVPp-TsO, Cell-1-Me-PVPCl, Cell-1-Me-PVPNO_3 (Bojanic et al., 1997). Cellulose loses adsorbed and chemisorbed water when heated in the temperature interval from 50^0C to 120^0C. By further heating this absolutely dry cellulose does not change mass until temperature reaches 300^0C, when very slow mass loss begins. This is referred as the first period of thermolysis of cellulose during which its depolymerization ocurs. Sudden loss of weight, the second period of thermolysis between 360 and 425°C, is followed by formation of series of gaseous products. At temperatures above 425°C, there is slow loss in weight, since already charred and probably crosslinked lignocellulosic residue material finishes its degradation and transfers to gaseous phase. This is known as the third period of thermolysis. In reaction between acryloyl chloride and cellulose and by its grafting with 4-vinylpyridine, the chemical composition and the degree of order in structures of the initial cellulose is changing. When heated in the inert atmosphere and after losing adsorbed and chemisorbed water at temperatures above 120°C synthesized cellulose acrylate and Cell-PVP show similar response as the initial cellulose. Perkin Elmer TGS-2 thermogravimetry apparatus was used to determine thermal stability of tested cellulose materials. Experiments were conducted in nitrogen atmosphere at gas flow of 45 cm^3/min, the heating rate of $10^0C/min$, and samples weights were 5 mg. The values for the mass residue were read off ($m/m_o x 100$) at 200, 270, 340, 410, and 4800C, temperatures where the mass loss ($m_o - m/m_o x100$) was 10, 50, 90 and 100 wt. %, and temperature fields in which I , II, and III period of cellulose thermolysis happened. These data for cellulose acrylate and Cell-PVP are shown in Table 5.

Materials	Rest of mass m/m_o%, T^0C					Loss of mass				T^0C, Thermolysis period		
	200	270	340	410	480	10	50	90	100	I	II	III
Cellulose	95	94.5	93	27	8	360	400	425	616	300-360	360-425	425-600
Cell-akrilate	93	90	80	33	25	285	375	580	682	200-340	340-380	380-640
Cell-PVP	94	91.5	73	31	29	290	355	620	739	130-330	330-360	360-730

Table 5. TGA results for cellulose, Cell- acrylate and Cell-PVP

Thermolysis of cellulose and Cell-PVP occurs in three stages after loss of adsorbed and chemisorbed water at temperatures up to 120^0C. The first period of thermolysis starts at lower temperatures than cellulose, at $270\text{-}300^0C$. The second period of thermolysis begins and ends at lower temperatures, $330\text{-}380^0C$, with lower mass loss of 50% than with cellulose: $340\text{-}400^0C$, loss of 80 wt%. Period of slow weight loss up to complete degradation, unlike cellulose in which this is happening in the temperature range of $410\text{-}600^0C$, is extended to $340\text{-}730^0C$. The reduction of thermal stability of cellulose acrylate and Cell-PVP compared to cellulose in the first period of thermolysis is reflected in smaller quantity of non-reacted sample at the same temperature. The second period of thermolysis takes place in more narrow interval $330\text{-}380^0C$ than with cellulose $360\text{-}425^0C$. This is due to greater accessibility of amorphous structures of cellulose derivatives and weaker hydrogen bonds in cellulose macromolecule, and it is assumed to be limiting step of the overall reaction of thermal decomposition. In terms of reaction in the third period, an increased amount of charred cellulose acrylate and significantly slower weight loss can be attributed to crosslinking of the structure, which increases the pyrolytic stability and that can be observed with non-substituted cellulose, as well. In the case of cellulose acrylate additional crosslinking of charred cellulose residue can be expected due to reactions of vinyl groups and their derivatives during thermolysis. Crosslinking of Cell-PVP structure contributes to more difficult diffusion of gaseous degradation products and their retention, i.e. binding with charred residue, hence the weight of the residue is declining even slower and the temperature at which thermolysis was completed is greater, 730^0C. This indicates increased stability of the cellulose after grafting (Barsbay et al. 2007). With non-substituted cellulose, but also in the case of thermolysis of cellulose derivatives in ionic form, limiting step is process of tearing off glycosidic bonds (Simkovic, et al., 1985). Although chemical composition compared to cellulose is altered, observed copolymers show three-step mass loss characteristic for cellulose and Cell-PVP. Cell-1-Me-$PVPCF_3$ and Cell-1-Me-$PVPBF_4$ show deviation from this type of reaction. They have multiple-response. Table 6. shows the results of thermolysis of the observed materials. Data were collected from corresponding TGA-curves.

Materials	Rest of mass m/m_o %					T^0C					Thermolysis period		
	T^0C					Loss of mass							
	130	200	270	340	410	480	10	50	90	100	I	II	III
Cell-1Me-$PVPCF_3COO$	96	95	85	64	31	26	247	350	600	760	143-275	275-340	340-355
Cell-1Me-$PVPBF_4$	93	93	90	71.5	46	39	270	367	633	760	120-325	325-377	377-660
Cell-1Me-PVPCI	95	92	90	66	33	27	270	345	595	650	120-270	270-377	377-660
Cell-1Me-PVPJ	95.3	95	94	82	32	27	287	350	610	710	250-350	350-520	520-710
Cell-1Me-$PVPNO_3$	96.5	92	90	64	33	28	272	344	605	710	140-270	270-380	380-710
Cell-1Me-$PVPPF_6$	95	95	93	54.2	49	44	290	405	740	760	140-270	270-310	310-600
Cell-1Me-PVP_P-T_SO	95	94	91	51	43	37	275	346	625	700	140-270	270-380	380-509
Cell-1Me-$PVPClO_4$	96	95	93	24	22	18	300	305	445	655	130-270	270-35	305-665

Table 6. TGA results of grafted cellulose copolymers in ionic form

Non-reacted residue of all cellulose copolymers in ionic form at 270^0C is 90-94 wt.% and it is similar to Cell-PVP which equals 91wt.%. The exception is very unstable Cell-1-Me-$PVPCF_3COO$ which has smaller non-reacted residue of 85.4wt.%. However, the mass loss of the first 10% of initial air dried cellulose sample takes place at lower temperatures than that of Cell-PVP. Out of these 10wt.%, around 5-7wt.% goes to moisture. The temperature of the 10% mass loss declines in the following manner: Cell-1-Me-$PVPCIO_4$>Cell-PVP>Cell-1-Me-PVPp-TsO >Cell-1-Me-$PVPNO_3$>Cell-1-Me-PVPCI = Cell-1-M-$PVPBF_4$>Cell-1-Me-CF_3COO. This also presents thermal stability of cellulose copolymers in ionic form. All observed samples are more susceptible to degradation during heating then Cell-PVP, except for Cell-1Me-$PVPCIO_4$ and Cell-1Me-$PVPBF_4$, which show increased stability. The first period of thermolysis takes place in a narrower temperature interval than Cell-PVP for all ion forms without exception. The second period of thermolysis with rapidly declining mass is the fastest in case of Cell-1Me-$PVPCIO_4$, which loses 40% of weight at temperatures of 300-305^oC, while other samples are losing mass slower. According to the amount of non-reacted residue at 340^0C they can be ranked in order: Cell-1-Me-BF_4 >Cell-1-Me-PVP CI > Cell-1-Me-PVP CF_3OO = Cell-1-Me-

PVPNO$_3$ > Cell-1-Me-PVPPF6 > Cell-1-Me-PVPp -T$_s$O > Cell - PVP >Cell-1-Me-PVPCIO$_4$. According to the mass of charred residue at this temperature, all observed grafted cellulose copolymers in ionic form, apart from the most unstable Cell-1-Me-PVPCIO$_4$, have higher amount of residue of 51-71wt.%. In comparison, Cell-PVP has non-reacted residue of 31wt.%. The rate of mass loss of Cell-1-Me-PVPPF$_6$, Cell-1-Me-PVPBF$_4$ and Cell-1-Me-PVPp-TsO$_4$ is smaller and of Cell-1-Me-PVPCF$_3$COO, Cell-1-Me-PVPNO$_3$, Cell-1-Me-PVPCl is similar to the one of charred Cell-PVP. The amount of charred residue at 480°C decreases in the following order: Cell-1-Me-PVPPF$_6$ > Cell-1-Me-PVPp-TsO > Cell-PVP > Cell-1-Me-PVPNO$_3$ = Cell-1-Me-PVPCl > Cell-1-Me-PVPCF$_3$COO > Cell-1-Me-PVPClO$_4$. Introduction of ion types in Cell-PVP decreases the thermal stability of new cellulose materials.

4.2 The application of new grafted cellulose and lignin copolymers for the selective sorption of noble metals

The application of synthesized grafted cellulose and lignin copolymers for selective extraction of gold, palladium and platinum from the solution with other metals was done (Bilba et al., 2010; Bojanic et al., 1998c, 2001; Dubiella-Jackowska et al., 2007; Liu et al., 2000; Masllorens et al., 2006; Nastasovic et al., 2006; Navarro et al., 2006; Othman et al., 2005; Sandic & Nastasovic, 2009; Tavlarides et al., 2006). Following copolymers synthesized for this purpose were used: Cell-PVP, Cell-1-Me-PVPJ, Cell-1-Me-PVPCF$_3$COO, Cell-PVIm, Lig-PVP and Tan-PVP. Three series of experiments were conducted and tests were performed at the Institute for Copper-RTB Bor Serbia. The determination of the amount of gold and palladium related to new cellulose and lignin materials has been carried out with Perkin Elmer Company, model 703 atomic absorption spectrophotometer with measurement accuracy of 0,001μgAu/dm^3. In the first series of experiments, the sorption of gold from the solution has been examined on: Cell-PVP, Cell-1Me-PVPJ, Cell-1Me- PVPCF$_3$COO. Gold solution for these tests was obtained by dissolving pure gold in aqua regia. After dissolution of gold, excess of nitric acid was eliminated from the solution by evaporation. HAuCl$_4$ solutions have been made with different content of gold and different pH values by dilution with bidistilled water. All new materials based on cellulose and lignin were treated the same way. Gold solution with concentration of 4.32g/dm^3 and pH=1.55 was used to study sorption of gold in these cellulose copolymers. 0.1g of cellulose copolymers were placed in 3 cm^3 of initial gold solution and the content of gold was measured before the placement, after 1h and 24h. All experiments were performed at room temperature. Since the degree of Au sorption on all the samples exceeded the value of 99 wt.%, sorption was good and it was practically carried out in 1 hour time. The capacity, g Au/g ion-exchanger for all cellulose copolymers, as well as for lignin and tannin copolymers. The solution originating from the industrial processing of anodic sludge from the stage of obtaining gold by the process of electrolytic refining in RTB Bor-Serbia was used. An electrolyte that can still be used, which contains 120g of gold, platinum, palladium, copper and iron/dm^3 was used to determine the capacity of the samples. In this solution, gold is present in the form of HAuCl$_4$, platinum and palladium in the form of H$_2$PtCl$_6$ and H$_2$PdCl$_6$ and the capacity was determined in the same manner as the degree of sorption: Cell -PVP-0.2g Au/1g ion-exchanger, Cell-1Me-PVPJ-0.24g Au/1g ion-exchanger Cell-1Me-PVPCF$_3$COO-0.22g Au/1g ion-exchanger and Cell-PVIm-0.23g Au/1g ion-exchanger Lig-PVP-0.4Au/1g ion-exchanger and 0.2g Pd/1g ion-exchanger. Regeneration of ion-exchanged Lig-PVP with HCl (1:1) was complete. However, characteristics after regeneration were not tested. Sorption of platinum is very good. It was not possible to examine Tan-PVP for the sorption of Au, since it is water-soluble. In the third series of experiments the aim was to examine certain characteristics of Cell-PVP such as capacity, level of sorption, period of sorption and selectivity in details. The solution originated from the industrial production in RTB Bor-Serbia, as in the second series, so Cell-PVP could be used for commercial purposes. New sorption materials for selective extraction of noble metals were applied in a continuous and batch mode. In batch experiments, calculated quantities of grafted copolymer were continuously mixed with the solution from which it was necesary to selectively extract certain noble metals, primarily gold. Sorbent was than separated and the noble metal was eluted in the same way, but with much smaller quantity of hydrochloric acid. Columns with the sorbent were used in a continuous mode. Sorption and desorption were conducted principally in the same manner as in batch experiments. 0.1M solution with gold concentration of 19.25g Au/dm^3 and pH=1.2, as well as 0.1 M solution with palladium concentration of 10.64g Pd/dm^3 and pH=0.28 were used to test the degree of sorption, capacity, period of sorption and selectivity. The results of such tests have been confirmed in the same way as in first and second series of experiments. The regeneration with HCl (1:1) has been studied and it was found that 0.1g of Cell-PVP ion-exchanger sorbed 0.1g of gold, which is twice less than the actual capacity, hence confirming that there has been a decline in its sorption power by 50 wt.%. The selectivity of Cell-PVP in the solution of gold, platinum and palladium has been studied with the presence of copper and iron, and it was shown that there is no sorption of non-noble metals. Grafted cellulose and lignin copolymers have, on one hand, the characteristics of the electron exchangers and, on the other hand, characteristics of complexes with high degree of selectivity. They have been used as new cellulose and lignin materials for noble metal ions containing wastewater treatments for chemical and electronic industries (Alguacil, 1998; Al-Merey et al., 2003; Azarudeen et al., 2009; Chang &Chen, 2006; Farang et al., 2007; Hussain & Khan 2011; Ladhe, 2008; Liu, et al., 2009; Nguyen et al., 2010; O Malley, 2002; Ran et al., 2002; Zhong et al., 2007). Chemical and electrochemical modification of cellulose and lignin is based on synthesis of new materials with 4-vinylpyridine or similar vinyl derivatives of heterocyclic skeleton with at least one nitrogen atom, and they were used as specific chemisorbents of noble metals ions from dissolved aqueous solutions. New technology for the synthesis of new materials based on cellulose and sorption of noble metals is confirmed in industrial production in RTB Copper Institute, Bor-Serbia. Sorption of noble metals is quick and complete. Au, Pd and Pt, as nanoparticles, directly implement into the matrix of biopolymers, and they are producing chelating complexes and ligands by chemisorption with a new structure of biopolymers in this way. Capacities are 20 wt.% higher than standard commercial products of world famous companies. They are outstandingly selective in relation to Fe and Cu and they can sorb very small concentrations of Au, Pd and Pt from the solution which gives them a complete ecological and economic feasibility and the advantage compared to existing technologies. Their special value in fundamental research is that bionanonoblemetals polymers are synthesized for use in nanoelectronics (Bloor et al., 2005, 2006; Burda et al., 2005; Cao et al., 2009; Dai et al., 2006; Feng et al., 2010; Hoppe et al., 2006; Ingrosso et al., 2010; Kim et al., 2007; Lai et al., 2008; Lazzari & Lopez-Quintela, 2003; Li et al., 2010; Liu, T et al., 2003; Liu, P. et al., 2009; Sarkar et al., 2010; Sathishkumar et al., 2009; Thompson, 2007; Yaghi et al., 2003; Yin et al., 2009; Yoon et al., 2008; Zhang et al., 2009; Zubarev et al., 2006) and cancer nanotechnology (Cai et al., 2008; Chen et al., 2008; Daniel & Astruc, 2004; De et al., 2008; Patraet al., 2007; Popovtzer et al., 2008; Xia et al., 2003; Yang et al., 2008; Love et al.,2005).

5. Future researches

For centuries, electrochemistry has played a key role in technologically important areas such as electroplating or corrosion. Electrochemical methods are receiving increasing attention in rapidly growing fields of science and technology, such as nanosciences (nanoelectrochemistry) and life sciences (organic and biological electrochemistry). Characterization, modification and understanding of various electrochemical interfaces or electrochemical processes at the nanoscale level have led to a huge increase of scientific interest to novel technologies. Electrochemical methods carried out at the nanoscale level lead to exciting new science and technology. Hence, trends in electrochemistry are leading to experiments obtaining ever smaller particles in seminano and nanoscale range.

Some applications of cellulose-based materials are either more economically profitable or require its use in the form of conductive composites. Natural polymers based on renewable materials with addition of chosen materials can be directly used as contemporary materials by electrochemical methods (Bojanic et al., 1996, 1998b, 2000). Tailoring new composites within a perspective of sustainable development is applied to more and more materials. Ecological concerns have resulted in a renewed interest in natural, renewable resources-based and compostable materials, and therefore issues such as materials elimination and environmental safety are becoming important. For these reasons, material components such as natural fibers, biodegradable polymers obtained from biomass can be considered as "interesting"–environmentally safe–alternatives for the development of new biodegradable composites (Bojanic, 1994, 2010). These polymers show a large range of properties and at present, they can compete with non-biodegradable polymers in different industrial fields (Bojanic, et al., 1988a, 1998c, 2001). Lignocellulosic feedstocks are composed primarily of carbohydrate polymers (cellulose and hemicellulose) and phenolic polymers (lignin). Lower concentrations of various other compounds, such as proteins, acids, salts and minerals, are also present. Both cellulose and hemicelluloses have favorable properties for potential use in the biomedical area, as they have the ability to pass through the digestive tract unchanged. Owing to their resistance to digestion, they are eligible as potential excipients that could be used in the pharmaceutical industry. Information about numerous existing possibilities of polymers containing dispersed conductive fillers and various methods of manufacture of such materials have been reported widely in the literature (Pinto et al., 2011). Also they found numerous technological applications as self regulating heater, photothermal optical recording, direction finding antennas, chemical detecting sensors used in electronic noses, chemical and electrochemical catalysts and adsorbents . The electrolytic powder production method usually allows products of high purity which can well be pressed and sintered. Besides, in recent years it has been shown that by different electrolysis regimes it is possible not only to obtain powders with a wide range of properties, but to predict the decisive characteristic of powders which are of vital importance for the powder quality and for the appropriate purpose. (Pavlović, M.G. & Popov, K.I., 2005; Popov, K.I. & Pavlović, M.G., 1993).

For metal powder application, a series of their properties are of interest; the size and shape of the particles, the bulk weight, flow rate, the corrosion resistance, the specific surface area, the apparent density and the quality of the sintered products. Finally, the properties mentioned depend on the shape and the size of the particles which can be influenced by electrolysis regimes (Pavlović et al., 1998, 2010; Pavlović, M.G. & Popov,

K.I., 2005; Popov & Pavlović, M.G, 1993). Generally speaking, the larger the powder specific surface the lower its apparent density, and all the more so the smaller the particle size. On the other hand, it seems to be that particle structure has the vital importance on apparent density and on the powder quality. The method most often employed to alter the electrical properties of a polymer is an extrinsic approach whereby the insulating polymer is combined with a conductive additive. The electrical conductivity of polymer composites does not increase continuously with increasing electroconductive filler content. Hence, in general, the percolation theory is used to describe the nonlinear electrical conductivity of extrinsic conductive polymer composites. The conducting additive is incorporated into polymers at levels that allow the composite to maintain its electrically insulative qualities, as well as at higher levels, which allow the composite to become electrically semiconductive. As the volume fraction of the conducting filler particles increases, the particles come into contact with one another to form the conduction paths through the composite. As the result there is a critical composition (percolation threshold) at which the conductivity increases by some orders of magnitude from the insulating range to values in the semiconductive or metallic range (Pinto et al., 2011). The conductivity of filled polymers is strongly dependent on the nature of the contact between the conductive filler elements. Therefore, the copper powder was galvanostatically produced since it has distinct dendritic morphology and large specific area (Pavlović, M.G. et al., 1998, 2010; Pavlović, M.G. & Popov, K.I. 2005, Popov & Pavlović, M.G, 1993). The electrical conductivity measurements as a function of filler content (copper powder-Fig. 5) showed typical *S*-shaped dependency with three distinct regions: dielectric, transition and conductive (Pavlović, M.M et al., 2011).

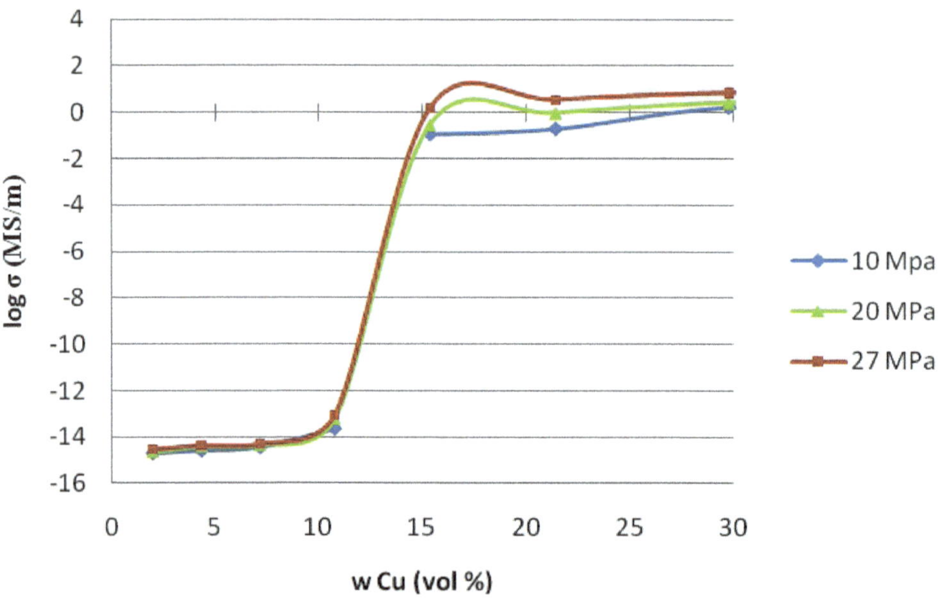

Fig. 5. Variation of electrical conductivity, as a function of filler content, of lignocellulose composites filled with copper powder under different processing pressures (Pavlović, M.M. et al., 2011).

Clearly, the samples with low filler content are practically nonconductive. Naturally, the electrical conductivity of the composites increases with the increase of the conductive filler content. The significant increase of the electrical conductivity can be observed as the copper content reaches the percolation threshold at 14.4% (v/v) for all the processing pressures. The value of the percolation threshold was obtained from of the maximum of the derivative of the conductivity as a function of filler volume fraction (Fig. 5). However, in the conductive region, composites with the same volume fraction of copper powder prepared under higher pressure have higher values of conductivity. Above the percolation threshold, the conductivity of composite increased by much as fourteen orders of magnitude (Pavlović, M.M. et al., 2011). In any case, when considering electroconductive polymer materials, it is desirable to thoroughly examine not only galvanostatically obtained copper powder as filler, but also copper powder obtained by other electrochemical procedures, as well as the effect of different deposition regimes, on the powdered metal electrodeposits morphology in order to obtain nanoscale powders. Above all, electrodeposition of metal powders, suchs as silver, gold, palladium, platinum, zinc, tin, etc., at a periodically changing rate (pulsating overpotential and reversing, and pulsating currents) should be examined. The purpose of this future researches will be to present a possibility of electrodeposition of metal powders with contolled grain size, morphology, and crystal structure of the particles. Morphology is probably the most important property of electrodeposited metals. It depends mainly on the kinetic parameters of the deposition process and the deposition overpotential or current density. In general, they depend on the shape and the size, which can be influenced by appropriate electrolysis regime. The assumption for the silver powder (Dimitrov et al., 1998; Maksimovic et al., 2007; Pavlović, M.G. et al., 1978; Popov et al., 1978, 1991, 1996, 1998; Radmilovic et al., 1998; Strbac et al., 1999) is that all effects which can be obtained by changing parameters determining the deposition regime in direct current deposition conditions can be obtained by changing the shape of the current or overpotential wave only in electrodeposition at a periodically changing rate. Furthermore, at the same time, the surface crystal structure of powder particles can be varied from polycrystalline to one characterized by well-defined crystal planes.Noble metals powder particles obtained in this way used with natural polymers based on renewable materials can be directly used as contemporary materials by electrochemical methods, and they will be able to satisfy different requirements. Such new materials obtained by the new technology are bionanonoblemetals biopolymers with nanoparticles of noble metals, and they can be used in nanoelectronics and cancer nanotechnology which is the subject of current and future researches. This will be of great benefit to many emerging technologies involving molecular electronics, miniature fuel cells, chiral catalysis and biomaterials with hybrid properties.

6. Conclusions

Procedure for the synthesis of cellulose acrylate has been developed and it has been successfully applied for the synthesis of lignin and tannin acrylate. The impact of the relations of the reactants: cellulose, potassium-t-butoxide, acyiloyl chloride and reaction period on the yield of cellulose acrylate has been thoroughly examined. The optimum synthesis conditions under which the cellulose acrylate is obtained with DS 2.4 and Y-80.7 is achieved when the ratio of reactants cellulose/potassium-t-butoxide/acryloyl chloride

= 1:3:10, and the reaction time is 10 hours. Radical copolymerization of synthesized cellulose, lignin and tannin acrylates with 4-vinylpyridine, 1-vinylimidazole, 9-vinylcarbazole and 1-vinyl-2-pyrrolidine yields grafed copolymers and polymers: Cell-PVP, Cell-PVIm, Cell-P9VK, Cell-P1V2P, Lig-PVP, Tan- PVP. In reactions of quaternization of grafted Cell-PVP, Cell-PVIm, Lig-PVP and Tan-PVP copolymers with methyl iodide following polymers were synthesized: Cell-1Me-PVPJ, Cell-3Me-PVImJ, Lig-1Me-PVPJ, Tan-1Me-PVPJ. Cyclic voltammetry method was used to investigate electrochemical behavior of synthesized: Cell-1Me-PVPJ, Cell-3Me-PVImJ, Lig-1-Me-PVPJ and Tan-1Me-PVPJ. Constant current regime electrolysis was used for substitution of iodide anion with anion present in the basic electrolyte from formerly mentioned copolymers. New materials based on cellulose were synthesized: Cell-1-Me-PVPClO_4, Cell-1-Me-PVPCl, Cell-1-Me-PVPCF_3COO, Cell-1-Me-PVPNO_3, Cell-1-Me-PVPp-TsO, Cell-1-Me-PVPBF_4, Cell-1-Me-PVPPF_6, Cell-3-MePVImClO_4, Cell-3-Me-PVImCl, Cell-3-Me-PVImCF_3COO, Cell-3-Me-PVImNO_3, Cell-3-Me-PVImp-TsO, Cell-3-Me-PVImBF_4 and Cell-3-Me-PVImPF_6. In the same manner new materials based on lignin were synthesized: Lig-1-Me-PVPClO_4, Lig-1-Me-PVPCl, Lig-1-Me-PVPCF_3COO, Lig-1-M-PVPNO_3, Lig-1-Me-PVPp-TsO, Lig-1-Me-PVPBF_4 and Lig-1-Me-PVPPF_6 as well as new tannine based materials: Tan-1-Me-PVPClO_4, Tan-1-Me-PVPCl, Tan-1-Me-PVPCF_3COO, Tan-1-Me-PVPNO_3, Tan-1-MePVPpTsO, Tan-1-Me-PVPBF_4 and Tan-1-Me-PVPPF_6. Thermal stability of some synthesed cellulose copolymers in ionic form was tested by thermogravimetry method and it was compared with the stability of pure cellulose acrylate, cellulose, and Cell-PVP. Cellulose copolymers in ionic form have lower thermal stability than pure cellulose and the type of anion present in the copolymer has a decisive influence on their behavior at elevated temperatures. The application of the following synthesized new materials was studied: Cell-PVP, Cell-1Me-PVPJ, Cell-1Me-PVPCF_3COO, Cell-PVIm and Lig-PVP as ion-exchangers for sorption of noble metals from aqueous solutions. These copolymers can selectively extract gold, platinum and palladium from solutions containing copper and iron. The degree of gold sorption from clean solutions for Cell-PVP, Cell-1Me-PVPJ and Cell-1Me-PVPCF_3COO is from 99.03 to 99.89 wt.%, capacity for Au from 0.20 to 0.40 g Au/g ion-exchanger. Capacity for Pd is 0,20g/ Pd g ion-exchanger and sorption of platinum is good. These values are 20 wt. % higher than standard commercial products of world famous companies, which are used to bind ions of noble metals from aqueous solutions. Selectivity of obtained ion-exchangers is of particular iterest from the theoretical and technical point of view. They are completely indifferent to copper and iron ions. New technology has been successfully applied for the synthesis of new materials based on lignin and tannin. It can be used for obtaining acrylate biopolymers. Biopolymers are copolymerized with a 4-vinylpyridine or similar vinyl derivative with heterocyclic skeleton and at least one nitrogen atom. Nitrogen atom is quaternized with methyl iodide and obtained material is electrochemicaly transformed whereby the iodide anion is replaced by another. Synthesized new materials are used as selective sorbents for the extraction of noble metals from diluted aqueous solution, especially gold. Such new materials obtained by the new technology are bionanonoblemetals biopolymers with nanoparticles of noble metals, and they can be used in nanoelectronics and cancer nanotechnology which is the subject of current and future researches. This will be of great benefit to many emerging technologies involving molecular electronics, miniature fuel cells, chiral catalysis and biomaterials with hybrid properties.

7. References

Achilleos, D.S & Vamvakaki, M. (2010). End-Grafted Polymer Chains onto Inorganic Nano-Objects , *Materials,* 3: 1981-2026.

Akelah, A. & Sherrington, D.C.(1981). Syntesis of some vinyl derivatives of cellulose and their grafting copolymerization vith styrene, *Journal of Appllied Polymer Science,* 26 (10): 3377-3384.

Alguacil, F.J. (1998).Recovery of tetrachloroaurate through ion exchange with Dowex 11 resin, *Revista de Metalurgia,* 34: 376-380.

Al-Merey, R., Hariri, Z. & Hilal, A.(2003). Selective seperation of gold from iron ore samples using ion exchange resin, *Microchemical Journal,*75: 169-177.

Anderson, J.L.; Coury, L.A. & Leddy, J. (2000). Dynmic electrochemystry. Methology and application,*Analytical Chemistry,* 72: 4497-4520.

Azarudeen, R.S., Ahamed, M.A.R., Jeyakumar, D, & Burkanudeen, A (2009). An Eco-friendly Synthesis of A Terpolymer Resin. Characterization and Chelation ion-exchanger Property, *Iranian Polymer Journal,* 18 (10): 821-832

Barsbay, M.; Guven, O.; Stenzel, M. H.; Davis, T.P.; Barner-Kowollik, C. & Barner,L. (2007). Verification of Controlled Grafting of Styrene from Cellulose via Radiation-Induced RAFT Polymerization, *Macromolecules,* 40: 7140-7147.

Bhattacharya, A. & Misra, B.N. (2004). Grafting: a versatile means to modify Polymers Techniques, foctors and applications, *Progress in Polymer Science,* 29: 767-814. Bilba,D.; Bejan, D. & Tofan, L.(1998). Chelating Sorbents in Inorganic Chemical Analysis, *Croatica Chemica Acta,* 71 (1): 155-178.

Bilba, D., Panduraru, C. & Tofan,L. (2010). Macroporous Anion Exchanger Purolite A-500 Loaded with Ferron for Palladium (II) Recovery, *Journal of the Iranian Chemical Society,7* (3): 608-614.

Bloor, D., Donnelly, K., Hands, P.J., Laughlin, P. & Lussey, D. (2005). Ametal-polymer composite with unusual properties,*Journal of Physics D-Applied Physics,*38: 2851-2860.

Bloor, D., Graham, A. & Williams, E.J. (2006). Metal-polymer composite with nanostructured filler particles and amplified physical properties,*Applied Physics Letters,*88: 102103.

Bojanic,V.; (1994) Synthesis and electrochemical modification of cellulose and lignine grafted copolymers with 4-vinylpyridine, *Ph.D. Thesis,* University of Beograd.

Bojanic, V.; Jovanovic, S.; Tabakovic, R. & Tabakovic, I. (1996). Synthesis and electrochemistry of grafted of cellulose with 4-vinylpyridine,1-vinylimidazole,1-vinyl-2-pyrrolidone, and 9-vinylcarbazole, *Journal of Applied Polymer Science* 60: 1719-1725.

Bojanic, V.; Jovanovic, S. & Teodorovic,M.; (1997). Thermogravimetric analysis of copolymers of cellulose grafted with 4-vinylpyridine in ionic form, *Chemical industry* 51 (4): 158-161.

Bojanic, V.; Jovanovic, S. & Tabakovic, I. (1998a). The Synthesis of new materials by the chemical modification of cellulose, *Chemical industry,* 52(5): 191-198.

Bojanic, V.; Tabakovic, I. & Jovanovic, S. (1998 b). The synthesis and electrochemistry of copolymers of lignin and 4-vinylpyridine, *Chemical industry,* 52 (7-8): 290-294.

Bojanic, V. & Jovanovic, S. (2000). The sinthesis and electrochemistry of graft polymers of tanin and 4-vinylpyridine, *Chemical industry,* 54 (1): 33-36.

Bojanic, V.; Jovanovic, S. & Jovanovic, L.(2001). The application of new graft copolymers of cellulose and lignin for the selective sorption of gold and palladium, *Chemical industry*, 55 (4):163-166.

Bojanic. V.; Tabakovic, R.; Tabakovic, I.; Dvornic, P.; Govedarica, M. & Dabovic,M.; (1988c). The procedure for obtaining a new graft copolymers based on cellulose and lignin and their application for selective sorption of noble metals, Serbia *Patent* 48 558 B.

Bojanic,V. (2010). Optimization of cellulose acrylate and grafted 4-vinylpyridine and 1-vinylimidaoze synthesis, *Chemical industry*, 64 (6): 529-535.

Bonne, M. J.(2008). Electrochemical Studies of Cellulose Matrices. Absorption, Diffusion, Reactivity Detection, *Ph.D.Thesis*, University of Bath.

Burda, C., Chen, X., Narayanan, R. & El-Sayed, M.A. (2005). Chemistry and Properties of Nanocrystals of Different Shapes, *Chemical Review*, 105 (4): 1025-1102.

Cai, W., Gao, T., Hong, H. & Sun, J. (2008). Applications of gold nanoparticles in cancer nanotechnology, Nanotechnology, Science and Applications, 1: 17-32.

Cao, M., Wang, M. & Gu, N. (2009). Calculated Optical Properties of Dielectric Shell Coated Gold Nanorods, *Chinese Physics Letters*, 26 (4): 045201.

Cao, Y.; Wu, J.; Meng, T.; Zhang, J.; He, J, Li, H. & Zhang, Y. (2007). Acetone- soluble cellulose acetates prepared by one-step homogeneous acetylation of cornhusk cellulose in an ionic liquid 1-allyl-3-methylimidazolium choride (Amim Cl), *Carbohidrate Polymers* 69: 665-672.

Chang, Y-C. & Chen, D-H. (2006).Recovery of Gold 3 ions by a Chitosan-coated Magnetic Nano-adsorbent,*Gold Bulletin*,39 (3): 98-102.

Chauhan, GS. & Lal, H. (2003). Novel grafted cellulose- based hydrogels for water technologies , *Desalination,* 159 (2): 131-138.

Chauhan, GS; Guleria, L. & Lal, H. (2003). Synthesis of graft copolymers of acrylamide and comonomers on to cellulose: A study of the effect of comonomer on polymer yields, structure and prpoperites , *Polymers & Polymer Compozites,* 11 (1): 19-29.

Chauhan, GS; Dhiman, SK; Guleria, LK; Misra, BN. & Kaur, I. (2000). Polymers from renewable resurces: kinetices of 4-vinyl pyridine radiochemical grafting onto cellulose extracted from pine needles,*Radiation Physics and Chemistry*, 58 (2): 181-190.

Chauhan, G.S; Singh, B. & Kumar, S. (2005). Synthesis and characterization of N-vinyl pyrrolidone and cellulosics based functional graft copolymers for use as metal ions and ions and iodine sorbents , *Journal of AppliedPolymer Science*, 98 (1): 373-382.

Chen, P.C., Mwakwari, S.C. & Oyelere, A. (2008). Gold nanoparticles: From nanomedicine to nanosensing,*Nanotechnology,Science and Applications*, 1: 45-66.

Chmielewska, D.K.; Sartovska, B.; Starosta, W & Wolo. M. (2010). Radiation synthesis of silver nano and microparticles in cellulose fibres, *Nukleonika,* 55 (3): 345-349.

Cohen Stuart, M.A.; Husk, W.T.S.; Genzer, J.; Muller. M.; Ober, C.; Stamn, M.; Sukhorukov, G.B.; Szleifer, I.; Tsukruk, V.V.; Urban, M.; Winnik, F.; Zauscher, S.; Luzinov, I. & Minko, S. (2010). Emerging applications of stimuli-rensponsive polymer materials, *Nature Materials* 9: 101-113.

Coskun, M. &Temuz, M.M. (2005). Grafting studies onto cellulose by atom-transfer radical polymerization, *Polymer International*, 54 (2): 342-347

Crini, G. (2005). Recent develompment in polysaccharide -based materials used as adsorbents in wastewater treatment, *Progres in Polymer Science,* 30: 38-70.

Dai, C-A.; Wu,Y-L.; Lee,Y-H.; Chang, C.J. & Su,W-F. (2006). Fabrication of 2D ordered structure ofself-assembled block copolymers containing gold nanoparticles, *Journal of crystal Growth*, 288:128-136.

Daniel, M-C. & Astruc, D. (2004). Gold Nanoparticles.Assembly ,Supramolecular Chemistry, Quantum-Size-Related Properties, and Applications toward Biology, Catalysis and Nanotechnology, *Chemical Reviews*, 104(1): 293-346.

De, M., Ghosh, P.S. & Rotello, V.M. (2008). Applications of Nanoparticles in Biology,*Advanced Materials,20*: 4225-4241.

Dimitrov, A.T., Hadži Jordanov, S., Popov, K.I., Pavlović, M.G. & Radmilović, V. (1998). Electrodeposition of Ag from nitrate solutions. Part I. Effect of phosphate ions on morphology, *Journal of Applied Electrochemistry*, 28: 791.

Dubiella-Jackowska, A., Polkowska, Z. & Namiesnik, J. (2007). Platinum Group Elements:A Challenge for Environmental Analytics, *Polish Journal of Environmental Studies*, 16.(3): 329-345.

Farang, A. B.; Soliman, M.H.; Abdel-Rasoul, O. S. & El-Shahawi, M.S. (2007). Sorption characteristics and chromatographic separation of gold (I and III) from silver and base metals ions using polyurethane foams, *Analytica Chemica Acta*, 601: 218-229.

Feng, L., Gao, G., Huang, P., Wang, K., Wang, X., Luo, T. & Zhang, C. (2010). Optical properties and cataytic activity of bimetallic gold-silver nanoparticles,*Nano Biomedicine and Engineering*, 2 (4): 258-267

Filcento-Olteanu, A., Dobre, T., Radulescu, R., Panturu, E. & Panturu, R. (2010). Selective recovery of gold from hydrochloric solutions with Amberlite XAD-7 ion-exchanger resin,*UPB Scientific Bulletin,Series B*: *Chemistry and Materials Science*,72 (3): 233-242.

Gandini, A. (2008). Polymers from Renewable Resources: A Challenge for the Future of Macromolecular Materials, *Macromolecules*, 41 (24): 9491-9504.

Granja, P.L.; Jeso, B.D.; Bareille, R.; Rouais,F.; Baquey, C. & Barbosa, A. (2006). Cellulose phosphates as biomaterials .In vitro biocompatibility studies,*Reactive Funtional Polymers*, 66: 728-739.

Gunic, E. & Tabakovic, I.; (1988). Electrochemical synthesis of compounds.17. Anodic oxidation of hydrazoues in the presence of heterotomic and Schiff bases, *Journal of Organic Chemistry*, 53: 5081-5087.

Gupta, KC. & Sahoo, S. (2001). Co (III) acetylacetonate-complex-initiated grafting of N-vinyl pyrrolidone on cellulose in aqueous media, *Journal of Applied Polymer Science*, 81 (9): 2286-2296.

Hetnze, T. (1988). New Ionic polymers by cellulose functionalization, *Macromolecular Chemistry and Physics* , 199 (11): 2341-2364.

Hoppe, C.E., Lazzari, M., Pardinas-Blanko, I. & Lopez-Quintela, M.A. (2006). One-Step Synthesis of Gold and Silver Hydrosols Using PolyN-vinyl-2-pyrrolidone as a Reducing Agent, *Langmuir*, 22: 7027-7034.

Hu, G.; Heitmann, J.A. & Rojas, O.J. (2009). In situ Monitoring of cellulase activity by microgravimetry with a quartz crystal microbalance, *Journal of Physical Chemistry*, 113: 14761-14768.

Hubbe, M.A.; Rojas, O.J.; Lucia L.A. & Sain, M. (2008). Cellulosic nanocomposites: a review, *BioResources*. 3 (3): 929-980.

Hubbe, M.A.; Hasan, S.H. & Ducoste, J.J. (2011). Cellulosic supstrates for removal of pollutants from aqueous systems : a review .1.metals, *BioResources*, 6 (2): 2161-2287.

Hubicki, Z., Wawrzkiewicz, M. & Wolowicz, A. (2008). Application of Ion Exchange Methods in Recovery of Pd(II) Ions-a Review,*Chemia Analityczna*,53: 759-78

Hussain, G. & Khan, M. A. (2011). Adsorption of Gold (III) from Aqueous Solutions on Baggase Ash, *Journal of the Chemical Society of Pakistan*, 33 (3): 317-323.

Ibrahim, S.F.; El-Amoudy, E.S. & Shady, K.E. (2011). Thermal Analysis and Characterization of Some Cellulosic Fabrics Dyed a New Natural Dye and Mordanted with Different Mordants ,*International Journal of Chemistry*, 3 (2): 40-54.

Ingrosso, C., Panniello, A.M., Comparelli, R., Curri, M.L. & Srriccoli, M. (2010). Colloidal Inorganic Nanocrystal Based Nanocomposites.Functional Materials for Micro and Nanofabrication,*Materials*,3: 1316-1352.

Jovanovic S.; Stojanovic, Z. & Jeremic, K.(2002). Polymers based on Renewable raw materials, *Chemical industry*, 56 (11): 447-460.

Kalia, S; Kumar, A. & Kaith, B.S. (2011). Sunn hemp cellulose graft copolymers polyhydroxybutyrate composites: morphological and mechanical studies, *Advaced Materials Letters*, 2 (1): 17-25.

Kamel, S. (2007). Nanotechnolgy and its applications in lignocellulosic composites, a mini review, *Express Polymer Letters*, 9: 546-575.

Kaur. J.& Dhiman, P.K. (2011) Sinthesis, Characterization of cellulose graffed N-oxde reagent and its applcation in oxidation of alkyl aryl halides, *International Journal of Organic Chemistry* 1: 6-14.

Kim, B.J., Bang, J., Hawker, C.J., Chiu, J.J., Pine, D.J., Jang, S.G., Yang, S-M. & Kramer,E.J. (2007). Creating Surfactant Nanoparticles for Blok Copolymers thtough Surface Chemistry,*Langmuir*,23: 12693-12703.

Kim, J.; Yun, S.; Mahadeva, S.K.; Yun, K.; Yang, S.Y.; & Maniruzzman, M. (2010), Paper Actuators Made with cellulose and hybrid materials, *Sensors* 10: 1473-1485.

Kononova, N.O., Goryaeva, G.N. & Dychko, V.O. (2009). Ion exchange recovery of palladium(II) from nitrate weak acidic solutions,*Natural Science*,1: 166-175.

Ladhe, A.R. (2008). Surfactant and Metal Sorption Studies by functionalized membranes and quartz crystal microbalance,*Ph.D.Thesis*, University of Kentucky.

Lai, W-H., Su, Y-H., Teoh, L-G. & Hon, M-H. (2008). Commercial and natural dyes as photosensitizer for a water-based dye-sensitized solar cell loaded with gold nanoparticles,*Journal of Photobiology A.Chemistry*,195: 307-313.

Lazzari, M. & Lopez-Quintela, M.A. (2003). Block Copolymers as a Tool for Nanomaterials Fabrication,*Advanced Materials*,15 (19): 1583-1594.

Li, D.; Lee, J.Y & Kim, D.H. (2011). Responsive polymer/gold nanopartilcle composite thin films fabricated by solvent-inducet self-assembly and spin-coating, *Journal of Colloid and Inerface Science* 354: 585-591.

Li, Y., Schluesener, H.J. & Xu,S. (2010). Gold nanoparticle-based biosensors, *Gold Bulletin*,43.(1):29-41.

Liu, T., Burger, C. & Chu,B. (2003). Nanofabrication in polymer matrices, *Progress in Polymer Science,28*: 5-26.

Liu, P.; Liu, G.F.; Chen, D.L.; Cheng, S.;Y. & Tang, N. (2009). Adsorption properties of Ag (I), Au (III), Pd (II) and Pt (IV) ions on commercia 1 717 anion-exchanger resin, *Transactions of Nonferrous Metals Society of China*, 19: 1509-1513.

Liu, P,. Pu, Q. & Su, Z. (2000). Synthesis of silica gel immobilized thiourea and its application to the on-line preconcentration and separation of silver,gold and palladium, *Analyst*, 125: 147-150.

Liu, X-Y., Cheng, F., Liu, Y., Liu, H-J. & Chen,Y. (2009). Preparation and Characterization of novel thermorensponsive gold nanoparticles and their responsive catalysis properties, *Journal of Materials Chemistry*,20: 360-368.

Love,J.C.;Estroff,L.A.;Kriebel,J.K.;Nuzzo,R.G.&Whitesides,G.M.(2005).Self-Assembled Monolayers of Thiolates on Metals as a Form of Nanotechnology,*Chemical Reviews*,105 (4):1103-1169.

Lu, J.Z., Duan, X., Wu, Q. & Lian, K. (2008). Chelating efficiency and thermal ,mechanical and decay resistance performances wood-polymer composites,*Bioresourse Technology*, 99:5906-5914.

Maliyekkal, S.M., Lisha, K.P. & Pradeep,T. (2010). A novel cellulose-manganese oxide hybrid material by in situ soft chemical synthesis and its application for the removal of Pb(II) from water,*Journal of Hazardous Materials*,181: 986-995.

Maksimović, V.M., Pavlović, M.G., Pavlović, Lj.J., Tomić, M.V. & Jović, V.D. (2007). Morphology and growth of electrodeposited silver powder particles, *Hydrometallurgy*, 86: 22-26.

Masllorens, J., Roglans, A., Antico, E. & Fontas,C. (2006). New applications of azamacrocyclic ligands in ion recognition,transport and preconcentration of palladium, *Analytica Chimica Acta*,560: 77-83.

Nada, A. M. A.; Alkady, M.;Y. & Fekry, H. M. (2007). Synthesis and characterization of grafted celluose for use in water and metal ions sorption, *BioResources*, 3 (1): 46-59.

Nastasovic, A.B.; Jovanovic, S.M.; Onjia, A. E.; Sandic, Z.P.; Malovic, Lj.T.; Jakovljevic, D.M.&Vukovic, Z.M. (2006). The apllication of macroporous copolymers in the sorption of heavy and precious metals from aqueous solutions, *Chemical inustry*, 60 (11-12): 306-310.

Navarro, P., Vargas, C., Alonso, M. & Alguacil, F.J. (2006). The Adsorption of Gold on Activated Carbon from Thiosulfate-Ammonical Solutions, *Gold Bulletin*, 39 3.93-97.

Nguyen, N.V.; Lee, J.C.; Kim,; S.K., Jha, M.K., Chung, K.S. & Jeong, J. (2010). Adsorption of gold (III) from waste rinse water of semiconductor manufacturing industries using Amberlite XAD-7HP resin, *Gold Bulletin*, 43 (3): 200-208.

O Malley, G.P. (2002). Recovery of Gold from Thiosulfate Solutions and Pulps with Anion-exchanger Resin, *Ph.D.Thesis*, Murdoch University.

Othman, N., Goto, M. & Mat, H. (2005). Solvent extraction of metals from liquid photographic waste using acidic extractants, *Journal Teknologi*, 42 (F): 25-34.

Ozdemir, S.S., Buonomenna, M.G. & Drioli, E. (2006). Catalytic polymeric membranes: Preparation and application,*Applied Catalysis A.General*,307: 167-183.

Patra, H.K., Banerjee, S., Chaudhuri, U., Lahiri, P. & Dasgupta, A.K. (2007). Cell selective response to gold nanoparticles, *Nanomedicine. Nanotechnology, Biology and Medicine*,3:111-119.

Panyala, N.R., Pena-Mendez, E.M. & Havel, J. (2009). Gold and nano-gold in medicine. Overview, toxicology and perspectives, *Journal of Applied Biomedicine*, 7:75-91.

Parajuli, D. (2006). Development of some novel lignin derivatives for adsorptive removal of heavy metals and of precious metals, *Ph.D. Thesis*, Saga University.

Pavlović, M.M., Ćosović, V., Pavlović, M.G., Talijan, N. & Bojanić, V. (2011). Electrical conductivity of lignocellulose composites loaded with electrodeposited copper powders", *International Journal of Electrochemical Science*, 6: 3812-3829.

Pavlović, M.G., Maksimović, M. D., Popov, K.I. & Kršul, M.B. (1978). The effect of pulsating overpotential on the morphology of electrodeposited silver powder particles, *Journal of Applied Electrochemistry*, 7: 61.

Pavlović, M.G., Pavlović, Lj.J., Maksimović, V.M., Nikolić, N.D. & Popov, K.I. (2010). Characterization and morphology of copper powder particles as a function of different electrolytic regimes, *International Journal of Electrochemical Science*, 5: 1862-187.

Pavlović, M.G. & Popov, K.I. (2005). Metal Powder Production by Electrolysis, *Electrochemistry Encyclopedia*, http://electrochem.cwru.edu/ed/encycl/.

Pavlović, M.G., Popov, K.I. & Stojilković, E.R. (1998). The effect of different deposition conditions on the morphology and grain size of electrodeposited metal powder, *Bulletin of Electrochemistry*, 14: 211-217.

Petrovic, Z.S.; Milic,J.; Xu,J. & Cvetkovic, I. (2010). A Chemical Route to High Molecular Weight Vegetable Oil-Based Polyhydroxyalkanoate, *Macromolecules*, 43: 4120-4125.

Pinto, G. & Maaroufi, A. K., (2005). Conducting Polymer Composites of Zinc-Filled Urea-Formaldehyde, *Journal of Applied Polymer Science*, 96: 2011-2015.

Pinto, G., Maaroufi, A.K., Benavente, R. & Perena, J.M. (2011). Electrical Conductivity of Urea–Formaldehyde–Cellulose Composites Loaded with Copper, *Polymymer Composites.*, 32: 19.

Popov, K.I. & Pavlović, M.G. (1993). In *Modern Aspects of Electrochemistry,* Electrodeposition of metal powders with controlled particle grain size and morphology, B.E. Conway, J.O'M. Bockris and R.E. White, Eds.,Vol. 24, Plenum, NewYork, 299–391.

Popov, K.I., Pavlović, M.G. & Maksimović, M.D. (1978). The mechanism of formation of a surface film of silver on a platinum electrode in galvanostatic deposition, *Journal of Applied Electrochemistry*, 8: 531.

Popov, K.I., Pavlović, M.G., Grgur, B.N., Dimitrov, A.T. & Hadži Jordanov, S. (1998). "Electrodeposition of Ag from nitrate solution. Part II. Mechanism of the effect of phosphate ions", *Journal of Applied Electrochemistry*, 28: 797.

Popov, K.I., Pavlović, M.G., Mitrović, B.A. & Toperić, B.V. (1991). Electrodeposition of silver powder by pulsating and reversing currents, *Journal of Applied Electrochemistry*, 21: 50.

Popov, K.I. Pavlović, M.G., Stojilković, E.R. & Radmilović,V. (1996). Silver powder electrodeposition by constant and pulsating overpotential, *Journal of the Serbian Chemical Society*, 61 (1): 47.

Popovtzer, R., Agrawal, A., Kotov, N.A., Popovtzer, A., Balter, J., Carey, T.E. & Kopelman, R. (2008). Targeted gold nanoparticles esnable molecular CT imaging of cancer, *Nano Letters,8* (12): 4593-4596.

Puoci, F; Iemma, F, Spizzirri, U.G, Cirillo, G, Curcio, M. & Picci, N. (2008). Polymer in Agriculture. a review, *American Journal of Agricultural and Biological Sciences*, 3(1): 299-314.

Radmilović, V., Popov, K.I., Pavlović, M.G., Dimitrov, A. & Hadži Jordanov, S. (1998). The mechanism of silver granular electrodeposits formation, *Journal of Solid State Electrochemistry*, 2 (3) : 162 .

Ran, Y.; Fu, J.; Rate, A.W. & Gilkes, R. J. (2002). Adsorption of Au (I, III) complexes on Fe, Mn oxides and humic acid,*Chemical Geology*, 185: 33-49.

Rodriguez, F.; Castillo-Ortega, M.M.; Encinas, J.C.; Sanchez-Corrales, V.M.; Perez-Tello, M. & Munive, T. G. (2009). Adsorption of a Gold-Iodide Complex AuI_2 onto Cellulose

Acetate-Polyaniline Membranes. Equilibrium Exsperiments, *Journal of Applied Polymer Science*, 113: 2670-2674.

Roy, D; Semsarilar, M; Guthrie, J.T. & Perrier, S. (2009). Cellulose modification by polymers grafting: a review, *Chemicol Society Reviews*, 38: 2046-2064.

Saarinen, T. (2008).Adsorption studies of polyelectrolytes and enzymes on lignocellulosic model surfaces,*Ph.D.Thesis*,Helsinki University of Technology.

Sarkar, P., Bhui, D.K., Bar, H., Sahoo, G.P., Samanta, S., Pyne, S. & Mirsa, A (2010). Aqueous-Phase Synthesis of Silver Nanodiscs and Nanorods in Methyl Cellulose Matrix: Photophysical Study and Simulation of UV-Vis Extinction Spectra Using DDA Method,*Nanoscale Researh Letters*,5: 1611-1618.

Sandic, Z.P. & Nastasovic, A.B. (2009). Functionalized macroporous copolymer of glycidyl methacrylate: The type of ligand and porosity parameters influence on Cu(II) ions sorption from aqueous solutions, *Chemical industry*, 63 (3):269-273.

Sathishkumar, M., Sneha, K., Kwak, I.S., Mao, J., Tripathy, S.J. & Yun, Y-S. (2009). Phyto-crystallization of palladium through reduction process using Cinnamom zeylanicum bark extract, *Journal of Hazardous Materials*, 171: 400-404.

Sharma, R.K. & Chauhan,G.S. (2009). Synthesis and characterization of 2-hydroxyethyl methacrylate and some comonomers onto exstraced cellulose for use in separation Technologies ,*BioResources*,4 (3): 986-1005.

Sharma, R. K.; Singh, J. & Chauhan, G. S. (2010). Study of Immobilization of protease and sorption of bsa on cellulose ,cellulose derivatives, and graft copolymers, *BioResources*, 5 (4): 2547-2555.

Schwanninger, M., Rodrigues, J.C., Pereira, H. & Hinterstoisser, B. (2004). Effect of short-time vibratory ball milling on the shape of FT-IR spectra of wood and cellulose, *Vibrational Spectroscopy, 36*: 23-40.

Simkovic, I.; Antal, M.; Mihalov,V.; Konigstein J. & Micko, M. M. (1985). Influence of anionic form on thermal degradation of TMAHP-cellulose, *Journal of Applied Polymer Science*, 30:4707-4711.

Simkovic, I. (2007). Trends in Thermal Stability Study of Chemically Modified Lignocellulose Materials, *Nova Science Publisers, Inc.;* ISBN.987-1-60021-827-9,217-236.

Spiridon, I.; Teaca, C.A. & Bodirlau, R. (2011). Structural changes evidenced by FTIR-spectroscopy in cellulosic materials affer pre-treatment with ionic liquid and enzymatic hydrolysis, *BioResources* 6 (1) : 400-413.

Srivastava, S.; Haridas, M. & Basu, J. K. (2008). Optical properties of polymer nanocomposites, *Bulletin of Materials Science*, 31 (3): 213-217.

Štrbac, S., Rakočević, Z., Popov, K.I., Pavlović, M.G. & Petrović, R. (1998). The role of surface defects in HOPG on the electrochemical and physical deposition of Ag, *Journal of the Serbian Chemical Society*, 64 (7-8) : 483 .

Tabakovic, R.; Gunic, E.; Tabakovic, I. & Zupan, M. (1992). Electrochemical synthesis of new polymeric reagents,*Electrochimica Acta*, 37 (4): 751-75.

Tabakovic, R. & Tabakovic, I.(1999). Catalysis of Nifuroxazide Formation by Cross-Linked Poly viylpyrdine Supported acids, *Reaktive Functional Polymers*, 39 (3): 263-268.

Tavlarides, L.L., Lee, J.S., Nam, K.H. & Agarwal, N. (2006). Sol-Gel Synthesized Adsorbents for Metal Separation,*Tsinghua and Technology*, 11(2): 233-240.

Thomson, D. (2007). Michael Faradays Recognition of Ruby Gold: The Birth of Modern Nanotechnology, *Gold Bulletin*, 40 (4): 267-269.

Vainio, U. (2007). Characterisation of Cellulose-and Lignin-based materials using X-ray scattering,*Ph.D.Thesis,*University of Helsinki.

Vlasankova, R. & Somme ,L.(1999). Solid Phase Extraction and Preconcentration for the Determination of Trace Amounts of Patinum Group Metals in Environmental and Biotic Material A Critical Review, *Chemical Papers,* 53 (3): 200-209.

Vitz, J.; Yevlampieva, N.P.; Rjumtsev, E.& Schubert, U.S. (2010). Cellulose molecular properties in 1-alkyle-3-methylimidazolium-based ionic liquid mixfures with pyridine, *Carbohydrate Polymers* ,82: 1046-1053.

Wang, W. (2005). A study on the adsorption properties of quaternized cellulose, *Ph.D.Thesis,* Auburn University.

Wang, B. (2008). Dispersion of Cellulose nanifibres in biopolymer based nanocomposites, *Ph.D.Thesis,*University of Toronto.

Wu, Q., Henriksson, M., Liu, X. & Berglund, L.A. (2007). A High Strenght Nanocomposite based on Microcrystalline Cellulose and Polyurethane, *Biomacromolecules,* 8: 3687-3692.

Xia, B.Y., Yang, P., Sun, Y., Wu, Y., Mayers, B., Gates, B., Yin, Y., Kim, F. & Yan, H. (2003). One-Dimensional Nanostructures. Synthesis, Characterization and Applications, *Advanced Materials,*15 (5) 353-389.

Xin, T.T.; Yuan, T.; Xiao, S. & He, J. (2011). Synthesis of cellulose-graft-(poly methylmetacrylate) via homogeneous Atrp, *BioResources,* 6 (3): 2941-2953.

Xu, T. (2005). Ion exchange membranes. State of their development and perspective,*Journal of Membrane Science,* 263: 1-29.

Yaghi, O.M., O Keeffe, M., Ockwig, N.W., Chae, H.K., Eddaoudi, M. & Kim, J. (2003). Reticular synthesis and the desing of new materials,*Nature,*423: 705-714.

Yang, J., Eom, K., Lim, E-K., Park, J., Kang, Y., Yoon, D.S., Na, S., Koh, E.K., Suh, J-S., Huh,Y-M., Kwon, T.Y. & Haam, S. (2008). *In situ* Detection of Live Cancer Cells by Using Bioprobes Based on Au Nanoparticles,*Langmuir,*24: 12112-12115.

Yin, G., Xue, W., Chen, F. & Fan, X. (2009). Self-reparing and superhydrophobic film of gold nanoparticles and fullerene pyridyl derivative based on the self-assembly approach, *Colloids and Surfaces A: Physicochemical and Engineering Aspects,* 340: 121-125.

Yoon, H., Choi, M., Lee, J.K. & Jang, J. (2008). Versatile Strategies for Fabricating Polymer Nanomaterials with Controled Size and Morphology, *Macromolecular Research,* 16 (2): 85-102

Zhang, X., Guo, Q. & Cui, D. (2009). Recent Advances in Nanotechnology Applied to Biosensors, *Sensors,*9: 1033-1053.

Zhang, B.T.; Wang, W.; Zhang, D.; Zhang, X.; Ma,Y.; Zhou,Y. & Qi, L. (2010). Biotemplated Synthesis of Gold Nanoparticle -Bacteria Cellulose Nanofiber Nanocomposites and Their Application in Biosensing, *Advanced Functional Materials,* 20: 1152-1160.

Zhong, H., Wang, S., Qiu, Y. & Wang, A. (2007). Synthesis of Chelating Resin PETU and Its Adsorption to Ag 1 ,*The Chinese Journal of Process Engineering,* 7 (4): 689-693.

Zhou, C.; Wu,Q.; Yue,Y. & Zhang, Q. (2011). Apllication of rod-shaped cellulose nanocristals in polyacrylamide hydrogels, *Journal of Colloid and Interface Science,* 353: 116-123.

Zubarev, E.R., Xu, J., Sayyad, A.& Gibson, J.D. (2006). Amphiphilic Gold Nanoparticles with Shaped Arms, *Journal of the American Chemical Society,* 128 (15): 4958-4995.

9

Atomic Layer Deposition of Noble Metals – New Developments in Nanostructured Catalysts

Junling Lu, Yu Lei and Jeffrey W. Elam*
Argonne National Laboratory, Argonne, IL USA

1. Introduction

This chapter will review recent progress in the use of atomic layer deposition (ALD) to prepare noble metal nanoparticle catalysts in which the nanoparticle size, composition, and local environment can be tailored on the atomic scale to tune the reactivity, selectivity, and thermal stability of the catalysts. ALD is a thin film growth technique that uses alternating cycles of self-limiting chemical reactions between gaseous precursors and a solid surface to deposit material in an atomic layer-by-layer fashion[1]. By combining ALD processes for metal oxides, noble metals, and other materials relevant to catalysis, it is possible to engineer nanostructured catalysts with unique properties by depositing a sequence of discrete layers or particles which each perform a specific function. Although the idea of using ALD to prepare catalysts is not new[2], recent advances in ALD technology, coupled with innovative ideas for nanofabrication, have rekindled this field and now offer potential solutions to long-standing problems in catalyst synthesis.

We begin with a brief introduction to ALD. A broad variety of materials can be deposited by ALD using alternating exposures of two chemicals. For example, aluminum oxide (Al_2O_3) ALD uses sequential exposures to trimethyl aluminum (TMA) and H_2O as illustrated in Fig. 1[3]. In reaction A, the substrate to be coated is exposed to TMA vapor which reacts with surface hydroxyl groups (OH) to form new O-Al linkages and liberate methane (CH_4) into the gas phase. This reaction proceeds until all of the OH groups are consumed at which point the reaction terminates because the TMA is inert towards the methyl-terminated surface. In reaction B, H_2O vapor is introduced which reacts with the surface methyls producing CH_4 gas and again forming new Al-O linkages. Reaction B stops when all of the methyl groups are gone because the H_2O does not react with surface hydroxyls. The net result of one A/B cycle is to deposit a monolayer of Al_2O_3 on the surface and to regenerate the original starting surface so that the process can be repeated.

ALD reactors typically incorporate a chamber heated to ~100-400°C containing the substrate to be coated[4]. A constant stream of inert gas such as N_2 flows through the chamber at a pressure of ~0.1-10 Torr. The precursor vapors are injected into the carrier gas using pulsed valves and transported to the substrate where they react, and any excess precursor and

* Corresponding Author

byproduct gases are swept away. Desirable characteristics for ALD precursors include high volatility, good thermal stability at the ALD growth temperatures, and high reactivity with the other compound used for the film growth. A purge period is introduced between each precursor dose to prevent mixing of the chemicals which would cause non self-limited growth. Each ALD cycle can last seconds to hours depending on factors such as the reactivity of the precursors and the surface area and porosity of the substrate. ALD research and process development are greatly facilitated using in-situ measurements. For instance, an in-situ quartz crystal microbalance (QCM) with sub-monolayer sensitivity and 10-20 Hz update rate, is a valuable tool for monitoring film growth in real time, and can also provide details for understanding the ALD surface chemistry. Similarly, an in situ quadrupole mass spectrometer (QMS) installed in the reactor can help to understand the growth mechanism by identifying the gas phase products of the ALD surface reactions.

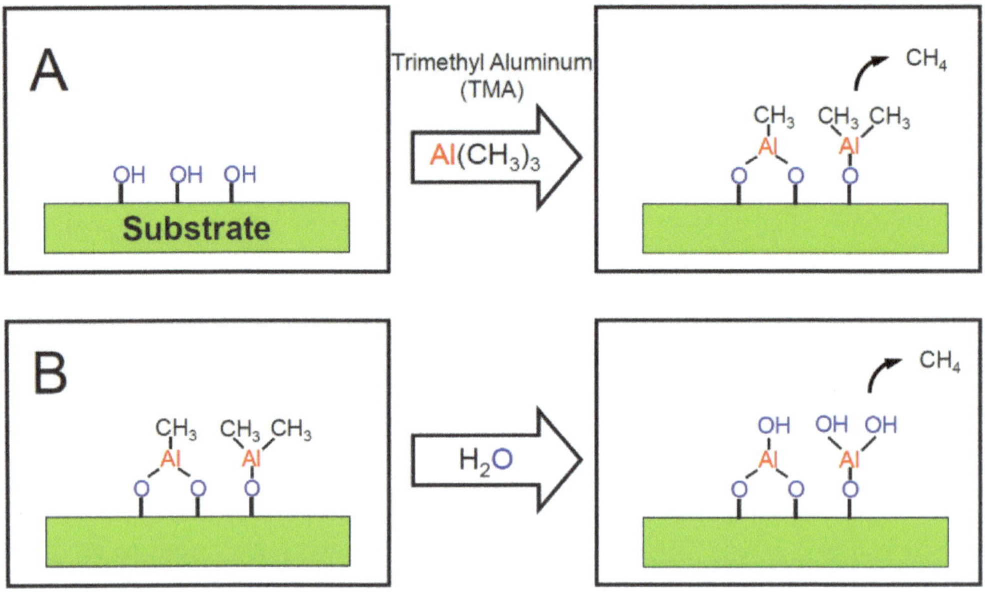

Fig. 1. Schematic illustration of surface chemistry for Al_2O_3 ALD.

One of the hallmarks of an ALD process is that the reactions saturate after a certain minimum exposure, after which additional precursor flux does not deposit more material. For instance, Fig. 2a shows that the Al_2O_3 ALD saturates at 1.2 Å/cycle following a TMA exposure of ~0.05 Torr s. A natural consequence of this self-limiting behavior is that the film thickness increases by a fixed amount with each A/B cycle. This behavior is displayed in Fig. 2b where the ALD Al_2O_3 thickness was measured by ellipsometry over a wide range of A/B cycles and the growth is exceptionally linear[5]. A second consequence of the self-terminating surface chemistry is that the precursor molecules can diffuse into narrow pores until they encounter an empty site, and in this way nanoporous materials can be coated with great uniformity. This capability is extremely useful for catalyst synthesis[6].

Figure 2b demonstrates that the Al_2O_3 ALD growth rate does not change with the number of cycles, and in particular the earliest cycles deposit virtually the same amount of Al_2O_3 as the later cycles. This behavior is typical of most ALD metal oxides and results from the prompt nucleation or initiation of metal oxide ALD on most surfaces (due to the ubiquity of OH groups), and usually generates very uniform, continuous films. In contrast, the ALD of noble metals such as palladium often shows a long incubation period on oxide surfaces[7]. This nucleation delay stems from the difficulty in removing the ligands from the noble metal precursor bound to an oxide surface. This poor nucleation, coupled with the natural tendency for noble metal atoms to diffuse and agglomerate, results in the formation of discrete metal islands which grow with increasing numbers of ALD cycles until finally converging to form a continuous film. Although this tendency is undesirable for microelectronics applications, it is ideal for preparing catalytic nanoparticles which must be highly dispersed and of nanometer size. Smaller nanoparticles are typically more active on a per-atom basis because the surface area to volume ratio increases with decreasing thickness. Moreover, the lower radius of curvature for smaller nanoparticles produces a larger fraction of under-coordinated surface atoms, which often have the highest catalytic activity. For these reasons, the following sections will highlight efforts to control the nanoparticle size, and in particular to achieve ultra-small (<1nm) noble metal nanoparticles.

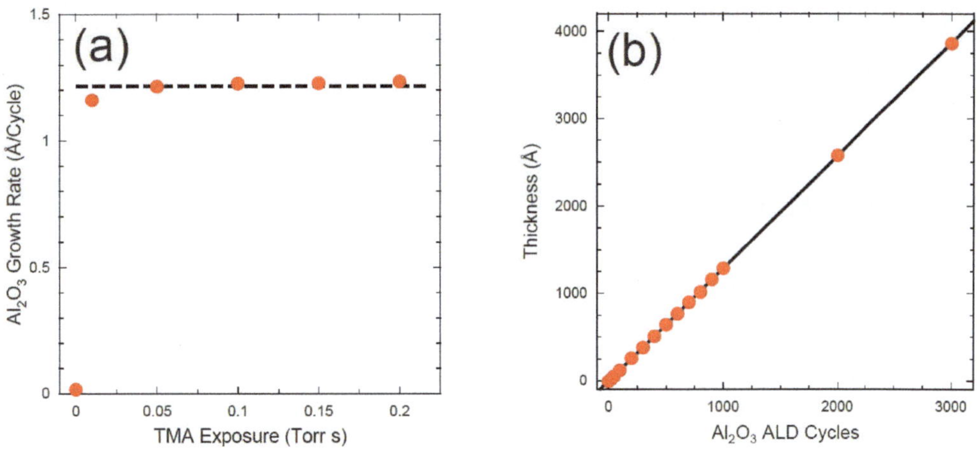

Fig. 2. (a) Saturation of ALD Al_2O_3 growth rate versus TMA exposure; (b) Linear growth of ALD Al_2O_3 versus number of Al_2O_3 ALD cycles.

In addition to depositing simple binary compounds such as Al_2O_3 and elements such as Pd, ALD can also synthesize ternary and more complex compounds such as $ZnAl_2O_4$[8], as well as mixed metals such as Pt-Ir[9]. The ALD of mixed materials is accomplished by alternating between the binary reaction sequences of the individual components in the mixture, and the composition is controlled by adjusting the ratio of ALD cycles for the different components. The ability to control the composition of a metal oxide support layer allows properties such as the surface acidity to be adjusted in a manner analogous to the Al/Si ratio in zeolites, and mixed-metal nanoparticles such as Pt-Ru often exhibit superior catalytic properties.

The capability to produce both thin, continuous layers as well as discrete nanoparticles makes it possible to engineer an "ideal catalyst" using a sequence of ALD processes as illustrated in Fig. 3. Starting from a substrate with the desired porosity, surface area, and nanostructure, the support layer is first applied by ALD in the form of a physically thin but chemically thick layer (~1 nm). Next, catalytic nanoparticles are deposited by ALD on top of the support layer with the desired size, composition, dispersion, and nanostructure. A final ALD process can be used to stabilize the nanoparticles against sintering, or to modify their catalytic behavior. The following sections will provide examples for all of the steps in this synthesis scheme. In particular, we will describe ALD of the noble metals Pt, Pd, and Ir, and the mixed-noble metals Pt-Ir, Pt-Ru, and Pt-Pd. We will show the effects of different ALD support layers (Al_2O_3, TiO_2, ZrO_2, and ZnO) as well as the influence of ALD over-layers on the stability and reactivity of catalytic nanoparticles.

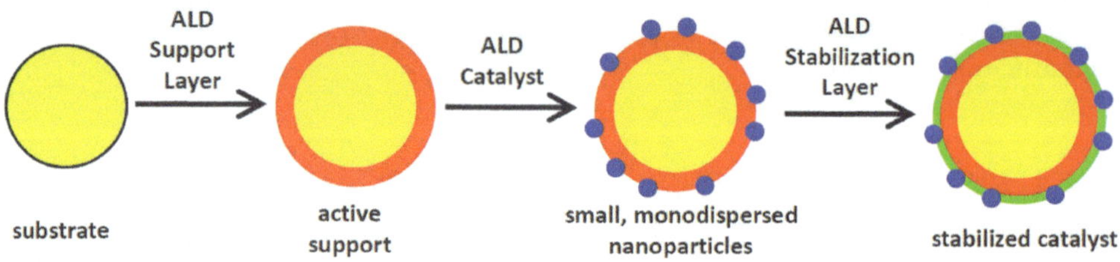

Fig. 3. Schematic illustration of catalyst synthesis by ALD.

2. Palladium ALD

Palladium (Pd) is a common metal for heterogeneous catalysis with applications in methanol partial oxidation and the water-gas shift reaction. Palladium ALD can be accomplished using Pd(II) hexafluoroacetylacetonate ($Pd(hfac)_2$) which has a relatively high vapor pressure compared to other β-diketonate derivatives of Pd.[10] Senkevich et al. demonstrated Pd ALD at 80 °C on Ir metal surfaces, where atomic hydrogen, formed from the dissociation of H_2 on the Ir surface, served as the reducing reagent.[11] To increase the substrate generality, Ten Eyck et al. used a remote hydrogen plasma for Pd ALD on Ir, W, and Si surfaces.[12] However, plasmas are not well suited to catalyst synthesis because they require line-of-site between the plasma and the substrate surface which is impractical for nanoporous catalyst templates. The lack of a suitable reducing reagent for Pd ALD on porous oxide surfaces has limited the development of ALD Pd catalysts.

This problem was solved by using formalin as the reducing reagent.[7] Pd ALD was investigated using amorphous ALD Al_2O_3 as the support material. The reaction mechanism during the first Pd ALD cycle was suggested to be:

$$\text{Al-OH}^* + \text{Pd(hfac)}_2 \rightarrow \text{Al-O-Pd(hfac)}^* + \text{hfacH} \quad (1)$$

$$\text{Al-O-Pd(hfac)}^* + \text{HCOH} \rightarrow \text{Al-O-Pd-H}_x^* + \text{hfacH} + \text{CO} + 0.5(1\text{-}x)\text{H}_2 \quad (2)$$

Here hfacH represents hexafluoroacetylacetone, and the asterisks designate surface species. In this mechanism, the $Pd(hfac)_2$ precursor first reacts with the surface hydroxyl and is then reduced by atomic hydrogen produced from HCOH. The formation of Pd nanoparticles could occur during reaction 1, 2, or both. *In situ* quartz crystal microbalance (QCM), measurements revealed a nucleation period of slower Pd growth on the Al_2O_3 surface before reaching a steady-state Pd growth rate of 0.2 Å/cycle at 200 °C. The Pd nucleation could be accelerated using larger initial $Pd(hfac)_2$ and formalin exposures (Fig. 4). (Contrast the delayed Pd nucleation in Fig. 4 with the prompt Al_2O_3 nucleation in Fig. 2b). The slow nucleation period was believed to result from slower kinetics for hfac removal by HCOH on the oxide surface, or from surface poisoning by readsorption of the hfacH product on the Al_2O_3 surface. The later hypothesis was supported by X-ray photoelectron spectroscopy studies that identified residual fluorine on the surface of thin ALD Pd films. George et al. suggested $Pd(hfac)_2$ adsorbs on hydroxylated Al_2O_3 surfaces primarily through dissociative adsorption based on only a slight loss of hydroxyl coverage during $Pd(hfac)_2$ absorption as observed by *in situ* Fourier transform infrared (FTIR) spectroscopy.[13] Surface poisoning by adsorbed hfacH was found to be much less pronounced on TiO_2 and ZnO surfaces as evidenced by the much shorter nucleation delay of Pd ALD on these surfaces.[14]

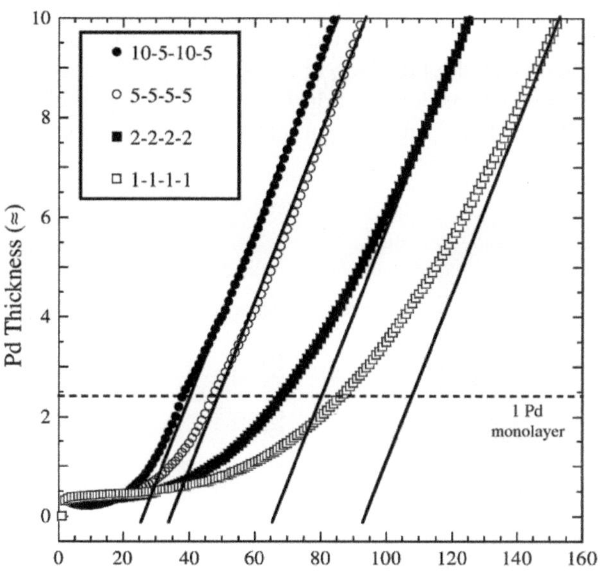

Fig. 4. Nucleation of Pd on Al_2O_3 using different Pd ALD timing sequences measured by *in situ* QCM. The timing sequence numbers designate the Pd precursor dose time, Pd precursor purge time, HCOH dose time, and HCOH purge time, respectively, in seconds. [7]

2.1 Control of ALD Pd particle size

In many heterogeneous and electro-catalytic reactions, particle size plays a major role in determining catalyst performance.[15] Therefore, a synthesis method which can precisely tune the size of noble metal nanoparticles while maintaining a narrow size distribution can improve the catalytic performance significantly. ALD provides fine control over noble metal nanoparticle size. For instance, the size of Pd nanoparticles on an Al_2O_3 support was found to vary linearly between 1.1 (± 0.5) nm and 2.9 (± 0.9) nm as the number of Pd ALD cycles was increased from 1 to 25.[14a]

The size of noble metal nanoparticles often depends on the metal loading since the nanoparticles form through diffusion which is influenced by the inter-particle spacing. Clearly the Pd loading can be controlled by the number of ALD Pd cycles, but the self-limiting and site-selective nature of ALD provides additional means for controlling the loading. Equation 1 implies that the density of surface hydroxyls dictates the number of adsorbed $Pd(hfac)_2$ precursor molecules during Pd ALD. Therefore, decreasing the surface hydroxyl density should reduce the Pd loading. In agreement with this idea, when the hydroxyl density on an ALD Al_2O_3 surface was reduced prior to the Pd ALD through thermal treatment, reaction with alcohols to form surface alkoxides, or reaction with TMA, the Pd loading decreased[16].

Temperature can also affect nanoparticle size in ALD. *In situ* QCM measurements demonstrated that $Pd(hfac)_2$ chemisorption on oxide surfaces can occur at temperatures as low as 80°C, far below the minimum temperature of 200°C required to strip the remaining hfac ligands using HCOH .[14a, 17] This finding suggested that one route to smaller Pd nanoparticles is to chemisorb the $Pd(hfac)_2$ at lower temperatures to reduce agglomeration. Indeed, low-temperature deposition of $Pd(hfac)_2$ yielded a particle size of 0.8±0.2 nm, significantly below the 1.4±0.4 nm size obtained using the standard 200°C $Pd(hfac)_2$ adsorption (Fig. 5).[16]

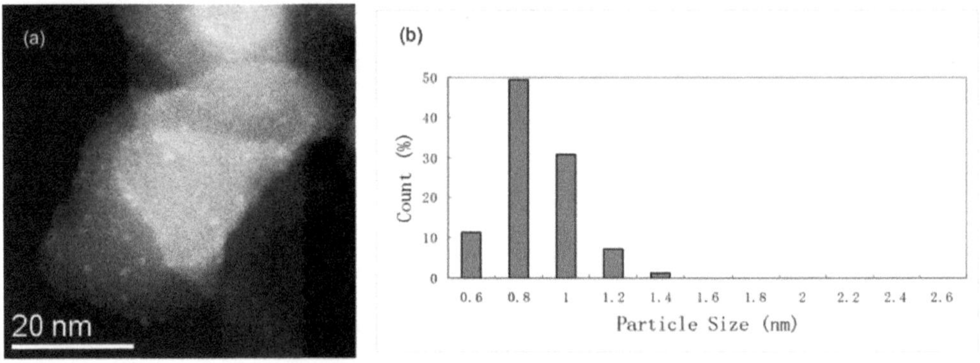

Fig. 5. Scanning transmission electron microscopy (STEM) image (a) and corresponding particle size distribution (b) of ALD Pd sample prepared using $Pd(hfac)_2$ at 100°C, followed by 2-cycle ALD Al_2O_3 to stabilize the $Pd(hfac)_2$ species before removing the hfac ligands at 200°C using HCOH.[16]

One unique advantage of ALD is the capability to tune the metal loading without affecting the particle size. For instance, when an ethanol exposure was used to remove a fraction of the Al_2O_3 surface hydroxyls by conversion to ethoxide species prior to a low-temperature $Pd(hfac)_2$ exposure, the Pd loading dropped by 36%, but the Pd particle size was unchanged.[16] In contrast, the Pd loading was found to increase substantially when the conventional $Pd(hfac)_2$/HCOH reaction scheme was replaced with a novel $Pd(hfac)_2$/TMA/H_2O sequence for Pd ALD. In this case, the Al_2O_3 that formed during the interposed TMA/H_2O exposures continuously provided new nucleation sites for $Pd(hfac)_2$ chemisorption. Simultaneously, the TMA acted as a reducing agent to remove the hfac ligands. In this way the Pd metal loading from 15 $Pd(hfac)_2$ exposures increased from 0.7 % to 6.6 % using this novel pulsing sequence while maintaining a Pd nanoparticle size of only ~1 nm.[17]

2.2 Activity of ALD Pd catalysts

The catalytic activity of ALD Pd nanoparticles was evaluated using the methanol decomposition reaction. The low-temperature decomposition of methanol to form carbon monoxide and hydrogen may provide a valuable source of hydrogen for transportation fuel,[18] and Pd is one of the most active species for catalyzing this process. It is generally believed that smaller Pd particles are more active for methanol decomposition.[19] In keeping with this trend, Fig. 6 shows that at the same space velocity (mol of methanol per gram of Pd), the activities of the ALD Pd catalysts increase in the sequence: Pd (2.2 nm) < Pd (1.4 nm) < Pd (0.8 nm). It is worth noting that the hydrogen production rate of the 1.4 nm ALD Pd catalyst at a space velocity of 120 000 mL/h/g-cat was significantly higher than some of the most active Pd-based catalysts reported for this reaction which is likely due to the highly uniform dispersion and small particle size.

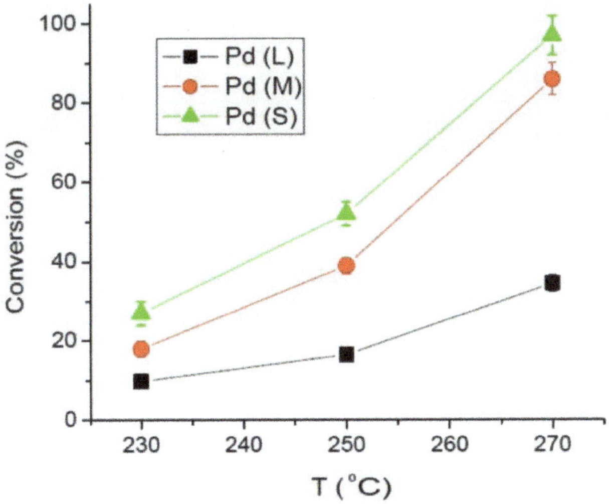

Fig. 6. Conversion of ALD Pd containing small (0.8 nm), medium (1.4 nm), and large (2.2 nm) nanoparticles measured in methanol decomposition.[16]

2.3 Support effects for ALD Pd catalysts

The catalytic behavior of noble metal nanoparticles can be greatly influenced by the underlying support material. For instance, cationic Pd species (most likely Pd^{+}) formed through the strong metal-support interaction are sometimes invoked to explain the higher activities of smaller Pd particles supported on CeO_2 or ZrO_2.[20] However, support effects are difficult to isolate using traditional catalyst synthesis methods. These methods rely on bulk supports which often have different physical properties such as surface area, pore size distribution, or particle size that can also affect performance. In contrast, ALD can be used to prepare the support as a thin (~1 nm), conformal layer on a template (Fig. 3). In this way, the same template can be used for a variety of ALD support layers allowing the composition to be varied while keeping the physical structure constant. ALD Pd nanoparticle catalysts were prepared on ALD Al_2O_3, ZrO_2, and TiO_2 layers on high surface area silica gel templates. The Pd loading was found to vary somewhat with the support layer (1.69% on Al_2O_3, 0.82% on ZrO_2, and 1.17% on TiO_2) and this probably reflects the different reactivity of the $Pd(hfac)_2$ precursor on these different surfaces. The relative rates of the three catalysts were evaluated for the methanol decomposition reaction, and the absolute rates were divided by the catalyst loading. The relative rates were found to increase as: $Al_2O_3 < TiO_2 < ZrO_2$ as shown in Fig. 6[21]. On a per-atom basis, the ALD Pd is 7x more active on the ALD ZrO_2 support compared to the ALD Al_2O_3 support.

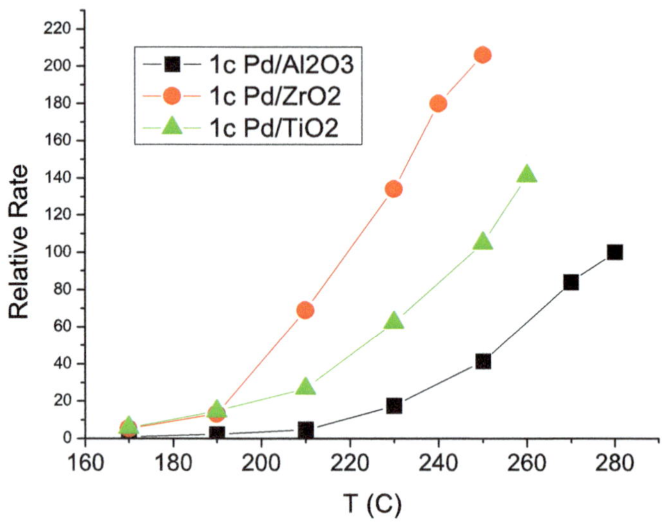

Fig. 7. Influence of metal oxide support on relative rate of methanol decomposition for ALD Pd catalysts.

3. Platinum ALD

Platinum (Pt) nanoparticles were first used as industrial catalysts by Vladimir Haensel at UOP in the 1940s [22]. Since then, Pt nanoparticles have become one of the predominant industrial precious metal catalysts and have received significant academic and industrial attention because of their unique catalytic properties. Supported Pt nanoparticles in the range of sub-nm to a few nm are effective for oxidative dehydrogenation of propane [23], water-gas shift reaction [24], fuel cells [25], direct conversion of methane [26], bio-mass reforming [27], and emission control [28]. The high market price of platinum and high activity of under-coordinated surface Pt atoms have motivated efforts to synthesize small Pt nanoparticles with a narrow size distribution using conventional preparation methods such as wet impregnation [29] and colloidal chemistry [30]. In light of the results presented above for Pd, ALD is a promising technique to prepare monodispersed Pt nanoparticles on high surface area supports with precise size control.

Pt ALD is typically conducted using alternating exposures of (trimethyl) methylcyclopentadienyl platinum (IV) ($MeCpPtMe_3$) and oxygen at 300 °C with dose and purge times of several seconds required to coat planar surfaces. To prepare Pt nanoparticles on high surface area supports, longer dose and purge times of several minutes are needed to allow complete saturation and precursor removal. Similar to Pd, Pt ALD on oxide surfaces can be divided into two regimes: nucleation and growth. The growth regime is characterized by a constant growth per cycle because the surface reactions have achieved a steady state. The mechanism for Pt ALD in the steady-state growth regime has been well studied and is given by [31]:

$$2(MeCp)PtMe_3 + 3\,O^* \rightarrow 2(MeCp)PtMe_2^* + CH_4 + CO_2 + H_2O \tag{3}$$

$$2\,(MeCp)PtMe_2^* + 24\,O_2 \rightarrow 2\,Pt^* + 3\,O^* + 16\,CO_2 + 13\,H_2O \tag{4}$$

In reaction 3, $MeCpPtMe_3$ reacts with oxygen adsorbed on the Pt surface from the preceding O_2 exposure. In reaction 4, O_2 serves to combust the remaining ligands on the adsorbed precursor, and some oxygen remains adsorbed on the surface. Pt ALD in the steady state regime is facilitated by the dissociation of O_2 on Pt to produce the necessary O* species. In contrast, we expect that Pt nucleation on oxide surfaces should be hindered because most oxides are not capable of dissociating O_2 at these temperatures.

3.1 Pt on $SrTiO_3$

To study Pt nucleation and growth on a catalytically relevant metal oxide surface, Bedzyk and coworkers prepared Pt nanoparticles on $SrTiO_3$ (001) single crystals [32] and high surface area $SrTiO_3$ nanocubes [33]. Contrary to expectations, the Pt growth during the nucleation period on $SrTiO_3$ (001) was ~2x higher than in the steady-state regime (Fig. 8). The higher initial growth was attributed to a greater density of reactive sites on $SrTiO_3$ compared to Pt. In contrast to the Pd study which used relatively short exposure times and showed a lower initial growth (Fig. 4), the Pt nucleation in Fig. 8 used much larger precursor exposures that were sufficient to saturate the surface reactions despite a lower reaction rate. Atomic force microscopy and scanning electron microcopy (SEM) revealed that the Pt nucleation occurred by island coalescence on the $SrTiO_3$ (001) surface, and the diameter of the Pt nanoparticles increased linearly with Pt ALD cycles before achieving film closure at ~40 cycles. This observation matches the transition between the nucleation and growth regimes in Fig. 8, and supports the hypothesis that the Pt grows faster on the $SrTiO_3$ (001) substrate. Grazing-incidence small angle X-ray scattering (GISAXS) was utilized to study the evolution in surface morphology with increasing Pt ALD cycles and showed that both the Pt nanoparticle size and inter-particle distance increased linearly with the number of cycles.

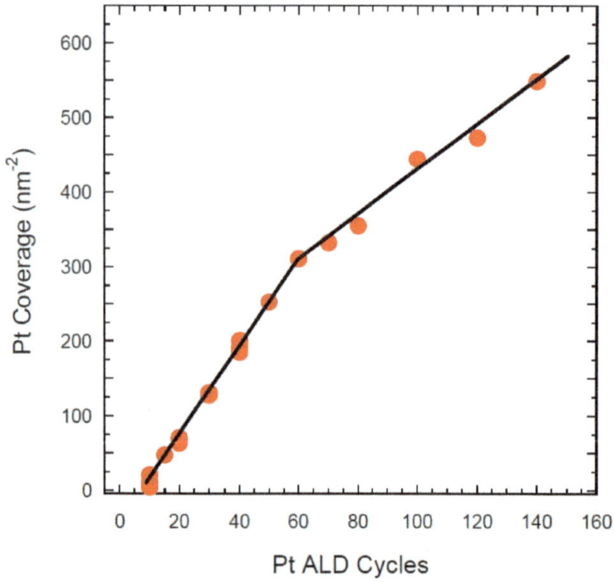

Fig. 8. ALD Pt coverage on $SrTiO_3$ (001)versus Pt ALD cycles.

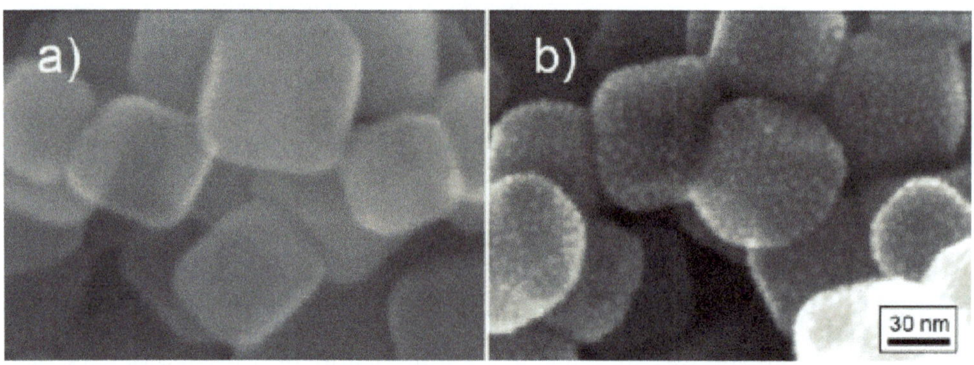

Fig. 9. SEM images of $SrTiO_3$ nanocubes before (a) and after (b) three Pt ALD cycles.

The surface area of the single crystal $SrTiO_3$ (001) substrate was far too low to facilitate catalytic testing by conventional methods. However, the surface reactions for Pt ALD should be independent of substrate morphology. Using this assumption, Pt was deposited on high surface area (20 m^2/g) $SrTiO_3$ nanocubes with a size of ~ 60 nm that were predominantly terminated with (001) facets. SEM images of the resulting materials revealed monodispersed Pt nanoparticles supported on the $SrTiO_3$ nanocubes (Fig. 9). During the early stages of Pt ALD on the $SrTiO_3$ nanocubes the mass was found to increase by 4.4 wt% per cycle, 2x higher than the value predicted using the steady-state growth rate for Pt of ~0.5 Å per cycle [33,31b]. Evidently, the enhanced Pt uptake observed during nucleation on the $SrTiO_3$ single crystal also occurs on nanophase $SrTiO_3$, suggesting that the Pt ALD is indeed independent of substrate morphology. X-ray scattering showed that both the Pt nanoparticle size and inter-particle spacing increased linearly with the number of Pt ALD cycles, consistent with the single crystal results and supporting a diffusional processes for nanoparticle formation. Figure 10 shows the size of the ALD Pt nanoparticles as determined from line broadening of the Pt(111) peak using wide angle X-ray scattering measurements. As with Pd, the ALD Pt nanoparticle size could be controlled precisely in the nm regime by adjusting the number of ALD cycles. The ALD Pt nanoparticles supported on $SrTiO_3$ nanocubes were significantly more active for propane oxidation than comparable catalysts prepared on a γ-Al_2O_3 support, emphasizing the important role played by the support[36].

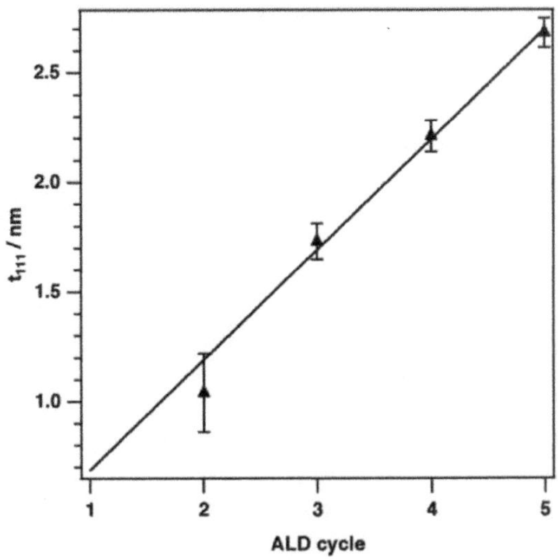

Fig. 10. Pt nanoparticle size on $SrTiO_3$ nanocubes versus Pt ALD cycles as determined using X-ray diffraction.

3.2 *In-situ* X-ray absorption studies of Pt ALD

A detailed understanding of the surface chemistry for noble metal ALD can lead to better methods for controlling the nanoparticle size, composition, and dispersion. X-ray absorption spectroscopy (XAS), is a powerful tool for determining the local geometric and electronic structure of materials under typical ALD process conditions. Marshall and co-workers used in situ XAS to study Pt ALD over γ-Al_2O_3, TiO_2, and $SrTiO_3$ surfaces[34]. In situ XAS measurements performed at the Pt L_3 edge (11.56 KeV) identified adsorbed $MeCpPtMe_2$, PtO, and metallic Pt as the predominant surface species present during Pt ALD. Time-resolved XAS measurements revealed the detailed evolution of these species during Pt nucleation (Fig. 11a) allowing a mechanism to be proposed for the surface reactions (Fig 11b). During the first Pt precursor exposure (Pt dose 1), $MeCpPtMe_3$ reacts with surface hydroxyl groups but remains essentially intact. The subsequent O_2 exposure at 100-300 °C converts all of the adsorbed $MeCpPtMe_2$ into PtO (O_2 dose 1). In the second Pt ALD cycle, $MeCpPtMe_3$ chemisorbs first on surface hydroxyl groups, and then later on the PtO as evidenced by the decrease in PtO and simultaneous increase in Pt metal during the later portion of Pt dose 2. The following O_2 exposure again converts all of the adsorbed species to PtO (O_2 dose 2). Apparently, no surface hydroxyls remain for the third and final Pt ALD cycle, because the $MeCpPtMe_3$ exposure (Pt dose 3) immediately causes a decrease in PtO concentration and a simultaneous increase in Pt metal in a 1:1 ratio.

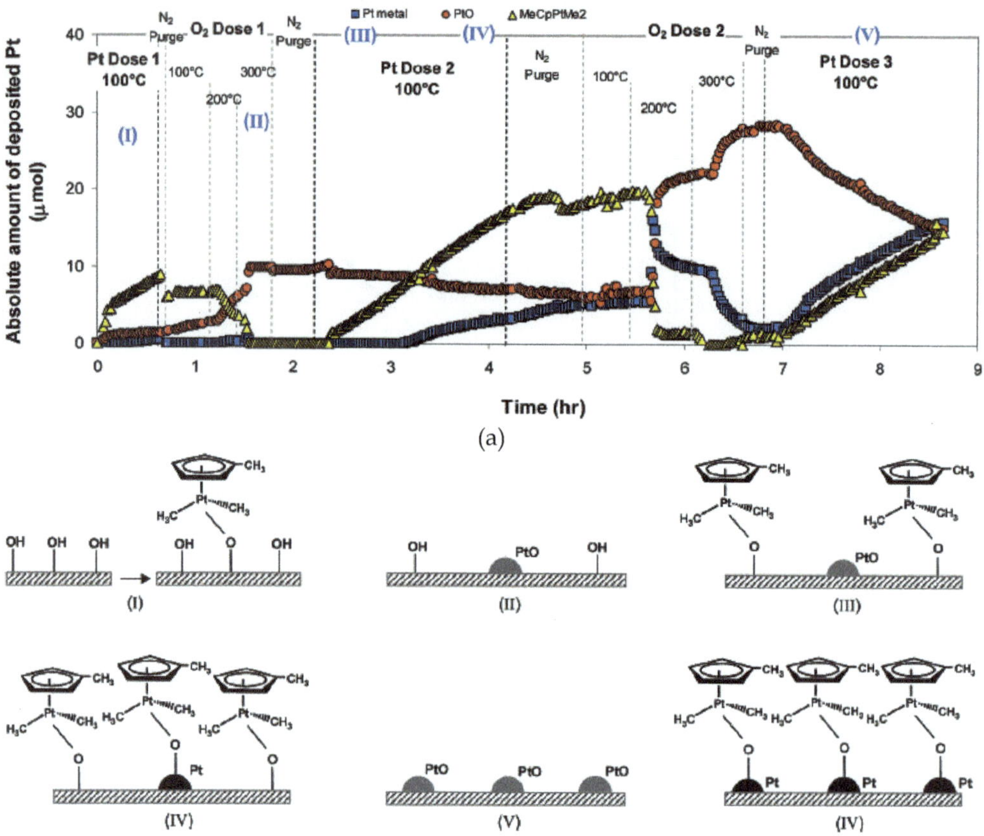

Fig. 11. (a) Concentration of Pt species on γ-Al_2O_3 during 3 Pt ALD cycles as determined using in situ XAS measurements. (b) Mechanism proposed for Pt ALD on γ-Al_2O_3.

4. ALD of mixed-noble metal catalysts

Bimetallic catalysts offer the possibility to combine the unique advantages of each component, allowing the catalytic properties to be tuned by adjusting the nanoparticle composition and structure. However, the design and economical synthesis of catalysts at the atomic scale represents a scientific Grand Challenge[35]. ALD offers a potential solution to this challenge. As with the pure Pd and Pt nanoparticles described above, the size of bimetallic particles should be controlled by the total number of ALD cycles performed. Moreover, the composition will be dictated by the relative number of ALD cycles used for each component. Finally, the structure of the bimetallic nanoparticles might be controlled by the order in which the individual cycles are executed. One of the appeals of ALD for the atom scale synthesis of bimetallics is that once a procedure is developed to create the desired nanoparticles, the same process can be used to deposit them on virtually any substrate, and at any quantity.

4.1 ALD of iridium-platinum films

To demonstrate the viability of mixed-noble metal ALD, Elam and coworkers synthesized thin-film mixtures of iridium and platinum using iridium (III) acetylacetonate – O_2 cycles for Ir ALD and $MeCpPtMe_3$ – O_2 cycles for Pt ALD at 300 °C [9]. In situ QCM measurements were performed using various Ir:Pt dosing ratios. These measurements found that the growth rates of Pt and Ir remained constant regardless of the dosing ratio, indicating that both the Ir and Pt ALD proceed equally well on either metal surface. This allowed the composition of the Ir-Pt mixed metal films to be easily controlled using the relative number of Ir cycles, and predicted using a simple rule-of-mixtures formula (Fig. 12). The lattice parameter for the Ir-Pt thin films was determined using X-ray diffraction measurements and was found to vary smoothly between the values of the pure Ir and Pt as the composition was adjusted, consistent with well-mixed, bimetallic alloys. Using this method, conformal Ir-Pt films were deposited successfully on high aspect ratio surfaces, suggesting that coating porous materials using this method should be feasible.

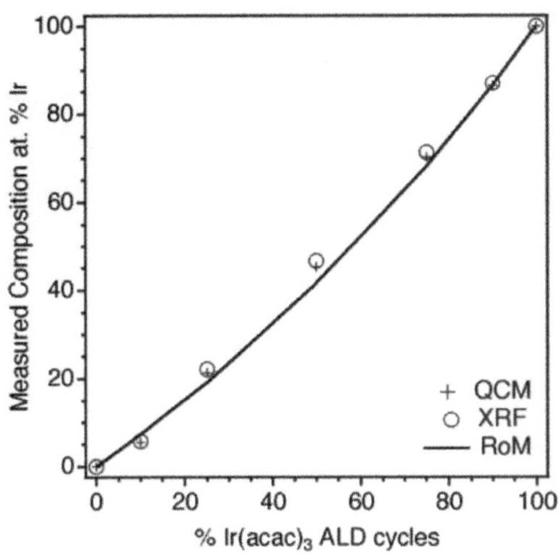

Fig. 12. Composition of ALD Ir-Pt films versus percentage of Ir ALD cycles measured by quartz crystal microbalance (QCM) and X-ray fluorescence (XRF), and predicted using the rule-of-mixtures (RoM).

4.2 ALD of platinum-ruthenium nanoparticles

Pt-Ru mixed-metal films were prepared using $MeCpPtMe_3$ – O_2 cycles for Pt ALD and 2,4-(dimethylpentadienyl)(ethylcyclopentadienyl) ruthenium (II) (Ru(DER)) – O_2 cycles for Ru ALD at 300°C [36]. Similar to the bimetallic Ir-Pt thin films, the growth rates of the Pt and Ru measured by QCM during the Pt-Ru mixed-metal ALD were in close agreement with the growth rates of the pure ALD Pt and Ru. Next, Pt-Ru nanoparticles were synthesized using 2 cycles Ru(DER)/O_2; 1 cycle $MeCpPtMe_3/O_2$; and 1 cycle Ru(DER)/O_2 over spherical Al_2O_3 nanopowder with a surface area of ~35 m^2/g yielding an average particle size of 1.2 ± 0.3 nm. Figure 13 shows the results of XAS measurement performed at the Ru K-edge (22.12 KeV). The Ru K-edge of Ru-Pt nanoparticles is shifted by ~3 eV to higher energy compared to the Ru/SiO_2 standard, suggesting the formation of bimetallic Pt-Ru. Furthermore, the magnitude of the Fourier transform of the Ru K-edge for the Ru-Pt nanoparticles clearly shows different features compared to the Ru nanoparticle standard, indicating the existence of Pt atoms in the first coordination shell of the Ru. Ru K-edge fitting yielded coordination numbers of 3.8 for Ru-Ru and 4.5 for Ru-Pt. These measurements were sufficient to confirm that the nanoparticles were bimetallic, but structure determination (core-shell versus alloy) would require XAS data for the Pt K-edge. The Ru-Pt nanoparticles supported on spherical Al_2O_3 nanopowder exhibited a much higher reactivity in the methanol decomposition reaction compared to a physical mixture of Ru and Pt monometallic catalysts, consistent with the unique atomic arrangement of a bimetallic catalyst.

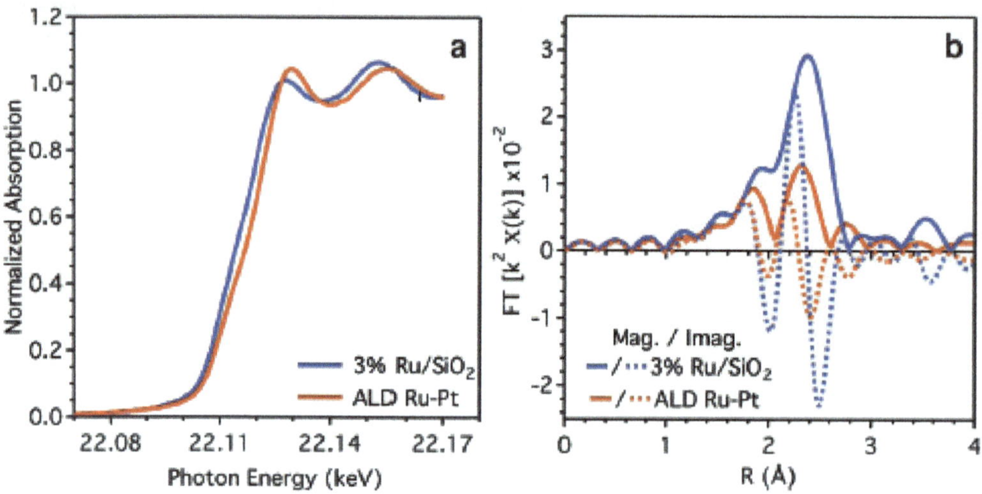

Fig. 13. XAS measurements showing (a) Ru K-edge for ALD Ru-Pt/Al_2O_3 and a reference 3% Ru/SiO_2; (b) Fourier transforms of the absorption spectra where the solid lines show the magnitude and the broken lines shows the imaginary part.

4.3 ALD of platinum-palladium nanoparticles

Pt-Pd mixed-metal nanoparticles ~1 nm diameter were synthesized over ALD modified SiO_2 gel with a surface area of 100 m^2/g using $MeCpPtMe_3$ – O_2 cycles for Pt ALD and $Pd(hfac)_2$ – HCOH cycles for Pd ALD. The composition of the Pt-Pd nanoparticles could be adjusted in two ways. Due to the different reaction kinetics for the Pt and Pd ALD processes, the Pt metal loading could be varied by adjusting the deposition temperature while the Pd loading remained constant with temperature. Consequently, the Pt-Pd composition could be controlled through the deposition temperature. In addition, the composition could be tuned using the ratio of Pt to Pd ALD cycles. The scanning transmission electron microscopy (STEM) image in Fig. 14a illustrates the homogeneity of the as-prepared Pt-Pd mixed-metal nanoparticles prepared using a 1:1 molar ratio supported by ALD Al_2O_3-coated SiO_2 gel. The mean size of the as-prepared Pt-Pd nanoparticles synthesized using one Pt ALD cycle followed by one Pd ALD cycle was 1.1 ± 0.2 nm. After 3.5% H_2 treatment at 250 °C for one hour, the particle size did change appreciably (Fig. 14b), indicating a high thermal stability for the Pt-Pd nanoparticles.

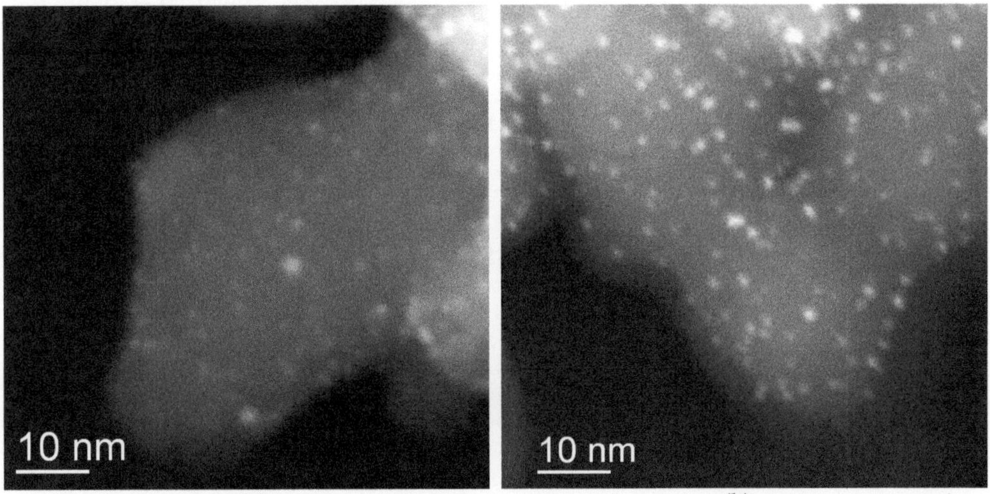

(a) (b)

Fig. 14. STEM images of Pt-Pd nanoparticles supported on ALD Al_2O_3-coated SiO_2 gel. (a) as-prepared, (b) after H_2 treatment at 250 °C for one hour.

The structure of the ALD Pt-Pd nanoparticles was investigated using XAS measurements at both the Pt L_3 edge (11.56 KeV) and the Pd K edge (24.35 KeV). The as-prepared samples were reduced in H_2 at 250 °C for one hour to obtain a fully metallic state prior to performing the XAS measurements. Small shifts in the edge position and changes in shape for both the Pt and Pd edges were clearly observed. Significant changes in the magnitude and imaginary part of both the Pt and Pd extended X-ray absorption fine structure (EXAFS) measurements indicated second scatterers in the first shell structure, i.e., a bimetallic nanoparticle. Furthermore, EXAFS model fitting showed that Pt-Pd forms a Pt-core, Pd-shell nanostructure, independent of the preparation temperature and ALD pulse sequence. Density functional theory (DFT) was used to calculate the lowest energy configuration for the Pt-Pd binary system. In vacuum, there was no clear preference for either a Pt- or Pd-rich surface. However, the Pd-rich surface was found to be the most stable with one monolayer of hydrogen adsorbed, in good agreement with EXAFS results[37].

5. Stabilization of noble metal nanoparticles

The stability of small particles against sintering is a serious problem limiting the application of metal nanoparticles. This problem is particularly severe for supported noble metal catalysts where sintering at high temperatures is a major contributor to catalyst deactivation.[38a-d] Various methods have been developed to encapsulate noble metal nanoparticles in porous materials using techniques such as chemical vapor deposition, grafting, micro-emulsion, and dendrimer encapsulation to form core-shell structures.[39a-h] Some of these encapsulated nanoparticles showed good sintering resistance up to 800 °C. However, in nearly all cases the catalytic activity was greatly reduced due to the mass transfer resistance imposed by the protective shell (usually tens of nm thick). In contrast to these previous methods, ALD should enable precise control over the thickness of the protective layer to achieve an optimum balance between enhanced stability and increased mass transfer resistance.

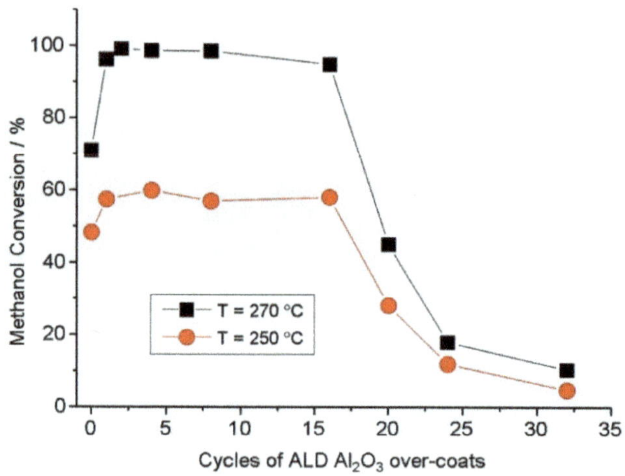

Fig. 15. Methanol conversion for Pd catalysts with 0 to 32 cycles of ALD Al_2O_3 over-coating in the methanol decomposition reaction carried out at 250 and 270 °C.[40]

To explore this possibility, Al_2O_3 ALD was used to form protective "over-coats" on ALD Pd nanoparticle catalysts. The catalytic activity and stability of the over-coated catalysts were evaluated using the methanol decomposition reaction. Surprisingly, up to a certain thickness formed using 16 ALD Al_2O_3 cycles, the protective layers preserved or even slightly enhanced the catalytic activity compared to the same catalyst without Al_2O_3 over-coating (Fig. 15).[40] The reason for this behavior was revealed through infrared spectroscopic measurements of adsorbed CO designed to probe the Pd surface. These measurements showed that the ALD Al_2O_3 over-coats preferentially nucleate at Pd corners, steps, and edges while leaving the catalytically active Pd (111) facets accessible for methanol conversion. An additional benefit of the over-coating was that the defect sites where the Al_2O_3 preferentially formed were also responsible for the undesired catalytic reaction to form coke. Consequently, the selectivity of the Pd catalyst for methanol decomposition could be controlled by the site-specific over-coating treatment to preferentially form CO and H_2 versus coke.

In addition to these changes in catalytic behavior, the ALD Al_2O_3 over-coats virtually eliminated the nanoparticle sintering exhibited by the uncoated Pd catalysts at elevated temperatures. The fact that the Pd surfaces were still accessible to the methanol and the CO probe molecules even after 8 ALD Al_2O_3 cycles indicates that the ALD Al_2O_3 over-coats are porous. To understand this behavior, the mechanism for Al_2O_3 ALD on Pd surfaces was investigated by combing *in situ* QCM and QMS experimental studies with density functional theory (DFT) calculations.[37] Both the experimental results and DFT calculations indicated that when TMA reacts on the Pd surface, $-CH_3^*$ surface species are deposited that inhibit further TMA chemisorption. This inhibition likely introduces discontinuities in the Al_2O_3 film and gives rise to the porosity. The porosity of the ALD Al_2O_3 overcoats could be increased through high temperature treatment or prolonged reaction at elevated temperatures, suggesting that the Pd catalyst might be modified by tuning the porosity of the Al_2O_3 over-coats to optimize a particular reaction.

6. Conclusion

Atomic layer deposition is a powerful method for catalyst synthesis. The self-limiting surface chemistry of ALD provides digital, sub-nm control over the amount of material deposited and allows nanoporous, high surface area materials to be coated uniformly. Noble metal nanoparticles can be prepared by ALD including Pd, Pt, Ir, and Ru, that are highly dispersed and uniform, and the size can be controlled easily on the nm scale by adjusting growth parameters such as the number of ALD cycles. By combining the ALD processes for two different noble metals, bimetallic nanoparticles and films and can be synthesized where the composition is dictated by the relative number of ALD cycles used for the two components. The capability to form uniform, continuous, ultrathin films of metal oxide support materials such as Al_2O_3, TiO_2, and ZrO_2, as well as size-selected catalytic noble metal nanoparticles, allows a catalyst to be fabricated starting from nearly any convenient template through a sequence of ALD processes. ALD can be used to stabilize noble metal nanoparticles while preserving their catalytic activity by forming porous overcoats that limit agglomeration but still permit access to the metal surface. The scale-up of ALD synthesis is straightforward owing to the self-saturating surface reactions and gas-phase precursors, so that novel materials developed in the lab can be manufactured for industrial use. The application of ALD technology to catalyst synthesis is an active area of research, and new capabilities are certain to emerge in the near future.

7. Acknowledgment

This material is based upon work supported as part of the Institute for Atom-efficient Chemical Transformations (IACT), an Energy Frontier Research Center funded by the U.S. Department of Energy, Office of Science, Office of Basic Energy Sciences. Y. L. and J. L. were supported in part by the U.S. Department of Energy, BES-HFI, Chemical Sciences under Contract DE-AC-02-06CH11357. Argonne is managed by UChicago Argonne, LLC, for the U.S. Department of Energy under contract DE-AC02-06CH11357.

8. References

[1] George, S. M., Atomic Layer Deposition: An Overview. *Chemical Reviews* 2010, 110, (1), 111-131.

[2] Haukka, S.; Lakomaa, E. L.; Suntola, T., Adsorption controlled preparation of heterogeneous catalysts. In *Adsorption And Its Applications In Industry And*

Environmental Protection, Vol I: Applications In Industry, Dabrowski, A., Ed. Elsevier: New York, 1999; Vol. 120, pp 715-750.

[3] Ott, A. W.; Klaus, J. W.; Johnson, J. M.; George, S. M., Al2O3 thin film growth on Si(100) using binary reaction sequence chemistry. *Thin Solid Films* 1997, 292, (1-2), 135-144.

[4] Elam, J.; Groner, M.; George, S., Viscous flow reactor with quartz crystal microbalance for thin film growth by atomic layer deposition. *REVIEW OF SCIENTIFIC INSTRUMENTS* 2002, 73, (8), 2981-2987.

[5] Elam, J. W.; Sechrist, Z. A.; George, S. M., ZnO/Al2O3 Nanolaminates Fabricated by Atomic Layer Deposition: Growth and Surface Roughness Measurements. *Thin Solid Films* 2002, 414, 43-55.

[6] Elam, J. W.; Libera, J. A.; Trang, H. H.; Feng, H.; Pellin, M. J., Atomic Layer Deposition of Al2O3 on Nanoporous Silica Gel Powder. *Journal of Physical Chemistry C* 2010, 114, 17286–17292.

[7] Elam, J. W.; Zinovev, A.; Han, C. Y.; Wang, H. H.; Welp, U.; Hryn, J. N.; Pellin, M. J., Atomic layer deposition of palladium films on Al2O3 surfaces. *Thin Solid Films* 2006, 515, (4), 1664-1673.

[8] Elam, J. W.; George, S. M., Growth of ZnO/Al2O3 Alloy Films Using Atomic Layer Deposition Techniques. *Chemistry of Materials* 2003, 15, 1020-1028.

[9] Christensen, S. T.; Elam, J. W., Atomic Layer Deposition of Ir-Pt Alloy Films. *Chemistry of Materials* 2010, 22, (8), 2517-2525.

[10] Igumenov, I. K.; Belosludov, V. R.; Stabnikov, P. A., Frontier trends in the prediction of vapour pressure of metal-organic precursors. *Journal De Physique Iv* 1999, 9, (P8), 15-22.

[11] Senkevich, J. J.; Tang, F.; Rogers, D.; Drotar, J. T.; Jezewski, C.; Lanford, W. A.; Wang, G. C.; Lu, T. M., Substrate-independent palladium atomic layer deposition. *Chemical Vapor Deposition* 2003, 9, (5), 258.

[12] Ten Eyck, G. A.; Senkevich, J. J.; Tang, F.; Liu, D. L.; Pimanpang, S.; Karaback, T.; Wang, G. C.; Lu, T. M.; Jezewski, C.; Lanford, W. A., Plasma-assisted atomic layer deposition of palladium. *Chemical Vapor Deposition* 2005, 11, (1), 60.

[13] George, S. M.; Goldstein, D. N., Surface poisoning in the nucleation and growth of palladium atomic layer deposition with Pd(hfac)(2) and formalin. *Thin Solid Films* 2011, 519, (16), 5339-5347.

[14] (a) Lu, J. L.; Stair, P. C., Nano/Subnanometer Pd Nanoparticles on Oxide Supports Synthesized by AB-type and Low-Temperature ABC-type Atomic Layer Deposition: Growth and Morphology. *Langmuir* 2010, 26, (21), 16486-16495; (b) Feng, H.; Elam, J. W.; Libera, J. A.; Setthapun, W.; Stair, P. C., Palladium Catalysts Synthesized by Atomic Layer Deposition for Methanol Decomposition. *Chemistry of Materials* 2010, 22, (10), 3133-3142.

[15] (a) Astruc, D., *Inorg. Chem.* 2007, 46, 1884-1894; (b) Astruc, D., *Nanoparticles and Catalysis*. Wiley-VCH: 2008; (c) Bell, A. T., *Sci.* 2003, 299, 1688-1691; (d) Chen, M. S.; Goodman, D. W., *Catal. Today* 2006, 111, (1-2), 22-33; (e) Boutonnet, M.; Lögdberg, S.; Svensson, E. E., *Current Opinion in Colloid & Interface Sci.* 2008, 13, (4), 270-286; (f) Cuenya, B. R., *Thin Solid Films* 2010, 518, (12), 3127-3150; (g) Aiken, J. D.; Finke, R. G., *J. Mol. Catal.* 1999, 145, (1-2), 1-44; (h) Liu, W., *China Particuology* 2005, 3, (6), 383-394.

[16] Feng, H.; Elam, J. W.; Libera, J. A.; Stair, P. C.; Miller, J. T., Subnanometer Palladium Particles Synthesized by Atomic Layer Deposition. *Acs Catalysis* 2011, 1, (6), 665-673.
[17] Lu, J. L.; Stair, P. C., Low-Temperature ABC-Type Atomic Layer Deposition: Synthesis of Highly Uniform Ultrafine Supported Metal Nanoparticles. *Angewandte Chemie-International Edition* 2010, 49, (14), 2547-2551.
[18] Wilson, M. S., Methanol decomposition fuel processor for portable power applications. *International Journal of Hydrogen Energy* 2009, 34, (7), 2955-2964.
[19] Saitoh, Y.; Ohtsu, S.; Makie, Y.; Okada, T.; Satoh, K.; Tsuruta, N.; Terunuma, Y., Effect of Pd Dispersion on Methanol Decomposition over Supported Pd-Catalysts. *Bulletin of the Chemical Society of Japan* 1990, 63, (1), 108-115.
[20] (a) Matsumura, Y.; Okumura, M.; Usami, Y.; Kagawa, K.; Yamashita, H.; Anpo, M.; Haruta, M., Low-temperature decomposition of methanol to carbon monoxide and hydrogen with low activation energy over Pd/ZrO2 catalyst. *Catalysis Letters* 1997, 44, (3-4), 189-191; (b) Matsumura, Y.; Shen, W. J., Methanol decomposition and synthesis over palladium catalysts. *Topics in Catalysis* 2003, 22, (3-4), 271-275.
[21] Feng, H.; Elam, J., 2011, in preparation.
[22] Vladimir, H. Alumina-platinum-halogen catalyst and preparation thereof. 2479109 08/16/1949, 1949.
[23] Vajda, S.; Pellin, M. J.; Greeley, J. P.; Marshall, C. L.; Curtiss, L. A.; Ballentine, G. A.; Elam, J. W.; Catillon-Mucherie, S.; Redfern, P. C.; Mehmood, F.; Zapol, P., Subnanometre platinum clusters as highly active and selective catalysts for the oxidative dehydrogenation of propane. *Nature Materials* 2009, 8, (3), 213-216.
[24] Fu, Q.; Saltsburg, H.; Flytzani-Stephanopoulos, M., Active nonmetallic Au and Pt species on ceria-based water-gas shift catalysts. *Science* 2003, 301, (5635), 935-938.
[25] Wasmus, S.; Kuver, A., Methanol oxidation and direct methanol fuel cells: a selective review. *Journal of Electroanalytical Chemistry* 1999, 461, (1-2), 14-31.
[26] Periana, R. A.; Taube, D. J.; Gamble, S.; Taube, H.; Satoh, T.; Fujii, H., Platinum catalysts for the high-yield oxidation of methane to a methanol derivative. *Science* 1998, 280, (5363), 560-564.
[27] Cortright, R. D.; Davda, R. R.; Dumesic, J. A., Hydrogen from catalytic reforming of biomass-derived hydrocarbons in liquid water. *Nature* 2002, 418, (6901), 964-967.
[28] Burch, R.; Breen, J. P.; Meunier, F. C., A review of the selective reduction of NOx, with hydrocarbons under lean-burn conditions with non-zeolitic oxide and platinum group metal catalysts. *Applied Catalysis B-Environmental* 2002, 39, (4), 283-303.
[29] Miller, J. T.; Schreier, M.; Kropf, A. J.; Regalbuto, J. R., A fundamental study of platinum tetraammine impregnation of silica 2. The effect of method of preparation, loading, and calcination temperature on (reduced) particle size. *Journal of Catalysis* 2004, 225, (1), 203-212.
[30] Tao, A. R.; Habas, S.; Yang, P. D., Shape control of colloidal metal nanocrystals. *Small* 2008, 4, (3), 310-325.
[31] (a) Kessels, W. M. M.; Knoops, H. C. M.; Dielissen, S. A. F.; Mackus, A. J. M.; van de Sanden, M. C. M., Surface reactions during atomic layer deposition of Pt derived from gas phase infrared spectroscopy. *Applied Physics Letters* 2009, 95, (1); (b) Aaltonen, T.; Ritala, M.; Sajavaara, T.; Keinonen, J.; Leskela, M., Atomic layer deposition of platinum thin films. *Chemistry Of Materials* 2003, 15, (9), 1924-1928.

[32] Christensen, S. T.; Elam, J. W.; Lee, B.; Feng, Z.; Bedzyk, M. J.; Hersam, M. C., Nanoscale Structure and Morphology of Atomic Layer Deposition Platinum on SrTiO3 (001). *Chemistry of Materials* 2009, 21, (3), 516-521.

[33] Christensen, S. T.; Elam, J. W.; Rabuffetti, F. A.; Ma, Q.; Weigand, S. J.; Lee, B.; Seifert, S.; Stair, P. C.; Poeppelmeier, K. R.; Hersam, M. C.; Bedzyk, M. J., Controlled Growth of Platinum Nanoparticles on Strontium Titanate Nanocubes by Atomic Layer Deposition. *Small* 2009, 5, (6), 750-757.

[34] Setthapun, W.; Williams, W. D.; Kim, S. M.; Feng, H.; Elam, J. W.; Rabuffetti, F. A.; Poeppelmeier, K. R.; Stair, P. C.; Stach, E. A.; Ribeiro, F. H.; Miller, J. T.; Marshall, C. L., Genesis and Evolution of Surface Species during Pt Atomic Layer Deposition on Oxide Supports Characterized by in Situ XAFS Analysis and Water-Gas Shift Reaction. *Journal of Physical Chemistry C* 2010, 114, (21), 9758-9771.

[35] Basic Research Needs: Catalysis for Energy. In *Report from the U.S. Department of Energy, Basic Energy Science Workshop*, 2007.

[36] Christensen, S. T.; Feng, H.; Libera, J. L.; Guo, N.; Miller, J. T.; Stair, P. C.; Elam, J. W., Supported Ru-Pt Bimetallic Nanoparticle Catalysts Prepared by Atomic Layer Deposition. *Nano Letters* 2010, 10, (8), 3047-3051.

[37] Lei, Y.; Lu, B.; Greeley, J.; Elam, J., 2011, in preparation.

[38] (a) Bartholomew, C. H., *Applied Catalysis A: General* 2001, 212, 17-60; (b) Sault, A. G., and V. Tikare, *Journal of Catalysis* 2002, 211, 19-32; (c) Datye, A. K., Q. Xu, K. C. Kharas, and J. M. McCarty, *Catalysis Today* 2006, 111, 59-67; (d) Bernal, S., J.J. Calvino, C. LoÂpez-Cartes, J.M. Pintado, J.A. PeÂrez-Omil, J.M. RodrõÂguez-Izquierdo, K. Hayek, and G. Rupprechter, *Catalysis Today* 1999, 52, 29-43.

[39] (a) Park, J.-N., A. J. Forman, W. Tang, J. Cheng, Y-S. Hu, H. Lin, and E. W. McFarland, *Small* 2008, 4, (10), 1694-1697; (b) Kanazawa, T., *Catalysis Letters* 2006, 108, (1-2), 45-47; (c) Takenaka, S., H. Matsumori, K. Nakagawa, H. Matsune, E. Tanabe, and M. Kishida, *Journal of Physical Cheistry: C* 2007, 111, 15133-15136; (d) Zhao, M., L. Sun, and R. M. Crooks, *Journal of the American Chemical Society* 1998, 120, 4877-4878; (e) Arnal, P. M., M. Comotti, and F. Schuth, *Angewandte Chemie-International Edition* 2006, 45, 8224-8227; (f) Joo, S. H., J. Y. Park, C-K. Tsung, Y. Yamada, P. Yang and G. A. Somorjai, *Nature Materials* 2009, 8, 126-131; (g) Ott, L. S., and R. G. Finke, *Coordination Chemistry Reviews* 2007, 251, 1075-1100; (h) Seipenbusch, M., and A. Binder, *Journal of Physical Cheistry: C* 2009, 113, 20606-20610.

[40] Feng, H.; Lu, J. L.; Stair, P. C.; Elam, J. W., Alumina Over-coating on Pd Nanoparticle Catalysts by Atomic Layer Deposition: Enhanced Stability and Reactivity. *Catalysis Letters* 2011, 141, (4), 512-517.

10

Precise Temperature Measurement Using Noble Metal Thermocouples

Yasser A. Abdelaziz
National Institute for Standards (NIS), Giza
Egypt

1. Introduction

1.1 Physical and chemical properties of platinum

Platinum is a white - grayish metal with brilliant luster. It is harder than gold and copper, platinum has also the valuable quality that it is often before melting so that it can be welded. Platinum and rhodium do not volatilize appreciably at 900 °C but at 1300 °C volatilization can be detected. The high melting temperature, and the fact that platinum is not affected by air and strong acids, enables it to be used in the manufacture of temperature measuring sensors like platinum resistance thermometers and thermocouple. Platinum has nearly the same coefficient of expansion as glass, and platinum wires can be fused in glass so as to make gas-tight joints. This property makes it possible to prepare platinum. Resistance thermometers in glass outer sheath Platinum are also a good conductor of electricity which makes it suitable for making of temperature sensors.

The platinum indium allay (90:10) is used for making the prototype kilogram from 1889 to 1960 the same alloy was used to make the international prototype meter bar.

The physical properties of Pt which makes it suitable for thermometric sensors are given in table (1).

Physical properties	Value
Density in g/cm^3	21.45
Melting point °C	1768.30
Molar heat capacity in J mal^{-1} K^{-1}	25.86
Thermal conductivity W m^{-1} K^{-1}	71.6
Thermal expiation at 25°C µm m-1 K-1	8.80
Tensile strength Mpa	125.24
Electrical resistivity at 20°C n Ω .m	105.
Temperature coefficient of electrical resistance from 20 to 100°C	0.003925

Table 1. The physical properties of Platinum

1.2 Thermoelectric phenomena

A thermocouple consists of two dissimilar conductors or thermo-elements joined to form a circuit. T. J. Seebeck (1822) first discovered that a thermocouple would produce a current in a closed circuit when one junction is at a different temperature from the other (1). The emf that produced the current is refereed to as the thermocouple emf or as the Seebeck voltage E and its temperature derivative dE/dT is known as the thermoelectric power or Seebeck coefficient S. The one conductor A is said to be positive with respect to the other B, if current would flow from A to B at the cooler of the two junctions as shown in figure (1).

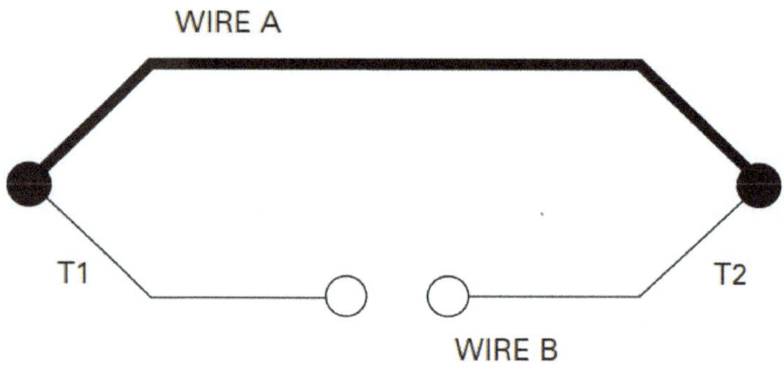

Fig. 1. A typical thermocouple

1.3 Mechanism of thermoelectricity

The purpose of this introduction is to show how the thermopower of metals and alloys depends so critically upon the composition, homogeneity and state of anneal of the material.

The origin of thermoelectricity in a metal can be demonstrated using the simple free-electron gas model of the conduction process. We can imagine a metal to be made up of a rigid lattice of atoms between which can move free electrons driven by electric, magnetic or thermal gradient in the conductor happen to be present. As a result of the temperature gradient in the conductor, the electrons at the hot end diffuse towards the cold end. The excess of electrons thereby produced at the cold end leads to a potential gradient being established at the same time. The negative charge at the cold end builds up until a state of dynamic equilibrium is established between the number of electrons having relatively high velocity, diffusing towards the cold end of the conductor driven by the temperature gradient and the number of electrons having relatively low velocity, diffusing towards the hot end of conductor driven by the electric potential gradient. This electric potential gradient arises whenever a temperature gradient exists in a conductor and is known as the thermoelectric emf. It follows that no thermoelectric emf can be present in the absence of a temperature gradient.

It is evident that the mechanism of electron scattering must play an important role in thermoelectricity. The electrons having a higher velocity would be scattered through smaller angles by the lattice atoms than those having a lower velocity; in this case an enhanced negative thermoelectric emf would develop. In other words the mean free path of the electron would depend upon its kinetic energy.

The complexity of the electron energy vs. scattering relations leads to the thermoelectric emf in different metals being very different from one another. Were they the same in all metals and alloys, then we would not be able to make use of it for temperature measurement, since the thermocouple is a device in which the difference between the thermoelectric emf in the two arms of the thermocouple gives an indication of the temperature difference between the hot and cold ends of the wires. The difference in emf is known as Seebeck effect.

The distribution of electric potential E(T) in a pair of conductors of different materials A and B having a junction maintained at a temperature T_2 and having free ends both maintained at lower temperature T_o is shown diagrammatically in figure (1) and (2). Both the junction and the free ends of the conductors are shown in regions of uniform temperature and both conductors pass through the same temperature gradient. T_o permit measurements to be made of the thermoelectric potential difference between the free ends of A and B. A further pair of identical conductor C is attached at T_o which lead to a detector at a temperature T_1.

$$S_{A(T)} = \frac{dE_{A(T)}}{dT} \quad and \quad S_{B(T)} = \frac{dE_{B(T)}}{dT} \tag{1}$$

Analysis of the circuit in figure (2) now reduces to determining the thermoelectric effects for the three wires A and B for the thermocouple and C for the instrument leads. Clearly it is essential that any connections must be isothermal.

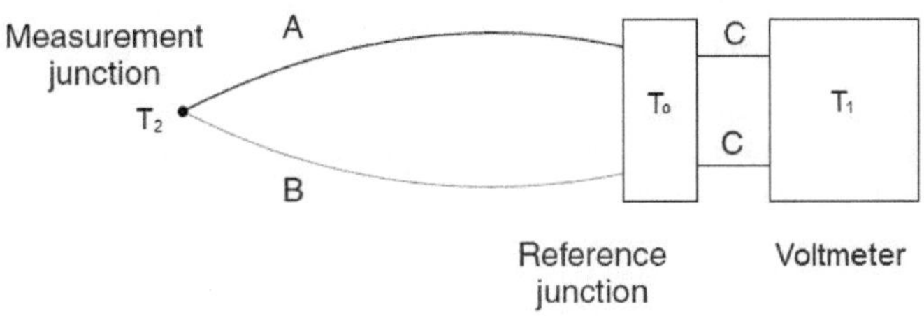

Fig. 2. Model of thermocouple measurement

The wires A, B and C should be kept as homogenous as possible by avoiding any chemical or mechanical treatment. The voltage E produced by the circuit is the sum of three parts.

$$E = E_{A(T2)} - E_{A(To)} + E_{C(To)} - E_{C(To)} + E_{B(T2)} - E_{B(To)} \tag{2}$$

Where for the wires A, B and C, EA, EB and EC are the voltages for the end points at T_2 and T_o which are the measurement and reference junction temperatures respectively.

From equation (2) it can be seen that the net contribution of the lead wires C is zero since they are both at the same temperatures T_o.

Under the above conditions the instrumentation for measuring a thermocouple voltage can be considered independent of the sensor. Also from equation (2) the voltage output from the pair of wires (A and B) is related to the difference between the thermoelectric effect for the wires A and B. It is common to consider only a relative Seebeck voltage E_{AB} and a relative Seebeck coefficient SAB and equation (2) becomes;

$$E = E_{AB(T2)} - E_{AB(T_O)} \tag{3}$$

simplification is made by choosing the reference temperature T_o= 0 °C and setting E_{AB} (0 °C) = 0, equation (3) becomes.

$$E = E_{AB(T2)} \tag{4}$$

Equation (4) is the basic equation for thermocouple, which relates the voltage produced to the temperature being measured.

However, the equation is based on several assumptions. The assumptions made are;

- The reference junction is 0 °C.
- The relative Seebeck coefficient is known.
- There are isothermal conditions for instruments and connecting wires.
- There are isothermal conditions for measurement and reference junctions.
- There are homogeneous wires connecting measurement and reference junctions.

Note

Various metals and metal alloys have been developed for use in the construction of thermocouples. The thermocouple is classified by a letter code for the particular combination of metals used, for example a Type S thermocouple comprises of platinum and platinum 10% rhodium alloys. The current international standard for thermocouples is IEC 584.

Thermocouples have sensitivities from a few microvolts per degree C to a few 10's of microvolt per degree C, varying with type and operating temperature. Sensitivities at 100 °C are shown in the table for various types.

Type	Materials	Max. Temp. °C	Sensitivity (μV/°C) at		
			20 °C	100 °C	500 °C
B	Platinum 30% rhodium/platinum 30% rhodium	1700	0	0.9	5.1
R	Platinum 13% rhodium/platinum	1400	5.9	7.5	10.9
S	Platinum 10% rhodium/platinum	1400	5.9	7.4	9.9
Au/Pt	Gold/Platinum	1000	6.2	8.1	16.2

Table 2. Noble metal thermocouple Sensitivities and Other Data

2. Noble metal thermocouples

In 1886, Le Chatelier discovered the platinum-10%rhodium vs. platinum thermocouple. Since that time, many investigators have made great contributions to the advancement of platinum group metals for temperature measurement, such as tungsten vs. iridium, palladium vs. platinum and gold vs. platinum etc.

In this series, the platinum–10%rhodium vs. platinum thermocouple is, from the scientific standpoint, the most important of all the thermocouples (base and noble metal) now employed. It was used to define the International Temperature Scale of 1948 (ITS-48) and the International Practical Temperature Scale of 1968 (IPTS-68) in the temperature range from the freezing point of antimony 630.5 °C to the melting point of gold 1063 °C on ITS-48 and 1064.43 °C on IPTS-68 (on ITS-90 the melting point of gold is 1064.18 °C). This thermocouple is noted for its high accuracy over a broad temperature ranges, its excellent mechanical and chemical properties and low electrical resistivity, it is readily obtained in uniform quality.

The work on the gold platinum thermocouples started at the National Institute of Standards and Technology (NIST) by the work of Roeser and Wensel (1). In 1987 McLaren and Murdock published a report (2) on the characteristics of Au/Pt thermocouple. His study had been undertaken to find a superior and practical sensor as an alternative to the standard Pt-10%Rh vs. Pt thermocouple for temperature measurements in the range (0 – 1000 °C). Subsequent to that study, Burns, Strouse and Mangum (3) undertook an investigation of such thermocouples to determine their feasibility to serve as rugged secondary reference thermometer at the 50 m °C level of uncertainty.

With the introduction of the International temperature Scale of 1990 (ITS-90) (4) emphasis shifted in 1990 to include the development of an ITS-90 based reference function to these thermocouples. Recently a reference function for Au/Pt thermocouples based on the ITS-90 was published by Gotoh et al (5) and by Burns, Strouse et al (3).

In the present investigation, the stability and reproducibility of Au/Pt thermocouples and of the Pt-10%Rh vs. Pt thermocouples have been carried out for comparison and to find the best mathematical formula for their calibration.

2.1 Assembly, annealing and stabilization of Pt-10%Rh vs. Pt thermocouple

Before any assembly began the Pt and Pt-10%Rh wires (0.5 mm diameter) were cleaned by boiling in distilled water followed in sequence by boiling in dilute nitric acid, distilled water, dilute hydrochloric acid, (approximately 2 N) and finally four times in distilled water (6). After drying in an oven in a dust free space the Pt and Pt-10%Rh wires, 120 cm long, were welded using a strongly oxidizing oxygen-natural gas flame to form the thermocouple junction. Surgical polyethylene gloves were worn during handling of the wires and thermocouple. The thermocouple was electrically annealed at 1100 °C using A.C for 1 hour. The purity of the clean, annealed platinum wires was within the requirements of the IPTS-68 relating to the emf output at the gold point and its relationship to that at the other two fixed points as will be shown in the results.

Twin bore recrystallized alumina ($Al_2 0_3$), 60 cm long was used to electrically insulate the wires. This alumina sheath was boiled in aqua-regia and distilled water in the same manner as the wires to remove any metallic contaminations on the surface, then dried in an oven at approximately 100°C for some hours and then heated in a furnace to 1200°C in air for 24 hours. The free ends of the thermocouple were passed through the bores in the insulation tube, the wires were gently eased through the twin bores of the $Al_2 0_3$ tube with the minimum amount of strain.

The completed thermocouple was stabilized by annealing in a furnace at which the temperature of the central 60 cm was at 1100 °C ± 10 °C. Cleaned single bore alumina tube was placed over the uninsulated wires and the thermocouple placed in a clean $Al_2 0_3$ sheath and positioned so that the tip and the adjoining 60 cm lay in the furnace uniform zone at 1100 °C. After 8 hours of annealing the freezing point of silver was measured with the thermocouple after which it was replaced in the annealing furnace for a further 2 hours and then another freezing point of silver was determined. The two successive silver points agreed to better than 1 μV. Cleaned Plastic tubing was placed over the uninsulated wires and was placed in a glass dust free container and stored in a position free from vibration.

2.2 Assembly, annealing and stabilization of Au/Pt thermocouple

The wire used for preparing the Gold/Platinum thermocouples with diameter 0.5 mm and nominal purity 99.99% (7). The method of this thermocouple preparation followed the recommendations reported in the extensive investigation by McLaren and Murdock (2). Their study established the technique required to obtain measurements of the highest precision with the Au/Pt sensor, including the recommended heat treatments to obtain optimum stability and thermoelectric homogeneity, a stress relieving technique to minimize the tensile emf arising from differential thermal expansion of the Au and platinum elements and the immersion depths required to minimize the sensor conduction loss.

For annealing the gold wire, 120 cm long segment mounted in a 2 bore alumina insulating tubes, 1.2 mm bore and 3 mm in diameter and of total length 120 cm. The $AL_2 O_3$ tubes were previously heated for 50 hours at 1000 °C. Then this assembly was heated in a 1.5 m long horizontal tube furnace for 10 hours at 1000 °C cooled over 3 hours to 450 °C held at 450 °C for 15 hours and then removed from the furnace.

For annealing the platinum wire an electric current was passed through the bare wire sufficient to heat it to 1300 °C, with this temperature sustained for 10 hours. Then the current was switched off and the wire was quenched in air. The electric current was then set to a value sufficient to heat the bare wire to 450 °C and this temperature was maintained for 1 hour to equilibrate the vacancies introduced by the annealing of the wire at 1300 °C. Again the current was switched off and the wire rapidly cooled in air so as not to oxidize the platinum. The annealed Pt and Au wires were assembled in a twin alumina bore ($AL_2 O_3$) sheath of length 75 cm and 4.7 mm diameter. The two wires were joined by a stress relieving coil consisting of five turns of 0.13 mm diameter Au wire at the measuring junction, and given a 20 hour furnace anneal at 450 °C to relieve strain introduced by the assembly process.

The wires emerging from the alumina insulating tube were insulated with flexible Polyvinyl tubing to within about 5 cm of their ends. To complete the assembly of the thermocouple a pair of insulated Cu wires (0.4 mm in diameter) were soldered to the Au and Pt wires to form the reference junction.

2.2.1 Stress relieving coil

Due to the big differential thermal expansion of the Pt and Au wires, in Au/Pt thermocouples with regular hot junction a significant stress thermoelectro-motive force was observed and amounted to the equivalent of < 0.04 °C at the silver point. This stress – thermo emf can be eliminated (2) simply by welding a weak stress-relieving Pt or Au coil between Au and Pt elements to form a stress- relieved junction for the Au/Pt thermocouple. The stress-relieving coil deforms easily as the Pt and Au elements expand in the furnace gradients and cannot transport the differential expansion stress to the portion of the wires in the thermo-emf generating temperature gradient zones. Avoidance of the expansion stress allows an Au/Pt thermocouple to reveal its superior intrinsic thermoelectric homogeneity and to attain nearly the performance level expected of an ideal thermocouple with truly homogeneous.

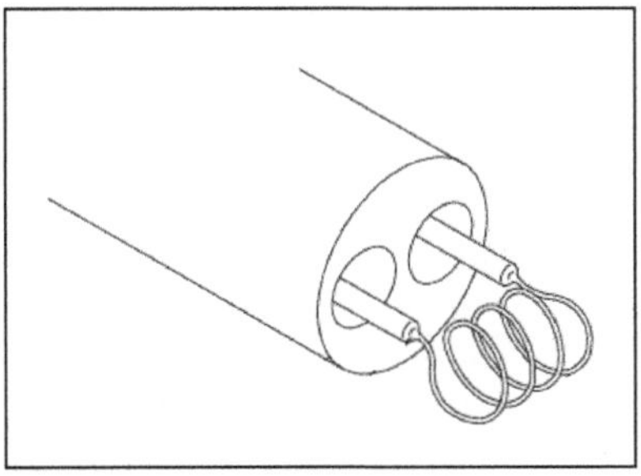

Fig. 3. The stress relieving coil of Au/Pt thermocouple.

2.3 Thermocouple reference junction

A thermocouple used to measure temperature is from its nature a differential sensor, which actually measures the thermal emf, generated as a result of the difference of temperature between its two junctions. It follows then that the temperature of one of its junctions must be known if the measured emf is to be related to the temperature being measured, as was mentioned before. This junction is known variously as the cold, ice, or reference junction. The other is the measuring junction that often is called the hot junction. The reference junction refers to the electrical connection between the thermocouple and copper wires leading to the measuring instrument. The two thermocouple wires directly connected to a copper leads, this assembly can inserted in a Pyrex tube immersed in a bath of water ice slush at 0 °C. The ice bath illustrated in figure 4 was made in a large dewar flask with a clear plastic cover drilled to receive the Pyrex reference junction tubes.

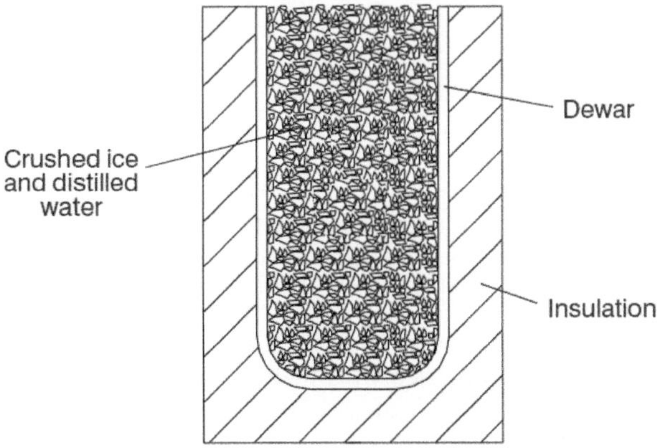

Fig. 4. Ice bath for reference junction

Clear shaved ice and distilled water is used, and water-ice slush filled the dewar flask at all times during observations. This latter precaution is important in the use of an ice bath because the water at the bottom of a bath in which the ice is merely floating may be as high as 4°C, the temperature of maximum density of water. The Pyrex tubes are immersed, so deeply in the ice bath that further immersion caused no detectable change in the thermal emf. This depth is found to be from 15 to 20 cm depending on the diameter of the wires.

3. Calibration of thermocouples and data fitting methods

Calibration of thermocouples is sometimes necessary in order to confirm or establish the emf-to-temperature relationship for the device concerned.

There are a number of ways of calibrating thermocouples. The emf of the thermocouple can be determined at a relatively small number of fixed points using special homogenous furnaces (Figure 5) and interpolation performed using agreed formulae or by considering the difference from a standard table. An alternative method is to compare the emf of the thermocouple being calibrated with that of the same type of standard thermocouple for a large number of temperatures and then fit a curve or algebraic relationship to the emf versus temperature data using a least squares method.

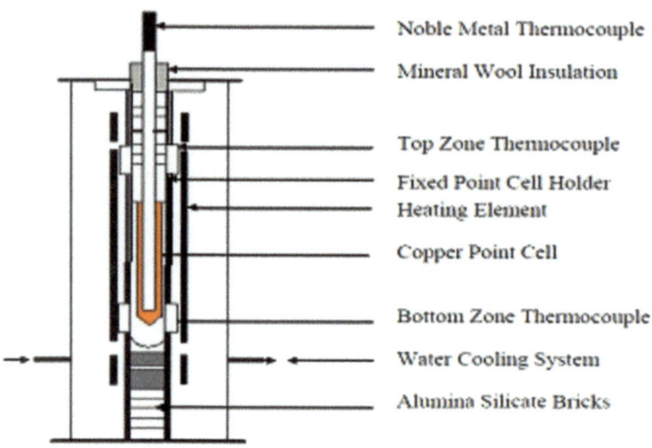

Fig. 5. The schematic diagram of metal fixed point cell and its furnace

Ideally a calibration should be traceable to the ITS-90. This can be achieved by a hierarchy of calibration activity between the end user and the national standards laboratory or accredited service. Calibration should be undertaken to a specified uncertainty and within a range of temperatures relevant to the application.

For calibration by comparison methods, a liquid bath can be used up to 600°C and a furnace, for higher temperatures up to a limit of about 1800°C (Figure 6). Fully automatic systems are commercially available that control a heater across a programmable temperature range and allow measurements to be taken from a calibrated temperature sensor and the thermocouple to be tested and provides the facility for the data to be logged in a data file.

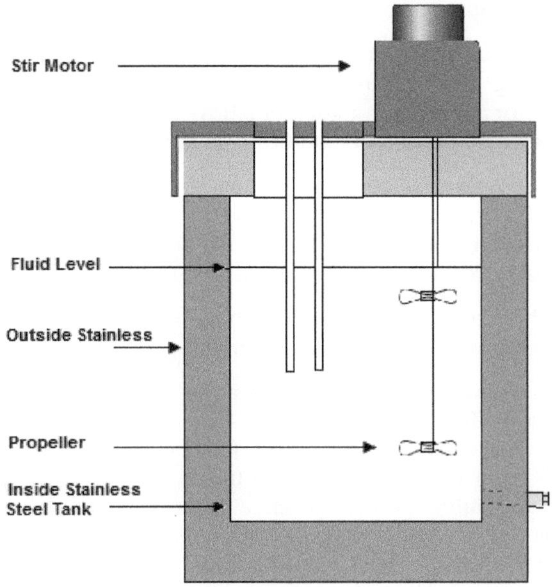

Fig. 6. Liquid bath used for thermocouples calibration

Despite the extensive use of thermocouples in scientific and industrial measurements for over half a century, theoretical studies have not yielded an adequate basis for predicting the electromotive force vs. temperature (ET characteristic) curve. Hence thermocouple ET characteristics are based on laboratory measurements, the data being fitted to an approximating formula. The approximating formula is at best an educated guess.

Platinum-10%Rhodium/Platinum thermocouples which satisfy the criteria stated in the text of the International Practical Temperature Scale of 1968 (IPTS-68) are used to realize temperatures in the range 630.74 °C to 1064 °C. The calibration of these thermocouples at the freezing points of Sb, Ag and Au resolves the quadratic relationship.

$$E = a + bt + ct^2 \quad (5)$$

which defines temperature in this range. However the thermocouple is used over a much wider range of temperature approximately from 100 °C to 1400 °C and it is desirable that it should be calibrated at that range with a minimum of uncertainty. There are two methods of thermocouple calibration generally referred to as the fixed point and the intercomparison methods. In the former the thermocouple is calibrated at the freezing points of metals specified by the ITS-90 whilst in the latter the output of the thermocouple is measured at temperatures determined by a calibrated thermocouple or resistance thermometer. In general calibration at the fixed points is more accurate than calibration by the intercomparison method. However there is very limited information on the estimated uncertainty of calibration in the region of interpolation. Before a rigorous comparison of the uncertainty in thermocouple calibration by the two methods can be made, the error associated with interpolation between fixed points must be evaluated.

3.1 Methods of interpolation

Methods which can be used for interpolation between measured data have been stated in many texts. Data fitting is a technique for constructing a mathematical model that characterizes the data or providing a mathematical formula by which the data can be reproduced. An extensive discussion of data fitting is given in Hamming's book(8).

When the intervals between the measured values are not equal the two methods most commonly used are those in which a curve is fitted to the points using a least squares method and the Lagrangian method (9). In forcing the curve to pass through the data points, Lagrangian method can introduce large deviations into the interpolation between the points.

The least squares method consists of fitting a polynomial function to minimize the sum of the squares of the deviations of all data points from the polynomial curve.

Functions of increasing powers are fitted and it is desired to find the lowest order curve, which gives a satisfactory fit. The criterion for satisfactory fit depends on the number of measured values. The most satisfactory criterion is to specify that the maximum deviation of the calibration points from the fitted curve should be less than the maximum uncertain associated with the calibration points. When a large number of calibration points are available, as the power of the fitted polynomial increases, the modulus of the maximum deviation of the calibration points from the fitted curve decreases rapidly at first and then becomes more nearly constant. The order of the polynomial one beyond which this modulus becomes nearly constant is the one chosen as a satisfactory fit.

Computer generated plots of emf vs. temperature can quickly reveal an adequate mathematical fitting of the approximating functions to the experimental data.

Another method used specifically for the calibration of thermocouples is to draw a curve of the differences between the measured values and a smooth thermocouple reference table for example the IEC reference tables. In this method, which has been detailed by Roeser and Lonberger (10), the difference between the measured values and the reference tables at the calibration points are determined, and smooth curve is fitted to these differences. The values of the difference curve are then added to the reference tables to give the final calibration table for the thermocouple. However any irregularities in the reference tables will produce irregularities in the calibration table.

3.1.1 Interpolation by using the standard reference function

Standard reference functions play a very important role in thermocouple thermometry and lead to considerable savings in time and effort. A standard reference function represents the behavior of typical example of particular type of thermocouple. The calibration of an individual example of this type of thermocouple is thus reduced to determining the difference between its behavior and that of the standard embodied in the reference function. Provided that the original work which led to the reference function was well done and that the thermocouple conform to the IPTS criteria mentioned before. It will be found that the differences are very small. The number of the calibration points necessary to determine adequately the differences from the standard reference function will be correspondingly small and the whole process will be simple and economic.

Figure (7) shows differences from the standard reference function measured for a number of type R thermocouples. The differences from the table were measured (Coats 1978a) at the freezing point of Zinc (≈419 °C), Silver (≈960 °C) and Gold (≈1064 °C). The accuracy of the measurement was estimated to be ±0.2 °C.

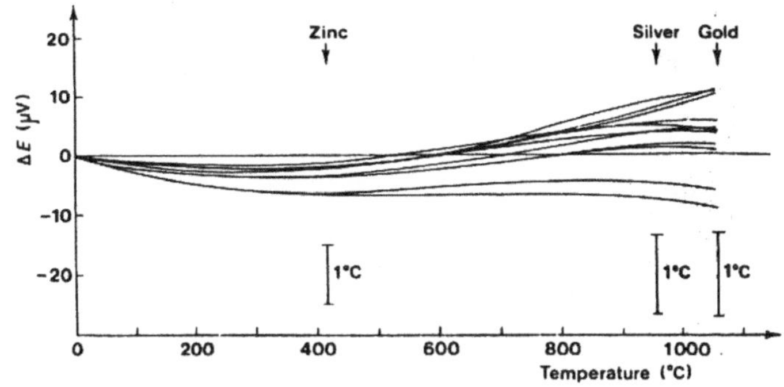

Fig. 7. Difference curves between the IEC Standard Reference Table and the emfs of a series of Pt-13%Rh/Pt thermocouples.

It is clear that a quadratic expression of the differences is perfectly adequate to give a calibration of each thermocouple within the limits of accuracy of measurement.

3.2 The reference functions for Au/Pt thermocouples

In that work Gotoh et al (5) established a reference function (reference table) for the Au/Pt thermocouple through the investigation of the emf-temperature relationship over the range 0-962 °C, with such a reference function calibration at a small number of fixed points to fit the deviation of particular Au/Pt thermocouple from the reference function should be sufficient to permit high precision temperature measurement over the full range 0-962 °C.

The thermocouples used in that project were initially annealed after annealing they were calibrated at the metal fixed points using the well developed techniques by McLaren and Murdock (11). The measurement sequence was indium, tin, zinc, antimony, Au-Cu eutectic (liquidus point) and silver. The emf was measured with a Keithley nanovltmeter.

Next a comparison measurement was performed in the stirred bath beginning in the water bath from 25 to 75 °C followed by the oil bath from 75 to 300 °C and finally the nitrate salt bath from 300 to 494 °C. For intercomparison a high temperature platinum resistance thermometer HTSPT was used.

The emf temperature data were analyzed by fitting polynomials of various degrees by the method of least squares. The preferred equation relating emf in millivolts to t90(°C) is

$$emf = \sum_{i=1}^{8} C_i (t_{90})^i \qquad (6)$$

Where:

c_1= 6.0310705X10^{-3}, c_5=-3.5645621X10^{-14},
c_2= 1.9377291X10^{-5}, c_6= 3.1611501X10^{-17},
c_3=-2.2104563X10^{-8}, c_7=-1.7549464X10^{-20},
c_4= 3.1103354X10^{-11}, c_8= 4.2602445X10^{-24}

The above reference function provides emf in μV as a function of temperature in degrees Celsius to within ±10 mK from 0 to 962 °C. The agreement between the intercomparison data and the fixed-point data was not as good as was expected by the authors.

3.3 The reference function for Pt-10%Rh vs. Pt thermocouple

In July 1990 the Comite Consultatif de Thermometrie requested its working group 2 to collaborate with national laboratories in the production of new reference tables and functions for the standard Pt-10% Rh vs. Pt thermocouples based on the International Temperature Scale of 1990 (ITS-90). Pursuant to this recommendation, eight national laboratories [12] have obtained new data for Pt-10%Rh vs. Pt thermocouples. The thermoelectric voltages of those thermocouples have been measured as a function of t(90) in the range -50 °C to 1070 °C with temperatures obtained from standard platinum resistance thermometer that were calibrated in accordance with the ITS-90 to 962 °C and extrapolated to 1070 °C. In addition the thermoelectric voltage of the thermocouples has been determined at various thermoelectric fixed points. From the results of these measurements, polynomials giving the thermoelectric voltage as a function of t(90) have been developed.

The new reference function for the Pt-10% Rh vs. Pt thermocouples for the temperature range from -50 to 1064.18 °C is of the form

$$E = \sum_{i=0}^{8} a_i \cdot (t_{90})^i \tag{7}$$

where t(90) is in degree Celsius and E is in microvolts. The coefficients of equation for the temperature range –50oC to 1064.18 °C are;

a_1=5.40313308631	a_5=-3.31465196389X10^{-11},
a_2=1.25934289740X10^{-2},	a_6=2.55744251786X10^{-14},
a_3=-2.32477968689X10^{-5},	a_7=-1.25068871393X10^{-17},
a_4=3.22028823036X10^{-8},	a_8=2.71443176145X10^{-21}

The eight laboratories involved in this study are:

- National Institute of Standards and Technology NIST, USA,
- Istituto di Metrologia G. Colonnetti IMGC, Italy,
- Korea Research Institute of Standards and Science KRISS, Korea,
- National Physical Laboratory NPL, UK,
- National Research Laboratory of Metrology NRLM, Japan,
- Mendeleyev Institute for Metrology VNIIM, Russia,
- Van Swinden Laboratorium VSL, Netherlands,
- And Shanghai Institute of Process Automation Instrumentation SIPAI, China.

4. Factors affecting the calibration constancy and reproducibility of Pt-10%Rh vs. Pt thermocouples

4.1 The effect of temperature gradient on annealed thermocouples

Stabilizing the gradient and the Seebeck coefficient can minimize the thermocouple drift. If the impurity concentration, (homogeneity) along the wires is reasonably uniform over the section of wires that are to be subjected to the main gradient, then as the average Seebeck coefficient is constant along the wire, the thermal emf generated will depend only upon the temperature difference and will be independent of the temperature gradient as the junction is displaced along the ingot. If on the other hand the impurity distribution is not uniform within the gradient zone then there should, obviously, be measurable gradient dependence.

4.2 The effect of soaking thermocouples in a high temperature gradient

The most likely reason why annealed thermocouples are unstable when introduced into a temperature gradient is that while uniform temperature homogenize impurities within the wire during annealing, temperature gradients induce segregation of impurities.

When a thermocouple wire is annealed, it is subjected to a uniform high temperature. If there are impurity concentration gradients within the wire they gradually disperse by diffusion as the wire approaches a stable thermodynamic equilibrium state. As the rate of approach to homogeneity depends on the mobility of impurities and as this mobility increases with increasing temperature, the higher the annealing temperature the faster the thermocouple wire will reach its annealed state. If however, an annealed thermocouple wire is inserted into a temperature gradient region then the temperature gradient will induce thermal diffusion of impurities. Thermal diffusion will cause segregation of impurities until a steady state is reached in which the flow induced by the temperature gradient is in balance with that induced by the gradient of concentration.

Due to the relatively low mobilities in the solid state, especially at lower temperatures ($T \leq 1000\ ^{o}C$), such diffusion process take a very long time to reach the steady state and hence thermoelectric stability. In the present study the thermocouples are gradient stabilized in the Ag furnace. They remained stable in all other fixed point furnaces because, although the temperature gradient was altered, the temperatures were also lowered thus lowering the mobility to low enough values that the altered gradients could not cause major impurity rearrangement within the short duration of our measurements. Such impurity gradient accumulation by thermal diffusion was reported by Darling and Selman (13). According to their findings a Pt vs. Pt-13%Rh thermocouple was used to monitor a temperature set at 625 ^{o}C. After several years of use the Pt-13%Rh wire showed a measurable depleted Rh concentration in certain sections of the wire exposed to high temperature and an increased Rh concentration towards the colder parts of the wire. Soaking the wire at a uniform temperature of 1200 ^{o}C the accumulated Rh concentration gradients dissipated (a tendency towards thermodynamic equilibrium) restoring the generated emf to its original value.

4.3 Factors affecting the calibration constancy and life of Pt-10%Rh vs. Pt thermocouples

The platinum vs. platinum rhodium thermocouple consists of a positive element of platinum 10 % rhodium and a negative element of platinum. The platinum from which the negative element is prepared is the purest metal produced commercially. This high purity is necessary in order to prevent unpredictable thermoelectric effects caused by impurities. Contamination of the pure platinum leg in service is perhaps the main reason for loss of calibration. In respect to the platinum vs. platinum 10 % rhodium couples, the following also may limit the life or constancy of calibration;

1. Diffusion of rhodium from the bead into the Platinum.
2. Volatilization of the Platinum or Rhodium.
3. Contamination of the thermocouple.

All of these are temperature dependent. The second is also dependent on the atmosphere in which the thermocouple is used. The third is dependent on the nature of the insulating and protection tubes and the atmosphere in which the thermocouple is used. These will be discussed in the following.

Diffusion of rhodium increases with the temperature and the time during which the thermocouple is exposed to the temperature. Its effect is negligible (14) if the furnace temperature is uniform, and if there is sufficient depth of uniform temperature, as then the zone containing the diffused rhodium acts only as a connector between the legs of the thermocouple and so does not interfere with the indications of the thermocouple. Should the diffused zone enter a region of non-uniform temperature, the result would be to reduce the emf output of the thermocouple, thus making the indicated temperature lower than the true temperature of the furnace. Diffusion of rhodium into the platinum leg may also occur by volatilization from platinum-rhodium wire and deposition on to the platinum. This only significantly occurs at temperatures over 1200°C. Its occurrence can be detected by an examination of the twin bore insulating tube, where small metallic crystals will be seen.

The loss of platinum and rhodium in air or oxidizing atmospheres is higher than in vacuum. This has been attributed to the formation of volatile oxides. The rate of volatilization depends upon the temperature, percent oxygen, and the velocity of the gas stream over the metal surface.

Contamination of a thermocouple is by far the easiest method of shortening its life. Should any metals (solid, liquid or vapour) come into contact with the thermocouple, alloying will take place and this will change the emf of the thermocouple and render it unfit for further use. Furnace gases or chemical fumes, particularly those carrying sulfur compounds may cause embrittlement of the wires in addition to affecting the emf output. It is very difficult to predict how long a couple may last. So many factors arc involved in this type of an evaluation that a standard answer cannot be given. Some experimental results show that volatilization appears to be the main problem.

5. General conclusion

- Temperature sensors constructed from high-purity noble metals exhibit the highest degree of interchangeability and stability in comparison to all other artifact thermometers.
- The Au/Pt thermocouple should be considered as a suitable and challenging alternative to the high temperature platinum resistance thermometers HTSPRT for many precise temperature measurements in the range from 660 °C to 950°C. The HTSPRT, however, is expensive about (10,000 US $) and fragile and requires either a DC resistance comparator bridge (60,000 US$) or an AC resistance comparator (50,000 US$) to measure electrical resistance. In contrast, a normal Au/Pt thermocouple with 170 cm long, 0.5 mm diameter thermo-element, has a material value of (1000 US$) and requires only a superior digital Nanovoltmeter (≅3000 US$) to measure the thermo-emf.

The Au/Pt thermocouple, following suitable high temperature preparatory annealing, is also capable of repeatability of few mK at the Ag point and the uncertainties in temperature measurements arising from the intrinsic thermoelectric inhomogeneity along the Au and Pt wires can be reduce to 10 or 20 mK.

Further on the basis of cost, simplicity of construction, ruggedness and reconstruction of damaged thermo-elements by re-annealing and re-assembly procedures.

- Au/Pt thermocouple should be given serious considerations as a superior replacement for the standard Pt-10% Rh vs. Pt thermocouple in the temperature range from 0 to 1000 °C and as a worthy alternative to the high temperature platinum resistance thermometer in the range 500 °C to 1000 °C in the future International Temperature Scale.
- Following careful annealing of Au and Pt wires forming the thermocouple to stabilize their thermopower for sustained exposure to high temperature, the Au/Pt thermocouple has distinct advantages over the standard Pt-10%Rh vs. Pt thermocouple for temperature measurements up to 1000 °C. Substituting Au element for the Pt 10%Rh element in the standard thermocouple eliminates the thermoelectrically degrading effect on the thermopower of the Pt-10% Rh element.

6. Acknowledgement

The author express his gratitude to Prof. Dr. M. M. Ammar, for his helpful discussion and advice during the progress of the work.

7. References

[1] Roeser, W.F. and Wensel H.T.: Temperature, its Measurement and Control in Science and Industry, 1, (1941) pp.1309.

[2] Mclaren E. H. and Murdock E.G., "Au/Pt thermocouple–27703", National Research Council of Canada, NRCC, (1987).

[3] Burns G. W., Strous G.F., Liu B.M. and Mangum B.W.: Temperature, its measurement and Control in Science and Industry, AIP, 5, pp. 541-548, (1982).

[4] Preston-Thomas, H.: Metrologia, 27, 3-10 and 107, (1990).

[5] Gotoh M., Hill K. D. and Mwdock E.G., Gold versus platinum thermocouples reference table: Rev. Sci. Instrum. 62, 2778-2791, (1991).

[6] Stability and Calibration of Platinum / Palladium Thermocouples Following Heat Treatment, Y. A. Abdelaziz, F. M. Megahed and M. M. Halawa. Measurement journal IMEKO-UK, Vol. (35), pp. 413-420, (2004).

[7] Ripple D. C., Burns G. W., "Thermoelectric properties of a selected lot of gold versus platinum thermocouples", TEMPMEKO, pp. 447-452, (2004).

[8] Hamming R.W., Numerical Methods for Scientists and EngineersP McGraw-Hill, pp. 81 (1962)

[9] T. P. Jones: TMCSI, 4 Part 3, pp. 1561-1567, (1972).

[10] Roeser, W.F. and Lonberger S.T.: National Bur. Standards Circular, 1, pp. 590, (1958).

[11] Mclaren E. H., Can, J. Phys., 35, pp 78 (1957).

[12] Burns G. W., Strous G.F., Liu B.M. and Mangum B.W.: Temperature, its measurement and Control in Science and Industry, AIP, 6, pp. 541-546, (1992).
[13] Darling A. S., Selman G.L.: Temperature, its measurement and Control in Science and Industry, 4, pp. 1633-1644, (1972).
[14] Mortlock A. G., J. Sci. Instr., Vol. 1, pp. 256-258, (1958).

Part 4

Advance Catalyst

11

Role of Precious Metal Catalysts

Takashiro Muroi
Industrial Catalysts Laboratory
Japan

1. Introduction

Precious metal catalysts have been used in many industries, such as refinery, petrochemicals, polymer, chemicals, pharmaceuticals, and environment, For example, automotive industry utilizes large amounts of precious metal catalysts as a part of auto exhaust gas purifier, because of their high activity and selectivity and stability under various reaction conditions. This article, introduces wide variety of applications on precious metal catalysts. We recognize that precious metal catalysts play very important role in our lives. Not only unique properties of precious metals but also advanced preparation technology allow us to use precious metal catalysts for wide range of applications. Principle of precious metal catalyst preparation is introduced, and various industrial applications of precious metal catalysts follow. Applications of precious metal catalysts were disclosed by some literatures. (Rylander 1967) (Bartholomew and Farrauto 2006)

2. Properties of precious metal catalysts

2.1 High activity and selectivity of precious metals in catalysis

Precious metal catalysts consist of highly dispersed nano-scale precious metal particles on supports with high surface area such as carbon, silica, and alumina. The nano scale metal particles easily adsorb hydrogen and oxygen in the atmosphere. The hydrogen or oxygen is very active due to its dissociative adsorption through *d*-electron of out of shell of precious metal atoms. Dissociatively adsorbed hydrogen and oxygen readily react with many substitutes at under the mild conditions. Although commercial plants are operated under heated and pressurized conditions, many hydrogenation and oxidation reactions proceed at room temperature. In such case, the product yield is relatively high because byproducts formation can be minimized under milder reaction conditions. For example, in hydrogenation of di-nitrotoluene to di-aminotoluene, reaction conditions of Ni catalyst requires 5 Mpa, 150 °C, and solvent such as methanol. However, such reaction with Pd/carbon proceeds under only 0.4 Mpa, and 90 °C without solvent. Quantity for Pd/carbon is one tenth of Ni catalyst.

Each precious metal catalyst shows unique characteristics, For instance, olefin hydrogenation can be accomplish with Pd/Al_2O_3 without hydrogenating aromatic bond at mild condition. In case of hydrogenation of phenol, Pd/Al_2O_3 gives produces cyclohexanone. Similarly, Pt/Al_2O_3 is highly selective to cyclohexane formation and Rh or Ru/Al_2O_3 gives cyclohexanol selectively. Furthermore, preparation technology of precious metal catalyst has been advanced in recent years based upon nano technology. As a result, energy saving, high productivity, production cost reduction were achieved. Precious catalyst is not almighty, low activity for hydrogenolysis of esters, low selectivity for selective oxidation in gas phase. However, many base metal catalysts changed to precious metal catalysts in history.

2.2 Stability

Precious metals are stable. They do not easily form oxides by oxidation. The oxides of precious metals are, on the other hand, relatively not stable. Precious metals do not easily dissolve in acid or alkaline solution. Thus for example, Pd/carbon can be used for hydrogenation of maleic acid in water which is acidic condition. However Ni catalyst cannot be used for the same reaction condition due to the leaching. Melting point of precious metals is higher than base metals. It corresponds to resistance to migration and sintering of precious metal catalysts. Because of high thermal stability, precious metal catalyst has been used as automotive exhaust gas purification catalysts.

2.3 Advance in catalyst preparation method

2.3.1 Control of metal particle size

Catalytic activity and selectivity largely depend upon metal particle size of metals of the catalysts for the some reactions. It is necessary to make sure if the development of reaction is influenced by metal particles size prior to selection of proper catalyst. (Bond 1968) Generally, hydrogenation and hydrogenolysis prefer to small particles, and oxidation prefers to large metal particles. However, there are many exceptions because of different carrier, preparation method and impurity of the catalysts. Smaller metal particles show high tolerance of sulfur poison because of high surface area for adsorption of sulfur. (Okada 1993) The size of metal can be controlled by selection of metal salts, conditions of impregnation and reduction conditions, including metal concentration and reduction temperature.

2.3.2 Metal distribution of catalysts: Ex. eggshell type

Diffusion of reactant to the catalyst surface controls reaction rate at mild condition. Hydrogen and oxygen penetrate into inside of micro pores in the catalysts. However, large molecules are difficult to diffuse into the smaller pores. BASF Catalysts., (Former Engelhard) developed and commercialized highly active hydrogenolysis catalyst which is eggshell type of Pd/carbon as a slurry type catalyst. The catalyst was applied for hydrogenolysis to produce aspartame. DuPont developed eggshell type of vinyl acetate catalyst. This is fixed-bed type catalyst, which is prepared by impregnation method using Pd and Au salt and SiO_2 carrier. Preparation of vinyl acetate catalysts consist of, 1) impregnation of metal solution, 2) metal fixation by alkaline, 3) keeping long times, 4) treated with reductive agent, 5) washing, 6) drying, 7) impregnate potassium acetate and 8) dry. Pd and Au deposited outer surface layer of the carrier, which is egg-shell type distribution. On the other hand, in case reduction is immediately employed after impregnation, Pd, and Au deposited more inner layer on the carrier. (DuPont 1977) Eggshell type catalysts have been applied for many other reactions, such as hydrogenation of polymer resin and hydrogenolysis of large molecular, and are expected to be used for other reactions as well. However, there is an example that highly dispersion, not eggshell type metal distribution, shows higher activity than eggshell type catalysts in the hydrogenation of benzoic acid under severe condition because of no diffusion control. (Grove 2002)

2.3.3 Unreduced catalysts

The reduced catalysts have been supplied commercially. Unreduced catalysts give higher activity for some hydrogenation reactions. Small particles are formed possibly due to some effect by solvent, which is in-situ reduction before the reaction starts. A combination of eggshell and unreduced catalyst also gives higher activity in hydrogenation at mild condition. But, unreduced catalysts cannot be used for dehydrogenation or oxidation. Unreduced $Pd(OH)_2$/carbon, called Pearlman's catalyst, is used widely in pharmaceuticals.

2.3.4 Bi-metalic catalysts

Bi-metalic Pd-Ru/carbon catalyst is highly active for hydrogenation and hydrogenolysis of 1,4-butyndiol. Rh-Ru/carbon shows high activity for hydrogenation of aromatics. (Rylander 1967) Adams catalyst, Pt-Ru shows highly active for aromatic nitrocompounds. (Bond 1964) Pt doped Pd/carbon gives excellent activity for debenzylation. (N.E.Chemcat 2002)

2.3.5 Modified catalysts: Doping

Modification effectively improves activity and selectivity of catalysts. Addition of alkaline or alkaline earth metal, such as Na, K, Ba, to Pd/carbon leads to high activity towards hydrogenation of aromatic nitro compounds because of electron transfer to Pd. Pd-Pt-Fe/carbon catalyst is commercialized for hydrogenation of di-nitrotoluene. Pd modified with Na is applied for selective hydrogenation of mesityl oxide to produce MIBK. Na prevents hydrogenation of carbonyl. Selective hydrodechlorination can be achieved by Pd/carbon modified Ag or Sn. Pd/Al_2O_3 modified with heavy metals, such as Pb, and Bi, is applied for hydrogenation of phenyl acetylene in styrene stream. Modification with heavy metals gives leads to high activity for liquid phase oxidation.

2.3.6 Control of migration and sintering

Pd-Au/SiO_2 is used for production of vinyl acetate by acetoxylation of ethylene. In case of Pd/ SiO_2 without Au addition, sintering is observed, which is caused through palladium acetate formation within a few days. Thus, Au addition prevents sintering of Pd by formation of solid solution with Pd. Fluorocarbon substitute, R-134a is introduced from R-114 by hydrodechlorination. Pd/carbon granular shows short life because of sintering in hydrogen chloride. Addition of small amounts of Re or Au to Pd/carbon gives lengthens catalyst life. (Asahi Glass 1989) High thermally stable catalysts have been developed for automotive exhaust catalysts. Precious metal loading amount was needed to be reduced. The technology by interaction with metal oxide such as ZrO_2, CeO_2, TiO_2, enables lower loading of precious metals.

2.3.7 Control of dissolution: Leaching

Although precious metal is stable for migration and/or dissolution, it occurs in acidic or alkaline condition with oxygen atmosphere. Pt ion is leached out from the catalyst a few %

at reflux condition. Pd/Al_2O_3 is highly active and selective in hydrogenation of vinyl acetylene to butadiene in order to recover butadiene after extraction of butadiene. However, Pd is dissolved with acetylene by producing acetylide in short time. Pd_4Te/Al_2O_3 was developed to prevent leaching by acetylene, and the catalyst life was significantly improved. Pd_4Te is a chemical compound formed on the alumina carrier. The preparation is, at first Pd/Al_2O_3 is prepared, and secondary $TeCl_4$ is impregnated to Pd/Al_2O_3. Subsequently, the impregnated catalyst is heated at 500 °C. (JSR 1987) Pd is dissolved in acetoxylation of butadiene producing 1,4-butandiol because of acidic condition and presence of oxygen. Pd_4Te/carbon is developed by Mitsubishi Chemical and commercialized. (Mitsubishi Chemical 1973) Small particle of Au, less than 5 nm shows excellent oxidation and esterification activity. Methyl glycolate can be produced by ethylene glycol and methanol in slurry bed, of Au/TiO_2-SiO_2. This catalyst gives high activity, however Au leached out at oxidation condition. Addition of Pd to Au prevents dissolution. (Nippon Shokubai 2004)

3. History/background of precious metal catalyst development

3.1 Sulfuric acid

Industrial revolution started in England in 19th century. Demand of sulfuric acid as basic chemicals, especially producing sodium carbonate for Leblanc process, was increasing for the purpose of bleaching cotton cloth. The first catalytic process is oxidation of sulfur dioxide by Pt/asbestos which was developed by P. Philips in England in 1831. Lead chamber process replaced by catalytic process. That is the beginning of industrial catalysts. Ammonium sulfate for fertilizer production started using sulfuric acid and ammonia from coal distillation plant in the late of 1830s. Production of explosive cellulose nitrate from niter using sulfuric acid has started in Switzer-land in 1845. Novel developed dynamite which is nitro glycerol impregnated clay in 1866. Pt/asbestos had been used until finding V_2O_5 catalyst by BASF in 1915.

$$SO_2 + 1/2\,O_2 \rightarrow SO_3$$

3.2 Nitric acid

W. Ostwald found production of nitric acid by oxidation of ammonia with Pt-plate, and the nitric acid plant was commercialized in 1908. He received Novel prize in 1909. Nitric acid is very important to produce explosives. Ammonia was produced in coal furnace at the time. Commercialization of ammonia synthesis using nitrogen in the air by Haber and Boch was in 1913. Nitric acid was produced by CuO-MnO_2 or Fe_2O_3-Bi_2O_3 once. But since base metal catalysts show lower activity, Pt-Rh-gauze has been used now. Vaporized Pt is collected by Pd at downstream.

$$NH_3 + 5/4\,O_2 \rightarrow NO + 3/2\,H_2O$$

$$NO + 1/2\,O_2 \Leftrightarrow NO_2$$

$$2\,NO_2 \Leftrightarrow N_2O_4$$

$$3\,NO_2 + H_2O \Leftrightarrow 2\,HNO_3 + NO$$

3.3 Acetic acid

Cellulose nitrate as fiber was replaced by cellulose acetate for flammability issue. Acetic acid, raw material of cellulose acetate was produced by oxidation of acetaldehyde which was produced hydration of acetylene by $HgSO_4$ in coal times. Coming petroleum time around after 1960s, acetaldehyde has started to be produced by oxidation of ethylene. The process was developed by Hoechst-Waker which uses $PdCl_2$ and $CuCl_2$ in homogeneous reaction.

$$CH_2{=}CH_2 + 1/2\,O_2 \rightarrow CH_3CHO \quad (PdCl_2\text{-}CuCl_2)$$

$$CH_3CHO + 1/2\,O_2 \rightarrow CH_3COOH \quad (Mn(AcO)_2$$

Most of acetic acid is produced by carbonylation of methanol in the world, which was developed by BP (Former Monsant) using Rh carbonyl complex. The product yield is more than 99%. Consumption of Rh is less than 1ppm.

$$CH_3OH + CO \rightarrow CH_3COOH$$

New development, called CATIVA process by BP, uses Ir and Ru carbonyl complex. The precursors are IrI_3 and RuI_3. Ru complex works as co-catalyst for Ir complex which consumption is said one tenth of Rh complex. In addition, Showa Denko in Japan developed direct acetic acid process from ethylene by gas phase Wacker process. The catalyst is Pd/ $H_4SiW_{12}O_{40}$-SiO_2. The process was commercialized once. (Showa Denko 1994)

$$C_2H_4 + H_2O + O_2 \rightarrow CH_3COOH + H_2O$$

Chiyoda Corporation has developed immobilized Rh complex with resin. It was made as fixed bed process producing acetic acid by carbonylation of methanol. The process was licensed to China. (Chiyoda Corporation 1993)

4. Refinery applications

4.1 Reforming

Due to increasing gasoline demand, reforming process, which produces gasoline with high octane number, was developed by Shell in 1949. The reactions are dehydrogenation of naphthene and isomerization of paraffin. The UOP reforming process is called "Platformer" because the catalyst constituent is Pt. At first, the catalyst life was short due to carbon deposition. In 1967, Chevron developed Pt-Re/Al_2O_3 bi-metalic catalyst which can be used almost 10 years with several times of regeneration. After that, continuous cyclic regeneration (CCR) process was developed by UOP and IFP. This process can produce gasoline with higher octane number due to higher aromatics yield under much severe condition or under low pressure. The catalyst is spherical Pt-Sn/Al_2O_3. More than three hundred plants including semi-regeneration process are running in the world.

4.2 Aromatization

To meet the demand of aromatics for in petrochemical industries, aromatization of n-hexane process was developed by Chevron. The catalyst for this application is Pt-F/L-zeolite which was improved by Idemitsu Kosan Co., Several plants in the world have been applying this technology.

4.3 Removing aromatics

Removing aromatics from diesel fuel is necessary for environmental protection of atmosphere for more severe environmental protection. Many hydrotreating processes are developed, and consist of two series reactors of reactors. First reactor contains Ni-Mo/Al_2O_3 or Co-Mo /Al_2O_3 and the second contains Pt/Al_2O_3 or Pt/SiO_2-Al_2O_3. Bi-metalic catalyst was also developed because hydrogenation by Ni needs high temperature which occurs dehydrogenation to minimize the negative influence by sulfur. Pt-Pd/SiO_2-Al_2O_3 shows high activity even containing several tens ppm of sulfur. (Vaarkamp 2000)

4.4 Isomerization

Isomerization of butane and pentane is difficult by SiO_2-Al_2O_3. Bi-functional catalyst, Pt/SiO_2-Al_2O_3 with halogen can progress isomerization because the reaction occurs through dehydrogenation and hydrogenation by Pt and isomerization by acid. Jet fuel production, for example Hysomer process developed by Shell uses Pt/zeolite under hydrogen pressure. Japan Energy developed Pt-SiO_2/ZrO_2-Al_2O_3 for isomerization of naphtha. Small content of sulfur in naphtha deactivate this catalyst. Hybrid catalyst, consisting of Pt/SiO_2/ZrO_2-Al_2O_3 with addition of Pd/Al_2O_3 was developed to improve sulfur tolerance. The mechanism of sulfur tolerance is decomposition of sulfur compound by Pd is explained by Watanabe et al. (Watanabe 2005)

4.5 Hydrocracking

Gas oil or heavy gas oil is hydrocracked to produce gasoline. The typical cracking catalysts are Co-Mo/Al_2O_3 or Ni-W/SiO_2 or zeolite. Unicracking process licensed by UOP uses Pd/Y-zeolite producing jet fuel.

4.6 FCC additives

Fluid Catalytic Cracking (FCC) processes produce gasoline or diesel fuel from oil from heavier oil. FCC processes are operated in fluid bed with continuous regeneration processes for removal of deposited carbon. The size of catalyst is 70~100 μm ϕ composed SiO_2-Al_2O_3 and USY zeolite. Some additives improve stability and propylene yield. Pt/Al_2O_3 powder is also used to accelerate removal of deposited carbon.

5. Application in petrochemical

5.1 Hydropurification

Ethylene and propylene are important raw materials for many chemicals and polymers. Ethylene and propylene contain acetylene and dienes as impurities from cracker. Acetylenes and dienes are selectively hydrogenated to ethylene and propylene respectively with 0.02-0.2% Pd/Al_2O_3 in gas phase. These catalysts life is more than several years with carbon removal regeneration. Purification of butadiene in presence of 1-butene and 2-butene is employed under hydrogen by Pd/Al_2O_3 in liquid phase. Unsaturated C_4 and C_5 molecules are recycled to ethylene cracker after hydrogenation to increase ethylene yield. Small amount of phenyl acetylene as an impurity is removed as polystyrene applying by hydrogenation by Pd/Al_2O_3 or modified Pd/Al_2O_3. (Table-1)

Stream	Removal	Catalysts	Phase	Remarks
Ethylene	Acetylene	0.01-0.02%Pd/Al_2O_3	Gas	
Propylene	Acetylene	0.1-0.2%Pd/SiO_2	Gas	
Butadiene	Vinyl acetylene	0.2%$Pd\text{-}Te/Al_2O_3$	Liquid	
1-Butene	Butadiene	0.5%Pd/Al_2O_3	Liquid	Addition of CO
2-Butene	Butadiene	0.1%Pd/Al_2O_3	Liquid	
C_4, C_5	Butene	0.3%Pd/Al_2O_3	Liquid	Return to cracker
Styrene	Phenyl acetylene	0.3%Pd/Al_2O_3	Gas	

Table 1. Hydropurification in ethylene plant

5.2 Hydrogenation of pyrolysis gasoline

Byproducts in liquid phase of ethylene cracker are aromatics and unsaturated compounds. In order to blend into gasoline, di-olefins are selectively hydrogenated to mono-olefin by Pd/Al_2O_3. Subsequent to the first reactor, $Ni\text{-}Co/Al_2O_3$ is utilized to remove mono olefins, sulfur, and nitrogen prior to production of benzene, toluene, and xylene before solvent extraction. Catalyst life of Pd/Al_2O_3 is generally several years with several times of regenerations. Every ethylene cracker equips this hydrogenation process.

5.3 Propylene

Propylene demand is increasing due to strong polypropylene market. Dehydrogenation of propane with Pt/Al_2O_3 is one of commercialized processes. Several plants apply this process.

$$CH_3CH_2CH_3 \rightarrow CH_2{=}CHCH_3 + H_2$$

5.4 Cyclohexane

Highly pure cyclohexane is intermediate of Nylon. It is produced by hydrogenation of benzene with Ni or Pt/Al_2O_3.

5.5 Oxo alcohol

2-Ethylhexanol, a raw material of di-octylphthalate (DOP), is used for vinyl chloride resin plasticizer. 2-Ethylhexanol is derived from n-butyl aldehyde is produced by hydroformylation of propylene with hydrogen and carbon monoxide by Rh complex catalyst. $RhH(CO)(PPh_3)_3$ with tri-phenylphosphine produces n and i butylaldehyde. Rh complex generates more than 10 of n/i ratio. (Table-2) In the same reaction, Co complex shows low n/i ratio. n-Butylaldehyde is separated by distillation from the boiling reactor. 2-Ethylhexanol is produced after condensation with NaOH and subsequent hydrogenation by Ni. (Fig. 1)

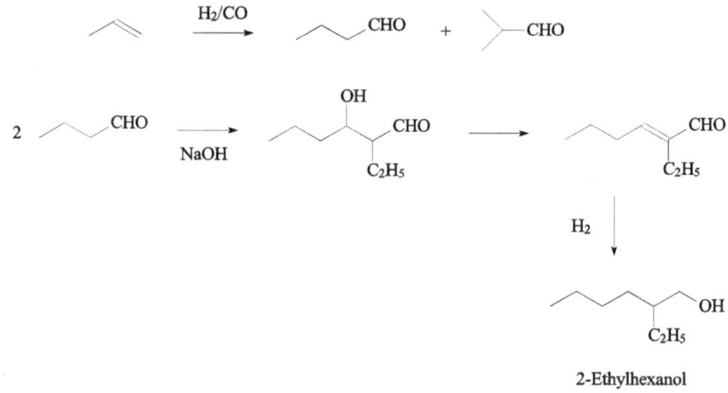

Fig. 1. 2-Ethylhexanol from propylene by hydroformylation

Catalyst	$RhH(CO)(PPh_3)_3$
Temp. °C	60～ 120
Pressure atm	1 ～50
Catalyst % Metal/Olefin	0.01～0.1
n / *i*	10 / 1
Conversion of Propylene %	30

Table 2. Hydroformylation by $RhH(CO)(PPh_3)_3$

6. Bulk chemicals

6.1 Ammonia

Fe-K catalysts have been used for last hundred years under various reaction conditions since Habar-Boch developed the catalyst. In addition, Ru-Cs/carbon graphite catalyst was developed by BP and Kellogge. (Czuppon 1993)

$$N_2 + 3/2\,H_2 \rightarrow 2\,NH_3$$

The process called KAAP process is commercialized. Two large plants are under operation in Trinidad and Tobago. Adsorption of nitrogen becomes weak due to electron transfer from Cs to Ru and Ru and from Ru to carbon graphite. The catalyst support, graphite is firstly, produced by thermal treatment of active carbon at 1,500 °C in argon, and in air at around 400 °C followed by subsequent thermal treatment at 1,700 °C in argon. Remaining catalyst preparation takes place as, impregnation of $RuCl_3$ solution to the graphite, reduction with hydrogen at 450 °C, drying, impregnation of Cs nitrate solution, and drying. (BP 1984)

6.2 Hydrogen peroxide

Hydrogen peroxide, H_2O_2, has widely been used as bleaching and oxidation agent for food, textile and paper. H_2O_2 has been producing by oxidation of hydroxyl alkyl anthraquinone after hydrogenation of alkyl anthraquinone by Pd. Small catalyst particles of Pd/Al_2O_3 or Pd/SiO_2 is generally used in slurry phase, and its alkaline modification prevents ring hydrogenation. Kemira process uses Pd black. Some fixed bed processes are commercialized.

These catalysts form smooth surface because decomposition of hydrogen peroxide occurs by leaked fine catalyst particles after separation. Carbon monoxide in hydrogen stream is removed to less than 1 ppm by methanation with Ni or Ru/Al_2O_3 since small quantity of carbon monoxide poisons Pd catalyst at lower reaction temperature, e.g. 40 °C. (Fig. 2)

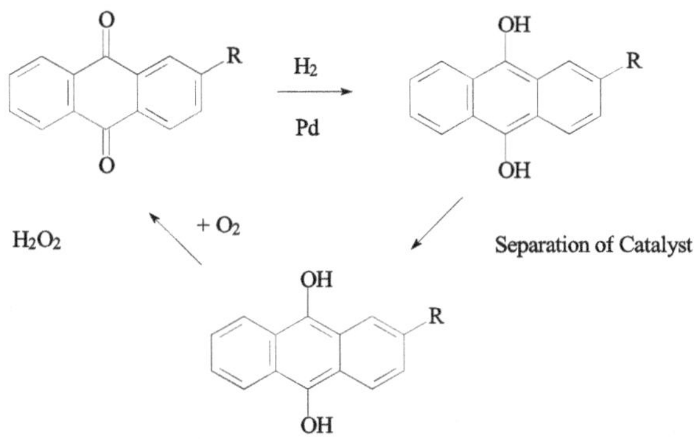

Fig. 2. Production of hydrogen peroxide by anthraquinone process

Direct hydrogen peroxide production, which is the process synthesize H_2O_2 from hydrogen and oxygen was developed by DuPont and commercialized several plants of paper plants in North America. This process uses Pd/carbon. HBr is used stabilizer of peroxide.

$$H_2 + O_2 \rightarrow H_2O_2$$

Head water has developed direct process at out of explosion range, e.g. H_2 3vol%, O_2 20vol% N_2 77vol% in H_2SO_4 solution with addition of small amount of NaBr and Pd-Pt/carbon black catalyst. The performance of this process is reported as 33% of the overall hydrogen conversion, 100% H_2O_2 selectivity with respect to hydrogen and 4.8% of hydrogen peroxide in a final liquid product at 45 °C and 10 MPa. (Headwater 2005)

6.3 Phenol

Phenol has been used as a raw material of phenol resin, nylon and others. Most phenol process uses cumene process in the world. Cumene is produced by alkylation of benzene with propylene by acid catalyst, such as H_3PO_4/Al_2O_3 or zeolite. Subsequently cumene hydroperoxide is produced by oxidation of cumene and phenol and acetone are converted from cumene hydroperoxide by sulfuric acid. In the step of phenol production, α-methyl styrene produced as byproducts is hydrogenated by Pd/Al_2O_3. Produced cumene by hydrogenation of α-methyl styrene is recycled to the feed. When α-methyl styrene demand increases, cumene hydroperoxide is hydrogenated to cumyl alcohol by Pd/Al_2O_3 and returned to stream to produce α-methylstyren more. Off gas treatment, such as oxidation of cumene, takes place by Pt/honeycomb catalyst at 350°C.

6.4 Aniline

Aniline has been produced by amination of phenol using Lewis acid such as Al_2O_3 and hydrogenation of nitrobenzene by Ni or Pd/carbon in slurry phase. Pd/carbon or modified Pd/carbon by Pt or Fe is used mainly for highly activity.

6.5 Chlorine

Hydrogen chloride is produced as byproducts in TDI process, polycarbonate and other plants. It has to be neutralized by NaOH before disposal. Oxidation process for Cl_2 production was developed using acid resistant $RuO_2/TiO_2\text{-}\alpha Al_2O_3$ by Sumitomo Chemical and was commercialized 3 plants. The commercialized Sumitomo fixed bed process is compact compare to Mitsui fluid bed process using Cr catalyst. In the preparation of $RuO_2/TiO_2\text{-}\alpha Al_2O_3$, the carrier is firstly prepared by mixing $\alpha\text{-}Al_2O_3$ and TiO_2 (100%rutile) powder followed by extrusion and calcination at 800 °C, Ru salt is impregnated on the prepared $TiO_2\text{-}\alpha Al_2O_3$, and impregnated carrier is calcined. (Sumitomo Chemical 2004) Commercialized 2 plants are under operation with TDI plant and polycarbonate plant.

$$2\,HCl + 1/2\,O_2 \rightarrow Cl_2 + H_2O$$

6.6 1,4-butanediol

1,4-butanediol(1,4-BG) is used for production of polyester, polyurethane and others. Poly butylenes terephthalate (PBT) is used as engineering plastics for automotive and electronics. PBT is used as performance textile for sport wears and swimsuits, for instance, known spandex. 1,4-BG can be synthesized from butadiene by acetoxylation. The process is developed and commercialized by Mitsubishi Chemical. The catalyst is bi-metalic catalyst which is resistant to acetic acid. The first process was consisted of trickle bed reactor with Pd_4Te/carbon granular. The bi-metalic catalyst was prepared by, for example, charging active carbon granular into mixed solution of $PdCl_2$ and TeO_2 followed by drying and heating with methanol at 450 °C.(Mitsubishi Chemical 1973) (Tanabe 1981) (Ohno 2000)

Improved process use Pd_4Te/SiO_2 in fixed bed in up flow mode.

$$H_2C{=}CHCH{=}CH_2 + 2AcOH + 1/2O_2 \rightarrow AcO\text{-}CH_2CH{=}CHCH_2\text{-}OAc + H_2O$$

Di-acetoxybutene is hydrogenated to di-acetoxybutane by Pd/carbon and hydrolysis to 1,4-BG by acid ion exchange resin.

Dairen Chemical in Taiwan commercialized homogeneous process, which produces hydroxyfuran by hydroformylation with Rh hydride complex. Hydrogenation to 1,4-BG is conducted by Ni catalyst.(Fig.3)

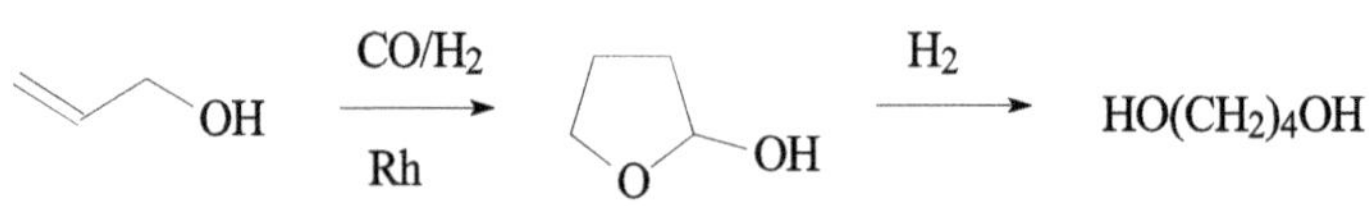

Fig. 3. Hydrofomylation of allylalcohol

6.7 MIBK

MIBK is widely used as a solvent. At first, diacetone alcohol is produced by condensation of acetone with $Ba(OH)_2$. And then, diacetone alcohol is dehydrated to mesityl oxide at second step. At last, mesityl oxide is hydrogenated to MIBK by Cu catalyst or Pd/carbon. Direct synthesis process of MIBK, Pd/acid ion exchange resin catalyst. The condensation and dehydration are with ion-exchange resin. Hydrogenation is done by Pd catalyst. The catalyst is prepared ion exchange of Pd salt and subsequently reduced by reductive agent. (Fig.4)

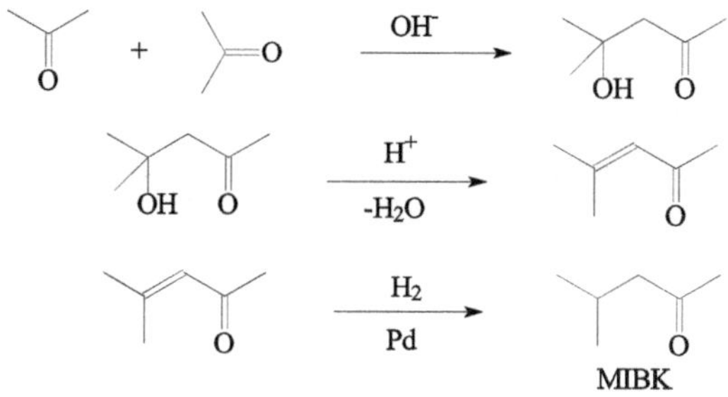

Fig. 4. MIBK production from acetone

6.8 Oxalic acid

Oxalic acid is used for hydration of starch, and it is a raw material of fertilizer and others. Di-methyl oxalate (DMO), an intermediate of oxalic acid, is synthesized from methanol and carbon monoxide. Methyl nitrate is produced at first under the mild condition without catalyst.

$$2\,CH_3OH + NO + 1/2\,O_2 \rightarrow 2\,CH_3ONO + H_2O$$

Carbon monoxide and methyl nitrate react to form DMO and NO by Pd/α-Al_2O_3. DMO is hydrated to oxalic acid. Produced NO is recycled in the system. For example, a reaction condition is at 0.2Mpa•G, and 110 ℃. (Yamamoto 2010) (Fig.5)

$$2\,CO + 2\,CH_3ONO \rightarrow (COOCH_3)_2 + 2\,NO$$

$$(COOCH_3)_2 + 4\,H_2O \rightarrow (COOCH_3)_2 \cdot 2\,H_2O + 2\,CH_3OH$$

$$\underline{2\,CH_3OH + 2\,NO + 1/2\,O_2 \rightarrow 2\,CH_3ONO + H_2O}$$

$$2\,CO + 3\,H_2O + 1/2\,O_2 \rightarrow (COOH)_2 \cdot 2\,H_2O$$

Fig. 5. Coupling of CO by methyl nitrate

Ethylene glycol (EG) can be produced from di-methyl oxalate by hydrogenolysis with Cu/SiO_2, e.g. at 3Mpa, 200 °C. China is expecting to produce EG form di-methyl oxalate using carbon monoxide which is originated from coal.

$$(COOCH_3)_2 + 4\,H_2 \rightarrow HOCH_2CH_2OH + 2\,CH_3OH$$

6.9 Di-methyl carbonate (DMC)

Di-methyl carbonate is a raw material of resin and others. DMC is expected to replace phosgene in production of polycarbonate and isocyanate. China started to use it diesel fuel additives. DMC is produced from methanol and carbon monoxide. Ube Ind. in Japan developed and commercialized a process applying oxy carbonylation of methanol by $PdCl_2$-$CuCl_2$/Al_2O_3 in gas phase. (Fig.6)

$$2\,CH_3OH + 2\,NO + 1/2\,O_2 \rightarrow 2\,CH_3ONO + H_2O$$

$$\frac{CO + 2\,CH_3ONO \rightarrow (CH_3O)_2CO + 2\,NO}{2\,CH_3OH + CO + 1/2\,O_2 \rightarrow (CH_3O)_2CO + H_2O}$$

Fig. 6. Dimethyl carbonate by oxy carbonylation of methanol

6.10 γ -butyrolactone

γ-butyrolactone (GBL) used as a resin solvent and an intermediate of N-methylpyrrolidone can be produced from maleic anhydride. Maleic anhydride is hydrogenated to GBL by $Ru(AcAc)_3$ with trioctylphosphine in tetraglyme in homogeneous reaction. This process is commercialized by Mitsubishi Chemical. Reaction rate increases using succinic acid anhydride after hydrogenation by Pd/Al_2O_3. Conversion of succinic acid anhydride is 79.2%, Selectivity of GBL is almost 100% at 200 °C, 1.0MPa, 14 h.(Fig.7)

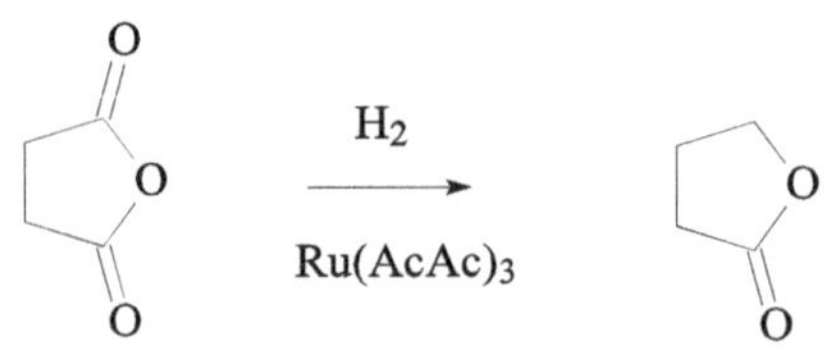

Fig. 7. Hydrogenation of succinic acid anhydride

7. Polymer

7.1 Vinyl chloride

Poly vinyl chloride is used very widely such as pipe, sheet, film, housing material for stability. Vinyl chloride monomer (VCM) is produced by thermal decomposition of ethylene dichloride (EDC) which is produced by oxychlorination with $CuCl_2/Al_2O_3$. After decomposition of EDC, 2,000-4,000 ppm of acetylene is produced in HCl stream. Acetylene in hydrogen chloride atmospher is removed by hydrogenation with acid resistant catalyst which is Pd/α-Al2O3 or Pd/SiC in order to reduce chlorine consumption in recycled oxychlorination process. (Kaneka 1987)

7.2 Methyl methacrylate

Methylmethacrylate (MMA) demand has been increasing in Asia. It is a transparent polymer and is used as glass substitute. A few routes are known to produce MMA. Isobutene process is developed in Japan. MMA is produced by esterification of methacryl acid which is introduced two step oxidation through methacroleine from isobutene. Asahi Kasei Chemicals commercialized direct process from methacroleine and methanol by $Pd\text{-}Pb/SiO_2$ using slurry bed. Oxygen is fed from the bottom of the reactor as very fine bubbles. The catalyst is Pd_3Pb/SiO_2. (Asahi Kasei Chemicals 1996) The size of silica carrier containing alumina and magnesia is about 60μm. The catalyst is prepared by starting mixture of HCl solution of $PdCl_2$ and $Pb(NO_3)$, immediately the metal solution mixture is impregnated 1hr and then reduced by hydrazine.

$$CH_2{=}C(CH_3){-}CHO + CH_3OH \xrightarrow{O_2} CH_2{=}C(CH_3){-}COOCH_3 + H_2O$$

MMA processes from ethylene are summarized as Fig.8. Lucite commercialized carbomethoxylation by Pd phosphine complex. BASF process uses hydroformylation by Rh complex. (Fig.8)

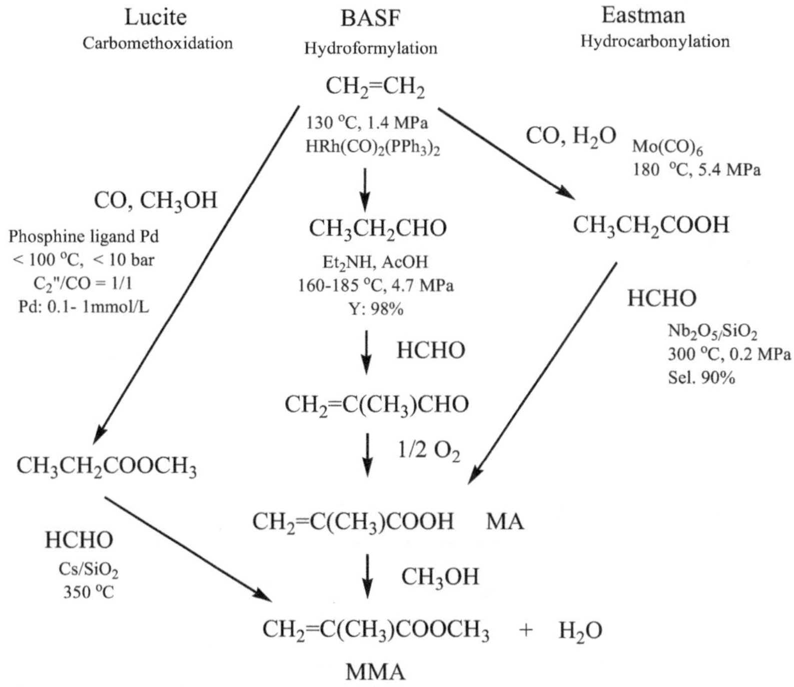

Fig. 8. MMA synthesis routs from ethylene

7.3 Toluene di-isocyanate

Toluene di-isocyanate (TDI) is used for polyurethane. TDI demand has been increasing due to the applications such as cushion of bed and sheet of car. TDI is produced from di-aminotoluene and phosgene. Di-aminotoluene is produced by hydrogenation of di-nitrotoluene with Pd/carbon or Pd-Pt-Fe/carbon in slurry bed.

7.4 Hydrogenated methaxylene diamine

Methaxylene diamine (MXDA) is used for heat resistant polyamide polymer. It is produced by ammoxidation of o-xylene with Fe catalyst. By Ru/Al_2O_3, a part of MXDA is hydrogenated to 1,3-di-amino cyclohexane which is highly heat resistant polymer material.

7.5 Cyclohexane di-methanol (CHDM)

Cyclohexane di-methyl is used heat resistant polyester resin. It is produced by hydrogenation of di-methylterephthalate. Pd or Ru/Al_2O_3 is used for hydrogenation of di-methyltrephthalate at first step, hydrogenolysis of cyclohexane di-methyl carboxylate is done by CuCrOx catalyst. (Fig.9)

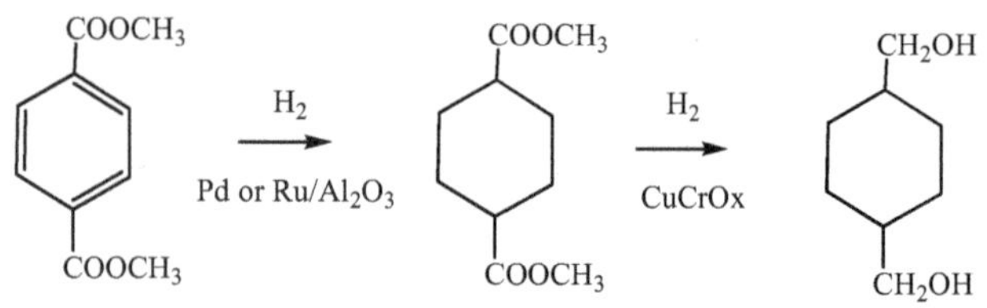

Fig. 9. DME synthesis from di-methyl trephthalate

7.6 Hydrogenated polymer

Hydrogenation of polymer is important to increase solubility to other polymer and transparency, which is applied such as adhesive for dipper. Hydrogenation of polybutene is conducted by Pd/Al_2O_3 in fixed bed. C_5, C_9 and terpentine resin are hydrogenated by Pd/Al_2O_3 and Pt/Al_2O_3. Hydroxypolybutadiene rubber is produced by hydrogenation with Ru/carbon prevention of hydrocracking of hydroxyl group. Nitril rubber and norbornene resin are hydrogenated with Rh complex in homogeneous phase. (Bayer 1980) Separation of catalyst from the system is very difficult. If the catalyst is very highly active, the loading quantity can be very small. Consequently, removal of catalyst is not necessary. (Shinohara 1997) It suggests that reaction system does not need filtration system like polymerization of polyethylene or polypropylene. Many resins are hydrogenated by precious metal catalysts. (Table-3)

Reactants	Catalysts	Reactor bed	Remarks
Polybutene	Pd/Al_2O_3	Fixed	
C_5 Petroleum resin	Pd/Al_2O_3	Fixed	
C_9 Petroleum resin	Pt/Al_2O_3	Fixed	
Hydroxy Polybutadiene Rubber	Ru/carbon	Slurry	
Nitril butadiene rubber	Pd/SiO_2	Slurry	
Norbornene resin	Ru complex	Homogeneous	
Terpentine resin	Pd/Al_2O_3	Fixed	Natural resin
Rosin	Pd/carbon	Fixed, Slurry	Hydrogenated rosin

Table 3. Commercialized hydrogenated resin

8. Synthetic fiber

8.1 Nylon

8.1.1 Inventa process

NO is produced by oxidation of ammonia with Pt-Rh gauze. NO is used for producing hydroxylamine which is reacted with cyclohexanone. NO is reduced by modified Pt carbon catalyst or Pt/carbon graphite in sulfuric acid solution at 40-60 °C. Anone oxime is converted to ε-caprolactam by Beckman rearrangement reaction. Cycrohexanone is produced by oxidation of cyclohexane or hydrogenation of phenol by Pd/carbon or Pd/Al_2O_3 in fixed bed reactor. (Fig.10)

$$4\,NH_3 + 5O_2 \rightarrow 4\,NO + 6\,H_2O$$

$$2\,NO + 3\,H_2 + H_2SO_4 \rightarrow (NH_2OH)_2 \cdot H_2SO_4$$

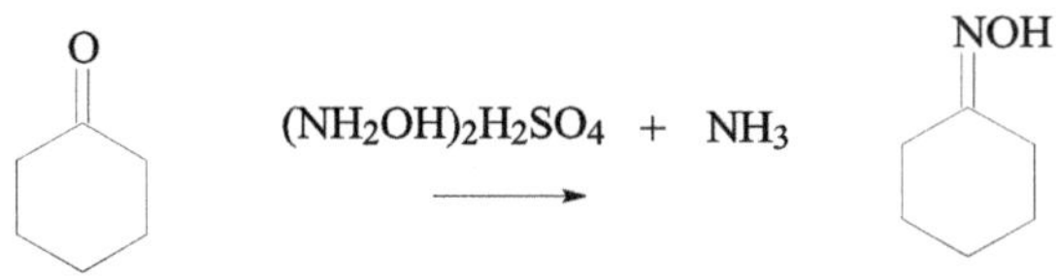

Fig. 10. Nylon-6 production route by Inventa process

8.1.2 DSM process

Ammonium nitrate is reduced by Pd/carbon or Pd-Pt/carbon in DSM process. Selectivity of Pd/carbon is higher than Pt/carbon. Mixed Pd/carbon and Pd-Pt/carbon gives highly activity and selectivity in H_3PO_4 aqueous solution with addition of GeO_2. Reaction temperature is 40-70 °C. The process is continuous slurry bed reactor installed swing filter system. Selectivity is effectively increased by small amount of halogen ion. (Fig.11)

$$H_3PO_4 \rightarrow H^+ + H_2PO_4^-$$

$$NO_3^- + 2\,H^+ + 3H_2 \rightarrow NH_3OH^+ + 2H_2O$$

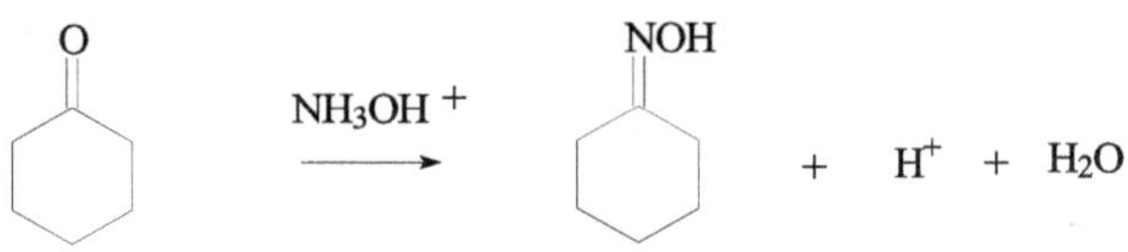

Fig. 11. Cyclohexanone oxime synthesis by DSM process

8.1.3 Snia viscosa process

Firstly toluene is oxidized to benzoic acid by metal acetate such as cobalt acetate in the aqueous phase. Subsequently, benzoic acid is hydrogenated to cyclohexane carboxylic acid by Pd/carbon in slurry bed. Hydrogenation condition is 170 °C, 1.0-1.7 Mpa. The production yield appears to be almost 100%. Cyclohexane carboxylic acid is reacted with $NOHSO_4$ and produces ε-caprolactam sulfate. (Fig.12)

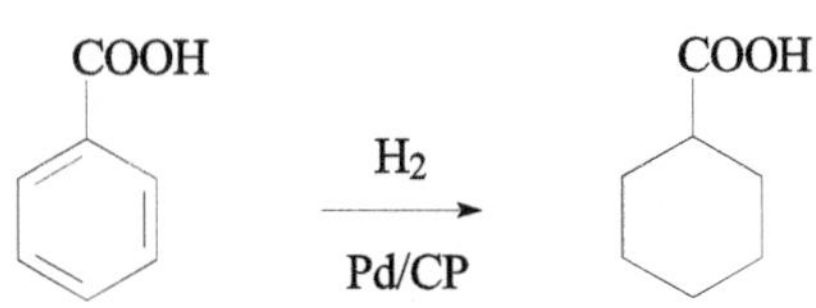

Fig. 12. Hydrogenation of benzoic acid

8.1.4 Asahi Kasei process

Asahi Kasei Chemical commercialized selective hydrogenation of benzene to cyclohexene. A 60,000Mton/y plant started to operate in 1990. Selective hydrogenation of benzene to cyclohexene takes place in water with Ru black catalyst. Cyclohexene is hydrated to cyclohexanol by high silica containing MFI. The process was licensed to China. (Fig.13)

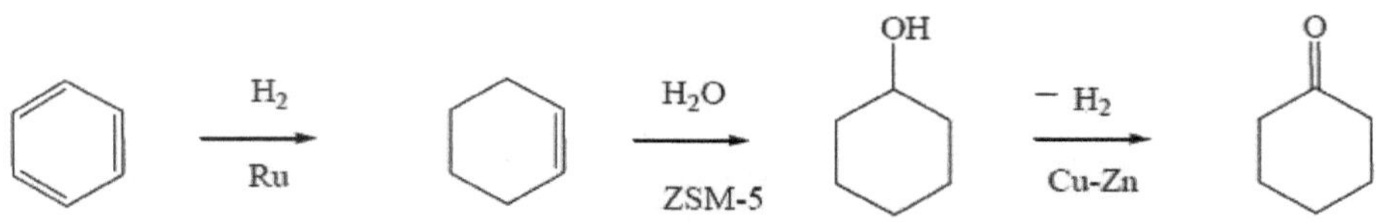

Fig. 13. Cyclohexanone production route by selective hydrogenation of benzene

8.1.5 Nylon 12

Nylon 12 is used as a hot melt adhesive. One of the Nylon 12 processes is hydrogenation of nitril compound. An intermediate of nylon-12, 12-Aminododecane acid is produced by hydrogenation of 11-cyanoundecane acid with Ru/SiC at 110-130 °C and 3.5 MPa, in mixed solvent of NH_4OH and n-propanol. Conv. 100%, Sel. 99.6%, yield of secondary amine is less than 0.4%. 11-cyanoundecane acid is produced by ammoxidation of cyclohexanone. (Ube Ind. 1976) (Fig.14)

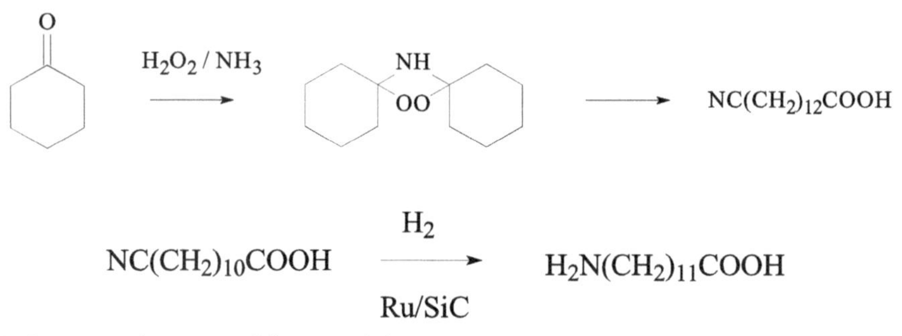

Fig. 14. Cyano undecane acid from cyclohexanone

8.2 Polyester

Most textile fiber has become polyester. Polyethylene terephthalate is produced ethylene glycol and terephthalic acid. Polyethylene terephthalate is very widely used to make bottles, for examples PET. Terephthalic acid is produced by oxidation of *p*-xylene introduced by isomerization of mixed xylene with ZSM-5 catalyst. Pt/ZSM-5 is used to convert ethyl benzene to *p*-xylene in hydrogen atmosphere. 4-carboxybenzaldehyde (4-CBA) is produced as a byproduct in oxidation of *p*-xylene by Co and Mn acetate in acetic acid solution. Producing pure terephthalic acid, 4-CBA is needed to remove to less than 25 ppm by hydrogenolysis by Pd/carbon granular in water solvent under severe condition. (Fig.15)

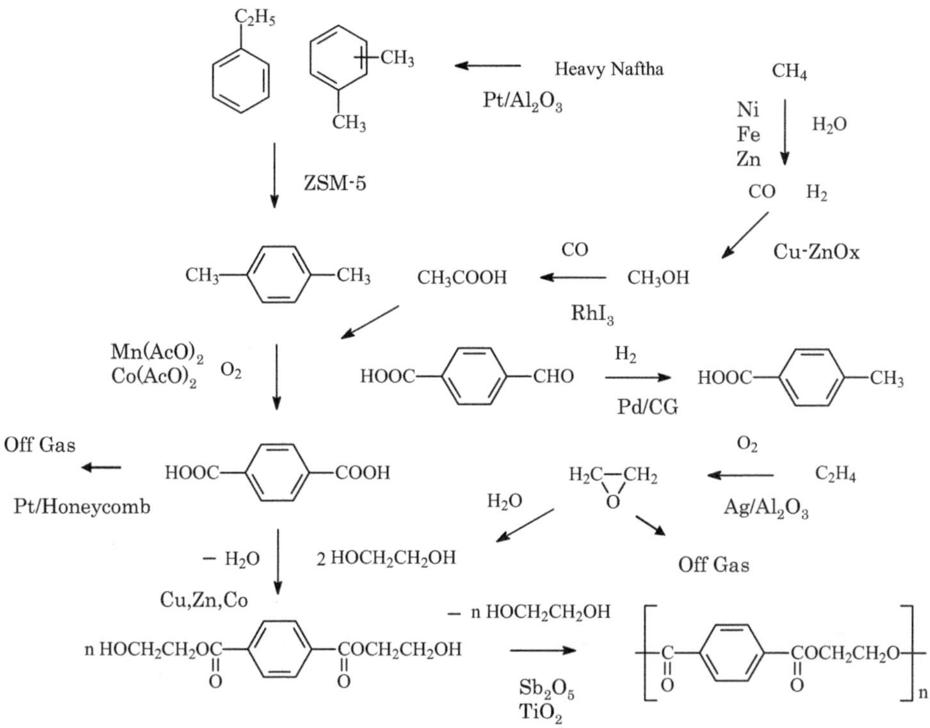

Fig. 15. Polyester production routs

8.3 Vinyl acetate

Vinylacetate is used for textile, paint, adhesive and others. Acetoxylation of ethylene in gas phase is commercialized developed by Bayer and ND (Millenium) with Pd-Au/SiO_2 or Pd-Au/alumina by multi tube reactor. Allylalcohol also commercialized by acetoxylation of propylene with Pd-Cu/SiO_2.

9. Gas production and purification

9.1 City gas

Producing city gas from coke oven gas contains hydrogen and carbon monoxide. Methanation is applied to produce methane by Ru/Al_2O_3 after removal of tar and sulfur compounds.

$$CO + 3H_2 \rightarrow CH_4 + H_2O$$

9.2 Purification of hydrogen

Removal of oxygen in hydrogen is proceeds with Pd/Al_2O_3. It is known as DEOXO process, and the catalyst was developed former Engelhard Ind., The technology has been used widely.

The catalyst is applied for purification of He, Ar with addition of hydrogen. CO causes catalyst poisoning for hydrogenation. Thus, methanation using Ru/Al_2O_3 plays important role to converts CO to non-poisoning CH_4. Selective oxidation of CO in hydrogen process is also known. It called "Select oxo process" using Pt/Al_2O_3 modified with Co or Fe developed by former Engelhard. The principle is originated from the facile adsorption of CO on Pt compared to hydrogen. Oxygen addition leads to reaction with CO on the catalyst surface. The catalyst is used for purification of hydrogen for ammonia synthesis and fuel cell.

9.3 Purification of nitrogen

Pure nitrogen is essential in fabrication of semiconductor in atmospheric gas. Pure nitrogen gas is generally produced in cryogenic system from the air. But, small content of CO is difficult to remove from N_2 in cryogenic system. Pt or Pd/Al_2O_3 can oxidize CO to CO_2 at 100-150℃ before cryogenic system.

9.4 Carbon dioxide

Carbon dioxide is used for as a coolant and for carbonated drinks. Quality of food grade CO_2 is regulated by government. Hydrocarbons and CO and others are main impurities in off gas of CO_2 from oxidation plant such as ethylene oxide plant. CO_2 is purified to food grade by addition of small amount of oxygen by Pt and Pd/Al_2O_3.

9.5 Reductive gas

Hydrogen gas is produced by decomposition of NH_3, a mixed catalyst of Pt/Al_2O_3 and Rh/Al_2O_3 is used at 700 °C. Transfer of H_2 is difficult, however it is feasible to transport as liquid NH_3.

Reductive gas for annealing furnace is produced by combustion of butane with $Rh/\alpha\text{-}Al_2O_3$ at 800 °C.

10. Specialty chemicals

10.1 Dyestuff and organic pigment

Many dyestuffs are produced by hydrogenation of aromatic nitro compounds. Halo nitro compounds are hydrogenated to halo amino compound by Pt/carbon or sulfur modified Pt/carbon. Monochloro acetic acid using as a raw material of dyestuff is produced selective hydrodehalogenation from di-chlrolo and tri-chloro acetic acid with Pd/carbon pellet in fixed bed reactor. The di- and tri-chloro acetic acid are produced by chlorination of acetic acid. *p*-Methoxy aniline, which is an intermediate of printer ink, is produced from nitrobenzene by Bamberger rearrangement reaction using Pt/carbon in methanol solvent.

Intermediate of an important yellow organic pigment is di-chlorohydrazone (DCH). It is produced from o-chlorobenzene using Pt/carbon.

10.2 Rosin

Rosin is gathered or extracted from pine tree. Disproportionated rosin is used as an emulsion polymerization agent for butadiene rubber. Disproportionation of rosin is conducted with Pd/carbon without hydrogen. Hydrogenated rosin is very stable. It is used as pavement paint and used as one of the ingredients of chewing gum. Hydrogenated rosin is produced by hydrogenation using Pd/carbon at severe condition.

10.3 Antioxidant

Antioxidant for rubbers has been produced by reductive alkylation of di-phenyl amine with MIBK. Pt/carbon or sulfur modified Pt/carbon has been used for reductive alkylation.

10.4 Liquid crystal

Liquid crystal is produced by Suzuki coupling reaction. Suzuki coupling reaction is coupling halo compound and boron acid compound by Pd with alkali, for example $Pd(PPh_3)_4$ or Pd acetate in homogeneous reaction. 5CB (4-pentyl-4'-cyanobiphenyl) is typical liquid crystal compound is synthesized by Suzuki coupling reaction. (Fig.16) BND and PBD are produced by Suzuki coupling. (Fig.17)

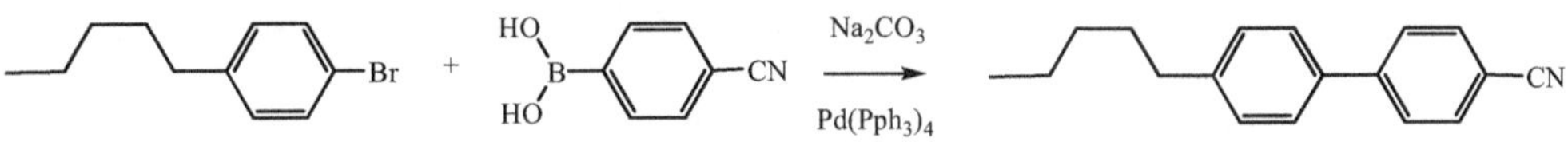

Fig. 16. 4CB synthesis by Suzuki coupling

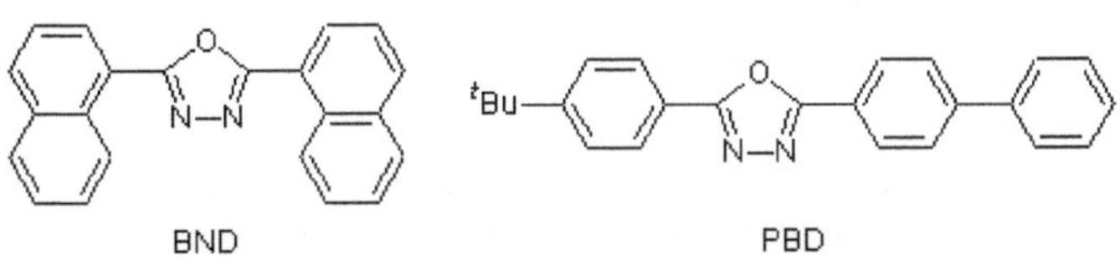

Fig. 17. Example of Liquid crystal produced by Suzuki coupling

Pd/carbon can be used instead of Pd complex. 4-Fluorobiphenyl acetic acid is produced by Pd/carbon.

10.5 Food industry

Sorbitol, known as a sweetener, is produced by hydrogenation of D-glucose. It is used for a moisturizing agent for ham and bacon, additive in teeth paste and others. Ru/carbon or Ni has been used as hydrogenation catalysts. Continuous process using Ru/carbon granular is developed. Succinic acid is produced by hydrogenation of maleic acid using Pd/carbon in slurry and fixed bed reactor.

Lecithin is hydrogenated to hydrogenated lecithin by Pd/carbon, which is used for food additives and cosmetics.

10.6 Synthetic perfume

Leaf alcohol (cis-3-hexene-1-ol) is produced by selective hydrogenation by Lindlar's catalyst which is 5%Pd-2.7%Pb/$CaCO_3$. (Fig.18)

$$CH_3CH_2C\equiv CCH_2CH_2OH \xrightarrow{H_2} CH_3CH_2\overset{H}{C}=\overset{H}{C}CH_2CH_2OH$$

Fig. 18. Leaf alcohol synthesis by Lindlar's catalysts

Many synthetic perfumes are derivatives of isoprene. Acetylene compound is selectively hydrogenated by Lindlar's catalyst after ethynylation. Linalool is derived from isoprene. (Fig. 19) Many synthetic perfume, such as jasmine, are synthesized by using Lindlar 's catalyst.

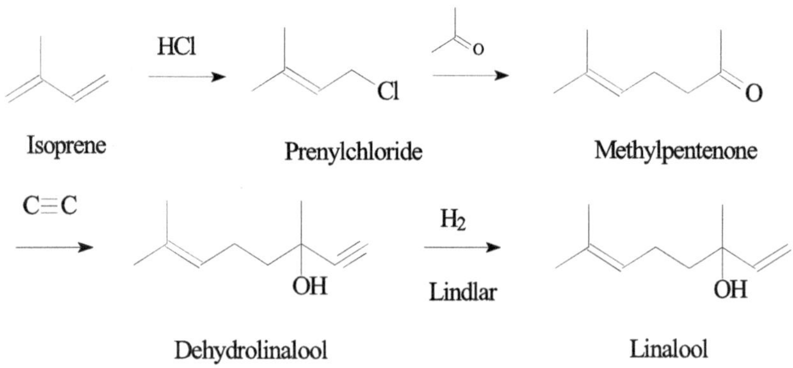

Fig. 19. Linalool synthesis by Lindlar's catalyst

10.7 Cosmetics

Squalene is extracted from liver of shark lives in deep sea. Squalene is hydrogenated with Pd/carbon to squarane which is used basic oil of cosmetics.

11. Medicines

11.1 Vitamin

Lindlar's catalyst was developed for producing vitamin A. (Fig.20)

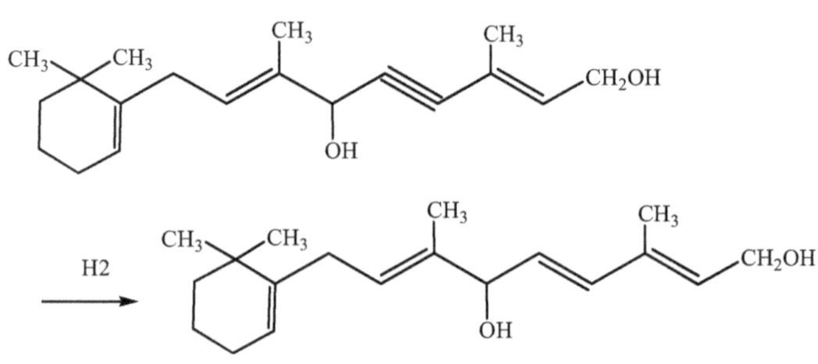

Fig. 20. Vitamin A synthesis by Lindlar's catalyst

Isophytol , an intermediate of vitamin E, is produced selective hydrogenation of carbonyl by Pd/carbon.

11.2 Medicine for pain and fever

Acetaminophen has been widely used as medicine for pain and fever. *p*-Aminophenol is an intermediate of acetaminophen synthesized from nitrobenzene by hydrogenation in sulfuric acid solution or directly by Bamberger rearrangement reaction with Pt/carbon.

Ibuprofen, which shows small side effect, is produced from *i*-butyl benzene using Pd/carbon and $Pd(PPh_3)_4$. (Fig.21)

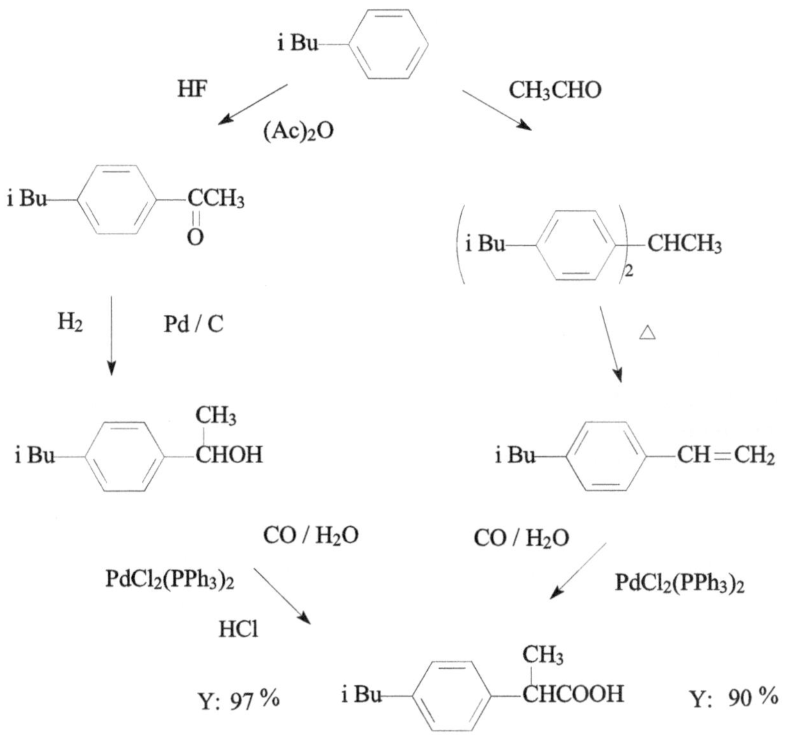

Fig. 21. Ibuprofen synthesis route by Pd

11.3 Hemostatic agent

Trans-ternexamic acid, used as a hemostatic agent, is produced by hydrogenation of nitril of *p*-cyanobenzoic acid with Pd/carbon at 100 °C and 5.0 Mpa, in NaOH aqueous solution. Consequently hydrogenation of aromatic conducts by Ru/carbon at the same condition.

11.4 Antibiotics

11.4.1 Di-hydroxystreptomycin

Streptomycin is hydrogenated to di-hydrostreptomycin which is more stable to alkaline and does not react with other carbonyl compounds. Hydrogenation is conducted by Adams PtO_2 in sulfuric acid solution. (Fig.22)

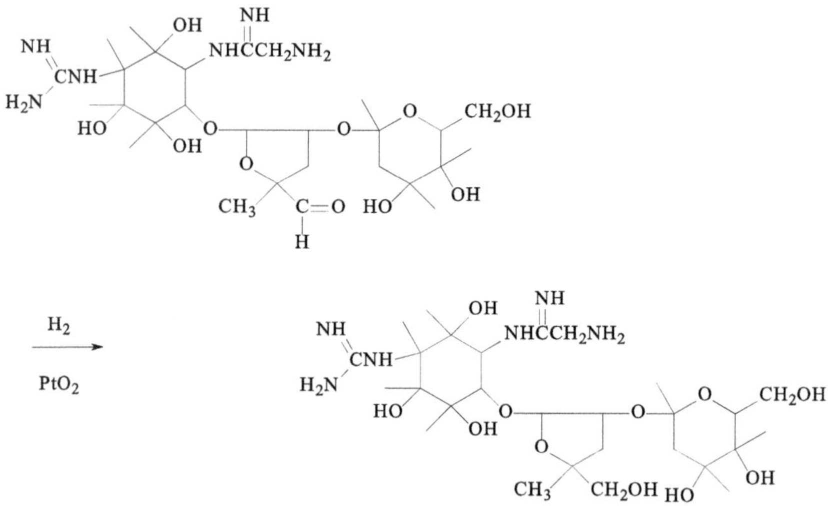

Fig. 22. Hydrogenation of streptmycin

11.4.2 Parasiticide

Wilkinson catalyst was discovered by Dr. Wilkinson in 1966. Wilkinson catalyst, $RhHCl_2(PPh_3)$ is used for hydrogenation of avermectin to produce ivermectin for parasiticide. This application is the largest use of Wilkinson catalyst. (Fig.23)

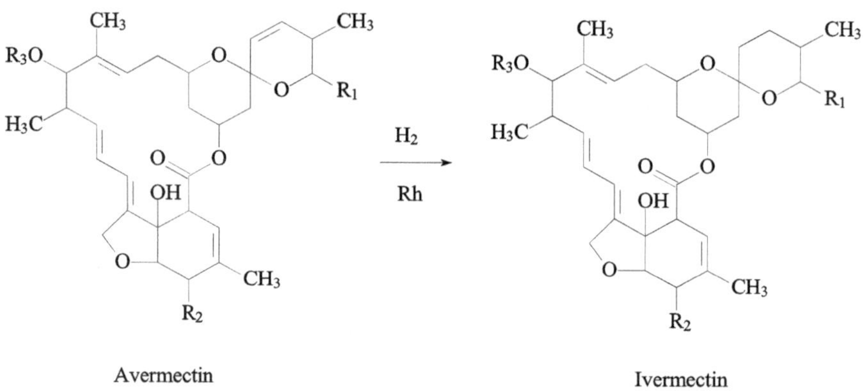

Fig. 23. Hydrogenation of avermectin

11.4.3 Tetracycline

Tetracycline type antibiotics are synthesized by Pd/carbon and Rh/carbon. Minocycline is produced by hydrodechlorination with Pd/carbon and employed enantio selective hydrogenation of methylene by Rh/carbon or Wilkinson Rh complex. (Fig. 24)

Doxycycline also produced through hydrodechlorination by Pd/carbon and hydrogenolysis by Rh/carbon.

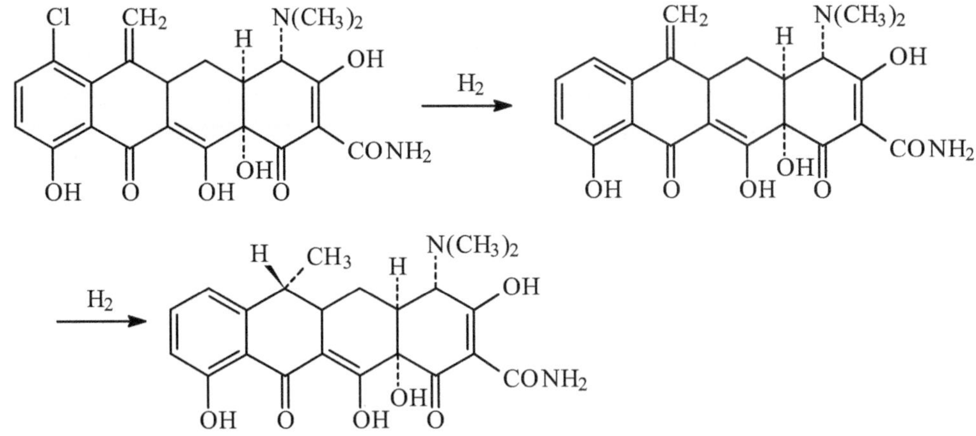

Fig. 24. Minocycline synthesis route

11.4.4 Carbapenem

Carbapenem antibiotics are produced by chiral Ru complex developed by Dr. Noyori.

β- lactam antibiotics were outstanding as next generation of antibiotics. It was intermediate of carbapenem antibiotics, 1β-methylcarboxylic acid. It is commercialized by chiral synthesis by Ru-BINAP. Target was synthesis of 4-acetoxy-2-azetizinone. Takasago International commercialized 40 ton/y of chiral 3-hydroxybutaneacid from α-acylaminomethyl acetic acid ester, at 50 °C, H_2 Press. 10Mpa, 20hrs, Yield is 99% ee, 95% de. (Fig.25) (Akutagawa 1987)

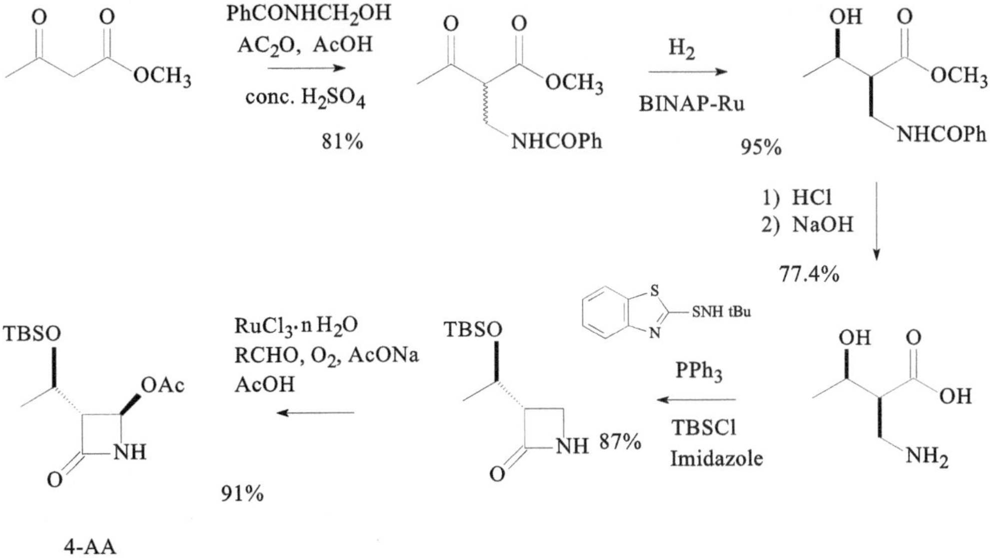

Fig. 25. 4-Acetoxy-2-azetizinone synthesis route

11.5 L-Dopa

A part of former Monsant developed a drug uses in the treatment of Parkinson's disease.

L-Dopa [3-(3,4-dihydroxyphenyl)L-alanine] is synthesized by Rh complex with chiral bisphosphine ligand such as 1,4-diphenylphosphinobutane, at ~50 °C, and ~40psig. (Fig.26)

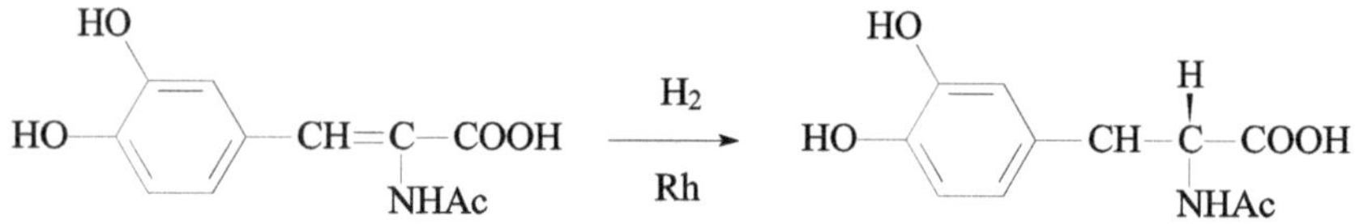

Fig. 26. L-Dopa synthesis

11.6 Indinavir (protease inhibitor)

Piperadine amide is one of the main intermediates of indinavir which is protease inhibitor developed by Merck. There are some synthesis routes to produce piperadineamide. One of synthesis routs is hydrogenation of piperadinecarboxylic acid t-butylamide hydrogenated by Pt/carbon and Ru/carbon and introduced rac-piperadine-2-carboxylic acid.

11.7 Anti-hypertensive

Many medicines are producing by Suzuki coupling using tetrakis-tri-phenyl Pd ($Pd(PPh_3)_4$) or Pd acetate for anti-hypertensive. (Table-4)

Name of medicine	Maker
Valsartan	Novartis
Losartan	Merk
Telmisartan	Boehringar-Ingelheim

Table 4. Example of anti-hypertensive medicine using Suzuki coupling

11.8 Other medicines

Suzuki coupling is applied for many anti-cancer medicines such as Vicenistatin. Negishi coupling is applied to produce anti-depressant. Many alkaloids are produced by the reaction using Pd complex.

11.9 Agricultural chemicals

Many agriculture chemicals are produced using Pt or Pd/carbon. Chloro pyridine is used for an intermediate of insecticide. Di- or tri chloropyridine which is over chlorination is hydrogenated to mono-chloro pyridine by Pd/carbon. Poly chloro amine is also hydrogenated to mono or di-chloroaniline by Pd-Sn/carbon. Reductive alkylation is general reaction to produce intermediates of herbicide by Pt/carbon. Heck carbonylation using $PdCl_2$ is applied for benzanilide type pesticide.

12. Nuclear power plant

12.1 Reduction of Uranium

One of extraction method of Plutonium (Pu) from spent Uranium (U) is conducted using U^{4+}. Spent U contains Pu $^{4+}$ is reduced to Pu^{3+} and separated by U^{4+}. U^{4+} is produced from U^{6+} by hydrogenation with Pt/SiO_2 in PUREX process.

$$U(NO_3)_6 + 2H_2 \rightarrow U(NO_3)_4 + 2HNO_3$$

12.2 Recombiner

Operating nuclear reactors mainly function with boiling water reactor (BWR) and pressurized water reactor (PWR) in the world. In BWR, water (light water) is used as coolant, and moderator is warmed to approx. 200 °C and pressurized to 7 Mpa to make steam in nuclear furnace. Steam is directly sent to gas turbine and recycled after cooled by water. In PWR, operates at approx. 15 MPa, 300 °C of liquid water is produced as the first water which makes steam of the second water for gas turbine. The first water is recycled. Water and nitrogen are decomposed to ^{3}H, ^{6}N, ^{19}O and radioactive Kr, and Xe from fuel rod by neutron. Gas in condenser after gas turbine is separated and introduced to recombiner, which installed Pd or Pt catalyst, after heated. H_2 including T is reacted with O_2 and separated as liquid. Gas components are introduced to gas hold up tower until radioactivity becomes lower. (Fig.27)

$$H_2 + 1/2\,O_2 \rightarrow H_2O$$

$$T_2 + 1/2\,O_2 \rightarrow T_2O$$

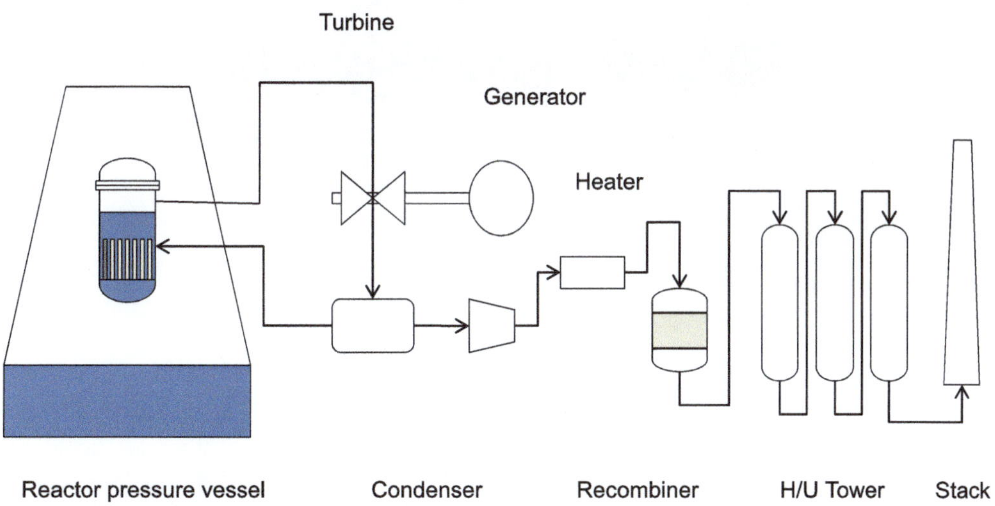

Fig. 27. Waste gas treatment in BWR

In PWR, the first water is recycled which is used warming the second water for gas turbine. Gas containing radioactive rare gases and H_2 containing T are purged and treated by recombiner, and treated gas holder as like as BWR (Fig.28)

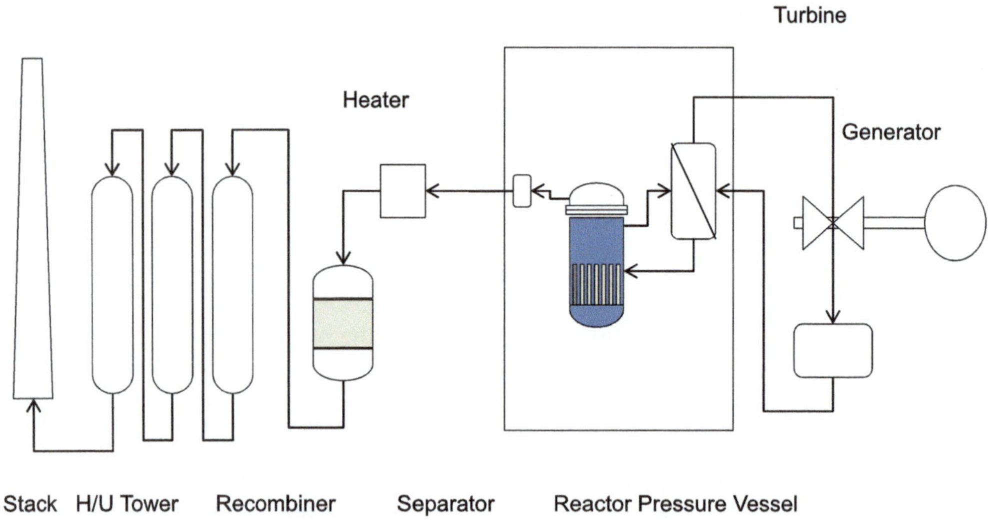

Fig. 28. Waste gas treatment in PWR

Pd/Al_2O_3 or Pt/metal form is used as a catalyst. Pd/Al_2O_3 sphere is charged SUS element. Water proof catalyst shows high activity for condensation of water. Pt/metal form is used as plate form, and charged 40 layers in the reactor. (Fig.29)

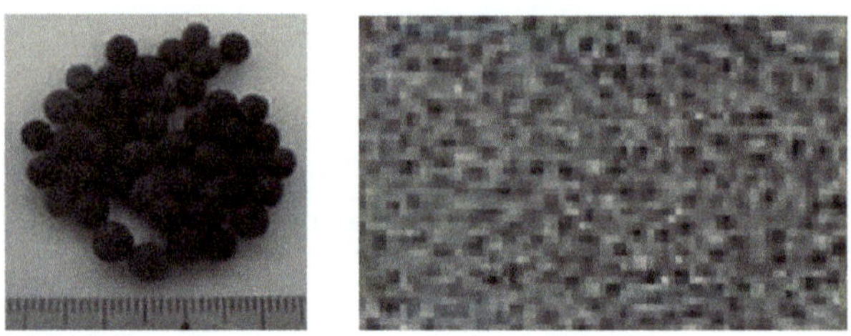

Pd/Al_2O_3 sphere Pt/metal form

Fig. 29. Recombiner catalysts

The example of reaction condition is described in Table-5.

		Inlet	Outlet
Gas Composition %	H_2	3.3%	< 1 ppm
	O_2	1.7%	
	Steam	Balance	Balance
Temp. °C		143	427

Press.0.12 MPa, S.V.: 1,500 hr^{-1}, O_2: Leak 0.5%

Table 5. Recombiner reaction condition

12.3 Flammability control system (FCM)

Three Mile Island accident taught us introducing FCM which is countermeasure of accidental coolant loss by breakage of cooling pipe and/or others. When cooling system is out of order, steam and Zr, which is cover metal of fuel rods, react and produces H_2. And H_2 is potentially generated from fuel rods even after shut down of nuclear furnace. It may hydrogen explosion.

$$H_2O + 1/2\,Zr \rightarrow 1/2\,ZrO_2 + H_2$$

FCS is installed recombiner catalyst which is the same catalyst as waste gas treatment system. FCS is fixed outside or inside of the vessel as shown in Fig.30.

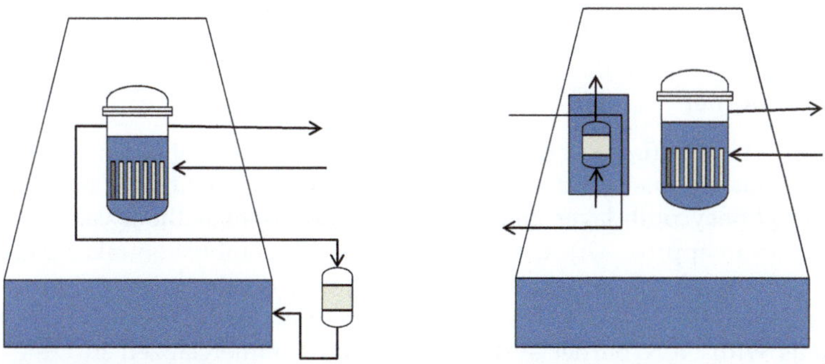

Fig. 30. FCS installed reactor

Radioactive I_2 are catalyst poison for Pd and Pt. I_2 adsorbent such as Ag/Al_2O_3 or Ag/zeolite are introduced. Such adsorbents are fixed on the front of the reactor.

13. Environment

13.1 VOC abatement

VOC (Volatile Organic Compounds) are removed by catalytic combustion system using Pt/honeycomb at lower than 350 °C. A lot of plants, not only chemical plants such as acrylic acid, maleic acid, phenol plants, but also printing, enamel wire, even food plant such as coffee plants, apply VOC removal. In case of high VOC content, heat recovery system equipped with Pt/honeycomb catalysts. The self combustion system after ignition without heating is running in some case. Off gas treatment of PTA plants and phenol plants use Pt/honeycomb, and generated heat during off gas treatment is recovered. In polyester film facilities, Pt/honeycomb reduce falling particle of organic compounds onto the film. Organic chloride compounds are completely oxidized by Pt/mordenite also.

13.2 Gas turbine off gas abatement

NOx can be reduced by adding water to fuel gas in gas turbine. NOx concentration can be reduced to 40 ppm from 150 ppm. However, addition of water increases CO from 10 ppm to 400 ppm because of lower flame temperature. In order to solve the problem, Pt/honeycomb is applied to oxidize CO completely. Since Pt/Al_2O_3 is poisoned by sulfur dioxide producing $Al_2(SO_4)_3$, sulfur resistant catalysts, Pt/TiO_2 are developed. Suggested reaction temperature and GHSV are 315-600 °C, and 200,000hr^{-1}, respectively. (Bartholomew 2006) (Fig.31)

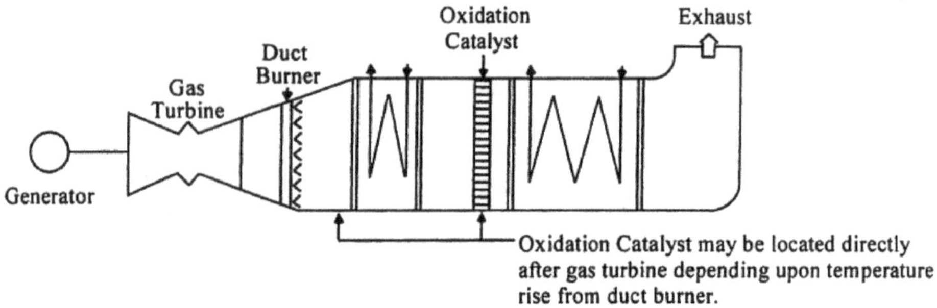

Fig. 31. Gas turbine installed CO oxidation catalyst

13.3 Low NOx burner

NOx is produced in the flame at high temperature. Production of NOx can be controlled by flam less combustion using Pd catalyst. Catalytica developed low NOx burner which consists of Pd/honeycomb layer and the combustion room without catalyst. Combustion gas is heated up to approx. 470 °C by burner. Catalytic combustion takes place in catalytic layer up to approx. 870 °C. At last, final combustion is conducted in flam less combustion room without catalyst. (Fig.32) NOx concentration of off gas is generally less than 5 ppm. Several plants with NOx burner system have been commercialized in USA. Ferrite type stainless containing Al honeycomb is selected for the honeycomb carrier instead of ceramics for prevention of heat destruction. (Shoji 1999) The system called Xonone Combustion System was developed by Catalytica Energy Co., NOx content is very low in the test results at power plant using gas engine. Small amount of NOx is produced at pre-burner, disclosed at Calfornia Air Resources Board Precertification Program in June 2002. (Table-6)

Off gas	Content ppm
NOx	1.13
CO	1.36
VOCs	0.16

Table 6. Power plant test result at O_2 15%

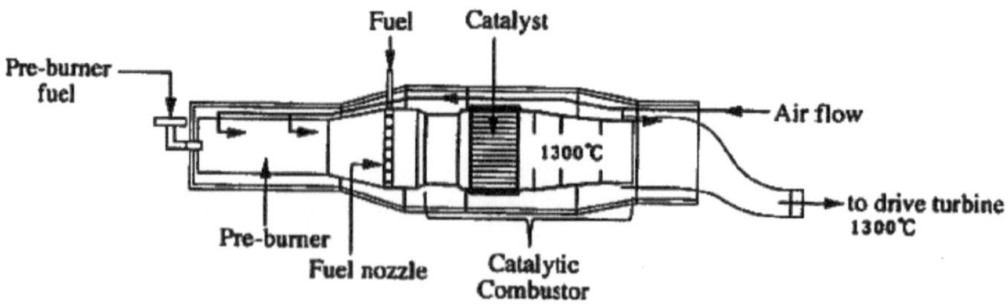

Fig. 32. Catalytic converter

14. Auto exhaust catalyst

14.1 Pt consumption

50% of Pt, 60% of Pd, and 85% of Rh in production from mining per year are consumed for automotive exhaust purification catalysts. Recovery of metals from spent automotive exhaust catalyst is currently 15% for Pt and Pd, and 30% for Rh so far.

14.2 Gasoline engine

Three way catalysts that remove CO, HC and NOx from exhaust gas of gasoline engine are used in the world. Contaminants of automotive exhaust gas of gasoline engine are reduced almost 99% by the catalytic converter. Pt and Pd are especially effective for combustion of hydrocarbon and carbon monoxide. Rh is effective for NOx reduction under stoiciometric condition. For example Pd-Pt/honeycomb is fixed in front part and Pd-Rh or Pt-Rh/honeycomb is fixed in the rear part of the converter which is installed under driver floor. (Fig. 33)

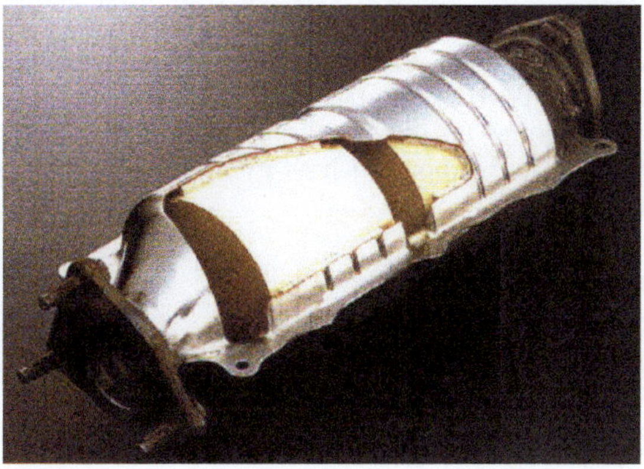

Fig. 33. Three way catalyst

Pt content cannot be reduced because of sintering which occurs at high temperature when reaches driving speed high. Some advanced technologies have been developed. Toyota Motor found Pt shows positive influence by strong interaction with metal oxide. Pt on CZY (CZY : 50%CeO_2, 46%ZrO_2, 4%Y_2O_3) does not show sintering at 800 °C, 5 hrs, and in the air. This is known as strong metal support interaction (SMSI). Pt is strongly fixed on oxide support by forming Pt oxide with carrier. By conducting XAFS analysis, Toyota Motor proved that Pt strongly bonds with carrier through surface oxygen atm. (Nagai 2007)

Daihatsu Motor developed heat resistant catalyst that consists of precious metal and peroveskite as support. PM is taken into lattice of peroveskite at oxidation atmosphere at high temperature. PM is out of lattice at reductive atmosphere at low temperature. Precious metal content of the catalyst can be reduced because sintering is significantly prevented. The catalysts are commercialized for light vehicle, and application of high exhaust gas temperature. The mechanism is investigated by XAFS study. (Daihatsu 2006) (Daihatsu 2008)

14.3 Purification of diesel engine gas

NO_2 is found to be oxidizing agent for soot. NO is oxidized to NO_2 by Pt/Al_2O_3 / honeycomb at first converter, and then soot is combusted to CO_2 at soot trap filter which Pt coated pores of wall through honeycomb consisting of SiC as a heat resistant material.

$$NO + 1/2\,O_2 \rightarrow NO_2$$

$$C + 3/2\,NO_2 \rightarrow CO_2 + NO$$

$$NO + 1/2\,O_2 \rightarrow NO_2$$

$$2\,NO_2 + C \rightarrow CO_2 + NO$$

15. Amenity

15.1 Catalytic heater

Catalytic heaters, Pt coated on alumina or silica wool, are widely used in Northern Europe. Natural gas or LPG is used as a fuel. Flameless infrared light heats the room. Portable catalytic heaters are sold for outside leisure, for instance, for camping, fishing and others. Honeycomb type of catalytic heaters is several pieces of Pt/honeycomb connected in series. It generates far infrared light which convey heat in distance. Thus, it is very useful to apply for large room such as gymnasium. Cord less catalytic heater is used for soldering iron using LPG.

15.2 Closed type battery

Lead batteries are used as a power source for emergency light, telephone system and traffic light and others. Small amount of water is decomposed as hydrogen and oxygen during charging battery. Waterproof Pd/Al_2O_3 particles are charged in porous ceramic cage. Hydrogen and oxygen are converted to water and dropped back into battery. Water level can be kept same for the long time.

$$2\,H_2 + O_2 \rightarrow 2\,H_2O$$

15.3 Ozone decomposition

Aircraft have to take air into the cabin at high altitude where air contains more ozone. This ozone uptake potentially makes from passengers airplane sick. Ozone converter is installed in the every plane now. Ozone is decomposed by Pd/honeycomb catalyst.

$$2\,O_3 \rightarrow 3\,O_2$$

16. Space

Decomposition of hydrazine by Ir/Al_2O_3 is propulsion power for control of satellite position in the space. Thrusters are fixed at the several sides of satellite wall.

$$N_2H_4 \rightarrow N_2 + 2\,H_2$$

17. Future

17.1 Fuel cell

17.1.1 Fuel cell system

Polymer electrolyte fuel (PEFC) is expecting to apply for residential and automotive use for at lower operation temperature. Target amount of Pt use is 0.5-0.7g/unit for residential use, 25-35 g/vehicle. Currently 5-10 times more of Pt is required. Since electric car cannot run long distance, fuel cell car is expected to be commercialized. Infrastructure of hydrogen supply remains as a technical challenge. (Table-7)

	Residential use	Automobile use
Power Capacity	0.7～1.0 kw	50 kw
Fuel	City gas, LPG, Kerocene	Hydrogen
Pt use amount	0.5～0.7 g/unit	25～35 g/car
Life of Cell	10 years	100,000 km

Table 7. Expecting performance of fuel cell

PEFC system for residential use equipped with purification, steam reforming, shift reaction, PROX, electrode and off gas treatment. It still appears to be complicated. (Fig.34)

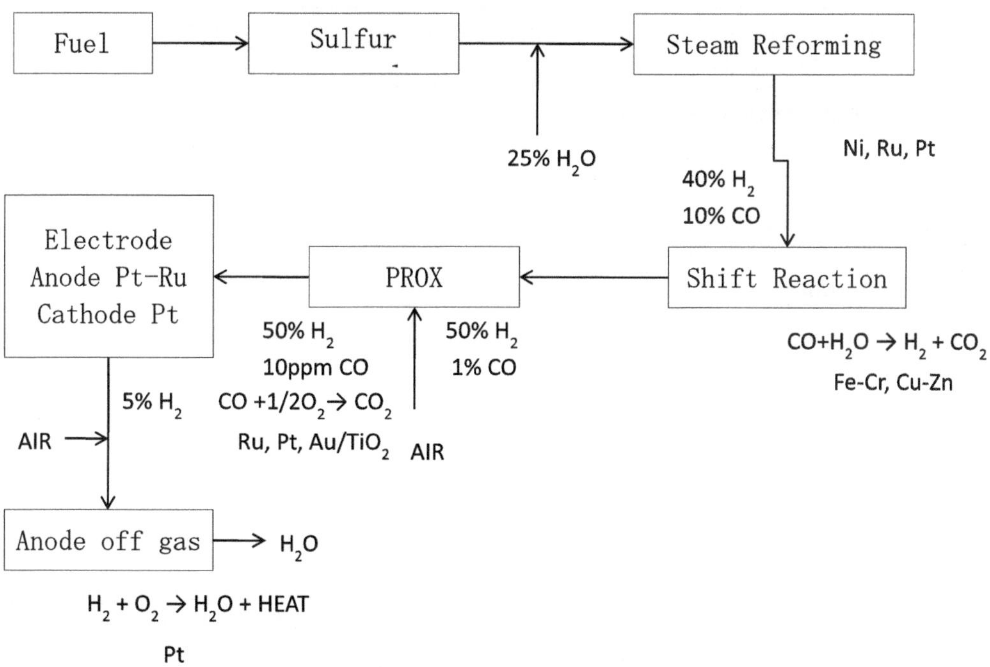

Fig. 34. Fuel Cell system flow

17.1.2 Reforming and hydrogen purification

City gas can be a fuel for a fuel cell of residential use. Containing small amount of sulfur is removable by Fe or Ag/Al_2O_3. Ni or Pt/Al_2O_3 is used for steam reforming to produce hydrogen. Since containing CO impurity in hydrogen poisons catalyst, CO is needed to remove by shift reaction with Cu-Zn. Remaining 1% of CO is removed by preferential oxidation (PROX) to less than 10 ppm by using Ru, Pt, or Au/TiO_2. Ni/Al_2O_3 causes carbon formation in case of kerosene as fuel. Ru/ZrO_2 is developed to overcome the issue.

17.1.3 Electrode

Purified hydrogen is introduced to anode, however small amount of CO (10 ppm) is remain which is catalyst poison, therefore CO resistant Pt-Ru coated carbon is used. Air is fed to Cathode which is Pt coated carbon. Carbon is high temperature treated graphite, however carbon oxidation would not be overcome by produced hydrogen peroxide for long use. Reduction of amount of Pt and Ru loading is the most important target. An example of typical Pt/carbon and Pt-Ru/carbon electrode is shown in Table-8.

	Cathode	Anode
Pt content	50 %	27 %
Ru content	—	13 %
Moisture	< 5 %	< 5 %
BET surface area	320 m^2 /g	120 m^2/g
CO uptake	40 ml/g·Pt	40 ml/g·Pt+Ru
XRD crystal diameter	3 nm	6 nm

Table 8. Typical catalyst for electrode

17.1.4 Off gas

Off gas containing hydrogen is combusted to H_2O by Pt/honeycomb catalysts.

17.2 Dry reforming

Dry reforming instead of steam reforming is coming up to reduce CO_2 when syngas is produced from natural gas. Syngas can be used to produce F/T oil and methanol which can be further converted to propylene. Ni/Al_2O_3 can be used as same as steam forming catalyst, However, Ru/MgO shows stable performance with low carbon deposition. This system is operated in pilot plant scale in Niigata as a national project in Japan. (Chiyoda 2006)

$$CH_4 + CO_2 \rightarrow 2\,CO + 2\,H_2$$

17.3 Hydrogen transfer

Electric power generation by utilizing wind power is effective in South America and North islands. Organic hydride is effective material as a hydrogen carrier. For example, toluene is hydrogenated to methyl cyclohexane by Pt/Al_2O_3 by hydrogen produced electrolysis by wind power. Methyl cyclohexane is transported to consumption area by vessel, and hydrogen is taken from methylcyclohexane by dehydrogenation with Pt/Al_2O_3,. Dehydrogenated toluene is again, transports to strong wind area by vessel and recycled.

17.4 Biomass

17.4.1 Propanediol from glycerol

1,2 -propanediol can be synthesized from glycerol which is expected byproduct of bio-diesel oil production. Mixed catalyst with Ru/carbon and amberlyst gives higher activity than Ru/carbon alon. It presumes that hydroxyacetone is formed as an intermediate with ion-exchange resin, and hydrogenation undergoes by Ru/carbon. At last, 1,2-propanediol is formed.(Furizuno 2006)

17.4.2 Hydrogenolysis of glycerol to methanol

Oxford University has disclosed production of methanol from glycerol by hydrogenolysis. Conversion is 50% and selectivity is 80% by Ru/graphite at 2.0 Mpa, and 100 °C, 24 hrs. (Fig. 35) (ISIS Inovation 2009) Produced methanol can be used to synthesize FAME by transesterification.

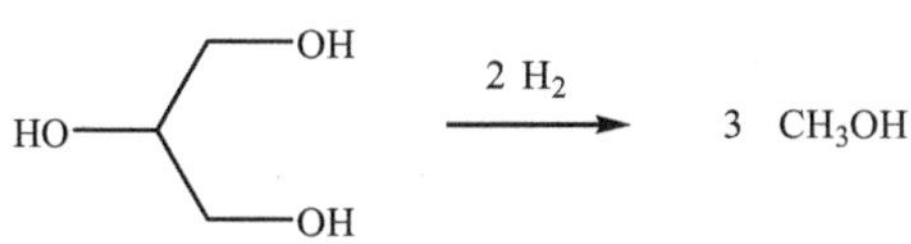

Fig. 35. Methanol by hydrogenolysis of glycerol

17.4.3 Hydrogen from glycerol

Hydrogen can be produced from diluted glycerol aqueous solution. Hydrogen is generated from 1-10wt% of glycerol at 220 °C and 2.9 Mpa by Pt/Al_2O_3. Conversion is 90% and selectivity is 90%. Small amount of alkanes are produced by Rh or Ni. (Boonyanuwat 2006)

18. Acknowledgement

The article could be completed thanks to the kind assistance of Dr. Shigo Watanbe.

He graduated Akita University in Japan and worked in Evonik Degussa as project leader living in USA.

19. References

Akutagawa, Kumobayashi, (1987) Catalysts & Catalysis Vol.29, No.5, 363
Asahi Glass (1989) JP H01-128942
Asahi Kase Chemicals (1996) JP H08-332383
BP (1984) JP Tokko S59-16816 BP
Bartholomew C.H., Farrauto R.J., (2006) Industrial Catalytic Processes, Wiley-Interscience.,
Bartholomew C.H., Farrauto R.J., (2006) Industrial Catalytic Processes, Second Edition, Wiley-Interscience

Bayer (1980)UK-PS 1,558,491
Bond G. C., (1964) D. E. Webster Proc. Chem. Soc. 398
Bond, G. C. (1968)Proc. 4th Intern. Congr. Catalysis Moscow, 266,
Boonyanuwat A., Jentys A., Lercher J.A., DGMK/SCI-Conference, Synthesis Gas Chemistry, Oct.4-6, Dresden, Germany (2006)
Chiyoda Corp., JOGMEC (2006) JP2006-055820
Chiyoda Corporation (1993) JP H05-3062254
Czuppon T.A., S.A. Knez, Robert V. Schneider, (1993) The 1993 American Institute of Chemical Engineers Ammonia Safty Symposium, Orland, Florida, Sep.
Daihatsu Motor (2006) JP 2006-210175
Daihatsu Motor, Hokko Chemical (2008) JP 2008-217461
DuPont (1977) USP4,048,096
Furizuno, Miyazawa, Kunimori, Tomishige, (2006) Catalysts & Catalysis, the 98th CASTJ, A, 414
Grove D. E. (2002) Platinum Metals Rev., 92, 46, (2)
Headwaters (2005) WO2005009611
ISIS Inovation (2009) WO2009/130452
JSR (1987) JP Tokko S62-23726
Kaneka (1987) JP S62-12771
Mitsubishi Chemical (1973) USP 3,755,423
Mitsubishi Chemical (1973) USP 3,755,423
N.E.Chemcat (2002) JP 2002-263490
Nagai Y. (2007) Catalysts & Catalysis, Vol.49, No.7, 591(2007)
Nippon Shokubai (2004) JP 2004-181357A
Ohno H (2000) Catalysts & Catalysis, 46, Vol.42 No.1
Okada, Tabata, Masuda, Matsui, (1993) Catalysts & Catalysis, Vol.35, No.4
Rylander P. (1967) Hydrogenation over platinum metal catalysts, Acadimic Press.
Rylander, P. N. (1967) Catalytic Hydrogenation over Platinum Metals, Academic Press. 326,
Shinohara H., (1997) The Chemical Daily-Books, 76, Dec.
Shoji T (1999) Catalysts & Catalysis, Vol.41, No.1 37
Showa Denko (1994) JP H06-293695
Sumitomo Chemical (2004) JP2004-181408
Tanabe Y (1981) Hydrocarbon Process 187, Sept.
Ube Ind. (1976) JPS51-127022
Vaarkamp M, (2000) Chemical Catalyst News, Engelhard, Nov.
Watanabe K., (2005) PETROTECH, Vol.28, No.10, 731
Yamamoto Y. (2010) Catal Surv Asia 14: 103-110

12

Nano/Micro-Patterning of Semiconductors by Site Selective Chemical Etching Using Noble Metals as Catalyst

Sachiko Ono* and Hidetaka Asoh
Department of Applied Chemistry, Faculty of Engineering, Kogakuin University, Tokyo Japan

1. Introduction

Controlled silicon structures on the micron to nanometer order have received much attention owing to their potential applications in various fields such as electrochemical, optoelectrical, and biological sciences. Although the techniques commonly used in fabricating nano-/microstructured silicon are conventional lithographic techniques using a resist mask with an optical, electron, or X-ray beam, chemical etching is also widely used in silicon micromachining. Three-dimensional silicon microstructures such as pillars, tubes, and macropores are fabricated by electrochemical etching in hydrofluoric acid (HF). This is a promising technique for the micromachining of silicon (Lehmann & Foll, 1990; Kleimann et al 2001; Matthias et al 2004).

Among three-dimensional silicon microstructures, silicon with a regular porous structure of the order of submicrons to nanometres, which is normally fabricated by electrochemical etching, has been studied intensively from the viewpoint of both basic research and commercial applications over the past few decades (Lehmann, 2002; Asoh and Ono, 2007, 2010). In the present chapter, nano-fabrication of semiconductor surfaces such as Si and GaAs by the combination of site selective chemical etching using various noble metals as catalyst and colloidal crystal templating is described with focusing on the difference in etching properties of each metal species. By applying this technique, ordered hexagonal arrays of crystalline Au nanodots on silicon were also obtained.

A novel technique for fabricating porous silicon without applying an external bias, so-called metal-assisted chemical etching, was proposed (Li & Bohn, 2000). According to their report, a porous silicon layer can be formed easily and efficiently by immersing silicon wafers coated with a noble metal (i.e. Au, Pt, or Au-Pd) in HF mixed with hydrogen peroxide (H_2O_2) solution. A mechanism involving a localized electrochemical process has been proposed to explain the mechanism of metal-assisted chemical etching as follows (Li & Bohn, 2000):

Cathode reaction (at noble metal surface as a local cathode):

* Corresponding Author

$$H_2O_2 + 2H^+ \rightarrow 2H_2O + 2h^+, \tag{1}$$

$$2H^+ \rightarrow H_2\uparrow + 2h^+. \tag{2}$$

Anode reaction (at silicon surface):

$$Si + 4h^+ + 4HF \rightarrow SiF_4 + 4H^+, \tag{3}$$

$$SiF_4 + 2HF \rightarrow H_2SiF_6. \tag{4}$$

Overall reaction:

$$Si + H_2O_2 + 6HF \rightarrow 2H_2O + H_2SiF_6 + H_2\uparrow. \tag{5}$$

In other words, when oxidants (H_2O_2) are reduced on the surfaces of noble metal catalysts, positive holes (h^+) are generated. After the removal of electrons from metal particles, the potential of the metal shifts towards a positive value to a level enabling the injection of h^+ into the silicon substrate. Finally, anodic oxidation and the dissolution of silicon take place in the chemical etchant containing HF.

Various types of silicon microstructures such as deep straight nanopores and helical silicon nanopores have been prepared by metal-assisted chemical etching (Yae et al, 2003; Tsujino & Matsumura, 2005a; Tsujino & Matsumura, 2005b). These microstructures are expected to be used for improving the light-emitting properties and conversion efficiency of solar cells. The formation of porous silicon by metal-assisted chemical etching can proceed in the presence of not only Au, Pt, and Au-Pd particles but also other noble metals such as Ag and Pd.

Recently, we have reported the fabrication of ordered Si, GaAs and InP microstructures such as convex arrays and nanopore patterns with regular periodicity of the order of micrometres by combining colloidal crystal templating and site-selective chemical etching using patterned noble-metal thin films as catalysts (Asoh et al, 2007a; Asoh et al, 2007b; Arai et al, 2008; Yasukawa et al, 2010). In addition, by using a shape-controlled Pt-Pd thin film as a catalyst, we have fabricated silicon microwells containing Pt-Pd thin films (Asoh et al, 2008a). We carry out a preliminary study to examine the effect of noble metal catalyst species on the morphology of macroporous silicon prepared by metal-assisted chemical etching (Asoh et al, 2009). We mainly focus on the correlation between the morphology of etched semiconductor microstructures, etching rate, and catalyst species in the present chapter.

2. Colloidal crystal templating

Concerning the application using two-dimensional (2D) colloidal crystals, "natural lithography", which has been proposed by Deckman and Dunsmuir in 1982, has attracted attention due to relatively easy process in comparison with conventional lithography. Based on such process, uniformly sized microstructures could be produced on a substrate using a monolayer coating of colloidal spheres instead of a conventional resist. In recent years, these techniques, which are often called "colloidal lithography" or "nanosphere lithography", have been reported on the nano-/micro-fabrication or nano-/micro-patterning on a wide variety of solid substrates including semiconductors, metals and ceramics. In the case of the use of 2D colloidal crystals as a physical mask, however, the target position for lithography is restricted to only interspaces among spheres. Although low ability to fabricate high-quality two-dimensional (2D) colloidal crystals, which limits the application fields for resultant patterns, is often pointed out as one of the drawbacks of such lithographic techniques, the regularity of colloidal crystals with wafer-scale sizes has been improving year by year (Jiang & McFarland, 2004). More recently, an efficient self-assembly technique for fabricating centimeter-sized single-domain 2D colloidal crystals under capillary forces within a wedge-shaped cell has been reported (Sun et al, 2010). For more details on the different strategies for the self-assembly into ordered 2D crystalline arrays and their applications, see the review papers (Velev and Kaler, 2000; Xia et al, 2000).

3. Formation of microstructures on silicon using colloidal crystal templating, hydrophobic treatment and electroless metal plating

The principle of pattern transfer for fabricating convex and hole arrays is schematically shown in Fig. 1 (Asoh et al, 2007a; Asoh et al, 2007b; Ono et al, 2007; Arai et al, 2008). Figures 2a and 2b show SEM images of Ag particles deposited on a silicon substrate (Asoh et al, 2007a). In the case without a mask, the fine Ag particles spread out over the whole silicon surface, as shown in Fig. 2a. The sizes and distribution of the particles on the silicon surface are in agreement with previous results (Tsujino & Matsumura, 2005a). On the other hand, when a 2D hexagonal array of polystyrene spheres of 3 μm diameter was used as a direct

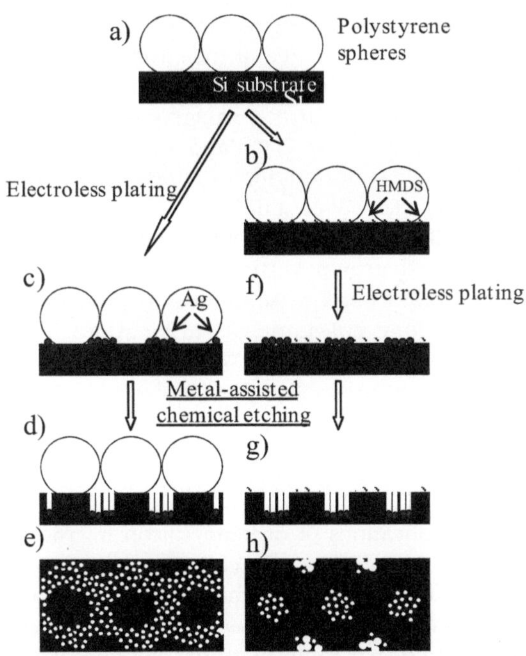

Fig. 1. Schematic model of site-selective chemical etching of silicon: (a) colloidal crystals on silicon substrate; (b) HMDS coating; (c), (f) electroless plating; (d), (g) chemical etching of silicon using Ag particles as catalyst; (e), (h) top view after removal of colloidal crystals.

mask during electroless plating (i.e., metal deposition which occur without the use of external electric source), a Ag honeycomb pattern was obtained, as shown in Fig. 2b. Using colloidal crystals as a mask, Ag particles were deposited selectively among the spheres. In other words, selective metal deposition can proceed only on the exposed parts of the silicon surface, which are located in the voids among the spheres on the silicon substrate. This result indicates that colloidal crystals can act as a mask for localized electroless plating in aqueous solution.

The details of the electroless plating process were reported in one of our previous papers (Asoh et al 2007c). In Fig. 2b, the center-to-center distance between the holes in the Ag honeycomb pattern, which was basically determined by the diameter of the polystyrene spheres, was approximately 3 μm. The framework of the Ag honeycomb pattern was composed of an aggregation of fine Ag particles. The sizes of the Ag particles were mostly between 50 nm and 100 nm but were scattered in the range from 10 nm to 250 nm.

3.1 Convex arrays of silicon using Ag particles as etching catalyst

After the deposition of the Ag particles on the silicon substrate, the specimens were immersed in a mixed solution of HF and H_2O_2 to form porous silicon by metal-assisted chemical etching. Figure 2c shows an SEM image of the etched silicon surface using a patterned Ag catalyst. The periodicity of the obtained silicon convex arrays was approximately 3 μm, corresponding to the diameter of the polystyrene spheres used as a mask for electroless plating. The configuration of silicon convex arrays, which were arranged hexagonally over the entire area of the specimen, has an inverse relation to the honeycomb pattern of the Ag particles, as shown in Fig. 2b. That is, chemical etching proceeds only on the Ag-coated Si surface, in agreement with the proposed mechanism (Li & Bohn, 2000; Chattopadhyay et al, 2002) and, consequently, the contact area between the polystyrene spheres and the underlying Si substrate has a disklike shape. However, prolonged chemical etching in a HF-containing solution caused the destruction of silicon microstructures owing to the excessive dissolution of the horizontal plane. Nevertheless, deep straight holes that had grown vertically downward from the surface and were deeper than 30 μm were found in a fractured cross section as shown in Fig. 3 (Ono et al, 2007). The diameter of the long holes appears to be approximately 100 nm, which is in agreement with the size of the Ag particles. Ag particles were found at the tip of straight long holes as shown in the insert in Fig. 3b, similarly to those described in an earlier report (Tsujino & Matsumura, 2005a).

3.2 Hole arrays of silicon using Ag particles as etching catalyst

To expand the range of applications of ordered silicon microstructures, it is necessary to control the silicon surface morphology. By colloidal crystal templating, it is possible to fabricate negative and positive patterns by changing the configuration of the Ag particles used as a catalyst. To fabricate a metal pattern that is the reverse of the Ag honeycomb pattern shown in Fig. 2b, two-step replication was applied. First, a colloidal crystal mask was formed on the silicon substrate, as described above. Second, specimens were placed in hexamethyldisilazane (HMDS) vapor overnight (Fig. 1b). HMDS is a popular reagent for forming hydrophobic surfaces based on the immobilization of trimethylsilyl groups on the surface (Ivanisevic & Mirkin, 2001; Maccarini et al, 2005), namely, areas of HMDS-coated silicon exhibit hydrophobicity and are thought to inhibit Ag deposition. Finally, electroless plating and chemical etching were conducted as described above (Figs. 1f-h).

Figure 4a shows the isolated patterns of Ag particles deposited by selective electroless plating using HMDS-coated silicon. The deposition is restricted to well-defined bare silicon surfaces, and does not occur on the HMDS-coated silicon. This result indicates that the HMDS-coated areas, which are located in the voids among the spheres on the silicon substrate, possess sufficient hydrophobicity and can act as a mask for localized electroless plating in aqueous solution. In Fig. 4a, the center-to-center distance between the island microarrays of Ag particles was approximately 3 μm. The isolated Ag patterns were composed of an aggregation of Ag particles with sizes in the range of 50-100 nm.

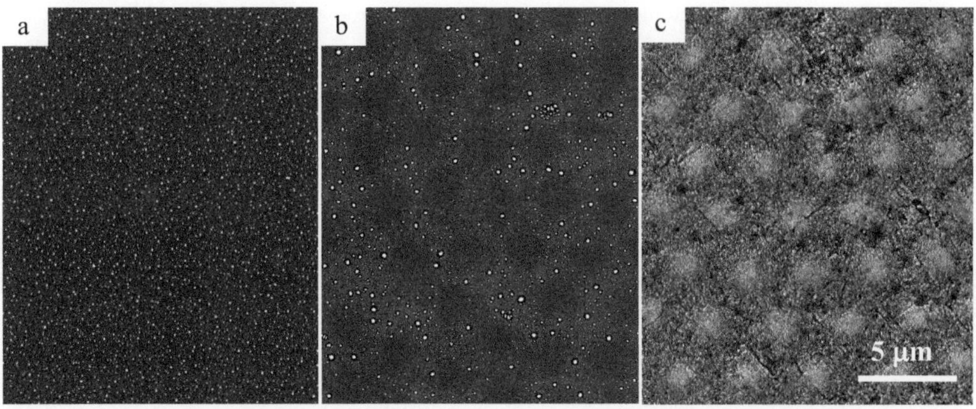

Fig. 2. SEM images of Ag particles deposited on silicon: (a) without mask and (b) with colloidal crystal mask. Electroless plating was conducted in $AgClO_4/NaOH$ for 20 min. (c) SEM image of surface of silicon etched in 5 mol dm^{-3} HF/1 mol dm^{-3} H_2O_2 for 1 min.

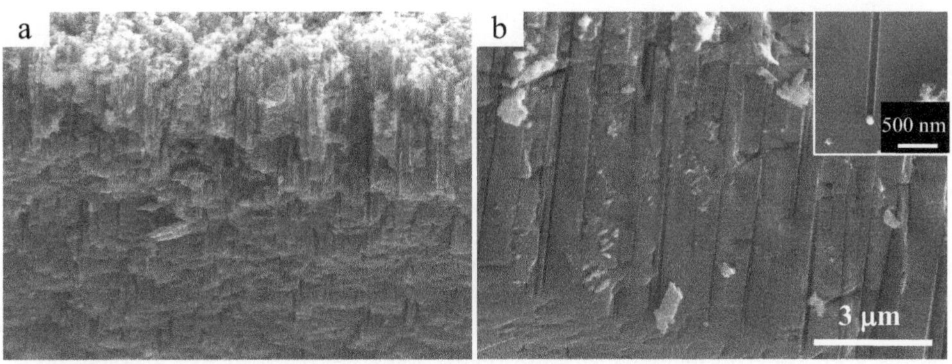

Fig. 3. SEM images of a fractured section of the silicon substrate obtained after etching with the Ag honeycomb pattern shown in Fig. 2 for 30 min: (a) near-surface region and (b) intermediate area revealing deep straight holes that had grown vertically downward from the silicon surface. The insert shows a deep hole with a Ag particle at the tip.

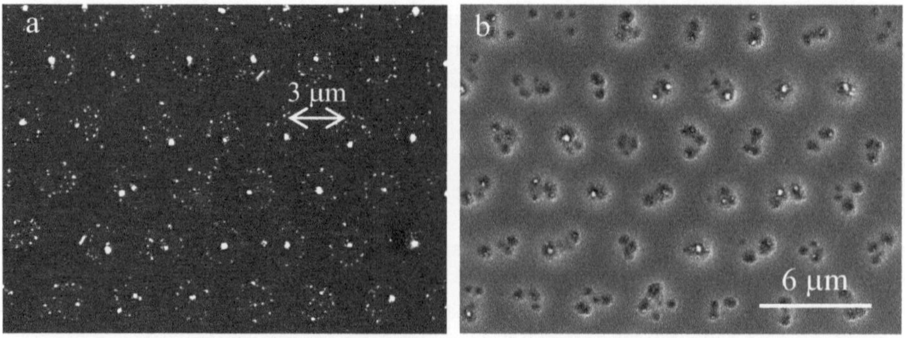

Fig. 4. (a) Isolated patterns of Ag particles formed on silicon and (b) silicon nanohole array. The electroless plating and etching conditions were the same as those for Fig. 2.

Figure 4b shows an SEM image of a nanohole array on a silicon substrate. The aggregation of nanoholes with sizes in the range of 50-100 nm was arranged hexagonally over the entire area of the specimen, corresponding to the 2D hexagonal array of polystyrene beads used as the original mask for the formation of the HMDS honeycomb pattern. Some Ag particles, which were detected as bright circular spots, were observed at the bottom of the pores due to the short etching time. The sizes of the particles observed in Fig. 4b coincided with those of the deposited Ag particles shown in Fig. 4a. These results indicate that chemical etching proceeds only on the Ag-coated silicon surface, and the consequently the HMDS-coated silicon parts remain in a honeycomb pattern.

Figure 5 shows silicon hole arrays with different periodicity after site-selective chemical etching using isolated Ag patterns (Asoh et al, 2007b; Arai et al, 2008). The magnification of each image was the same. The periodicity of the holes was basically determined by the diameter of the polystyrene spheres used as a mask. In each case, chemical etching proceeds only on the Ag-coated silicon surface, and consequently the areas of HMDS-coated silicon remain in a honeycomb pattern. The shortest hole periodicity, which was attained by the optimization of the etching time, was approximately 200 nm as shown in Fig. 5d. This indicates that the formation of silicon hole arrays with a periodicity of less than 1 μm can be achieved by the process described in this section.

4. Macroporous silicon formed by metal-assisted chemical etching

4.1 Silicon microwells containing Pt-Pd thin films

The fabrication process of Si microwells is schematically shown in Fig. 6. Silicon substrates were precleaned in 1 wt% HF to remove organic contaminants and native oxides. A mixed suspension consisting of equal volumes of a 0.2 wt% suspension of polystyrene (PS) nanospheres with a diameter of 200 nm (Polysciences, Inc.) and a 0.5 wt% suspension of silica microspheres with diameter of 3 μm (Bangs Laboratories, Inc.) was dropped on the substrates. The suspension on the substrates was dried in air for more than one day, during which the mixture containing spheres of two different diameters self-assembled into a close-packed structure due to the capillary forces. After the complete evaporation of the solvent, the silicon substrates with binary colloidal crystals formed from the spheres were heated at

100 °C for 1 h to combine the adjacent PS nanospheres [Fig. 6(a)]. After heating, the silica spheres, which were used as a template, were selectively removed by immersing the specimens in 10 wt% HF for 10 min [Fig. 6(b)] to obtain a PS honeycomb mask. SEM images of the binary colloidal crystals and the PS honeycomb mask are shown in Fig. 7.

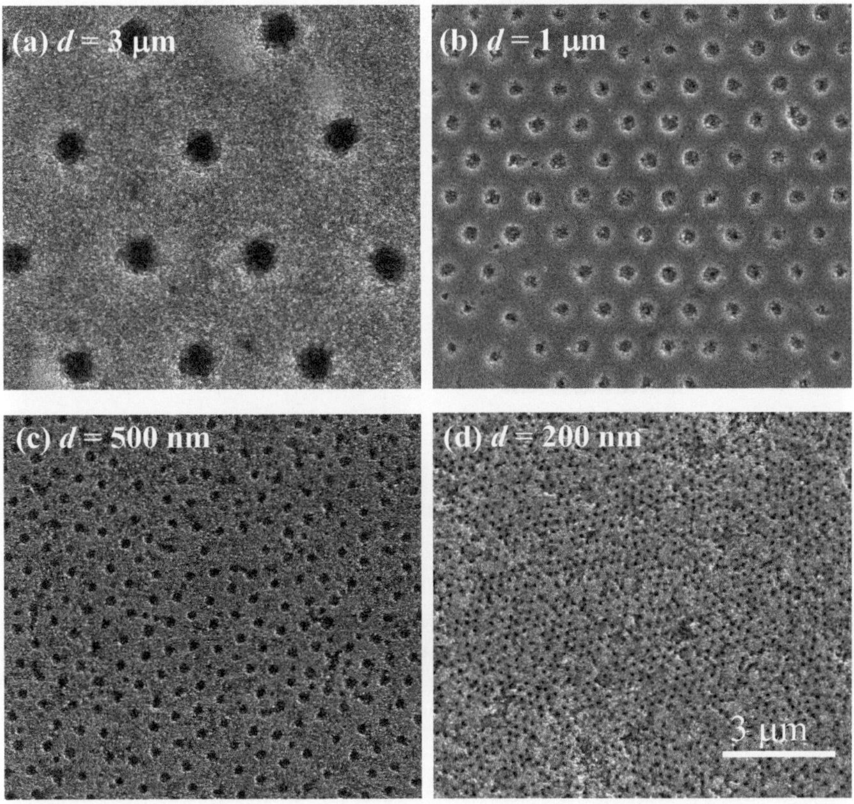

Fig. 5. SEM images of silicon hole arrays after removal of mask: (a) 3 μm periodicity, (b) 1 μm periodicity, (c) 500 nm periodicity, and (d) 200 nm periodicity. Electroless plating was conducted in 10^{-3} mol dm^{-3} $AgClO_4$ and 10^{-3} mol dm^{-3} NaOH for 20 min. The chemical etching times in 5 mol dm^{-3} HF/1 mol dm^{-3} H_2O_2 were (a) 5 min and (b)-(d) 30 s.

Metal thin films were deposited on the silicon substrates by ion sputtering (Hitachi E-1010) using a PS honeycomb mask composed of densely packed PS nanospheres. The sputtering was carried out at a discharge current of 15 mA in a vacuum with the pressure below 10 Pa [Fig. 6(c)]. Pt - Pd (80% Pt and 20% Pd), Au, and Pt targets were used as sputtering targets. The deposition rates of Pt - Pd, Au, and Pt were 6, 10, and 6 nm min^{-1}, respectively. The morphology and thickness of the deposited metal layers were examined by atomic force microscopy (AFM, Digital Instrument NanoScope IIIa). After sputtering, the specimens with locally deposited metal films were etched in a mixed solution of 5 mol dm^{-3} HF and 1 mol dm^{-3} H_2O_2 at room temperature [Fig. 6(d)]. Finally, the PS honeycomb mask was removed by immersing the specimens in 97 % toluene [Fig. 6(e)].

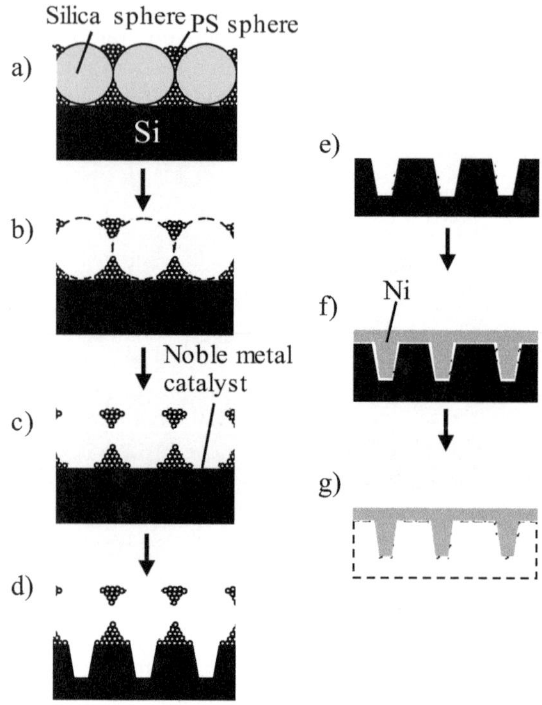

Fig. 6. Schematic model of fabrication of silicon macropore arrays: (a) formation of composite colloidal crystals on silicon substrate, (b) removal of silica sphere, (c) formation of metal catalyst layer, (d) chemical etching of silicon, (e) removal of PS honeycomb mask, (f) electrodeposition of nickel layer, and (g) removal of silicon by immersion of substrate in TMAH.

To examine the morphology of the silicon macropores in the direction of pore depth, a metal replica was prepared by metal plating. After coating the surface of silicon microstructures with metal catalyst layers by ion sputtering, nickel layers were electrodeposited using conventional nickel plating solution [Fig. 6(f)]. Finally, the silicon substrates were selectively removed by immersing the specimens in tetramethyl ammonium hydroxide (TMAH) [Fig. 6(g)]. The ordered geometric pattern formed on the silicon substrates was evaluated by scanning electron microscopy (SEM, Hitachi S-4200) and focused ion beam microscopy (FIB, Hitachi FB-2100).

When ion sputtering was carried out using the PS honeycomb mask, isolated circular metal thin films with a diameter of approximately 1.6 μm were deposited in the interspaces of the PS mask (Asoh et al, 2008a). Fig. 8(a) shows a typical AFM image of the Pt - Pd thin film arrays deposited on a silicon substrate (Asoh et al, 2009). From the cross-section analysis of the AFM image, the maximum thickness of the Pt - Pd layer deposited by using the PS honeycomb mask after a sputtering time of 5 min was estimated to be 10 - 15 nm. The thickness of the deposited layer was less than half the thickness estimated from the sputtering rate of Pt - Pd. The decrease in the deposition thickness is assumed to be caused by the use of the PS honeycomb with a thickness of approximately 2 μm during sputtering.

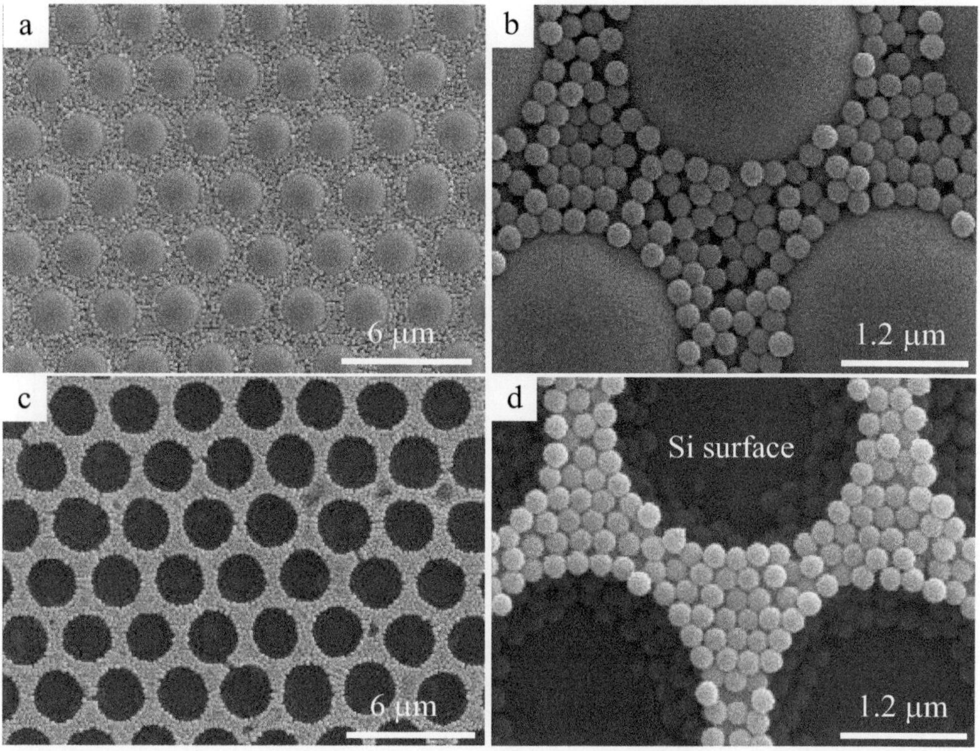

Fig. 7. SEM images of (a, b) binary colloidal crystals and (c, d) PS honeycomb mask after removing silica spheres by immersion in 10 wt % HF. The diameters of the large silica and small PS spheres were 3 μm and 200 nm, respectively. (a, c) Low-magnification view and (b, d) high-magnification view.

In addition, from a high-magnification AFM image of the same specimen, it was confirmed that the isolated circular Pt - Pd thin films were composed of clusters of Pt - Pd nanoparticles with a size range of 5 - 15 nm, as shown in Fig. 9(a) (Asoh et al, 2009). Ion sputtering, which is generally used for coating nonconducting materials with a metal layer for SEM observations, was used for depositing Pt - Pd in order to form a smoothly shaped metal thin film and to prevent the formation of a granular coating.

On the other hand, from the cross-section analysis of the AFM image of the deposited Au layer, the layer thickness was estimated to be 10 - 25 nm for a sputtering time of 3 min, as shown in Fig. 8(d). The isolated Au circular thin films were also composed of clusters of Au nanoparticles with a size range of 10 - 30 nm, as shown in Fig. 9(b). As compared to the isolated circular Pt - Pd thin films, the Au nanoparticles in the isolated circular Au thin films varied considerably in size, and the surface of the deposit was markedly uneven with granular particles.

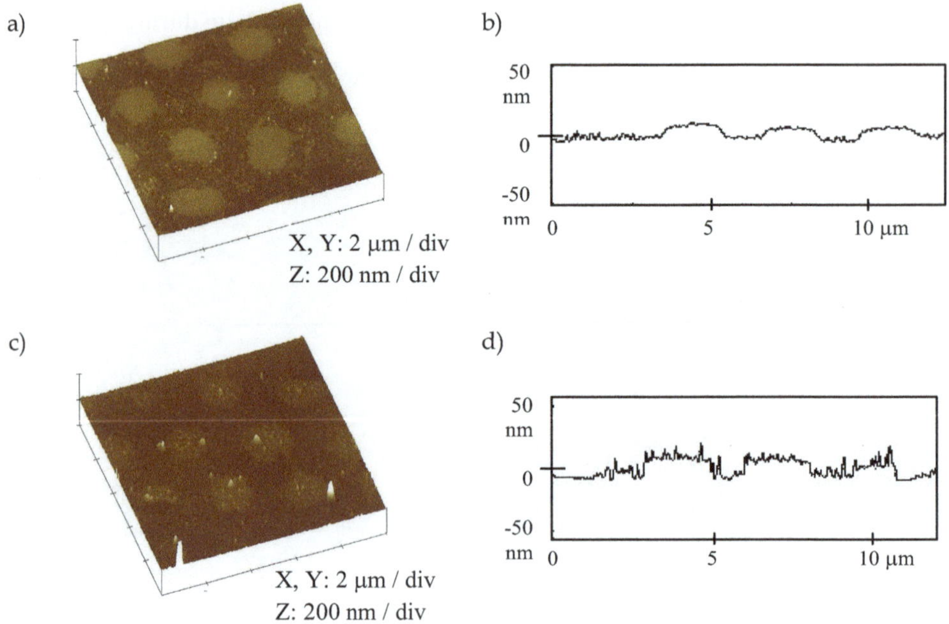

Fig. 8. AFM tapping-mode images of silicon surface after ion sputtering for (a, b) 5min using Pt - Pd and (c, d) 3 min using Au. (b, d) Typical line scan images of cross-section of each specimen.

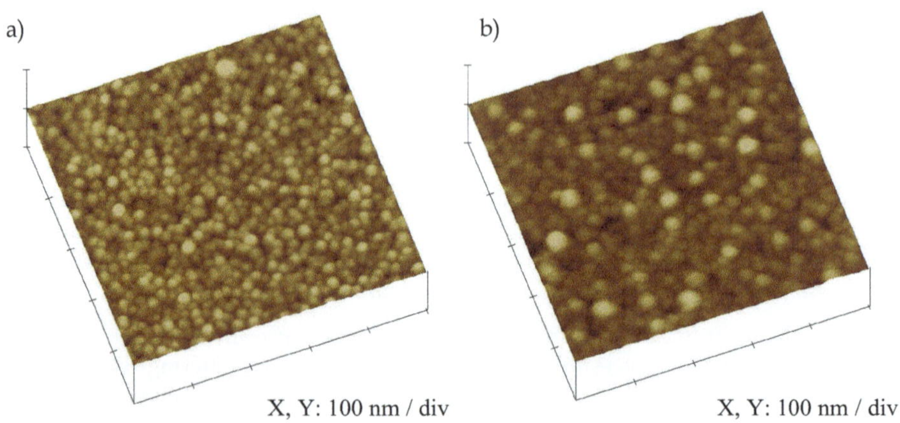

Fig. 9. High-magnification AFM images of (a) Pt - Pd and (b) Au films sputtered on Si substrate. Scan area was 500 nm^2.

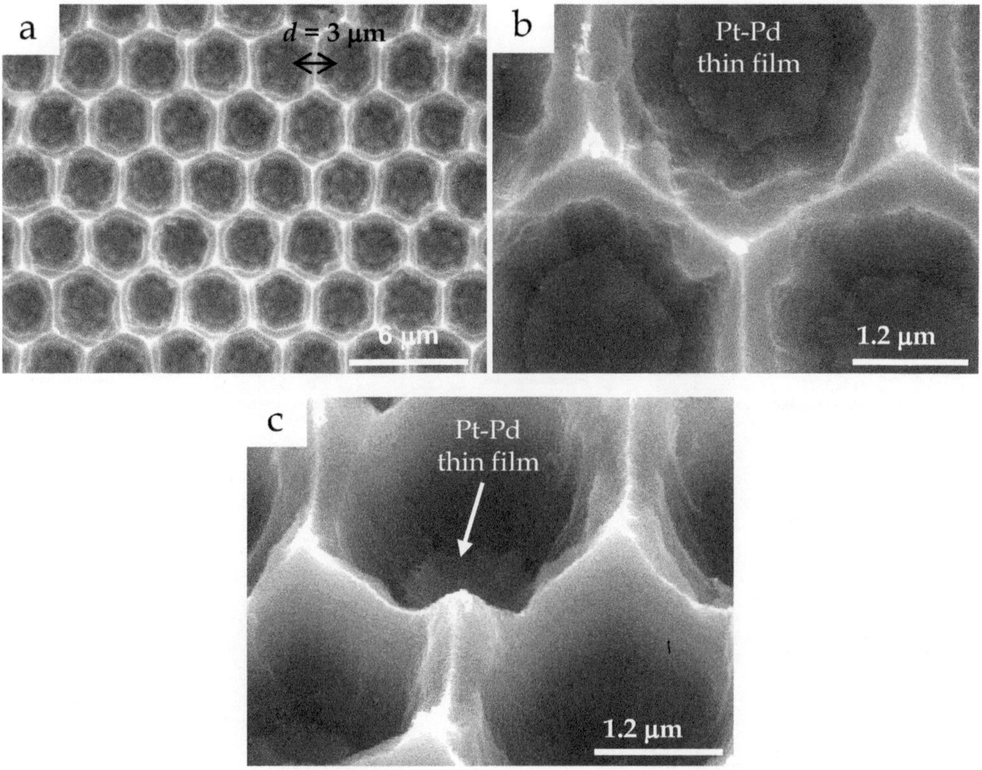

Fig. 10. (a) SEM image of macroporous silicon formed by metal-assisted chemical etching, (b) high-magnification view, and (c) tilted (45°). Chemical etching was conducted in 5 mol dm^{-3} HF/1 mol dm^{-3} H_2O_2 for 3 min using a Pt-Pd catalyst.

By using circular metal thin films as a catalyst instead of metal nanoparticles, the formation of silicon microwells with micrometer-scale openings was also achieved by metal-assisted chemical etching. The etching conditions were the same as those using Ag nanoparticles as a catalyst (Fig. 2). During chemical etching, the central part of the silicon substrate surrounded by the honeycomb mask gradually sagged downward. Figures 10a-c show plane-view and tilted-view SEM images of the silicon surface etched using the patterned Pt-Pd catalyst (Asoh et al, 2008a).

From the tilted view shown in Fig. 10c, it was confirmed that the pores were conical. The diameter of the opening of each silicon microwell was approximately 3 μm due to the chemical dissolution of the horizontal plane. The depth of each silicon microwell was estimated to be approximately 2 μm. The crest and side walls of the silicon microwells were extremely smooth. In addition, the most noteworthy point is that the circular Pt-Pd thin films used as the catalyst remained at the bottom of each well. The SEM images shown in Figs. 10b and 10c revealed that the residual detected as bright contrast was the Pt-Pd catalyst.

Fig. 11. SEM images of Au-coated silicon after chemical etching in HF/H_2O_2 for (a, b) 1min and (c, d) 5min. (a, c) Top view and (b, d) cross-sectional view of silicon prepared by FIB. The periodicity of the PS honeycomb mask was 3 μm.

4.2 Effect of noble metal catalyst species on morphology of macroporous silicon formed by metal-assisted chemical etching

To examine the effect of the catalyst species on the morphology of the etched silicon structure, chemical etching was also carried out using patterned Au thin films. Figure 11 shows the surfaces and cross-sectional images of silicon after chemical etching using the patterned Au thin films (Asoh et al, 2009). Before carrying out FIB processing to observe the cross section of the pores, tungsten was deposited on surface of the specimens to prevent damage to the etched silicon surface by gallium ion beam. Although metal-assisted chemical etching was carried out using Au catalyst under the same etching conditions indicated in Fig. 10, the morphology of the resultant porous structure was significantly different from that of the silicon macropores formed using Pt-Pd catalyst.

When the etching time was equal to 1 min, the localized dissolution of silicon was observed only on the silicon surface covered with Au thin films [Figs. 11(a) and 11(b)]. After carrying out metal-assisted chemical etching for 5 min, the pore depth increased up to 5 μm. The etching rate was estimated to be approximately 1 μm min^{-1}. One of the notable features of metal-assisted chemical etching using Au catalyst is the suppression of pore widening at the outermost surface of silicon. The diameter of the pores hardly increased during chemical etching. The dissolution of silicon was accelerated locally at the silicon/metal interface in the direction of the pore depth, resulting in the formation of macropores with a relatively high aspect ratio.

Fig. 12. SEM images of Pt-Pd-coated silicon after chemical etching in HF/H_2O_2 for (a, b) 1 min and (c, d) 2 min. (a, c) Top view and (b, d) cross-sectional view. Chemical etching was conducted after removing PS honeycomb mask.

The following factors are considered as the reasons behind the suppression of pore widening. (I) As discussed previously, the isolated Au circular thin films are composed of clusters of Au nanoparticles with a relatively large size. Thus, the configuration of metal catalysts affects the morphology of pores due to the catalytic action of the catalysts. (II) The diffusion area of h^+ during pore formation in the presence of Au catalyst is narrower than that of Pt-Pd catalyst. The injection/diffusion behaviour of h^+ is assumed to be affected by a Schottky barrier formed at the silicon/metal interface and the work function of each noble metal.

Figure 12 shows the SEM images of Pt-Pd-coated silicon after chemical etching in HF/H_2O_2 (Asoh et al, 2009). In this case, chemical etching was carried out after removing the PS honeycomb mask. A comparison with Fig. 10 showed that despite employing the same etching conditions, the preferential dissolution of silicon in this case remarkably proceeded in the direction of pore depth. It was also apparent that pore widening at the silicon surface was effectively suppressed. After metal-assisted chemical etching for 2 min, the pore depth increased up to 7 μm. The etching rate, which was estimated to be approximately 3.5 μm min^{-1}, was ~2 times faster than that of etching carried out using the PS mask and Pt-Pd catalyst, as shown in Fig. 10, and ~3.5 times faster than that of Au catalyst, as shown in Fig. 11. When metal catalysts were located in isolation on the silicon substrate and the PS honeycomb mask was not used, the diffusion of hydrogen produced by the reduction of protons and the dissolution of silicon probably proceeded effectively. As a result, straight macropores were formed within a relatively short etching time, and excessive pore widening at the outermost surface of macroporous silicon was prevented.

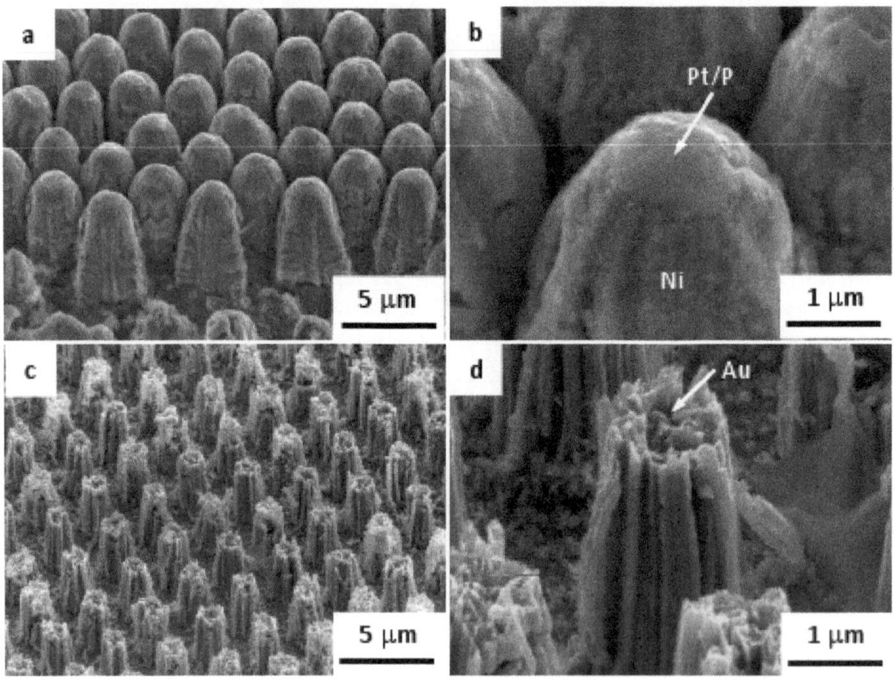

Fig. 13. SEM images of nickel replica of etched silicon. (a, b) Pt-Pd-coated silicon was etched for 2 min. (c, d) Au-coated silicon was etched for 3 min. (b, d) High-magnification view of each specimen. SEM observation was conducted from an angle of 45° to the surface.

To evaluate the morphology of the porous structure in the direction of pore depth, a nickel replica was prepared by metal plating using obtained macroporous silicon as a template. From the SEM image of the obtained nickel replica shown in Fig. 13, it was confirmed that the height of nickel rods, that is, the depth of the pores was almost uniform in both specimens (Asoh et al, 2009). These images of the nickel replica were basically compatible with the SEM images of macroporous silicon. However, the shape of the side surface of the nickel rods was found to differ in the case of each specimen. In the case of Pt-Pd catalyst, the side surface of nickel rods, that is, the side walls of the silicon macropores were relatively smooth. It should be noted that the obtained nickel rods had Pt-Pd thin films at their tips. This result also revealed that metal-assisted chemical etching proceeded only in the Pt-Pd-coated area on the silicon surface. On the other hand, in the case of Au catalyst, a number of channels were observed on the side surface of the nickel rods. Such structure is thought to be reflected in the traces of Au nanoparticles divided from circular thin film. In other words, the outer surface of silicon macropores was composed of the cluster of nanopores with sizes in the range of 50–100 nm. Suppression of pore widening during metal-assisted chemical etching using Au catalyst may be caused by the specific catalytic action of Au nanoparticles.

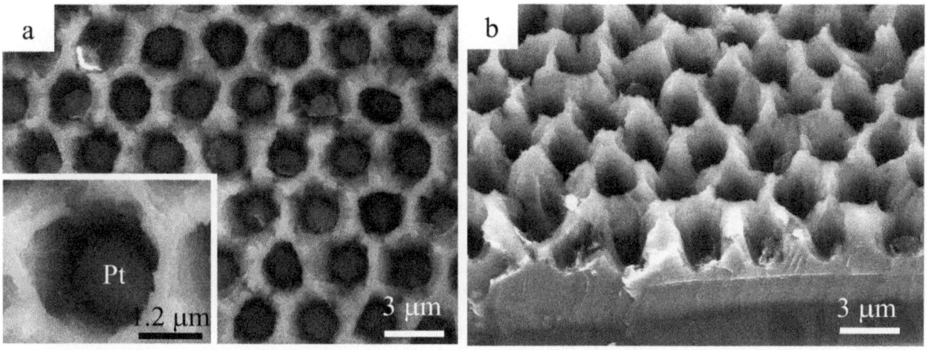

Fig. 14. SEM images of Pt-coated silicon after chemical etching in HF/H_2O_2 for 1 min. (a) Top view and (b) cross-sectional view. Chemical etching was conducted after removing PS honeycomb mask. Inset shows a high-magnification top view of the same specimen.

On the basis of the present method, ordered macropore arrays could be formed independent of noble metal catalyst species. Figure 14 shows the silicon macropore arrays formed by metal-assisted chemical etching using Pt single catalyst (Asoh et al, 2009). Metal-assisted chemical etching was carried out after removing the PS honeycomb mask. The etching behaviour when Pt catalyst was used was basically similar that in the case of Pt-Pd catalyst. From the tilted view of the specimen shown in Fig. 14(b), the pore depth after metal-assisted chemical etching for 1 min was estimated to be approximately 3 µm. A detailed study is under way for the clarification of the effect of Pd single catalyst.

The relationship between the pore depth and the etching time is summarized in Fig. 15 (Asoh et al, 2009). In addition to the results for the metal-assisted chemical etching carried out using the PS mask, the results obtained without using the PS mask are also plotted for each metal species. In the present study, etching rate increases in the following order: Au < Pt ≤ Pt-Pd. If the metal catalyst species are the same, the etching rate without using the mask is faster than that obtained using the mask. Further research on metal-assisted chemical etching using metal thin films as a catalyst would clarify the relationship between the mechanism for controlling the morphology of the resultant pattern and the etching conditions, such as the composition and concentration of an etchant, substrate parameters, resistivity, and doping density.

We described the differences in the catalytic action of Pt-Pd (or Pt) and Au catalysts on the morphology of etched silicon microstructures. In the case of Pt-Pd, although etching rate was faster than that of Au, there was excessive pore widening at the outermost surface due to the chemical dissolution of the horizontal plane. On the other hand, straight pores with relatively small diameter were obtained by using Au catalyst under the same etching conditions. Thus, the dissolution of silicon by using Au thin films as a catalyst was locally accelerated at the silicon/metal interface in the direction of the pore depth. The morphology of the resultant porous structure was assumed to be affected by the difference in the shape of metal catalyst and the diffusion behaviour of injected positive holes at the silicon/metal interface.

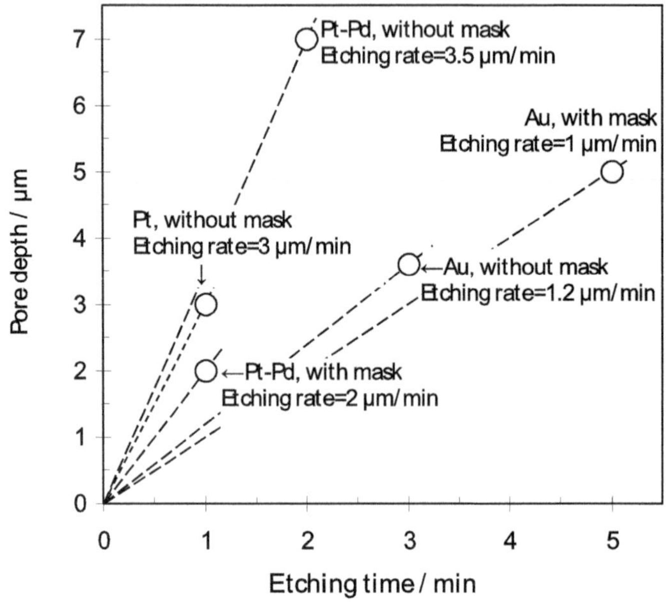

Fig. 15. Relationship between depth of pores formed by metal-assisted chemical etching using various noble metals and corresponding etching time.

4.3 Fabrication of macroporous silicon with a high aspect ratio

Figure 16 shows macroporous silicon with a high aspect ratio formed using circular Pt-Pd thin films as a catalyst. When metal-assisted chemical etching was conducted in HF with a high concentration of 10 mol dm^{-3}, the morphology of the resultant porous structure was significantly different from that of the silicon microwells formed in HF with a relatively low concentration of 5 mol dm^{-3} (Ono et al, 2009). In the case of low-concentration HF, injected positive holes are expected to diffuse into silicon bulk and oxidize silicon at locations away from the metal-coated silicon surface. On the other hand, in the case of high-concentration HF, the diffusion of positive holes is thought to be suppressed. One of the notable features of the obtained porous structure is that the diameter of each pore was hardly increased during chemical etching. Namely, the dissolution of silicon oxide is accelerated locally at the silicon/metal interface in the direction of the pore depth, resulting in the formation of macroporous silicon with a high aspect ratio. Further research on metal-assisted chemical etching using metal thin films as a catalyst would clarify the relationship between the mechanism for controlling the morphology of the resultant pattern and the etching conditions, such as the composition and concentration of etchant, substrate parameters, resistivity, and doping density.

Fig. 16. Cross-sectional SEM image of macroporous silicon formed by metal-assisted chemical etching. Chemical etching was conducted in 10 mol dm^{-3} HF/1 mol dm^{-3} H_2O_2 for 2 min using a Pt-Pd catalyst.

5. Periodic GaAs convex and hole arrays by metal-assisted chemical etching

5.1 Periodic micro GaAs convex arrays

GaAs nanostructures have been studied extensively, particularly from the viewpoint of their application to optoelectronic devices. For instance, the characteristic photoluminescence of GaAs porous structures that are produced by the polarization of the substrate in HCl solution has been reported previously (Schmuki et al, 1996). The design and fabrication of periodically ordered GaAs structures will lead to the development of new functional materials and devices in the future.

Studies have focused on the metal-assisted chemical etching of GaAs, and its effectiveness has been demonstrated (Yasukawa et al, 2008a, 2008b, 2008c, 2009). We fabricated GaAs convex arrays through electroless plating of Ag and Pd as metal catalysts and templating based on self-organized polystyrene (PS) spheres; however, convex structures with superior aspect ratios have not been obtained (Yasukawa et al, 2008a, 2008b, 2009). Herein, we describe the fabrication of GaAs convex and hole arrays using an ion-sputtered Pt-Pd catalyst to improve the aspect ratio of the etching structure. Pt-Pd was used because it markedly accelerates the etching rate of p-Si, and etching structures with a high aspect ratio were obtained (Asoh et al, 2009).

To fabricate convex arrays, Pt-Pd was selectively sputtered onto the substrate in the spaces among PS spheres through the PS mask so that a Pt-Pd honeycomb pattern was formed on GaAs. After Pt-Pd-metal-assisted chemical etching for 20 s in a mixed solution of 5 mol dm^{-3} HF and 1 mol dm^{-3} H_2O_2, ordered GaAs convex arrays were thus obtained as shown in Figs. 17(a) and 17(b) (Yasukawa et al, 2010). The etching depth was approximately 330 nm. GaAs was chemically dissolved using the etchant described above without the use of metal catalysts, resulting in two-dimensional striped etching patterns, i.e., anisotropic etching took place (Yasukawa et al, 2008b, 2009). The anisotropic etching in the present HF/H_2O_2 etchant was due to preferential etching of GaAs in the <010> crystallographic direction at an etching rate of ~ 0.2 nm s^{-1}. If the preferential etching direction is [010] or [010], then [001] or [001] is assumed to be the less preferential etching direction and *vice versa.*

The calculated etching rate of the convex arrays in the direction perpendicular to the substrate, [100], was ~16.7 nm s^{-1}. The etching rate using Pt-Pd was approximately two orders of magnitude higher than that in chemical dissolution. The metal-assisted chemical etching rates using Ag, Pd, and Au as catalysts were ~0.6, ~4.7, and ~8.1 nm s^{-1}, respectively (Yasukawa et al, 2008a, 2008b, 2008c, 2009); therefore, Pt-Pd-metal-assisted chemical etching was much faster and the aspect ratio of the resulting structure was improved. The difference between the work functions of each catalytic metal species could be one of the reasons for the difference between the metal-assisted chemical etching rates of Ag, Pd, Au, and Pt-Pd. In previous studies, empirically determined work functions were reported to be 4.26 eV for Ag (Dweydari and Mee, 1975), 5.12 eV for Au (Nieuwenhuys et al, 1974), 5.65 eV for Pt (Eastman, 1970), and 5.12 eV for Pd (Nieuwenhuys et al, 1974). In the metal-assisted chemical etching of a Si substrate, the fast etching rate originates from the higher work function, i.e., the relative ease of h^+ ejection (Asoh et al, 2009); the metal-assisted chemical etching rates of GaAs were highly consistent with this tendency. Therefore, it is reasonable for the Pt-Pd-assisted etching rate to be higher than those using other catalysts.

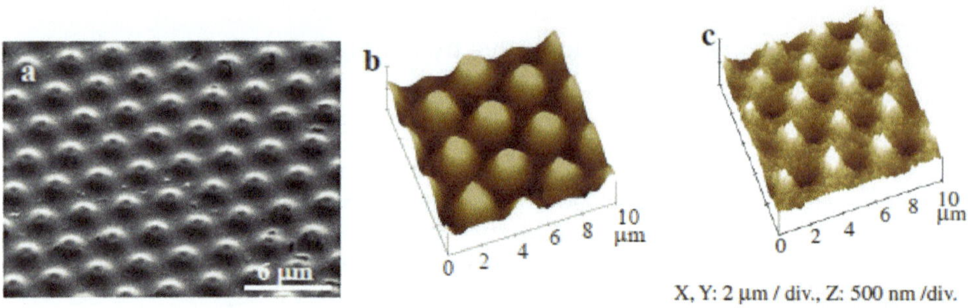

Fig. 17. GaAs convex structures obtained by Pt-Pd-metal-assisted chemical etching for 20 s in a solution of 5 mol dm^{-3} HF and 1 mol dm^{-3} H_2O_2. (a) SEM image at 45(to the surface and (b) AFM image. (c) AFM image of convex arrays formed by etching for 60 s.

Figure 17(c) shows convex arrays fabricated by etching for 60 s with a depth of ~ 430 nm. The etching rate was ~ 7.2 nm s^{-1}. The etching rate calculated from the structure was lower than that in the case of etching for 20 s. The rate of metal-assisted chemical etching is determined by the species of the substrate and the catalytic metal species, thereby the etching rate is constant as long as the same substrate and catalytic metal are used. The decrease in the etching rate with increasing etching time [Fig. 17(c)] was thus a consequence of the chemical dissolution of the convex structure itself (Yasukawa et al, 2010).

5.2 Periodic micro GaAs hole arrays

GaAs hole arrays [Figs. 18(a) and 18(b)] were fabricated by metal-assisted chemical etching for 60 s using the PS-honeycomb structural mask in the same etching solution as above (Yasukawa et al, 2010). Pt-Pd was deposited selectively in traces of SiO_2 spheres to form island-like patterns. GaAs was then etched to form hole arrays. The part of the substrate where Pt-Pd had not been deposited did not undergo metal-assisted chemical etching; however, striped etching patterns resulting from the anisotropic etching of the substrate surface were observed. Because Pt-Pd-metal-assisted chemical etching was performed after the removal of the mask, chemical dissolution occurred at the surface. The shapes of holes were approximately circular but somewhat elongated in the lateral direction in the images of samples with an average etch depth of ~670 nm. The metal-assisted chemical etching proceeded perpendicular to the substrate owing to h^+ being injected from the catalytic metal into GaAs. The diffusion of h^+ to the hole wall away from the catalytic metal/substrate interface occurs in all crystallographic directions in the substrate. However, anisotropic chemical dissolution depends on the substrate crystallography; therefore, the holes were elongated in the lateral direction (preferential etching direction).

Anisotropic etching is the most characteristic in GaAs because the crystallography of the underlying substrate strongly affects the resultant etching morphology. Such etching cannot be observed in the metal-assisted chemical etching of Si; therefore, different morphologies were observed between GaAs and Si fabricated by metal-assisted chemical etching even when the same metal catalyst was used.

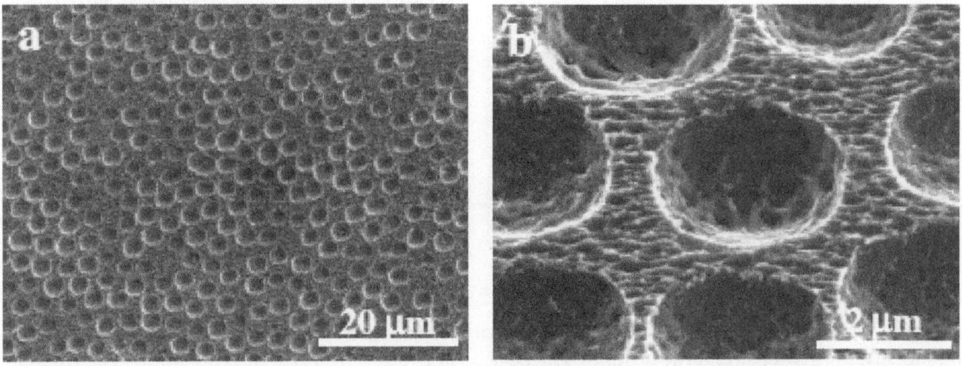

Fig. 18. SEM images of GaAs hole arrays fabricated by Pt-Pd-metal-assisted chemical etching for 60 s in a solution of 5 mol dm^{-3} HF and 1 mol dm^{-3} H_2O_2 observed from (a) top and (b) at 45° to the surface.

6. Ordered hexagonal array of crystalline Au nanodots on Silicon

Metal nanoparticles and nanodots have been a focus of intense research due to their novel physical properties in reference to their bulk counterparts (Daniel and Astruc, 2004). In addition, collective properties can arise due to the interactions of the individual nanoparticles in ordered arrays (Motte et al, 1997; Wang et al, 2007). This is why the control of individual dot properties (with controllable size and shape) and their relative arrangements (density, pattern shape) are crucial. Nanodots patterned in ordered arrays, especially gold nanodots, have been proposed for a wide range of applications, such as magnetic data storage, optoelectronic devices, biosensors and catalysts for the growth of aligned one-dimensional nanostructures.

In the past few years many methods have been developed for nanodot array patterning, such as focused ion beam or e-beam lithography, molecular beam epitaxy (MBE), chemical vapor deposition (CVD), selfassembly and template-based methods by using porous anodic alumina membranes as evaporation masks. However, most fabrication processes mentioned above are not satisfactory due to some drawbacks, such as low throughput, high cost of equipment and low uniformity of the shapes and sizes of dots.

We reported previously that metal patterns of two different types, that is, network-like honeycomb and isolated-island patterns with ordered periodicities (200 nm - 3 μm), were formed on an Si(100) substrate by a combination of colloidal crystal templating, hydrophobic treatment and subsequent siteselective electrodeless deposition (Asoh et al, 2007a, 2007c). Based on this strategy, it is possible to control the configuration of metal dot arrays by changing the mask structure for metal deposition.

We focus in particular on the controllability of the morphology and crystallinity of Au nanodots and try to control the heteroepitaxial growth of Au dots on Si surfaces by high temperature annealing above the eutectic temperature in an Au - Si system.

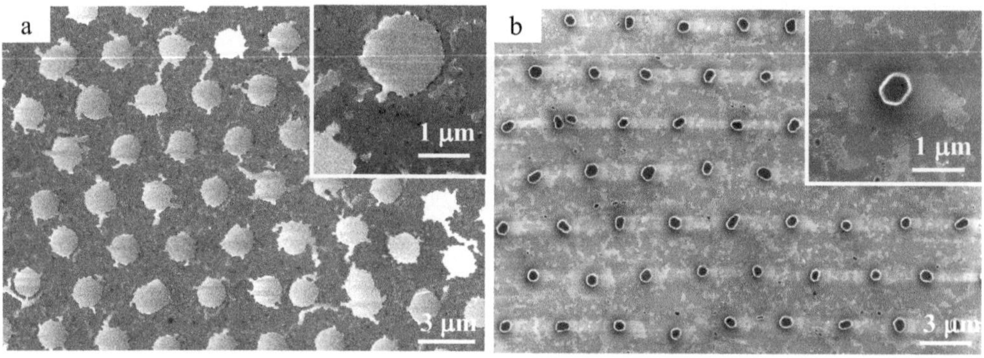

Fig. 19. (a) Au thin films pattern formed by sputter coating through PS honeycomb mask and removal of PS honeycomb mask. (b) Hexagonally-arranged Au nanodot arrays formed by annealing in Ar ambient.

In the present study, the PS honeycomb structure was applied as a mask for metal deposition. When sputtering was conducted through the PS honeycomb mask, isolated circular Au thin films with approximately 1.5 μm diameter and a density of ~1×10^7 cm^{-2} were deposited in the interspaces of the PS mask as shown in Fig. 19(a) (Sakamoto et al 2008). Each thin film consisted of polycrystalline granular Au particles with 20 - 40 nm grains. After the deposition of the Au thin films on the Si(111) substrate, the specimens were annealed under an argon atmosphere to control the morphology and the crystallinity of the deposited Au. Figure 19(b) shows an SEM image of the Au nanodot arrays after heat treatment for 1 h at 600 °C above the eutectic temperature (363 °C) in an Au - Si system. Au thin films composed of granular nanoparticles were self-assembled into spherical Au nanoislands by thermo migration. The diameter of each Au dot was approximately 650 nm (inset). Chemical analyses by EDX show that the heat treatment under argon does not change the chemical composition of the nanodots. The density of the nanodots is also still unchanged.

From the tilted view, it was confirmed that the shape of the Au nanodots was a truncated hexagonal dipyramid with defined facets (Fig. 20(a)) (Sakamoto et al 2008). The height of the thermally crystallized Au dot was approximately 260 nm. To identify the crystal structure of the Au dot, EBSD analysis was performed. In fact, electron backscatter diffraction (EBSD) is a technique which allows crystallographic information to be obtained from samples in the scanning electron microscope (SEM). In EBSD a stationary electron beam strikes a tilted crystalline sample and the diffracted electrons form a pattern on a fluorescent screen. This pattern is characteristic of the crystal structure and orientation of the sample region from which it was generated. The diffraction pattern can be used to measure the crystal orientation, measure grain boundary misorientations, discriminate between different materials and provide information about local crystalline perfection. When the beam is scanned in a grid across a polycrystalline sample and the crystal orientation measured at each point, the resulting map will reveal the constituent grain morphology, orientations and boundaries. This data can also be used to show the preferred crystal orientations present in the material.

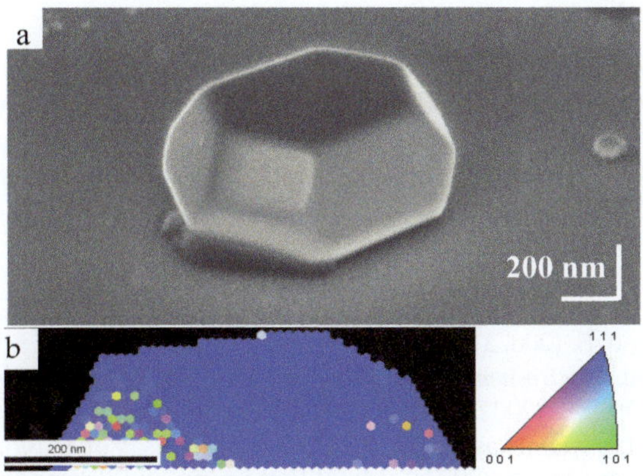

Fig. 20. (a) Tilted *(60°)* SEM image of Au nanodots. (b) Color mapping (colour online) of crystal orientation of Au nanodots by EBSD analysis (the area analyzed by EBSD is marked by the rectangle in (a)). A pattern quality map is plotted in the background and data is cleaned. (Sakamoto et al 2008)

Figure 20(b) shows EBSD mapping images of a side view of one dot. From EBSD analysis, it was ascertained that Au nanodots formed on Si(111) were almost single crystals with a (111) crystal face. We do not observe another crystal orientation. The shape difference observed in Fig. 19(b) can be related to the fact that all the nanodots under this heat treatment condition do not possess the perfect equilibrium structure (truncated hexagonal dipyramid with defined facets) (Shi and Stampfl, 2008). This specific growth of Au dots was assumed to be caused by the crystal orientation of the underlying Au/Si intermixed layer, that is, the Au silicide interfacial layer. Several studies have been carried out on interfacial phenomena between metals (Au) and semiconductors (Si) due to its importance in electronic devices and interconnects, and it is well known that Au silicide is formed due to diffusion of Si through the Au film during the annealing process (Ressel et al, 2003; Flammini et al, 2004; Negishi et al 2006). In addition, epitaxial growth of truncated hexagonal dipyramidal Au nanodots on the Au silicide phase of Au_7Si has been reported (Wu et al, 2003). According to their report, metallic Au was grown epitaxially on the Au_7Si*(111)* because the lattice spacing matches well, that is, the lattice mismatch is 3.4% with two unit cells of Au matching a single unit cell of Au_7Si in the (111) orientation. In the present study, it is inferred from the specific morphology of the Au nanodot that the epitaxial growth of Au nanodots could be controlled by the crystal orientation of the underlying Au silicide as well as Si(111).

7. Conclusion

The process presented in this paper is suitable for the large-scale production of ordered silicon macropores containing noble metal thin films, which cannot be fabricated by conventional lithographic techniques, because this patterning process involves colloidal crystal templating based on a relatively easy chemical treatment. On the basis of the present process, different types of noble metal films can also be embedded within a silicon substrate. Further research on the preparation of three-dimensional silicon microstructures by metal-assisted chemical etching would help us to determine their potential applications in optical devices, chemical sensors, and silicon-based biofunctional devices.

8. Acknowledgments

Part of this work was financially supported by a Grant-in-Aid for Scientific Research from the Japan Society for the Promotion of Science and the Light Metal Education Foundation of Japan. Thanks are also due to a grant of Strategic Research Foundation Grant-aided Project for Private Universities from Ministry of Education, Culture, Sport, Science, and Technology, Japan (MEXT).

9. References

Asoh, H.; Arai, F. & Ono, S. (2007a). Site-Selective Chemical Etching of Silicon Using Patterned Silver Catalyst. *Electrochemistry Communications*, Vol. 9, No. 4, pp. 535-539.

Asoh, H.; Arai, F. & Ono, S. (2007b). Micro Patterning of Silicon by Chemical Etching Using Patterned Noble Metals as Catalyst. *ECS Transactions*, Vol. 6, No. 2, pp. 431-437.

Asoh, H.; Sakamoto, S. & Ono, S. (2007c). Metal Patterning on Silicon Surface by Site-Selective Electroless Deposition through Colloidal Crystal Templating. *Journal of Colloid and Interface Science*, Vol. 316, No. 2, pp. 547-552.

Asoh, H. & Ono, S. (2007d). Fabrication of ordered anodic nanoporous alumina layers and their application to nanotechnology, In: Electrocrystallization in Nanotechnology, G. Staikov, (Ed.), Chapter 7 pp. 138-166: Wiley-Vch Verlag GmbH & Co. KGaA, ISBN 978-3-527-31545-4, Weinheim

Asoh, H.; Arai, F.; Uchibori, K. & Ono, S. (2008). Pt-Pd-Embedded Silicon Microwell Arrays. *Applied Physics Express*, Vol. 1, No. 6, pp. 067003/1-067003/3.

Asoh, H.; Arai, F. & Ono, S. (2009). Effect of Noble Metal Catalyst Species on the Morphology of Macroporous Silicon Formed by Metal-Assisted Chemical Etching. *Electrochimica Acta*, Vol. 54, No. 22, pp. 5142-5148.

Asoh, H. & Ono, S. (2010). Nanohole arrays on silicon, In: *Handbook of Nanophysics: Functional Nanomaterials*, K. Sattler, (Ed.), Chapter 28 pp. 28-1 – 28-14 : Taylor & Francis Books, Inc., ISBN 978-1-4200-7552-6, Boca Raton, FL, USA

Arai, F.; Asoh, H. & Ono, S. (2008). Electroless Deposition of Noble Metal Nano Particles as Catalyst and Subsequent Micropatterning of Silicon Substrate by Wet Chemical Etching. *Electrochemistry*, Vol. 76, No. 3, pp. 187-190.

Chartier, C.; Bastide, S. & Levy-Clement, C. (2008). Metal-assisted chemical etching of silicon in HF-H_2O_2. *Electrochimica Acta*, Vol. 53, No. 17, pp. 5509-5516.

Chattopadhyay, S.; Li, X. & Bohn, P. W. (2002). In-plane control of morphology and tunable photoluminescence in porous silicon produced by metal-assisted electroless chemical etching. *Journal of Applied Physics*, Vol. 91, No. 9, pp. 6134-6140.

Daniel, M. C. & Astruc, D. (2004). Gold Nanoparticles: Assembly, Supramolecular Chemistry, Quantum-Size-Related Properties, and Applications toward Biology, Catalysis, and Nanotechnology. *Chemical Reviews*, Vol. 104, No. 1, pp. 293-346.

Deckman, H. W. & Dunsmuir, J. H. (1982). Natural lithography, *Applied Physics Letters*, Vol. 41, No. 4, pp. 377-379.

Dweydari, A. W. & Mee, C. H. B. (1975). Work function measurements on (100) and (110) surfaces of silver. *Physica Status Solidi A*, Vol. 27, No. 1, pp. 223-230.

Eastman, D. E. (1970). Photoelectric Work Functions of Transition, Rare-Earth, and Noble Metals, *Physical Review B*, Vol. 2, No. 1, pp. 1-2.

Flammini, R.; Wiame, F.; Belkhou, R.; Taleb-Ibrahimi, A.; Gregoratti, L.; Barinov, A.; Marsi, M. & Kiskinova, M. (2004). Effect of annealing on the structure of the Au/Si(111)-H interface. *Surface Science*, Vol. 564, No. 1, pp. 121-130.

Ivanisevic, A. & Mirkin, C. A. (2001). "Dip-Pen" Nanolithography on Semiconductor Surfaces. *Journal of the American Chemical Society*, Vol. 123, No. 32, pp. 7887-7889.

Jiang, P. & McFarland, M. J. (2004). Large-Scale Fabrication of Wafer-Size Colloidal Crystals, Macroporous Polymers and Nanocomposites by Spin-Coating, *Journal of the American Chemical Society*, Vol. 126, No. 42, pp. 13778-13786.

Kleimann, P.; Linnros, J. & Juhasz, R. (2001). Formation of three-dimensional microstructures by electrochemical etching of silicon, *Applied Physics Letters*, Vol. 79, No. 11, pp. 1727-1729.

Lehmann, V. & Foll, H. (1990). Formation mechanism and properties of electrochemically etched trenches in n-type silicon. *Journal of The Electrochemical Society*, Vol. 137, No. 2, pp. 653-659.

Lehmann, V. (2002) *Electrochemistry of Silicon*, Wiley-Vch Verlag GmbH & Co. KGaA, ISBN 3-527-29321-3, Weinheim

Li, X. & Bohn, P.W. (2000). Metal-assisted chemical etching in HF/H_2O_2 produces porous silicon. *Applied Physics Letters*, Vol. 77, No. 16, pp. 2572-2574.

Maccarini, M.; Himmelhaus, M.; Stoycheva, S. & Grunze, M. (2005). Characterisation and stability of hydrophobic surfaces in water. *Applied Surface Science*, Vol. 252, No. 5, pp. 1941-1946.

Matthias, S.; Muller, F.; Jamois, C.; Wehrspohn, R.B. & Gosele, U. (2004). Large-area three-dimensional structuring by electrochemical etching and lithography. *Advanced Materials*, Vol. 16, No. 23-24, pp. 2166-2170.

Motte, L.; Billoudet, F.; Lacaze, E.; Douin, J. & Pileni, M. P. (1997). Self-Organization into 2D and 3D Superlattices of Nanosized Particles Differing by Their Size. The *Journal of Physical Chemistry B*, Vol. 101, No. 2, pp. 138-144.

Negishi, R.; Mochizuki, I. & Shigeta, Y. (2006). Fabrication of uniform Au silicide islands on the Si(111)-(7×7) substrate, *Surface Science*, Vol. 600, No. 5, pp. 1125-1128.

Nieuwenhuys, B. E.; Bouwman, R. & Sachtler, W. H. M. (1974). The changes in work function of group Ib and VIII metals on xenon adsorption, determined by field electron and photoelectron emission, *Thin Solid Films*, Vol. 21, No. 1, pp. 51-58.

Ono, S.; Oide, A. & Asoh, H. (2007). Nanopatterning of Silicon with Use of Self-Organized Porous Alumina and Colloidal Crystals as Mask. *Electrochimica Acta*, Vol. 52, No. 8, pp. 2898-2904.

Ono, S.; Arai, F. & Asoh, H. (2009). Micro-Patterning of Semiconductors by Metal-Assisted Chemical Etching through Self-Assembled Colloidal Spheres. *ECS Transactions*, Vol. 19, No. 3, pp. 393-402.

Ressel, B.; Prince, K. C.; Heun, S. & Homma, Y. (2003). Wetting of Si surfaces by Au-Si liquid alloys, *Journal of Applied Physics*, Vol. 93, No. 7, pp. 3886-3892.

Sakamoto, S.; Philippe, L.; Bechelany, M.; Michler, J.; Asoh, H. & Ono, S. (2008). Ordered Hexagonal Array of Au Nanodots on Si Substrate Based on Colloidal Crystal Templating, *Nanotechnology*, Vol. 19, No. 40, pp. 405304/1-405304/6.

Sato, T.; Fujino, T. & Hasegawa, H. (2006). Self-assembled formation of uniform InP nanopore arrays by electrochemical anodization in HCl based electrolyte, *Applied Surface Science*, Vol. 252, No. 15, pp. 5457- 5461.

Schmuki, P.; Lockwood, D. J.; Labbé, H. J. & Fraser, J. W. (1996). Visible photoluminescence from porous GaAs, *Applied Physics Letters*, Vol. 69, No. 11, pp. 1620-1622.

Shi, H. & Stampfl, C. (2008). Shape and surface structure of gold nanoparticles under oxidizing conditions, *Physical Review B*, Vol. 77, No. 9, pp. 094127/1-094127/9.

Sun, J.; Tang, C.; Zhan, P.; Han, Z.; Cao, Z. & Wang, Z., (2010). Fabrication of Centimeter-Sized Single-Domain Two-Dimensional Colloidal Crystals in a Wedge-Shaped Cell under Capillary Forces, *Langmuir*, Vol. 26, No. 11, pp. 7859-7864.

Tsujino, K. & Matsumura, M. (2005a). Boring deep cylindrical nanoholes in silicon using silver nanoparticles as a catalyst, *Advanced Materials*, Vol. 17, No. 8, pp. 1045-1047.

Tsujino, K. & Matsumura, M. (2005b). Helical nanoholes bored in silicon by wet chemical etching using platinum nanoparticles as catalyst, *Electrochemical and Solid-State Letters*, Vol. 8, No. 12, pp. C193-C195.

Wang, S.; Yu, G. J.; Gong, J. L.; Zhu, D. Z. & Xia, H. H. (2007). Large-area uniform nanodot arrays embedded in porous anodic alumina, *Nanotechnology*, Vol. 18, No. 1, pp. 015303/1-015303/4.

Wu, J. S.; Chen, Y. F.; Dhara, S.; Wu, C. T.; Chen, K. H. & Chen, L. C. (2003). Interface energy of Au_7Si grown in the interfacial layer of truncated hexagonal dipyramidal Au nanoislands on polycrystalline-silicon, *Applied Physics Letters*, Vol. 82, No. 25, pp. 4468-4470.

Yae, S.; Kawamoto, Y.; Tanaka, H.; Fukumuro, N. & Matsuda, H. (2003). Formation of porous silicon by metal particle enhanced chemical etching in HF solution and its application for efficient solar cells, *Electrochemistry Communications*, Vol. 5, No. 8, pp. 632-636.

Yae, S.; Abe, M.; Fukumuro, N. & Matsuda, H. (2008). Palladium enhanced etching of n-type silicon in hydrofluoric acid solution, *Electrochemistry*, Vol. 76, No. 2, pp. 144-146.

Yasukawa, Y.; Asoh, H. & Ono, S. (2008a). Site-Selective Chemical Etching of GaAs through a Combination of Self-Organized Spheres and Silver Particles as Etching Catalyst, *Electrochemistry Communications*, Vol. 10, No. 5, pp. 757-760.

Yasukawa, Y.; Asoh, H. & Ono, S. (2008b). Site-Selective Metal Patterning/Metal-Assisted Chemical Etching on GaAs Substrate through Colloidal Crystal Templating, *ECS Transactions*, Vol. 13, No. 3, pp. 83-92.

Yasukawa, Y.; Asoh, H. & Ono, S. (2008c). GaAs Microarrays by Noble-Metal Assisted Chemical Etching, *ECS Transactions*, Vol. 16, No. 3, pp. 253-258.

Yasukawa, Y.; Asoh, H. & Ono, S. (2009). Site-Selective Metal Patterning/Metal-Assisted Chemical Etching on GaAs Substrate through Colloidal Crystal Templating, *Journal of The Electrochemical Society*, Vol. 156, No. 10, pp. H777-H781.

Yasukawa, Y.; Asoh, H. & Ono, S. (2010). Periodic GaAs Convex and Hole Arrays Produced by Metal-Assisted Chemical Etching, *Japanese Journal of Applied Physics*, Vol. 49, No. 11, pp. 116502/1-116502/4.

Yokoyama, T.; Asoh, H. & Ono, S. (2010). Site-Selective Anodic Etching of InP Substrate Using Self-Organized Spheres as Mask, *Physica Status Solidi A*, Vol. 207, No. 4, pp. 943-946.

13

The Effect of Addition of ppm-Order-Pd to Fe-K Catalyst on Dehydrogenation of Ethylbenzene

Ryo Watanabe, Yasushi Sekine, Masahiko Matsukata and Eiichi Kikuchi
Department of Applied Chemistry, Waseda University
Japan

1. Introduction

Styrene, used for polymeric materials such as polystyrene resin, acrylonitrile–butadiene–styrene resin, and styrene–butadiene rubber, is an important monomer in the petrochemistry (Meima et al., 2001). The styrene production volume is about 23 million tons annually; more than 90% of styrene is produced by dehydrogenation of ethylbenzene, as in the following chemical equation (1) (Shekhah et al., 2004).

$$C_8H_{10} \rightarrow C_8H_8 + H_2 \quad (1)$$

Dehydrogenation of ethylbenzene is an endothermic reaction. High temperatures are necessary for high ethylbenzene conversion because of its thermodynamic limitation. Potassium promoted iron oxide (Fe–K) has been applied for dehydrogenation of ethylbenzene as an industrial catalyst (Lee, 1974). The Fe–K catalyst can be further modified by the addition of other promoters such oxides as MgO, Cr_2O_3, CeO_2 and MoO_3 (Eggertsen et al., 1947; Kearby et al., 1945; O'Hara, 1975; Pitzer, 1957; Riesser, 1979).

Generally, radial-type adiabatic reactors have been applied in commercial processes for dehydrogenation of ethylbenzene (Addiego, 2001). Superheated steam is supplied to the catalyst bed with ethylbenzene as a heating source for control of the reaction temperature. Steam also acts as a diluent for shifting the equilibrium conversion higher and suppresses over-reduction of the catalyst in the dehydrogenation atmosphere. Additionally, steam has the role of elimination of carbon deposition on the catalyst surface. The removal of coke from the catalyst surface using steam treatment improved the catalytic activity and that a low H_2O/ethylbenzene ratio caused coke deposition on the catalyst (Herzog et al., 1984; Devoldere et al., 1999). It was also reported that the increasing the feed molar-ratio of H_2O/ethylbenzene decreased the carbon deposition (Coulter et al., 1995).

The reaction mechanism of dehydrogenation of ethylbenzene on the Fe–K catalyst has been investigated widely. Langmuir–Hinshelwood (L–H) type including chemisorption of ethylbenzene and styrene proceeded on the Fe-K catalyst according to kinetic studies (Carrà et al., 1965). Analysis of infrared spectroscopy indicated that ethylbenzene adsorbed with the phenyl ring oriented to α-Fe_2O_3 (Busca et al., 1984). The styrene adsorption state on α-Fe_2O_3 catalyst was reported that styrene adsorbed via the vinyl group on the catalyst. (Addiego et al., 1994). The result of isotope exchange reaction revealed that the dissociation of α-hydrogen of ethylbenzene was the first step of ethylbenzene dehydrogenation, followed by the dissociation of β-hydrogen, which was the rate-determining step or post rate-determining step. (Miura et al., 1994)

Some researchers investigated the active phase of the Fe-K catalyst and showed $KFeO_2$ phase as the active site for dehydrogenation of ethylbenzene using a combination of XPS/UPS surface analysis (Hirano et al., 1986; Muhler et al., 1989, 1992). The active state of the Fe–K catalyst was analysed by transient response experiments and reported that initially high activity was correlated with Fe^{3+} state and the reduction from Fe^{3+} to Fe^{2+} state caused deactivation of the catalytic activity (Zhu et al., 2004). The reaction cycle was proposed that ethylbenzene adsorbed onto a terrace Fe^{3+} site; then C–H in the ethyl group was deprotonated by lattice oxygen at a step site. H_2O containing a lattice oxygen was postulated to desorb on the catalyst instead of H_2, meaning that oxidative dehydrogenation proceeded on the Fe-K catalyst (Weiss et al., 2000).

Recently, it was reported that the addition of small amounts of Pd or Pt enhanced the activity of Fe–K based catalysts (Williams et al., 2002). Previously we investigated hydrogen production by the steam–iron reaction using iron oxide modified with very small amounts of Pd at the temperature of 723 K and under atmospheric pressure. The addition of ppm-order Pd to iron oxide enhanced rates of the reduction by H_2 and the oxidation by H_2O (Urasaki et al., 2005). This work is intended to clarify the role of added precious metals such as Pd and Pt in relation to the reaction mechanism of dehydrogenation of ethylbenzene on the Fe–K catalyst (Sekine et al., 2008).

2. The activity for dehydrogenation of ethylbenzene over ppm-order Pd/Fe-K catalyst

For investigating the effect of Pd addition to the Fe-K catalyst on dehydrogenation activity and selectivity to styrene, we compared the result of activity tests on the Fe-K catalyst and Pd-promoted Fe-K catalyst. (Section 2.1) And also, the effect of Pd loading on dehydrogenation activity was investigated in the section 2.2.

2.1 Comparison of the activity and selectivity over Pd/Fe-K catalyst to those over bare Fe-K catalyst

Catalysts used in this study were commercially available Fe-K catalyst and Pd/Fe-K catalyst prepared by an impregnation method on the Fe-K catalyst. Fe-K mixed oxide which contained some promoters such as 10 wt%K_2O, 10 wt%CeO_2 and trace amounts of MoO_3, MgO and CaO immersed in an aqueous solution of the metal precursor salt of $Pd(NO_3)_2$. The amount of Pd was 10, 30, 300 and 1000 ppm. Then, the resulting slurry was dried on a hot plate under continuous stirring with subsequent calcination at 873 K for 6 h. Pt catalyst was prepared using the same impregnation method with an acetone solution of $Pt(CH_3COO)_2$.

The reactor used in this study comprised of a quartz tube (12-mm i.d.) containing the catalyst bed, which was fixed by quartz wool. A type K thermocouple enclosed in a quartz thermowell of 3-mm outer diameter was positioned inside the catalyst bed for accurate measurement of the catalyst temperature.

Catalytic activities, selectivities and stabilities were examined in a conventional fixed bed flow reactor. The weight of charged catalyst was 1.0 g. Reactions were carried out at 813 K at atmospheric pressure in the presence of steam; the molar ratio of steam to ethylbenzene was 12 and WHSV was 1.2 g h^{-1} g-cat^{-1}. Ethylbenzene and H_2O were fed separately using a micro feeder and a micro pump, respectively. Liquid products such as ethylbenzene, benzene, toluene and styrene were analysed using off-line flame ionization detection (FID) gas chromatography (GC8A; Shimadzu Corp.) with a capillary column of DB210 (0.25 mm×30 m) using N_2 as a carrier gas at 373 K. Tetralin was used as an internal standard substance. Gaseous products such as H_2, CO and CH_4 were analyzed using an off-line thermal conductivity (TCD) gas chromatography (GC8A; Shimadzu Corp.) with a 3 mm×2 m stainless steel column packed with active carbon (60/80 mesh) using Ar as a carrier gas at 393 K.

$$\text{Ethylbenzene conversion} = ([Sty] + [Bz] + [Tol])/([EB] + [Sty] + [Bz] + [Tol])\times 100 \quad (2)$$

$$\text{Styrene yield} = [Sty]/([EB] + [Sty] + [Bz] + [Tol])\times 100 \quad (3)$$

$$\text{Styrene selectivity} = [Sty]/([Sty] + [Bz] + [Tol])\times 100 \quad (4)$$

In these equations, [EB], [Sty], [Bz], and [Tol], respectively represent the yield of ethylbenzene, styrene, benzene, and toluene. Carbon balances of these investigations amounted to ≥95%.

For investigating the effect of Pd addition to the Fe–K based catalyst on dehydrogenation activity and selectivity, activity tests of the Fe–K and 10 ppm-Pd/Fe–K catalysts were carried out.

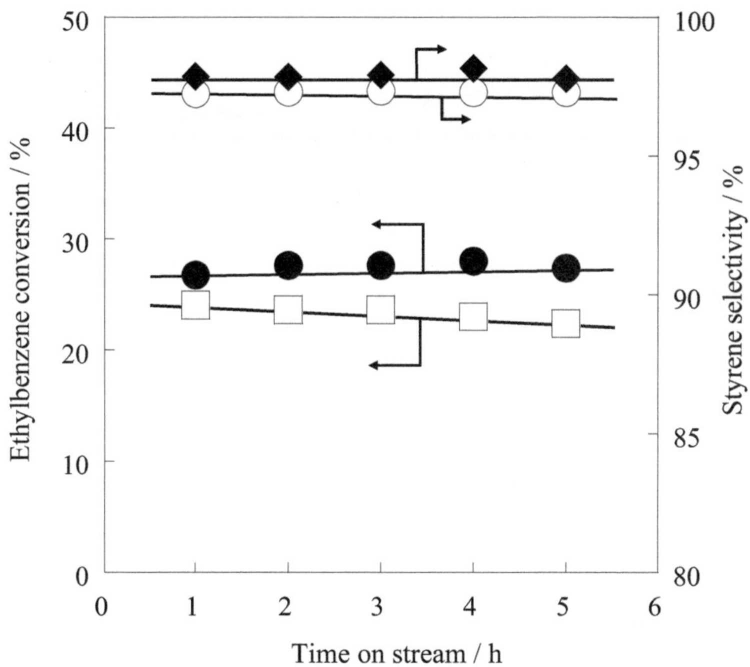

Fig. 1. Ethylbenzene conversion and styrene selectivity of (□,○) Fe-K catalyst and (●,◆) Pd/Fe-K catalyst.

Figure 1 shows the conversion of ethylbenzene and selectivity to styrene for these catalysts with time on stream. The Fe-K catalyst showed 24.1% of conversion at 1 h reaction. Ethylbenzene conversion was gradually decreased to 22.4% at 5 h reaction. On the other hand, 10 ppm-Pd/Fe-K catalyst showed 26.8% of conversion at 1 h reaction and 27.4% of conversion at 5 h reaction. The addition of small amount of Pd enhanced the activity for dehydrogenation of ethylbenzene over the Fe-K catalyst. Catalytic activity of 10 ppm-Pd/Fe–K catalyst was stable for 3 to 5 h at least. As for the selectivity to styrene, the Fe-K catalyst showed 97.2 and 97.3% at 1 h and 5 h, respectively. Benzene and toluene were formed as by-products with selectivities of 1.1% and 1.7%, respectively. While, 10 ppm-Pd/Fe–K catalyst revealed 97.8% selectivity to styrene at 1 h and 5 h. Selectivities to benzene and toluene were 1.1% and 1.1%, respectively at 1 h. The effect of addition of ppm-order-Pd to the Fe-K catalyst enhanced for dehydrogenation activity drastically and also improved selectivity to styrene.

2.2 The promotion effect of Pd on dehydrogenation activity of Fe-K catalyst

From section 2.1, we found that the trace amount of Pd addition to the Fe-K catalyst enhanced dehydrogenation activity. So, the promotion effect of Pd loading (the amount of Pd; 10, 30, 300 and 1000 ppm) on the Fe-K catalyst on dehydrogenation activity was investigated. Figure 2 shows the result of activity tests over various catalysts.

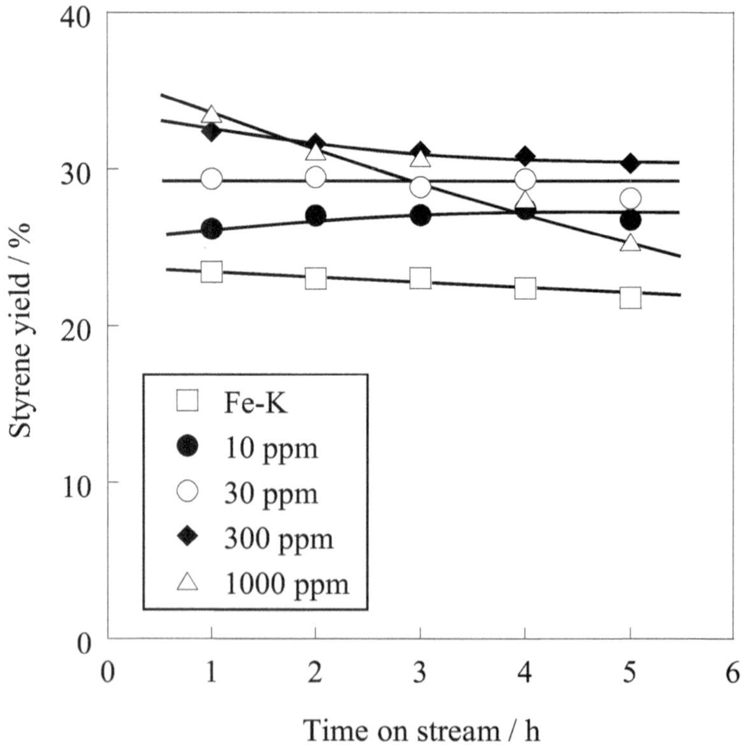

Fig. 2. The effect of Pd loading on styrene yield.

Figure 2 reveals that the conversion of ethylbenzene increases from 23.4% to 33.5% at 1 h reaction with increasing the amount of Pd. Figure 3 shows the relation between the initial catalytic activity and Pd concentration. As shown in fig. 2, catalysts of 10 ppm-Pd/Fe-K and 30 ppm-Pd/Fe-K showed high stability for dehydrogenation of ethylbenzene with time on stream. However, deactivation proceeded on catalysts of 300 ppm-Pd/Fe-K and 1000 ppm-Pd/Fe-K. From fig. 3, the effect of the addition of Pd seems to be saturated at a few hundred ppm of Pd on the surface of catalyst. Therefore, the optimum amount of Pd was less than 300 ppm in terms of activity and stability for dehydrogenation of ethylbenzene.

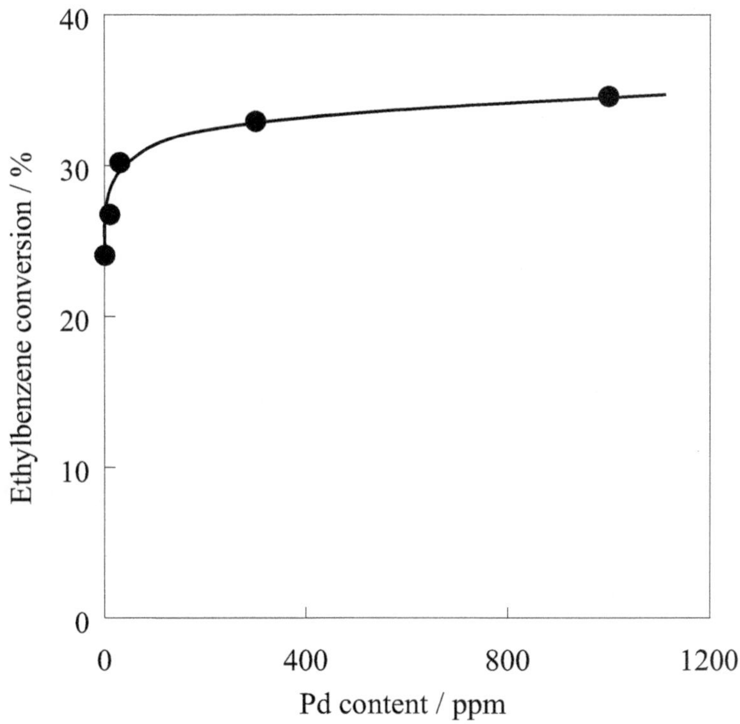

Fig. 3. Relation between the initial catalytic activity and Pd concentration.

For investigating the effect of the Pt addition to the Fe–K catalyst on dehydrogenation activity and selectivity, the activity test of Pt/Fe–K catalyst was carried out in the fixed-bed flow reactor at 813 K. Loading of Pt on the Fe-K catalyst was equivalent mole of 300 ppm-Pd/Fe-K catalyst. Figure 4 shows the result of activity and selectivity of the Fe-K, Pd/Fe-K and Pt/Fe-K catalysts. As shown in fig. 4, Pt/Fe-K catalyst showed 36.5 % and 31.6 % of conversion at 1 h and 5 h, respectively. Comparing conversion of Pt/Fe-K catalyst and Fe-K catalyst, trace amounts of Pt also enhanced the activity of the Fe-K catalyst. However catalytic activity of Pt/Fe–K catalyst gradually decreased with time on stream. Styrene selectivity was almost the same to that of the Fe-K catalyst. The result indicated the active site did not change by the addition of trace amounts of Pt to the Fe-K catalyst.

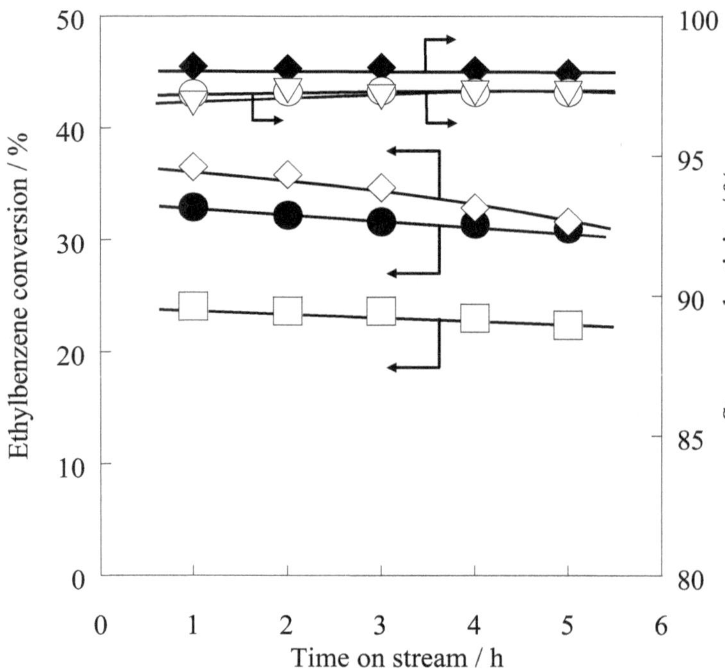

Fig. 4. Ethylbenzene conversion and styrene selectivity of (□,○) Fe-K catalyst, (●,◆) Pd/Fe-K catalyst and (◇,▽) Pt/Fe-K catalyst.

3. Reaction mechanism of dehydrogenation of ethylbenzene over Pd/Fe-K catalyst

In order to clarify the role of Pd, we investigated the reaction mechanism using periodical pulse experiments with reactants of ethylbenzene and H_2O supplied alternately to the catalyst bed. The period of each pulse was 600 s. The weight of charged catalyst was 1.0 g. Ethylbenzene and H_2O were supplied using a bubbler with He as a carrier gas. The molar ratio of ethylbenzene to He was 1/20; the total flow rate was 7.8×10^{-2} mol h^{-1}. The molar ratio of H_2O to He was 1/8, and the total flow rate was 8.4×10^{-2} mol h^{-1}. Reaction temperature was 783 K. The product composition at the reactor outlet was monitored continuously using an on-line quadruple mass spectrometer (HPR20; Hiden Analytical Ltd.). Calibration of mass spectrometer signals was based on prepared mixtures of a known composition. For all periodical experiments, the gas-phase composition was calculated from the mass signal at the following m/e = 2 (H_2), 4 (He), 18 (H_2O), 78 (styrene), and 91 (ethylbenzene).

Figure 5 shows the fraction of H_2O in produced hydrogen species ($[H_2O]/([H_2] + [H_2O])$) and the conversion of ethylbenzene when the pulse test was conducted on the Fe–K catalyst.

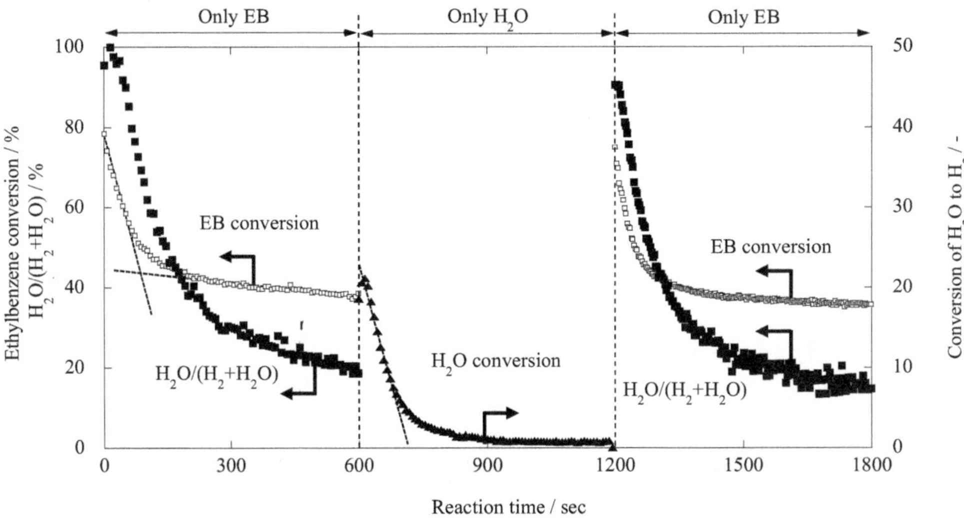

Fig. 5. Results for the periodical pulse reaction test over the Fe-K catalyst (Sekine et al., 2008).

In the case of feeding EB-alone to the catalyst bed, products were styrene and H_2O at the initial stage of the reaction. H_2 was gradually produced with decreasing of H_2O production. The fraction of H_2O and the conversion of ethylbenzene showed a high value at the starting time of the reaction. This steep and straight decrease of ethylbenzene conversion from 0 s to 100 s could be attributed to the domination of oxidative dehydrogenation of ethylbenzene (5) because the ratio of $H_2O/(H_2 + H_2O)$ was close to 1 at the early stage of the reaction.

$$C_8H_{10} + O^{2-}_{lat} \rightarrow C_8H_8 + H_2O + V_{ox} + 2e^- \quad (5)$$

Here, O^{2-}_{lat} denotes lattice oxygen in the Fe-K catalyst and V_{ox} shows the lattice vacancy. Oxidative dehydrogenation, by which styrene and H_2O were formed, competed with and progressed faster than, the simple dehydrogenation (6) (Langmuir-Hinshelwood type reaction) which formed styrene and H_2.

$$C_8H_{10} \rightarrow C_8H_8 + H_2 \quad (6)$$

Afterward, the conversion of ethylbenzene and fraction of H_2O decreased and reached a steady state. When H_2O supplied to the catalyst bed after feeding EB-alone, we confirmed the production of H_2. H_2 was considered to be produced through the following reaction (7).

$$H_2O + V_{ox} + 2e^- \rightarrow O^{2-}_{lat} + H_2 \quad (7)$$

Then, EB-alone was supplied to the catalyst bed again. We observed the formation of styrene and H_2O as products. The reactivity seemed to be recovered by the oxidation of the catalyst by H_2O. Hence, we concluded that the steady-state reaction mechanism of dehydrogenation of ethylbenzene on the catalyst was redox-type using the lattice oxygen in the catalyst.

We also investigated the reaction mechanism of 30 ppm-Pd/Fe-K catalyst using periodical pulse experiments with reactants of ethylbenzene and H_2O supplied alternately to the catalyst bed. Figure 6 shows the results of the modified Fe–K catalyst with small amount of Pd.

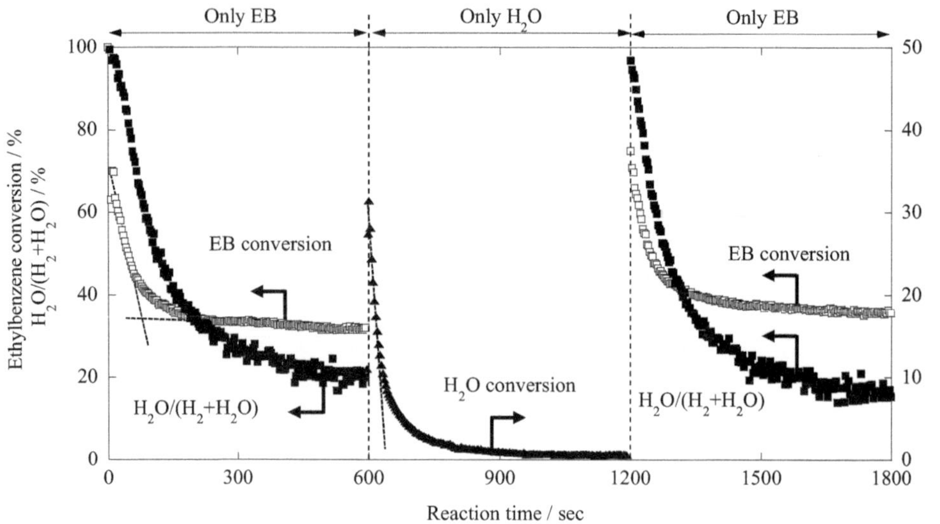

Fig. 6. Results for the periodical pulse reaction test over the 30 ppm-Pd/Fe-K catalyst.

Oxidative dehydrogenation of ethylbenzene (5) also proceeded on the catalyst by using lattice oxygen in the catalyst. And simple dehydrogenation (6) proceeded on the catalyst when lattice oxygen in the catalyst was consumed. After feeding EB-alone, H_2O supplied to the catalyst bed. H_2 was also produced by the oxidation of the catalyst by H_2O (7). When EB-alone was supplied to the catalyst again, the reactivity was recovered by the oxidation of the catalyst by H_2O. Therefore we concluded that redox-type mechanism using the lattice oxygen proceeded on the catalyst. Comparing results of periodical experiments of the Fe-K catalyst and Pd/Fe-K catalyst, there was a difference in the formation rate of H_2. Pd/Fe-K catalyst revealed higher oxidation rate at 600 s. So, the effect of ppm-order-Pd addition to the Fe-K catalyst on the enhancement dehydrogenation activity was investigated in terms of reaction mechanism in the next section.

4. The role of Pd on dehydrogenation of ethylbenzene in terms of reaction mechanism

For investigating the role of Pd for dehydrogenation of ethylbenzene in terms of reaction mechanism, we calculated the ratio of amounts of reactive lattice oxygen to total amounts of lattice oxygen in the catalyst and elucidated the effect of ppm-order Pd addition on redox property of catalysts (Section 4.1). And also, we considered the relation between the redox property of the catalyst and activity for dehydrogenation of ethylbenzene in this section 4.2.

4.1 Promotion of regenerating rate of lattice oxygen

We compared rates of release/regenerating of lattice oxygen over the Fe-K catalyst and Pd/Fe-K catalyst from figures 5 and 6. Here, the release rate of lattice oxygen was regarded as styrene formation rate at the time when feeding of ethylbenzene was started and regenerating rate of lattice oxygen was regarded as H_2 formation rate at the time when feeding of H_2O was started after feeding of EB-alone. Figure 7 shows the O^{2-}_{lat} balance in the catalysts during periodic pulse experiments.

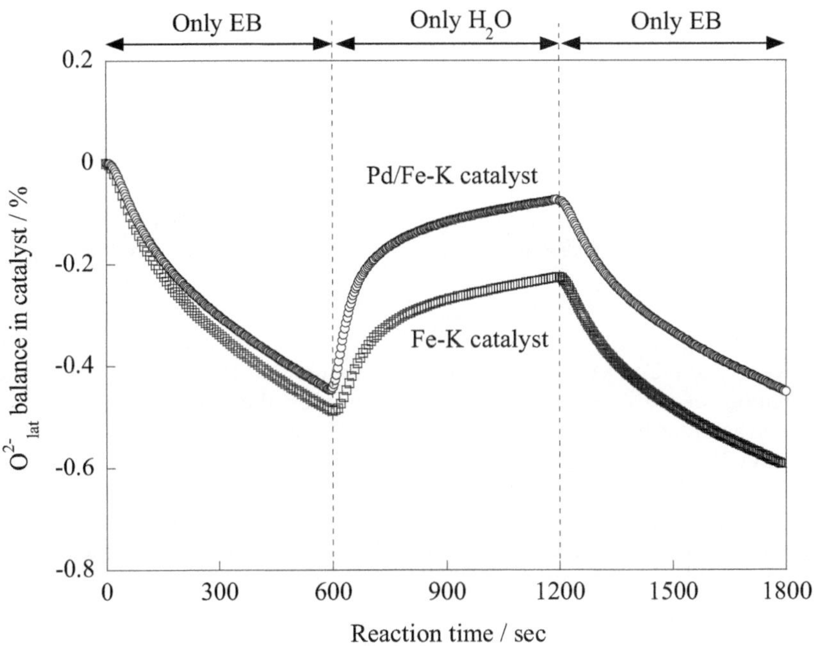

Fig. 7. Lattice oxygen balance in Fe-K and 30 ppm-Pd/Fe-K catalyst in the periodic pulse experiments (Sekine et al., 2008).

From figures 5, 6 and 7, we found the steep and straight decrease of ethylbenzene conversion from 0 s to 100 s could be attributed to the domination of oxidative dehydrogenation of ethylbenzene. So the amount of lattice oxygen was calculated by integration of releasing lattice oxygen from 0 s to 100 s. As a result, the amount of lattice oxygen showing high reactivity was inferred to be 1.7×10^{-5} mol g^{-1}, which was about 25%of surface lattice oxygen from the BET surface area (ca. 2 m^2 g^{-1}). The amount of lattice oxygen decreased in a straight trend in 100–600 s, implying that the concentration of surface lattice oxygen reached a steady state and dominated by the bulk diffusion. On the other hand, we found that the lattice oxygen that was consumed by ethylbenzene was regenerated by H_2O because H_2 was formed by the reaction with H_2O. The initial conversion of H_2O to H_2 was 20% at 600 s in fig. 5, subsequently, the conversion of H_2O decreased. The regeneration rate of lattice oxygen decreased from 1.7×10^{-7} mol s^{-1} g^{-1} to 4.1×10^{-8} mol s^{-1} g^{-1} at 700 s; then it reached the steady state. The amount of lattice oxygen that was regenerated before 700 s was 1.7×10^{-5} mol g^{-1}.

The total amount of lattice oxygen regenerated during the oxidation reaction with H_2O was 3.3×10^{-5} mol g^{-1}, which was 52% of the lattice oxygen consumed by ethylbenzene. The reactivity to ethylbenzene was mostly recovered by the oxidation with H_2O. As a result, we found that added Pd did not change the release rate of lattice oxygen, but enhanced regenerating rate of lattice oxygen in the catalyst. So enhancement of the regenerating rate of lattice oxygen might affect the activity for dehydrogenation of ethylbenzene.

Based on these results, we considered that the mechanism as follows; oxidative dehydrogenation was competitive with and faster than simple dehydrogenation; H_2O regenerated the surface lattice oxygen, showing high reactivity with ethylbenzene. The surface lattice oxygen was consumed by oxidative dehydrogenation of ethylbenzene and regenerated by H_2O. Previous reports showed that H_2O inhibited over-reduction of the catalyst. H_2O also acted as a diluent for shifting the equilibrium conversion higher and inhibited carbon deposition on the catalyst surface. However, in this work we considered that H_2O had the role of oxidant for the catalyst.

Comparing the result of periodic pulse experiment between the Fe–K catalyst (Fig. 5) and Pd/Fe–K catalyst (Fig. 6), there were no appreciable differences on the conversion of ethylbenzene and the fraction of H_2O. The rate of consumption of the lattice oxygen on each catalyst showed the same value, which was 1.7×10^{-7} mol g^{-1} s^{-1}. The amount of the lattice oxygen showing high reactivity was 1.7×10^{-5} mol g^{-1}, which was 25% of the amount of surface lattice oxygen. Consequently, we conclude that added Pd changed neither the rate of dehydrogenation of ethylbenzene nor the amount of the lattice oxygen giving high reactivity. On the other hand, the initial conversion of H_2O to H_2 was increased from 20% to 30% by adding Pd to Fe–K catalyst at 600 s in figs. 5 and 6 when H_2O supplied to the catalyst. We found that surface lattice oxygen was regenerated promptly within 70 s. The initial rate of regeneration of lattice oxygen was increased by adding Pd. At 670 s of the periodic pulse experiment, regeneration rate of lattice oxygen reached a steady state. H_2O regenerated 83% of lattice oxygen consumed by ethylbenzene after steam feed of 600 s, the amount of which was 5.6×10^{-5} mol g^{-1}. These results indicated that added Pd promoted the regeneration of lattice oxygen by H_2O although it did not affect the rate of dehydrogenation of ethylbenzene or the amount of the lattice oxygen giving a high reactivity. The amount of added Pd was 2.8×10^{-7} mol g^{-1}, which was 1.6% of the surface lattice oxygen giving a high reactivity. We found that the addition of Pd fewer than the surface lattice oxygen to Fe-K enhanced the catalytic activity.

4.2 The relation between the regeneration rate and steady-state activity

We showed that adding Pd improved the dehydrogenation activity by enhancement of regeneration rate of lattice oxygen. In this section, further experiments by changing the amount of Pd on the Fe-K catalyst were carried out for elucidating redox-type mechanism proceeding on a series of catalyst. The relationship among Pd content on the Fe-K catalyst, styrene formation rate in steady-state and releasing rate of lattice oxygen was investigated. Figure 8 shows the relationship between Pd content on the Fe–K catalyst and the rate of styrene formation/consumption of lattice oxygen. As a result, there was no relation among Pd concentration on the Fe-K catalyst, the rate of styrene formation and releasing rate of lattice oxygen. Further experiments indicated that added Pd did not affect the property of releasing lattice oxygen.

Next, the relation between the regeneration rate of lattice oxygen measured by the periodic pulse experiments and the steady-state dehydrogenation activity was examined. The result shows in figure 9. The regeneration rate of lattice oxygen was calculated by the periodic pulse experiments. The steady-state dehydrogenation activity was the conversion of ethylbenzene observed in the continuous reactor.

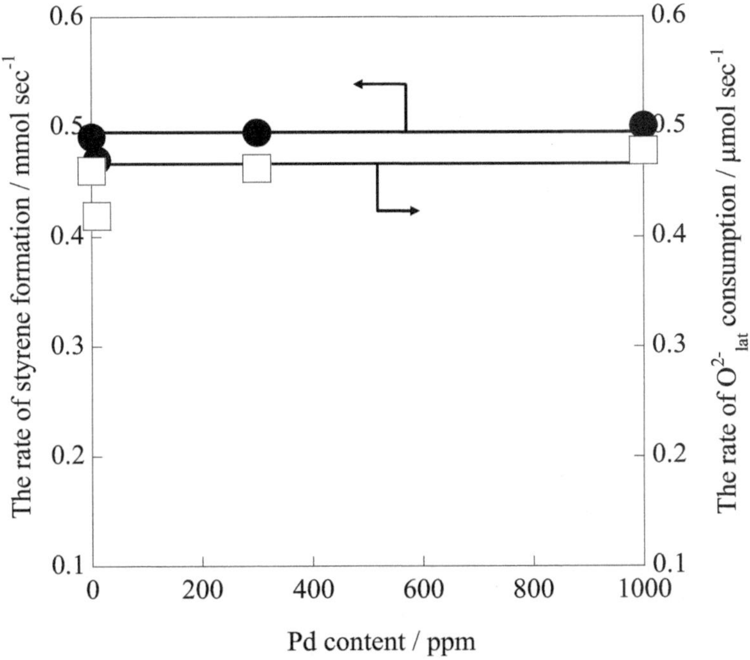

Fig. 8. The effect of Pd concentration on the styrene formation rate and O^{2-}_{lat} consumption rate (Sekine et al., 2008).

From figure 9, there was a linear relation between the regeneration rate of lattice oxygen and the steady-state dehydrogenation activity. Hence, the addition of Pd increased the regeneration rate of lattice oxygen on/in the Fe-K catalyst and steady-state activity for dehydrogenation of ethylbenzene. Therefore the promotion of regeneration of lattice oxygen with H_2O by the addition of small amount of Pd enhanced steady-state activity of the catalyst.

Through these investigations, we developed the precious metal-promoted catalyst based on Fe-K which showed high activity for dehydrogenation of ethylbenzene. And also we found that redox-type mechanism which ethylbenzene was oxidized by mobile surface lattice oxygen and consumed lattice oxygen was regenerated by H_2O proceeded on a series of catalysts. The effect of an addition of ppm-order-Pd changed neither the rate of oxidative dehydrogenation of ethylbenzene using lattice oxygen nor the amount of the lattice oxygen giving high reactivity. The regeneration rate of lattice oxygen was promoted by Pd addtion, and this result improved the dehydrogenation activity. We elucidated the effectiveness of utilization of steam as an oxidant of the catalyst and the control of activity by control of mobility of lattice oxygen in the catalyst.

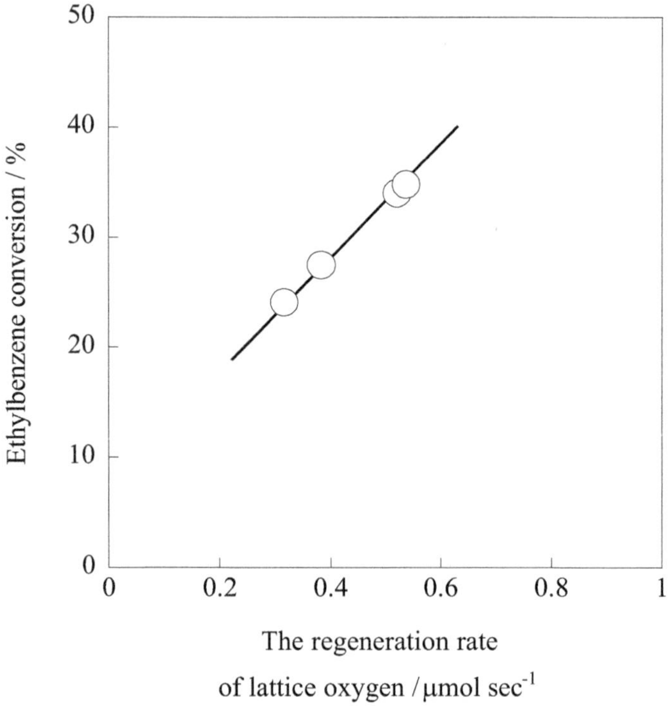

Fig. 9. The relation between regeneration rate of lattice oxygen and ethylbenzene conversion at steady-state (Sekine et al., 2008).

5. Conclusion

In this work, we examined the effect of ppm-order Pd addition on the activity of Fe-K mixed oxide catalyst for dehydrogenation of ethylbenzene. Results of periodical pulse experiments indicated that the oxidative dehydrogenation is competitive with and faster than simple dehydrogenation on the Fe-K mixed oxide. Consumed lattice oxygen was found to be regenerated by H_2O immediately. The amount of lattice oxygen showing a high reactivity for dehydrogenation of ethylbenzene was 0.14% of the total amounts of lattice oxygen in the catalyst, which meant reactive lattice oxygen was surface or subsurface lattice oxygen. We concluded that added Pd hardly changed the rate of dehydrogenation of ethylbenzene or the amount of surface lattice oxygen showing high activity. The surface lattice oxygen, which was regenerated by H_2O and Pd promoted the regeneration rate of surface lattice oxygen by H_2O.

6. References

Addiego, W.P., Estrade, C.A., Goodman, D.W. & Rosynek. P. (1994). An infrared study of the dehydrogenation of ethylbenzene to styrene over iron-based catalysts, *Journal of Catalysis*, 146: 407-414.

Addiego, W.P., Liu, W. & Boger. T. (2001). Iron oxide-based honeycomb catalysts for the dehydrogenation of ethylbenzene to styrene, *Catalysis Today*, 69: 25-31.

Busca, G., Zerlia, T., Lorenzelli, V. & Girelli, A. (1984). Fourier transform-infrared study of the adsorption of unsaturated and aromatic hydrocarbons on the surface of α-Fe_2O_3: I. Ethylene, *Journal of Catalysis*, 88: 125-130.

Carrà, S. & Forni, L. (1965). Kinetics of Catalytic Dehydrogenation of Ethylbenzene to Styrene, *Industrial & Engineering Chemistry Research*, 4(3): 281-285.

Coulter, K. & Goodman D.W. (1995). Kinetics of the dehydrogenation of ethylbenzene to styrene over unpromoted and K-promoted model iron oxide catalysts, *Catalysis Letters*, 31: 1-8.

Devoldere, K.R. & Froment, G.F. (1999). Coke formation and gasification in the catalytic dehydrogenation of ethylbenzene, *Industrial & Engineering Chemistry Research*, 38(7): 2626-2633.

Eggertsen, F.T. & Voge, H.H. (1947). Catalytic dehydrogenation, U.S. Patent No. 2,414,585.

Herzog, B.D. & Rase, H.F. (1984). In situ catalyst reactivation: used ethylbenzene dehydrogenation catalyst with agglomerated potassium promoter, *Industrial & Engineering Chemistry Product Research and Development*, 23(2): 187-196.

Hirano, T. (1986). Roles of potassium in potassium-promoted iron oxide catalyst for dehydrogenation of ethylbenzene, *Applied Catalysis*, 26: 65-79.

Hirano, T. (1986). Active phase in potassium-promoted iron oxide catalyst for dehydrogenation of ethylbenzene, *Applied Catalysis*, 26: 81-90.

Lee, E.H. (1974). Iron oxide catalysts for dehydrogenation of ethylbenzene in the presence of steam, *Catalysis Reviews*, 8(1): 285-305.

Kearby, K.K. & Elizabeth, N. J. (1945). Method for the catalytic dehydrogenation of hydrocarbon, U.S. Patent No. 2,370,797.

Meima, P.G. & Menon, P.G. (2001). Catalyst deactivation phenomena in styrene production, *Applied Catalysis A: General*, 212: 239-245.

Miura, H., Ansai, R. & Kawai, H. (1994). Deuterium exchange reaction of ethylbenzene over an Fe_2O_3-K_2CO_3-Cr_2O_3 catalyst, *Reaction Kinetics and Catalysis Letters*, 53(2): 323-329.

Muhler, M., Schlögl, R., Reller, A. & Ertl, G. (1989). The nature of the active phase of the Fe-K-catalyst for dehydrogenation of ethylbenzene, *Catalysis Letters*, 2: 201-210.

Muhler, M., Schlögl, R. & Ertl, G. (1992). The nature of the iron oxide-based catalyst for dehydrogenation of ethylbenzene to styrene 2. Surface chemistry of the active phase, *Journal of Catalysis*, 138: 413-444.

O'Hara F.J. (1975). Dehydrogenation catalyst, U.S. Patent No. 3,904,552.

Pitzer, E.W. (1957). Catalyst and dehydrogenation process, U.S. Patent No. 2,866,790.

Riesser, G.H. (1979). Dehydrogenation catalyst, US Patent 4,144,197.

Sekine, Y., Watanabe, R., Matsukata, M. & Kikuchi, E. (2008). High Performance of Fe–K Oxide Catalysts for Dehydrogenation of Ethylbenzene to Styrene with an aid of ppm-order Pd, *Catalysis Letters*, 125: 215-219.

Shekhah, O., Ranke, W. & Schlögl. R. (2004). Styrene synthesis: in situ characterization and reactivity studies of unpromoted and potassium-promoted iron oxide model catalysts, *Journal of Catalysis*, 225: 56-68.

Urasaki, K., Tanimoto N., Hayashi, T., Sekine, Y., Kikuchi, E. & Matsukata, M. (2005). Hydrogen production via steam–iron reaction using iron oxide modified with very small amounts of palladium and zirconia, *Applied Catalysis A: General*, 288: 143-148.

Weiss, W. & Schlögl, R. (2000). An integrated surface science approach towards metal oxide catalysis, *Topics in Catalysis*, 13: 75-90.

Williams, D., Mishima, Y., Andrei, R., Niiyama, K. & Danish, S. (2002). Jpn Patent No. 2002-509,790.

Zhu, X.M., Schön, M. Bartmann, U., Veen, A.C. & Muhler, M. (2004). The dehydrogenation of ethylbenzene to styrene over a potassium-promoted iron oxide-based catalyst: a transient kinetic study, *Applied Catalysis A: General*, 266: 99-108.

14

Organic Molecules on Noble Metal Surfaces: The Role of the Interface

Maddalena Pedio[1], Cinzia Cepek[1] and Roberto Felici[2]
[1]*IOM CNR, Laboratorio TASC Area Science Park, Basovizza Trieste*
[2]*ESRF Grenoble*
[1]*Italy*
[2]*France*

1. Introduction

Among the new developments in the use of noble metals, the organic-inorganic systems formed by Self Assembled Molecular (SAM) layering on noble metal surfaces represent a main topic. In this chapter we discuss the use of complementary surface science techniques Applied to these sysems. Our aim is to give an overview on our recent case studies of organic molecular layers interacting with noble metal surfaces. Functionalized molecular assembled films are important in many application fields, ranging from sensors and photovoltaic cells to nanostructured devices. In microelectronics adsorbed organic molecules on metal surfaces are a subject of intensive investigation, due to the fundamental interest in interfacial coordination chemistry and its many potential applications. Physical details and morphology, together with the grain size, orientation of the molecules in the film and the concentration of defects, play a crucial role in device performances, since they determine the carrier injection properties.

In particular we discuss how the adsorption of organic molecules can induce strong or weak perturbation of the noble metal substrates whose characterization is necessary for the control of the modified electronic properties with SAM-noble metal interfaces. These systems are complicated by the fact that the molecule-substrate system must be taken into account as a whole and it is not possible to simply transfer molecular functionalities, deduced in their isolated state, to the adsorbed case. The metal surface can force the molecule to rearrange its internal structure (leading to complex structural deformation of the adsorbed molecules) and the metal surface lattice plays an important role aligning the assembling with the main lattice directions. The mechanism of conjugated molecule ordering and (in some case) anchoring on surfaces is due to a balance of intermolecular binding forces and of molecule-substrate interaction involving a large number of sites. In case of strong bond, this can result in either displacive substrate reconstructions, involving extensive mass transport and the formation of ordered arrays of nanodimples.

Noble metal surfaces are valuable substrates for SAM because of their low chemical reactivity together with high atomic surface mobility resulting in the capability of reordering after molecular deposition. The high electron density at the crystal surface provides a charge reservoir. Comparisons of the different organic-metal surface combinations will be discussed, focussing on the role of the substrate moieties.

2. Research methods

The complexity of the organic molecules deposited onto noble metal surfaces deserves a multi-techniques approach. Our methodology is focussed on the interrelation between electronic properties and structural profiles. Electron Spectroscopies, Photoemission Electron Spectroscopy (PES) and Inverse Photoemission Spectroscopy (IPS) and X-ray Absorption Spectroscopy (XAS) are widely used, in campus and synchrotron radiation facitilies, to characterize the electronic properties of condensed matter and are successfully applied to low dimensional systems. The richness of information achievable with these techniques make them extremely useful for the fine characterization of organic-inorganic interfaces and Self Assembled Monolayer systems. These techniques are complementary to Scanning Microscopies (in particular Scanning Tunnelling Microscopy, STM) and Diffraction techniques Low Energy Electron Diffraction, (LEED), Helium Scattering, (HAS) and Surface X-ray Diffraction, (SXRD). Imaging through STM offers the access to details of the surface topography at atomic resolution and in particular cases to information about the Density of States (DOS) across the Fermi level by the Scanning Tunneling Spectroscopy (STS). Diffraction techniques provide structural details of the surface and (even buried) interfaces. In the following we shortly describe the main information provided by selected research methods. We address the reader to more complete reviews and text books (as for example Luth1994, Hufner1995, Feidenhans'l 1989, Woodruff 1988, Robinson1992, Besenbacher 1996, Chen1993).

2.1 Introduction to surface x-ray diffraction

Radiation can interact with matter in two ways: it can be diffused elastically and then the coherent sum of all the waves diffused by the single objects sum coherently giving diffraction, or it can induce (electronic, vibrational, and rotational) transitions in the atoms and molecules it interacts with. Diffraction is extremely important for the determination of the structural properties of matter being an interferometric technique sensitive to the relative positions of the scattering objects.

SXRD is usually applied to ordered samples. This technique provides information on the structure of both the terminating layers of the ordered substrate lattice or of an ordered new structure which is formed at the surface. In the first case, the basis of SXRD is the measurement of the diffracted intensity raising from the termination of the periodic bulk structure which is located along rods in the reciprocal space having their origin at the Bragg points of the bulk structure (Robinson 1986). These rods, which are perpendicular to the terminating surface, are usually referred as Crystal Truncation Rods (CTRs). In the second case the diffracted intensity is along continuous rods, still perpendicular to the surface with an in plane periodicity defined by the surface unit cell. If the surface unit cell is due to a reconstruction of the bulk cells and then it has in-plane dimensions which are commensurate with the bulk unit cell, the associated rods take the name of Fractional Order Rods (FORs) (Feidenhans'l1989). A measurement of the intensity of both the CTRs and FORs as a function of the continuous variable l, which is the reciprocal space coordinate perpendicular to the scattering surface, provides information on the expansion or contraction of the top layers of the substrate, on the substrate roughness and on the in plane position of all the atoms belonging to the surface unit cell.

X-rays interacts weakly with matter and the scattering process can be well described in the distorted Born approximation. This makes the data analysis quite simple because all the terms describing multiples scattering events can be neglected. In the case of ordered samples we can use the same approach normally employed in bulk crystallography. Among the different quantities it is important to highlight the role of the Patterson map obtained by Fourier transformation of the intensity of the fractional order rods at $l \approx 0$. In this case the map is the autocorrelation function of the electron density projected onto the surface. It usually provides hints on the main correlation vectors present in the unit cell. Because the unit cell is larger and commensurate with the bulk unit cell the atoms which have not moved from their ideal bulk termination position will not appear in the Patterson (Feidenhans'l, 1989).

The fitting of the data can be carried out using the ROD program (Vlieg, 2000). To gather sensitivity to a particular atomic species, all the above techniques can also be coupled to energy scans through absorption edges. In the simple case the change in the atomic scattering power at the resonance will give the possibility of determining the contribution of a particular atomic specie to the scattered intensity (Benfatto & Felici 2001). X-ray diffraction presents the advantage of being able to observe the buried interface even when it is covered by a thick layer of adsorbed molecules giving the opportunity of determining the substrate structure during the different growth phases.

2.2 Basic principles of scanning tunneling microscopy and scanning tunneling spectroscopy

Since the discovery the STM (Binnig and Rohrer) has revolutionised the field of surface science, making possible real space atomic resolution images of a sample surface. In STM measurements a fine pointed tip is brought extremely close to a surface and a voltage is placed between the tip and the sample surface. In this condition a tunneling current can exist between the tip and surface. The current flow is very sensitive to the distance between tip and surface. The tip can be rastered across the surface with the aid of piezo-electric actuators. The current travelling between the sample and tip is sensed and the tip is moved towards and away from the sample surface, keeping the current flow constant. Obviously, the sample needs to be electrically conducting, and to achieve atomic resolution for most samples the STM needs to operate in Ultra High Vacuum (UHV) conditions. Basics of STM and STS can be found in many reviews (Besembacher1996, Chen 1993, Gavioli2008 and refs therein).

STM is one of the most powerful tools for observing surfaces, adsorbates and molecular structures at nano scale level and a wealth of study using STM as main tool are present in the literature. STM and STS have found a wide range of applications, providing not only topographic details of the surfaces but the electron density distribution. The key of the measurement information is the tunneling matrix element, which depends explicitly from the sample and tip wave functions. Since the atomic structure of the tip is usually not known, one needs to model the tip wave function to understand the information provided by the tunneling current. Assuming a tip with radius R and an s-type only (quantum numbers $l \neq 0$ neglected) wave functions and small applied bias voltage, the current results to be

$$I \propto U \cdot n_t(E_F) \cdot e^{2xR} \sum_{\mu\nu} \left|\psi_\nu(r_0^2)\right|^2 \delta(E_\nu - E_F) \tag{1}$$

where E_F is the Fermi energy, r_0 is the center of curvature of the tip, $n_t(E_F)$ is the density of states at the Fermi level for the tip, and the decay rate $x = (2m\phi)^{1/2}/\hbar$ depends on the effective potential barrier height ϕ. The quantity

$$n_s(E_F, r_0) = \sum_{\mu\nu} \left|\psi_\nu(r_0^2)\right|^2 \delta(E_\nu - E_F) \tag{2}$$

is the charge density of the electronic states at the Fermi level (Tersoff1983, Tersoff1985 in Gavioli2008), i.e. the surface local density of states (LDOS) at E_F, evaluated at the center of curvature of the tip. The interpretation of the tunneling current as contour map of the surface LDOS is limited to small applied bias (less than 0.5 V) and ignores the angular dependence of tip wave functions.

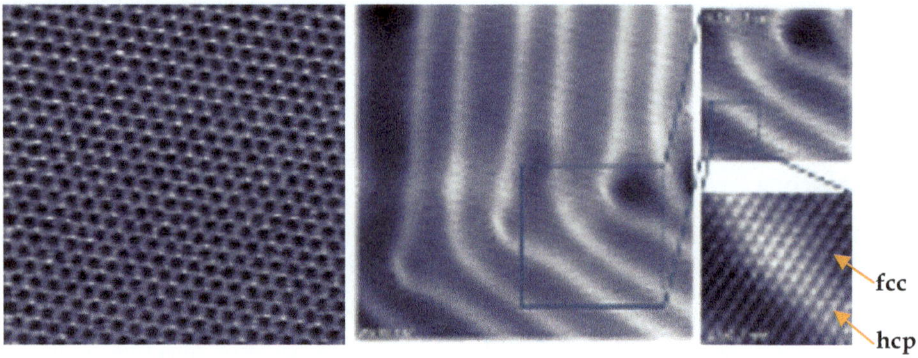

Fig. 1. a) Atomically resolved Cu(111) surface in constant height mode T=5K, 4.5x5 nm2 Omicron LT-SPM/Qplus Dr. B.Such, group Prof. E.Meyer, Univ. Basel, Swi. b) Ultrahigh resolution (2048x2048 pixels) topographical image of Au(111) reconstruction, 26x26 nm2, 40pA, 200mV, 4K SPECS Joule-Thomson STM Prof. W.Wulfhekel, Karlsruhe, Germany (web site http://www.specszurich.com/en/image_gallery_content---1--1122.html)

Figure 1 shows two examples the STM images of the Cu(111) and Au(111) surfaces. Gold presents the herringbone reconstruction with periodicity √3x23 (see description in section 3) formed by alternate hcp (clearer atoms) and fcc regions (dark). These topographic atomic resolved images give an idea of the power of high resolution of STM topography. Nevertheless STM images do not contain direct information on the absorption site and the configuration at the interface could only be inferred. More realistic assumption must be taken into account to interpret the imaging and the data of STM (Chen1990). The assignment of the atomically resolved structures should be taken with care, with a comparison to *ab-initio* calculations of the electronic structure or comparison to other experimental techniques.

2.3 Introduction to photoemission (valence photoemission UPS and core level photoemission XPS), inverse photoemission IPS and absorption (XAS)

Electron Spectroscopies, are precious tools for the fine characterization of organic-inorganic interfaces. They give access to the density of electronic empty and filled states (or bonding and antibonding orbitals of molecules) providing information on the electronic configuration, the molecule-metal surface bond, the effect of charge transfer, energy level alignment at interfaces and the orbital modification of the adlayer structures. This ensemble of information is fundamental, as we will discuss in section 4, for organic-inorganic systems, whose physical and structural fine details play a crucial role, in their properties and technological applications.

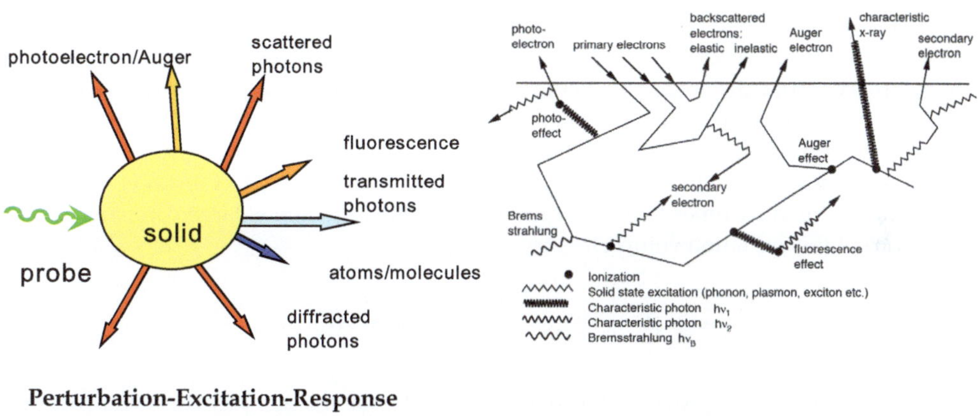

Fig. 2. Left: In a spectroscopic experiment the incoming probe (photons, electrons) interacts with the sample and produces different outgoing probes, depending on the different interaction processes each with its proper cross section σ; Right: possible processes inside a solid induced by an electron beam (Werner, 2005)

A spectroscopy experiment can be rationalized in three steps: Incoming probe onto a sample, Excitation inside the sample induced by the probe perturbation, detection of the outgoing response. When a photon (or electron) beam interacts with condensed matter, the products of such interaction compose different response channels. Therefore an electron spectroscopy experiment can be considered as a scattering (elastic or anelastic) experiment where a beam of electrons, or photons (UV, Soft or X-rays) impinging onto the system with prescribed energy and momentum, perturbs the sample, inducing its proper excitations or transitions that involve the eigenstate of the sample. Measuring the energy and momentum of the outgoing probes (emitted or absorbed X-rays or electrons) provides the "response" spectrum. Electron spectroscopies treated here, are in linear response regime, i.e., the interaction (photons-matter or electrons-matter) only weakly perturbs the system under study.

Examples of physical processes occuring when an electron (or a photon) interacts with a solid surface are presented schematically in the Figure 2 left (Werner, 2005). When the incoming electron is sufficiently energetic, a wealth of phenomena can take place leading to different types of emitted particles: elastically and inelastically backscattered electrons, exciton, phonon, plasmon excitations, photoelectrons, Auger recombination of a hole left by the photoelectron, fluorescence recombination and the so-called secondary electrons.

The information of the experiment is contained in the differential cross section $\frac{d\sigma(E)}{dVd\Omega}$ (V=volume, E energy, Ω solid angle of detection), and strongly depends on the experimental geometry adopted in the experiment, the properties of the incoming probe, the kind of detector, the angle of detection ecc., all quantities which can be modelled. The cross section σ of the excitation process, has the dimension of an *area* (namely is defined in barn= 10^{-24} cm^2) and is related to the transition probability of the interaction process. For each channel the σ (Hüfner1995, Woodruff1988, Bassani1983) can be expressed in terms of interaction probability W_{if} from the initial (wavefunction ψ_i, initial energy E_i) to the final (ψ_f initial energy E_f) states by using: a) the first order of the perturbation theory (with the assumptions of a weak perturbing probe), b) the single particle approximation (the final state approximate the ground state with only one particle passed from i to f state) c) the dipole approximation (that assumes that the wavelength of the radiation is large compared to the dimensions of the excitation volume).

$$\frac{d\sigma}{dVd\Omega} \propto \sum_{if} W_{if} \; ; \; W_{i\to f} \propto \frac{2\pi}{\hbar} \left| \langle \psi_f | r | \psi_i \rangle \right|^2 \delta(E_f - E_i - h\nu) \tag{3}$$

The quantity in bracket provides the density of states of the sample, the delta function implies the energy conservation, hν is the photon energy.

In this way, the measured flux of absorbed or emitted particles, as a function of energy and angle, can be compared with the calculated interaction probability, as for example the current density of photoelectrons, or the absorption of photons as a function of the photon energy etc. A rigorous treatment of the cross section calculation for the different electron spectroscopies is beyond the scopes of this chapter and we address the reader to manual and books for a deeper treatment (see for example Bassani1983, Hüfner1995, Lüth1995).

The schematics of the processes discussed in the following are shown in fig. 3: photoemission (PES, for historical reason the core level photoemission is called often X-ray Photoemission Spectrscopy XPS while valence band photoemission Ultraviolet Photoemission Spectroscopy UPS), Inverse Photoemission (IPS) and X-ray Absorption spectroscopy spectroscopies. PES corresponds to an the photoelectron initially in a core level (X-ray Photoelectron Spectroscopy XPS), or a valence band level (Ultraviolet Photoelectron Spectroscopy UPS), XAS from elemental edge corresponds to the absorption of the impinging photon by a core level electron, IPS technique involves the detection of photons emitted by the decay of an impinging electron from a level above the Vacuum Level to an empty electron state of the sample.

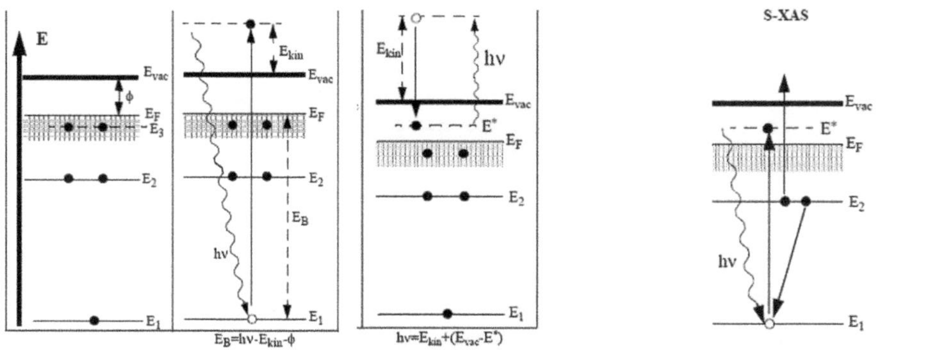

a) Ground State b) PES c) IPS d) XAS and Auger recombination

Fig. 3. Energy levels and processes of photoemission and absorption. A metal in a ground state (left panel) Evac indicates the vacuum level, ϕ the work function that is the energy between the Fermi level EF and the Evac, the E1 and E2 are proper energy levels of the sample. In Photoemission a photon with energy hv > E1-Evac promotes a photon above the Evac, leaving a hole in the state E1. In Inverse Photoemission an impinging electron makes a transition from the energy level above the Evac and an empty state E* inducing the emission of a photon, in absorption an impinging photon is absorbed leaving a core hole in the level E1 and the recombination produces either a fluorescence decay or an Auger electron.

2.3.1 NEXAFS

X-ray Absorption Spectroscopy (XAS) combined with synchrotron radiation is a well-established technique that provides information on the electronic, structural and magnetic properties of matter. In X-ray absorption, a photon is absorbed by the atom, leading to the transition of an electron from a core state to an empty state above the Fermi level. The absorption cross-section is proportional to the absorption coefficient and depends on the energy and on the measured element. To excite an electron in a given core-level, the photon energy has to be equal or higher than the binding energy of this core-level. This gives rise to the opening of a new absorption channel when the photon energy is scanned from below to above this core-level energy. The energies of the absorption edges therefore correspond to the transition from the core-level energies to the empty states and are characteristic for each element, making X-ray absorption an element-selective technique.

The fine structure above the x-ray absorption edge of a core level can be separated into a *n*ear-*e*dge region (Near Edge X-ray Absorption Fine Structure, NEXAFS, Another acronym used in the literature is XANES) (Stöhr1992), which is located within ≈ 30 eV above the edge, and an *e*xtended region at higher photon energies (Extended X-ray Absorption Fine Structure, EXAFS). Due to the small inelastic mean free path of the photoelectrons in the respective energy range, NEXAFS is dominated by multiple scattering, in contrast to EXAFS, where the photoelectron interacts with the surrounding atoms by single scattering events. On the other hand the near edge provides information of the cross section of the element selective density of final empty states, which for molecules are the directional antibonding π molecular states. An example is reported in figure 4 right. The photon absorption is controlled by dipole selection rules, which lead to general relations for the angle-dependence of the resonance intensities [Stohr, 1992].

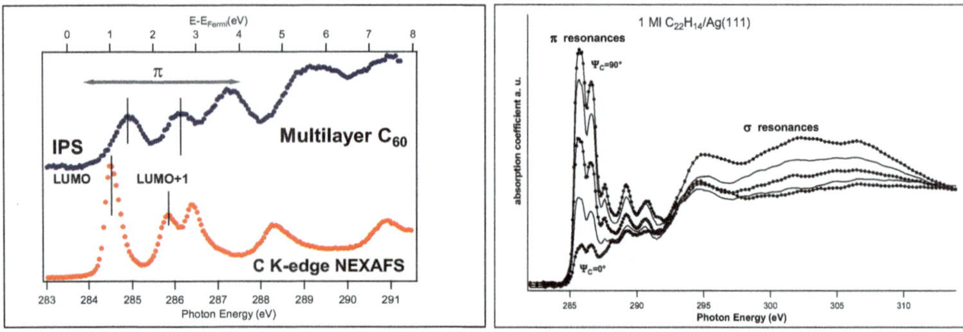

Fig. 4. Left: C Kedge NEXAFS of C_{60} multilayer compared with Inverse Photoemission (Felici 2009). Right: Example of C Kedge NEXAFS of the linear molecule Pentacene ($C_{22}H_{14}$) deposited (1ML) onto Ag(111) measured varying Electric field vector of the incident synchrotron light (Pedio 2007).

Making use of the polarized nature of the synchrotron radiation the orientation, relative to the surface of the molecular orbitals of an adsorbed molecule , can be determined. For this reason NEXAFS is mostly used to determine molecular orientations by observing the polarization dependence of the resonance intensity, which is mainly controlled by well-established symmetry selection rules. Figure 5 left) shows an example of the C kedge of C_{60} molecular films. The signals in the NEXAFS region can be divided into π and σ resonances. The lower energy π resonances are assigned to transitions from the core level into an unoccupied molecular orbital of (local) π symmetry that lies below the vacuum level, while the σ lie above. Thus, the NEXAFS spectroscopy probes empty levels below the ionization threshold, similarly to IPS (see next paragraph), but due to the absorption process, the final system has an electron less (N-1) and a hole in the core level. Note the differences of the energy separation of the peaks in the two spectra in Figure 4 left, due to the presence of the C1s core hole in the absorption process. Specifically Density Functional Theory (DFT) can reproduce the NEXAFS spectrum of the excited state (presence of the core hole) of molecules, either in the gas phase or in solid thin films, providing the oscillator strengths and the intensities (Triguero 1998,). The ground state is nicely reproduced by ground state DFT calculation (absence of the hole). For more details see (Brühweilr2002, Felici, 2009).

2.3.2 Photoemission

Photoemission spectroscopy (PES) is based on the photoelectric effect, which is the process of extracting an electron by means of an impinging photon with energy greater than the electron binding energy. PES measures, in its simpler version, the kinetic energy distribution of photoelectrons, which are emitted from a sample, when it is irradiated with monochromatic light. The energy of the photoelectrons leaving the sample is determined using an analyser that selects the kinetic energies of the photoelectrons and measures their intensity. This gives a spectrum with a series of photoelectron peaks. Within a single particle picture, the kinetic energy distribution of the detected electrons mimics the distribution in energy of the occupied electron states in the ground state (see Fig.5 left). In this case, the kinetic energy E_k of the photoelectron coming out from the sample is approximately given from the following relation:

$$E_k = h\nu - E_b - \Phi_A \quad (4)$$

where E_b is the binding energy of the photoelectron relative to the Fermi level, $h\nu$ the used photon energy and Φ_A is the work function of the electron analyzer.

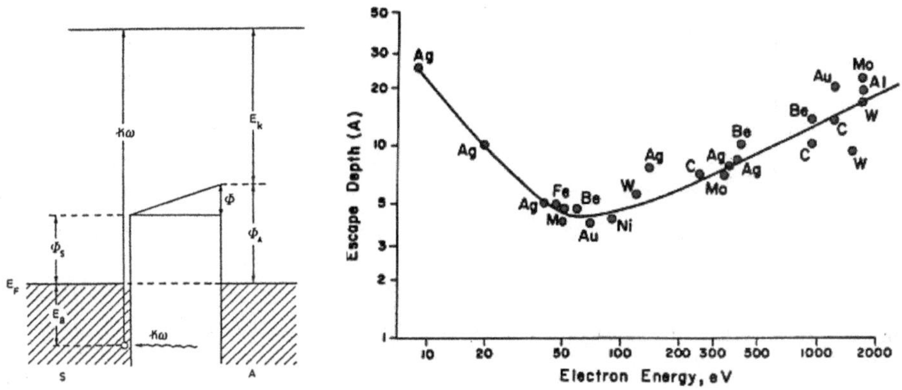

Fig. 5. Left: Level diagram of PES process in a metal. S indicates the sample, A the electron analyzer, E_F the common Fermi level. Right: Electron inelastic mean free path in different materials as a function of their kinetic energy.

Even if the photons penetrate into a solid for several μm, photoemission is a surface sensitive technique. The sampling depth in the photoemission process is determined by the mean free path of the photoelectrons in the solid. This quantity follows the so-called universal curve Shown in Fig. 5 right, and it changes as a function of the electron kinetic energy from about 4Å to about 50Å (Weber, NIST database).

Detecting electrons, XPS must be carried out in Ultra High Vacuum (UHV) conditions. The intensities of the photoemission structures reflect the density of states and depend furthermore on the cross sections of the photoemission processes involving electrons of different energy levels.

XPS sources in campus are normally anode emission of Al Kalpha (1486.6eV) or Mg Kalpha (1253.6eV); synchrotron radiation provide photon energies from approximately 10 eV up to the soft-X ray and hard X ray region. The binding energies of the peaks are characteristic of each element.

A measured XPS spectrum of Au(111) sample is shown in Figure 6a. It is characterized by sharp peaks, superimposed to a mostly featureless background, which exhibits a huge peak (not shown) in the low kinetic energy region (≈5-10eV). This background is ascribed to the almost continuum distribution of the secondary electrons which are produced after inelastic scattering within the solid, thus having lost memory of their primary energy. The excitation of most core levels requires at least soft X-rays, which explains the more usual acronym of the technique. Because of the discreteness of the core levels energies, each core level distribution is a fingerprint of a specific chemical element, thereby making XPS an atom-specific technique.

The peak areas can be used (with appropriate sensitivity factors) to determine the composition of the materials surface. The width of a photoemission core level depends on many different factors: mean lifetime of the photo-hole, electronic excitations, phonon coupling, instrumental broadening. A core level binding energy can shift, depending on the chemical bonds and local environment of a specific atomic site. That implies that the shape of each peak and the binding energy can be slightly altered by the chemical state of the emitting atom. Therefore the analysis of the core level lineshape (taken at high resolution, i.e. resolving power $\Delta E/E \sim 10^{-4}$) provides key information on the chemical bond. Thereby the identification of the chemical state of an atom in an unknown solid system is made possible, consistently with the energy resolution and the actual size of the core level shift. XPS is not sensitive to hydrogen or helium, but can detect all other elements.

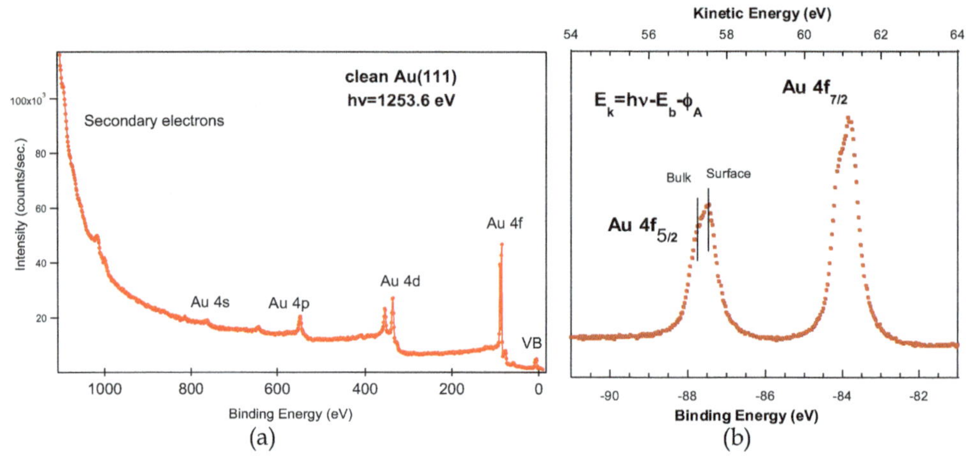

Fig. 6. a) Example of a photoemission spectrum from a Au(111) single crystal (hv=1253.6eV). Core levels and Valence Band emission are indicated; b) high resolution of the Au 4f core level of the Au(111) taken in surface sensitive conditions. The Au 4f emission is a doublet due to the spin orbit splitting of the 4f 7/2 and 4f 5/2. The components ascribed to bulk gold atom and surface atoms are labeled B and S, respectively.

In the photoemission spectra also other structures are visible, typically Auger peaks, shake up losses and correlation satellites due to relaxation processes, i.e. excitation and de-excitation of the system after the creation of the hole left by the photoelectron.

2.3.3 Valence band photoemission and inverse photoemission

The broad structure labeled VB at the low binding energies of the photoemission spectrum in Fig. 6a corresponds to the energy distribution of the valence band states. The typical feature of these shallow states is their delocalized character, compared to the local nature of core levels. They are thus expected to markedly change their distribution in energy as the chemical bonds are changed.

Direct and inverse photoemissions permit a direct measurement of the band structure. With the measurement of *photon* and *Ekin*, PES can determine the occupied state energy. If the sample is a monocrystal and the electron detector has a finite acceptance angle, one can also use momentum conservation law. Angle Resolved UPS (ARUPS) measures the angular distribution of the photoelectrons as a function of the impinging light angle, i.e. the momentum **k** dependence of the filled electronic states. ARUPS is used to determine the band structure of the material under investigation. As we have seen PES measurement is a mapping of the occupied states of a solid as a function of the binding energy. Once the electron is removed (it is sent above the vacuum level *Evac*), the solid remains in some excited state for a while, with a hole somewhere in the valence states. If the solid used to contain *N* electrons before the absorption of the photon, it only contains $N - 1$ of them at the end of the process.

UPS can also be used to identify molecular species on surfaces by identifying characteristic the electron energies associated with the bonding orbital of the molecules.

An inverse photoemission experiment is simply the opposite process. This allows a mapping of the empty states as a function of the energy and of the **k**-point of the solid. Here an extra electron is injected with energy *Ekin* into the solid initially in the ground state. It goes down the surface potential barrier φ and arrives in some electronic empty state (eigenstate) of the solid, with emission of a photon of energy $h\nu$. The system of electrons ends in an $N + 1$ particle excited state. As in the direct case, the energy conservation law permits to evaluate the energy of the empty states. Finally, direct and inverse photoemission experiments are the most direct experimental way to measure the band structure of crystals.

A variation of the angle of incidence of the electron beams leads to angle resolved measurements and thus the momentum dependence of the bands. In the isochromat mode, the incident electron energy is ramped and the emitted photons are detected at a fixed energy that is determined by the photon detector. The cross section of the IPS process is five orders of magnitude weaker than photoelectric effect, implying the use of highly efficient photon detectors (Dose, 1985, Luth, 1982).

Figure 7 shows the PES –IPS spectra of a metal (Au(111)) and a thin Ni-octaethyl porphyrin (NiOEP) film. In case of metal surface there is emission at the Fermi level (EF), while in the case of semiconductor there is a gap (transport gap, Et) between the peaks identified as the

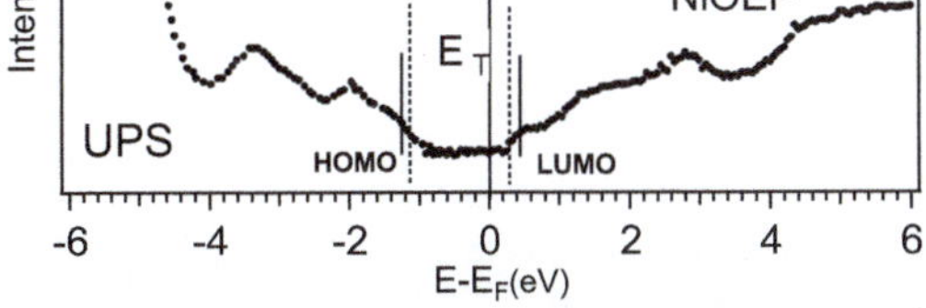

Fig. 7. UPS-IPS spectra of the Au(111) surface (upper red curves) and the multilayer NiOEP (lower black curves). The S and S1 labels indicate the surface filled and empty states, respectively. ET is the transport measured gap.

Highest Occupied Molecular Level (HOMO) and the Lowest Unoccupied Molecular Level (LUMO). In electronic the transport gap, Et, is the energy necessary to create a separated electron–hole pair. The charge injection processes requires promotion of an electron or a hole from the electrodes into one of the charge transport (HOMO or LUMO) states of the organic film. It exceeds the optical gap, Eopt, by 1 eV. (for details see Zahn2006).

2.4 Summary and comparison among the techniques

In section 4 we will discuss our results on organic molecules deposited onto noble metal surfaces, obtained by the described experimental techniques. Table 1 reports the list of typical applications.

Method/Name	Probe/ Signal	Detection limits	Typical applications
Photoemission/X-ray Photoelectron Spectroscopy XPS	Photons/Photoelectrons N-1 final system	0.01-1 at% Surface-bulk sensitive	Surface analysis and composition, Elemental sensitivity, Chemical shifts, stoichiometry
Photoemission/Valence Band UPS	Photons/Photoelectrons N-1 final system	UPS surface sensitive	DOS filled states Energy alignments Transport Gap
Inverse Photoemission/IPS	Elecrtrons/photons Radiative decay N+1 final system	Surface sensitive	DOS empty states Energy alignments Transport Gap
Near edge X ray Absorption Fine structure/NEXAFS	Photons/hole recombination Transitions form core to empty states, N-1 final system	Surface sensitive for adsorbates	Structure Molecular orientation TOGETHER with elemental local density of empty states.
Low Energy Electron Diffraction	Elastic back scattering of low energy electrons	Sub- single monolayer	Surface structure adsorbate structure. Long range order
Surface XRD	Elastic scattering of Hard X-ray photons	From submonolayer to bulk, depending on the incidence X-ray angle and experimental geometry	Atomic structure of surface and interfaces atoms and adsorbates Long range order
Scanning microscopy/ STM	Tunneling current	Atomic resolution in UHV	Topography of the surface
Scanning microscopy/ STS	**Tunneling current versus bias voltage**	Atomic resolution in UHV	Local Density Of States, absorption gap with atomic resolution

Table 1. Research methods and the typical applications to measure surfaces composition, electronic properties and structures.

3. Main topics of surfaces of noble metal surfaces

Flat surfaces of single crystal samples correspond to a single Miller Index plane and, as we have seen, each individual surface has a well-defined atomic structure. Low index Miller flat surfaces (Ashcroft & Mermin 1976) are used in most surface science investigations. Depending upon how a single crystal is cleaved or cut, flat surfaces of macroscopic dimensions are obtained. As many of the technologically most important metals, noble metals exhibit a face centered cubic (*fcc*) structure. The single crystal surfaces (100), (110) and (111) represent the most frequently studied surface planes of the fcc system (BALSAC web site) - however, they are also the most commonly occurring surfaces on such metals. The knowledge gained from studies on this limited selection of surfaces contributes in propagating the development of our understanding of the surface chemistry of these metals.

By splitting the metal crystal along a specific crystallographic plane, the number of nearest neighbour atoms (coordination number, CN[1]) is reduced. This implies that the energy of the surface atoms is higher than that of an atom inside the bulk. These bulk terminated surfaces rearrange the atoms in order to minimize the surface energy. This produces either relaxation, i.e. vertical rearrangement of surface layers (as for example in fcc (111) surfaces), or reconstruction namely the lateral rearrangement of surface atoms (as in fcc (110)). The minimization of the free energy has the tendency to reduce the surface area, altering the geometry of the surface atoms. In particular relaxation takes place because the surface has a high surface tension and the electrons near the surface tend to smooth their surface electronic charge.

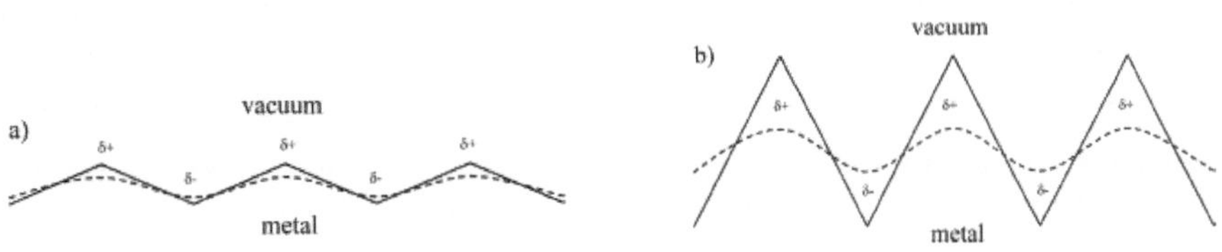

Fig. 8. Illustrations of the Smoluchowski smoothing model due to the redistribution of the charge density at the surface when the density relaxes. a) smooth, closely packed, and b) corrugated loosely packed surface. The lines are related to the contour of the electron density at the surfaces. The dashed lines show the contours after the redistribution of the electron charge density.

The electronic structure of noble metals is due to the simultaneous presence of s and d states in the valence band. The s electrons are essentially free and localized at the Fermi level, while the d states are spatially localized and yield to complex hybridization both in the bulk and at the metal surface. This complexity defines the electronic properties of the surface: the work functions, the surface energies, the reconstructions and the reactivities under molecular absorption. One of the physical observables more used to evaluate the electronic

[1] The Coordination Number of atoms in the bulk of the *fcc* structure is 12. To calculate the CN of a surface we should subtract to the bulk CN (12) those atoms which have been removed from above in forming the surface plane.

properties of surfaces and interfaces is the work function, (Figure 5 left, section 2.3.2) i.e. the energy that is necessary to promote an electron from the Fermi level to the vacuum. The work function (WF) depends on the element and on the structure of the surface. The interplay between the ionic potential (made by the atomic structure) and the electronic charge (filled electron valence states) determines the macroscopic properties of the surface and has strong relevance on technologic applications.

Accurate work functions have now been measured in ultrahigh vacuum for most common metals in various crystallographic orientations. The dependence of the WF on the surface orientation (WF anisotropy) in noble metals has an increasing trend from (110)→(100)→(111). For WF anisotropy the simpler model is the Smoluchowski (Figure 8). This effect (Smoluchowski,1941) consists in the redistribution of the electron cloud on a metal surface with a strong corrugation. Consider a closed packed and an "open" surface of some material. On the open surface we find the Smoluchowski effect of charge smoothing. This smoothing leads to a dipole moment which opposes the dipole created by the flow-out of the electrons. Hence, the work function of a closed packed surface will be higher than that of an open surface. The closed packed (111) surface has the highest workfunction.

Recently (Fall2000, Singh-Miller 2009) ab initio calculations by more accurate microscopic models improved the comprehension of the interrelation between structure and electron density at surfaces.

Theoretical simulations provide trends on the main surface properties. The literature, even recent, is quite rich. For example the evaluation of defect formation, surface relaxation, adhesion at surface, surface energy have been performed for surface atomic scale by effective medium theory (Stoltze 1994) and density functional theory (Vitos et al., 1998, Singh-Miller et al. 2009). Experimental values and example of theoretical results are reported in table 2.

The low values of Adatom/vacancy formation of the Au low Miller index surfaces (Table 2) reflect the high mobility of Gold surfaces (surface energy, E_{ad}/Vacancy) for which it is relatively easy to induce structural changes. The most stable surfaces present high surface atom density and surface atoms with high coordination number fcc(111)>fcc(100)>fcc(110).

The physical definition of a noble metal requires that the d-bands of the electronic structure are filled. Taking this into account, only copper, silver and gold are noble metals, as all d-like band are filled and don't cross the Fermi level (EF, fig. 5) that is the highest energy level that is occupied by the electrons in the metal. Other catalytically important precious metals (Pt, Rh, Pd) shows a more complicated configuration: for platinum two d-bands cross the Fermi level, changing its chemical behaviour. Note the reduced 6s-5d separation of ≈ 2 eV in gold as compared to ≈ 3.5 eV for the 5s-4d separation in the silver atom. This suggests the higher reactivity of silver.

The structure has a strong influence on the electronic properties and reactivity of the surfaces. Reactivity variations to different gases and deposited molecules on the surface even of the same metal are commonly measured. Reactivity strongly depends on the electronic properties, and hence on the crystallographic orientation of the surface (Hammer, 2006). The Hammer-Nørskov d-band model correlates changes in the energy center of the valence d-band density of states at the surface sites with their ability to form chemisorption bonds. A reactivity change is characterized as an electronic structure effect. In this way the model can explain the reactivity change from flat Au surfaces, over Au thin films to Au edges. The different reactivity can easily be seen while preparing clean metal surfaces in ultra high vacuum; surfaces of "physical defined" noble metals (e.g., gold) are easy to clean and stay clean for a long time, while those of platinum or palladium, for example, are covered by carbon monoxide very quickly.

Surface	a_0 / Å	Area unit cell a 2D / Å^2	2DBrillouin zone	Surface density/ (at/cm2)/ CN	WF (eV) Fall,2000/ Singh-Miller 2009	Surface Energy γ (J/m^2) Vitos 98	Theory Adatom/ vacancy Energy (eV) Stoltze 1994
Cu(111)	3,61	$\sqrt{3}/4a_0^2$	hexagonal	$1.78 \cdot 10^{15}$ /6	4.95	1.79	1.329
Cu(100)		$1/2a_0^2$	Square	$1.09 \cdot 10^{15}$ /4	4.6		0.984
Cu(110)		$\sqrt{2}/2a_0^2$	Rectangular P	$1.53 \cdot 10^{15}$ /2	4.45		0.466
Ag(111)	4.09	$\sqrt{3}/4a_0^2$	hexagonal	$1.38 \cdot 10^{15}$ /6	4.74	1.246	1.027
Ag(100)		$1/2a_0^2$	Square	$8.45\ 10^{14}$ /4			0.694
Ag(110)		$\sqrt{2}/2a^2$	Rectangular P	$1.20 \cdot 10^{15}$ /2			0.327
Ag poly					4.26		
Au(111)	4.08	$\sqrt{3}/4a_0^2$	hexagonal	$1.39 \cdot 10^{15}$ /6	5.3	1.5	1.009
Au(100)		$1/2a_0^2$	Square	$8.45\ 10^{14}$ /4	5.22		0.556
Au(110)		$\sqrt{2}/2a_0^2$	Rectangular P	$1.20 \cdot 10^{15}$ /2	5.2		0.273
Au poly	4.08				5.1		
Pt(111)	3.92	$\sqrt{3}/4a_0^2$	hexagonal	$1.51 \cdot 10^{15}$ /6	6.08	2.49	1.429
Pt(100)		$1/2a_0^2$	Square	$9.20\ 10^{14}$ /4	5.82		0.818
Pt(110)		$\sqrt{2}/2a_0^2$	Rectangular P	$1.30 \cdot 10^{15}$ /2	5.35		0.370

Table 2. Comparisons of low index metals surfaces of Cu, Ag, Au and Pt: structural symmetries, surface density, experimental work function, calculated surface energies and vacancy-adatom energy formations.

3.1 Two extremes: Comparison between (111) and (110) surfaces

In case of the [111] planes noble metal (Cu, Ag, Au and Pt) have a close packed structure, while the surfaces having [110] orientation are more "open" and a (2x1) reconstruction can take place at RT.

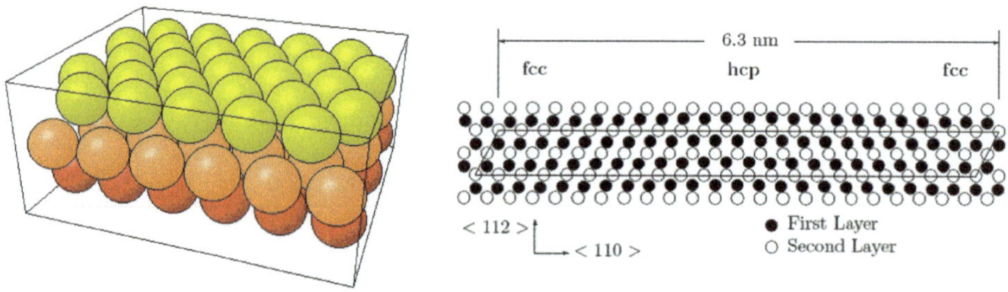

Fig. 9. Left: Structure bulk terminated fcc (111)[2]. Right: side view of the Au(111) herringbone reconstruction

The (111) surface is obtained by cutting the fcc metal in such a way that the surface plane intersects the x-, y- and z- axes at the same value - this exposes a surface with an atomic arrangement of 3-fold (apparently 6-fold, hexagonal) symmetry. This layer of surface atoms actually corresponds to one of the close-packed layers on which the fcc structure is based. This is the most efficient way of packing atoms within a single layer.

In a fcc (111) surface all surface atoms are equivalent and have a relatively high CN and the surface is almost smooth at the atomic scale.

The surface offers the following adsorption sites: On-top sites, Bridging sites, between two atoms, Hollow sites, between three atoms.

The effect of surface formation induces relaxation of the topmost metal layer. For example the clean Pt(111) surface does not show any reconstruction, however a S-XRD analysis of its CTRs (see section 2.1) shows that the surface top layer is slightly relaxed with respect to the nominal bulk termination and their fitting allows to determine an expansion of 4.5 ± 0.5 pm (+2.0%) with respect to the ideal bulk termination (Felici&Pedio 2009). Au(111) clean surface shows a Au (111) 22 $\sqrt{3}$ "herringbone" reconstructed surface (van Hove 1881, Sandy 1991), that involves a great number of surface atoms (Fig. 9 left , Herringbone side view, STM image Fig. 1 right).

The (110) surface is obtained by cutting the fcc unit cell in a manner that intersects the x and y axes but not the z-axis - this exposes a surface with an atomic arrangement of 2-fold symmetry i.e. a rectangular symmetry of the surface layer atoms. In one direction (along the rows) the atoms are in contact i.e. the distance between atoms is equal to twice the metallic(atomic) radius, but in the orthogonal direction there is a substantial gap between

[2] The pictures of figure 10 left and 11 left have been made from NIST SSD output and processed with BALSAC by K. Hermann. http://www.fhi-berlin.mpg.de/~hermann/Balsac/SSDpictures.html SSD is the NIST Standard Reference Database no. 42 by P. R. Watson, M. A. Van Hove, and K. Hermann

the rows. This means that the atoms in the underlying second layer are also, to some extent, exposed at the surface. Normally the bulk terminated (110) plane is energetically not favourable. The (110) surfaces of Ag and Au metals are characterized by the strong surface diffusion anisotropy energies. Typically this surface can be unstable and shows reconstructions ranging from the 1x2 to the 1x5 unit cell.

The clean Au(110) presents the missing row 1x2 reconstruction, while Cu(110) remains unreconstructed (Africh, 2010) as shown in Figure 10.

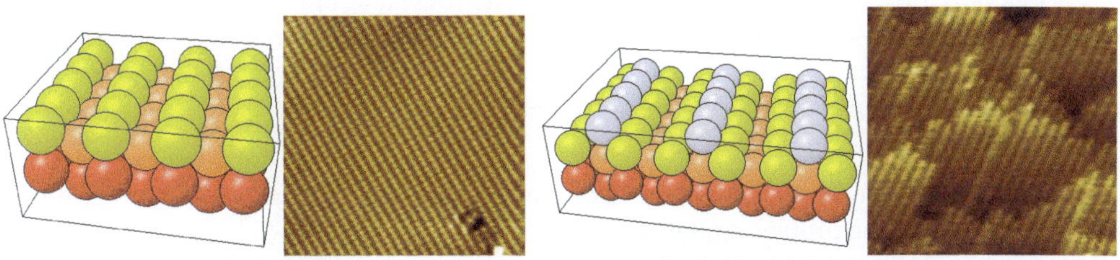

a) Cu(110)b) Au(110) (2x1) Missing row

Fig. 10. a) bulk terminated fcc (110) picture and STM image of Cu(110) surface (C. Africh). B) (110) missing row 2x1 reconstruction picture and topographical image of Au(110) 2x1 reconstruction ref. http://www.redespm. org.br/grupo10.html

The (110) substrates have higher mobility and reactivity. They offer a wide variety of possible adsorption sites, including : On-top sites, Short bridging sites between two atoms in a single row, Long bridging sites between two atoms in adjacent rows. Moreover higher coordination sites are possible (in the troughs that are facets with (111) orientation). The Au(110)-(1×2) surface shows an order-disorder transition with critical temperatures, Tc, between 650 K and 695 K depending on finite-size effects (Clark, 1986).

A large number of experimental and theoretical works have been envisaged to single out the basic aspects of molecules interacting with noble metal. In particular, the Au(111) surface is widely used as support of SAM providing an almost uniform reserve of electron charge. Differently from the (111) case, in a (110) 2x1 reconstructed fcc surface, the surface is atomically rough, and highly anisotropic. The surface atoms on a more open ("rougher") surface have a lower CN - this has important implications when it comes to the chemical reactivity of surfaces, as we discuss in the next section.

4. Survey of ordered molecular layers: From weak interaction to anchoring

Organic molecules can provide complex functionalities at a metal surface. The engineering of the functionalities of organic-inorganic interfaces has important technological applications, as cathalysis, sensors, optoelectronics, molecular spintronics, photovoltaic and so on. The development in these fields implies a deep understanding of the bonding and the lateral interactions that govern the organization of the molecular thin films and ultimately the design for the optimization of their functionalities. In order to accomplish this goal a rigorous surface science approach is necessary: dosing of known molecular fluxes on defined surfaces under ultra high vacuum conditions and fine characterization of these systems, including sophisticated data analysis, theoretical modelling and simulation.

The molecular ordening onto surfaces strongly depends on the delicate balance of the molecule-molecule and molecule-surface interactions (see discussions in ref. Rosei et al. 2003). This is reflected in the complexity of the phase diagrams of these systems. Small differences in the properties of the noble metal substrates (discussed previously) can lead to marked differences in the molecular deposited layers.

Depending on their structure the molecular semiconductor ordered films have important properties for microelectronics: integrability with inorganic semiconductor, low cost, large area bulk processing, tailoring for specific electronic and optical properties, systems more tolerant of defects (no electronically active), no dangling bonds that leads to great flexibility on the choice of substrates, charged carrier mobility increases with increasing pz-pz orbital, anisottropic mobility (for axample Pentacene *μ~1 cm^2/Vs direction in-plane, μ ~10^{-4} cm^2/Vs stacking).*

The disadvantages concern mainly the fact that a high degree of molecular ordering is necessary, to avoid the problems related to low carrier mobility.

The growth procedures are relevant in the bond, ordening and structure of the monolayer of organic semiconductors. All the cases discussed in the following are related to vapor deposition in UHV conditions.

We compare different single organic layers deposited onto ordered noble metal surfaces. The interest and the strong effort in the characterization of the first MONOLAYER[3] of organic molecules on single crystal surfaces adsorption is due to the general accepted result that structural and chemical properties of the monolayer are keys for controlled multilayer growth. The nucleation, growth, structure and interaction molecules strongly depend on the substrate nature and on the particular experimental protocol used during the deposition. The geometry of the first organic layer is exploited to drive the orientation of the next growing layers. From the electronic point of view, the first layer acts as a buffer layer, enabling the following layers to be exploited for the intrinsic electronic properties of the free molecule. Case studies of molecules that differ in size, symmetry and morphology, deposited on different low index noble metal substrates are here presented, on the basis of structural and electronic properties determined by the above techniques (section 2). Figure 11 shows the structural formulae of the molecules discussed in the following. Due to the extension of the literature on SAM deposited onto noble metal surfaces, the list of systems treated in this chapter cannot be complete. We focus our attention on our results obtained on ordered molecular layers mainly deposited onto Au(111) and Au(110) substrates, considering gold as the prototypical noble metals. Other specific systems involving other metal surfaces will be taken into account and discussed.

The large dimension of these molecules (scale 10 Å) with respect to the unit cell of metal surfaces (Table 2), implies that the local molecule-substrate interaction involves several unit cells of the substrate. The properties of these low dimensional systems depend, for example, on the nature of the chemical bond (ionic *versus* covalent *versus* Van der Waals), on the partial filling of the lowest unoccupied molecular orbital (LUMO) as a result of a charge

---

[3] For example 1 ML on Au(111) is defined as the packing density compared to the Au(111) plane, $1.39 \cdot 10^{15}$ atoms/cm^2, which is equivalent to an effective thickness of 2.35 A ° at the density of bulk gold.

transfer from the substrate and/or from dopant atoms (influenced by the substrate work function, the molecule-substrate interaction and possible screening effects), on the geometry of the molecular film which depends on the substrate surface chosen and on the thermal history of the film) , on possible adlayer-induced substrate reconstruction and on defects. In the next section we will describe the research methods with some detail.

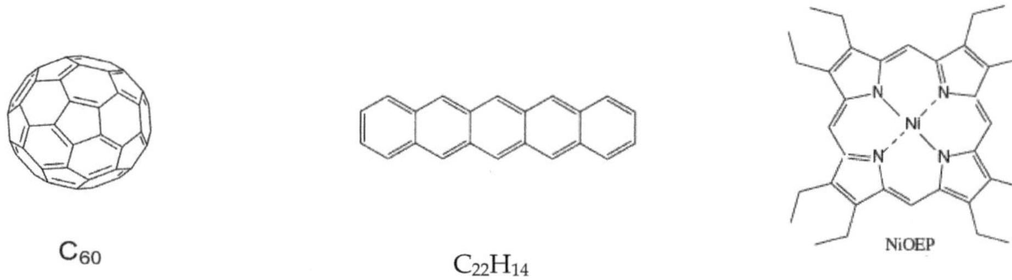

Fig. 11. Structures of different molecules. Left: Fullerene C_{60}, Center: Pentacene $C_{22}H_{14}$, Right: Ni Octaethylporphyrin NiOEP.

4.1 C_{60} on noble metals

Fullerenes (C_{60}) can be considered as model systems of large molecules. This kind of soccer-ball shaped molecule with its three dimensional character along with the strong modulation of the valence charge distribution over the molecular cage. This unique electronic structure leads to several possible interaction mechanisms with the host surface. The geometrical constraints imposed by the three dimensional cage allow only a few carbon atoms per molecule to be in contact with the substrate. The distribution of electronic charge (interface region) and the electronic properties of the first C_{60} ML can be tuned. When deposited on a single crystal surface, fullerene molecules are proving useful to encapsulate surface reconstructions and/or are functioning as templates for the formation of novel structures. They are widely used to create complex architectures on surfaces for their stability, electronic properties (doping), functionalization, and capability of forming reconstructed surfaces (Rudolf 1996 as cited in Gavioli &Cepek, 2008).

Monolayer (ML) and sub-monolayer films based on fullerene molecules deposited on top of metal or semiconductor surfaces, therefore, form a class of systems that could exhibit a wide range of physical behaviours. An increasing number of works is aimed at studying the interaction of single layers of C_{60} with nobel metal surfaces. Due to its large work function (7.6 eV) and electron affinity (2.7 eV), C_{60} has the propensity to accept electrons from many metallic surfaces. However, charge transfer from the metallic substrate to C_{60} is not always observed and the character of the interaction depends mainly on the metal, almost independently of its work function. Thin films of C_{60} adsorbed on crystalline substrates usually tend to form hexagonal or quasi-hexagonal ordered structures with a large variety of the substrate-adsorbate interactions, ranging from covalent to ionic character. An important issue is the investigation of the consequences of the C_{60} adsorption on the underlying substrate structure. According Maxwell et al. (Maxwell 1994) the bonding interaction of C_{60} in direct contact with a metal surface can be divided in three categories:

strong predominantly covalent bonding, intermediate predominantly covalent bonding and intermediate predominantly ionic bonding. Strong predominantly covalent bonding has been observed for a single layer of C_{60} on Ni(110), Pt(111), Rh(111) and Ta(110) where the strength of the interaction is able to catalyze the fullerene cage decomposition prior to the desorption of the monolayer. Interaction of intermediate strength with bonds of mixed ionic character has been observed for the C_{60} monolayer on Au(110), Au(111), Ag(110), Ag(111) and polycrystalline Ag Cu(110) and Cu(111),. Charge transfer has also been observed for C_{60} monolayers deposited on polycrystalline Au and Cu samples, but the stability of the C_{60} monolayer and the strength of the interaction onto these surfaces was not investigated in detail. The trend of these bond strengths reflects the Hammer- Nørskov model (par. 3).

A particular case has been observed on Ag(100). By using the appropriate sample preparation conditions, this system shows the reversible opening of a gap at the Fermi level at temperatures 25 ≤ T< 300K (see Fig. 12), whose origin is still under debate (Cepek et al. 2001, Goldoni et al., 1998).

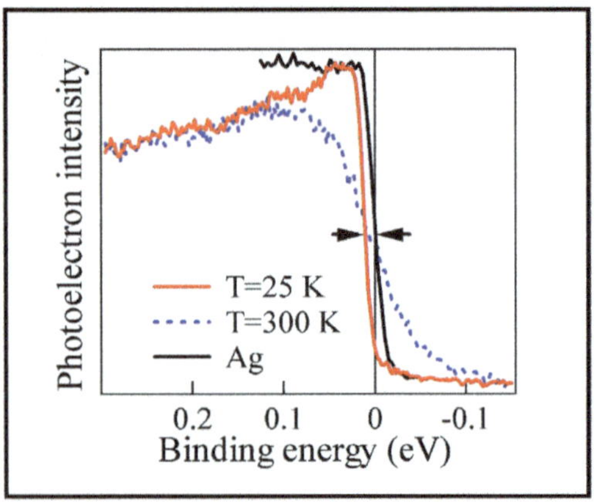

Fig. 12. Fermi Level photoemission spectra of the C_{60}/Ag(100) system at varying the temperature.

The C_{60} layer may not present long-range order, but in any case the quasi-hexagonal geometry is locally preserved. The C_{60} molecules adsorbed on Ag(100) follow this general trend. The growth mode and morphology of the C_{60} molecules adsorbed on Ag(100) at room temperature (RT) or above has been widely studied (Goldoni, 1998, Giudice,1998), Cepek,2000a), Lu, 2003). It has been observed that the C_{60} molecules form a strong bond on the Ag(100) surface at any growing temperature investigated (for STM, ranging from 100K to ≈670K). When deposited at room temperature or below, and by using an evaporation rate of ≈0.2ML/min.), the growth mode is not layer by layer, but the second layer start to grow before the first is completed (Fig. 13, a, b) (Cepek, 2000 a). When the C_{60} molecules are deposited at a temperature higher than the C_{60} sublimation temperature, or by annealing a C_{60} multilayer film, only one single layer of molecules are left on the surface. When the annealing or growing temperature is higher that ≈600K the molecules form a well ordered overlayer, characterized by long range order and by a sharp quasi-hexagonal LEED pattern, where the C_{60} - C_{60} distance is ≈10.4Å (see Fig. 13 c).

The most direct experimental technique, giving detailed information regarding interaction between the fullerene molecules and the substrate, is PES valence band (VB). Fig. 14 left compares the VB spectrum of the ordered monolayer with the clean substrate and the C_{60} multilayer. The spectra were measured at room temperature (RT) (Cepek, 2000 a). The C_{60} multilayer spectrum is the typical molecular-like spectrum of this compound. The five-fold degenerate highest occupied molecular orbital (HOMO) derived band lies at energy of ≈2.2eV, while the HOMO-1 derived band has energy of ≈3.6eV. The clean Ag(100) spectrum does not show any features in the HOMO and HOMO-1 energy range, so all the structures observed in the ML spectrum in this region are only due to the C_{60} adsorbed. We note that the ML spectrum shows an extra emission at the Fermi level compared to the multilayer spectrum, indicating a metallic behaviour. We want to remark that this extra emission cannot be due to the metallic Ag substrate, if the strong attenuation by the C_{60} layer is considered. So it results that the presence of the peak at the Fermi level is the typical fingerprint of a partially filling of the C_{60} lowest unoccupied molecular orbital (LUMO) derived band and it means that charge transfer from the Ag substrate to the C_{60} molecules takes place. By K doping, it has have found that the molecules adsorbed on the monolayer are in a charge state ≈-2 (Cepek, 2000 b)). Finally we want to note that the bond is activated by temperature.

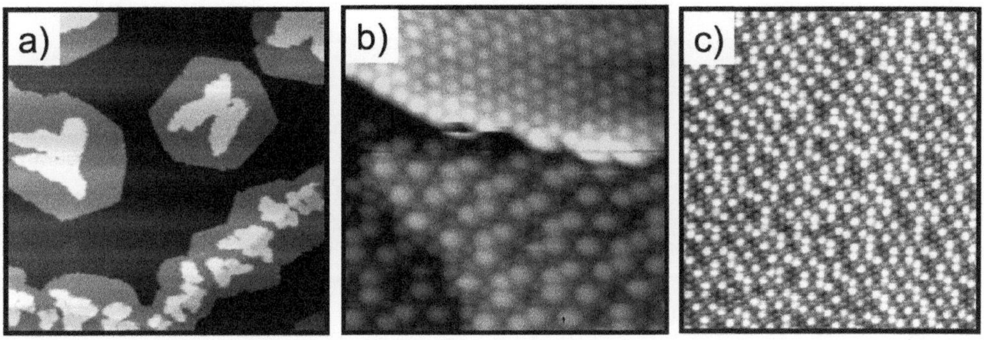

(Cepek et al., 2000 a)

Fig. 13. STM images of different C_{60} coverages deposited on Ag(100). See text for discussion.

Fig. 14, right shows the VB photoemission spectra of 0.7 ML of C_{60} adsorbed on Ag(001) at 150 K and subsequently annealed to different temperatures (RT and 600 K for ≈ 5 minutes, compared to a thick C_{60} multilayer (Cepek, 2000 a). The 150 K spectrum has been measured at 150 K, all the others at RT. The deposition of the same quantity of C_{60} molecules at RT produces a photoemission spectrum (not shown) identical to that of the 150 K deposited film annealed at RT. The same is true for the spectrum of a 600K deposited or annealed film.

Also in this case, the adsorption of C_{60} induces an increase of the photoemission intensity at the FL with respect to the clean surface thereby indicating that in all samples the bond is characterized by charge transfer from silver atoms to C_{60} molecules. The 150 K and RT VB photoemission spectra show that the HOMO and HOMO-1 derived bands consist of two components: one centered at the same BE of the multilayer spectrum, the other at the BE of one ordered C_{60} ML. This, combined with the STM images of Fig. 13 a, allows us to attribute the former component to C_{60} cages in the second layer and the latter to C_{60} molecules in direct contact with the substrate. Data fittings (continuous line of Fig. 14, right. See Cepek2000 a, b for more detail) confirms this hypothesis, showing that the percent of the molecules of the second layer is $x \cong 24\%$ for the 150K spectrum and $x \cong 20\%$ for the RT spectrum. The very good agreement of these results with the value of 17% obtained by STM, confirms that the molecules in direct contact with the substrate present the same bond at 150 K, RT, and 600 K while the second layer molecules do not interact with the substrate, as already observed for different systems (Cepek et al., 2000a). Taking into account that core level spectra indicate that no C_{60} molecules desorbs passing from 150 K to RT, this indicates that the increased thermal energy enables a fraction of the second layer molecules to descend the step edge and to incorporate in the first layer islands. From a careful inspection of the fittings in Fig. 14 right) (continuous line), one can notice that the 150 K deposited film spectrum is slightly sharper than the simulation result. This is due to the specific phonon broadening of the experimental data (measured at 150 K), which is lower than that used in the simulation, performed using spectra measured at RT. However it presents also lower photoemission intensity at the Fermi level with respect to the simulation, which can not be due to the different temperatures and may indicate a lower C_{60} charge state, or a higher disorder. In spite of these minor differences, passing from 150 K to RT, there is no evidence for the appearance of two different electronic states corresponding to the two different kinds of molecules (bright and dim) observed in STM images, which may be easily explained with the occurrence of the two molecular orientation revealed by the high resolution STM images (Lu, 2003) and by XPD.

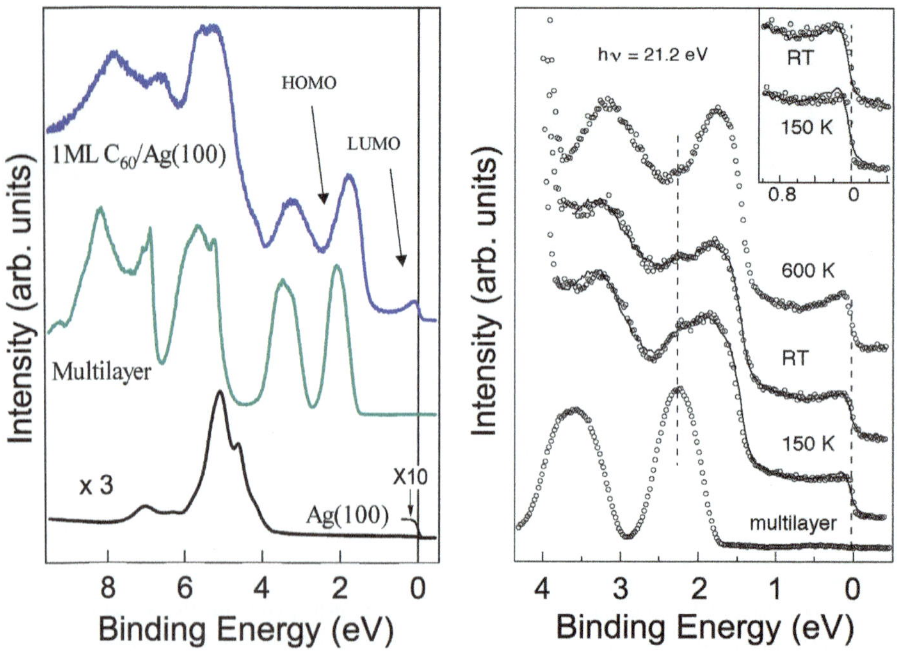

Fig. 14. Left: Valence Band spectra of C_{60} deposited onto Ag(100) . Clean Ag(100) and C_{60} Multilayer are compared. Right: Valence Band spectra for different Temperatures. See text for details.

We discuss two cases systems: the "covalent" C_{60}/Pt(111) (Felici, 2005) with the partially "ionic" C_{60}/Au(110) (Pedio et al., 2000, Hinterstein et el., 2008) systems.

Fig. 15 shows the phase diagram of the C_{60} deposited onto Pt(111) surface. The C_{60} deposited on Pt(111) at room temperature weakly interacts with the substrate atoms and molecular features measured by electron spectroscopies are similar in intensity and energy separation to the isolated molecule or to C_{60} multilayer. After annealing the fullerene bond with Pt(111) substrate become strongly covalent. The deposition of one monolayer of C_{60} on the Pt(111) surface at T < 580°C leads to an ordered double domain hexagonal reconstruction, that is present also for coverages exceeding 1 ML (Pedio et al., 1999, Felici et al., 2005].

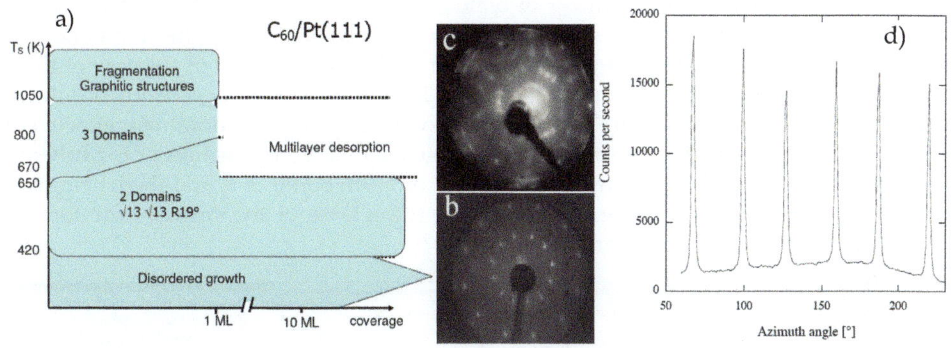

Fig. 15. a) Phase diagram of the C_{60}/Pt(111) systems obtained with C_{60} flux of 0.05 ML/min and slow temperature variations (1 K/min). The peaks belong to the two hexagonal domains separated of 27.8°, b) LEED pattern of the single layer of $\sqrt{13}\ \sqrt{13}R13.9°$, c) LEED pattern of the hexagonal 3-domain reconstruction. d) Azimuthal scan of the x-ray scattered intensity for an exchanged vector modulus corresponding to the first order of the $\sqrt{13}\ \sqrt{13}R13.9°$ reconstruction.

The long range molecular ordering formation is a kinetically controlled process. The double hexagonal pattern was observed even for 1 ML adsorbed at 100 K and annealed above 770 K (Cepek et al., 1996). When the sample is heated above 800 K or if less than a complete monolayer is adsorbed at T>670 K, a third domain becomes visible in the LEED pattern.

About 10 MLs of C_{60} were deposited at room temperature. Then the system was slowly annealed while monitoring with the x-rays the appearance of the $\sqrt{13}\ \sqrt{13}R13.9°$ reconstruction. At about 420 K sharps peaks corresponding to this reconstruction appear. This temperature is much lower than the multilayer desorption of the fullerenes and, in fact, x-ray reflectivity scans show the presence of a well ordered C_{60} multilayer. In Fig.15 d) we show an azimuthal plot of the intensity of the fist order reconstruction. The two P6 domains separated of 27.8° with a practical equal intensity are clearly observable. Further annealing at 650 K removes the multilayer and leaves only one layer bonded to the surface. The analysis of the CTRs provides the following information: a) the vertical expansion of the topmost substrate layer reduces to 1.1 +/- 0.4 pm, b) the increase in surface roughness is compatible with the presence of one vacancy per reconstruction unit cell in the substrate surface.

From the x-ray diffraction analysis this vacancy helps enormously the structural solution of the system. In our case the vacancy can be considered as a heavy atom with a negative number of electrons and the Patterson must then be interpreted by looking at the negative peaks (Fig.16). The dashed rhombus in the figure shows the surface unit cell of the Pt(111) substrate while the large continuous line rhombus is the surface unit cell of the √13 √13*R*13.9° reconstruction. The dashed contour lines represent the negative values of the map and the deepest minima are localized around the vacancies and at the Pt surface atom positions. A fitting of the data, assuming only 13 structural parameters, leads to a very simple solution of this complicated system. The surface Pt atoms displace slightly from their bulk terminated positions while the C_{60} lies on top of the vacancies pointing to the substrate surface with one of its hexagonal surfaces. The fullerene enters as much as possible in the vacancy trying to maximize the number of bonds which are formed with the Pt atoms and about 12 Carbon atoms form directs bonds with the Pt substrate. The C_{60} lies with one of its hexagonal faces on top of the vacancies at a height slightly larger than in the case of the Pt substrate. Similar results have been also observed in the case of the C_{60} adsorbed onto the Ag(111) (Li etal. 2009) and the Au(111) surface. Also in these cases the dominant effect is the formation of vacancies at the substrate surface which stabilize the position of the fullerene molecules. In case of Au(111) substrate the number of C-Au bonds is 6, and the herringbone substrate reconstruction is disappeared indicating the release of the stress of the topmost layer, induced by the C_{60} monolayer.

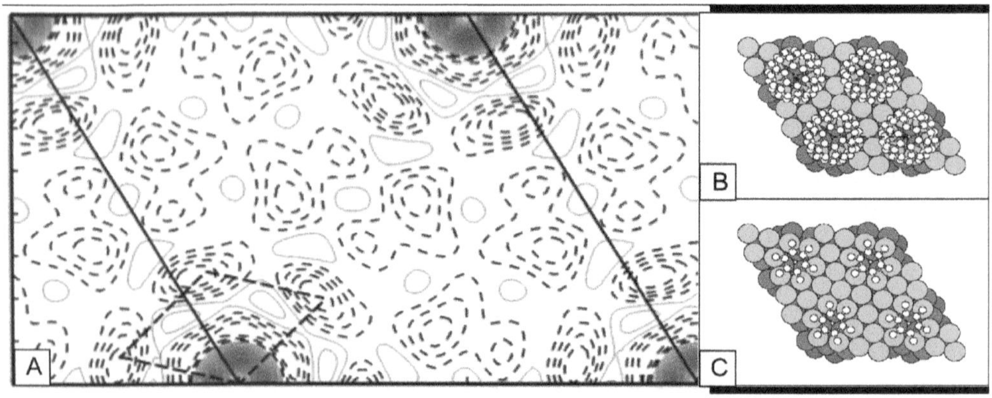

Fig. 16. a) the Patterson map of the √13 √13*R*13.9° reconstruction. Because the pattern is dominated by the signal due to a vacancy in the surface unit cell this map has an inverted sign usual. The negative peaks (dashed lines) correspond to the positions of atoms participating in the reconstruction. The dashed rhombus is the Pt(111) surface unit cell. In (b) we show a top view of the best model fit and in (c) a detail on the C atoms forming bonds with the substrate.

Figure 17 shows the spectroscopy results of C_{60} monolayers deposited onto different noble metal surfaces. On the left the photoemission of the C1s core levels of the ordered monolayers (Maxwell et al 1994, Pedio et al. 1999 private communication) are compared (bottom curve) with that relative to the pristine C_{60}. Note that the bulk C_{60} is a molecular solid tìwith the fullerene that interact by weak, Van der Waals interactions and the spectra are comparable (even though broader) with those obtained on the free molecule in gas phase. The experimental width in all monolayers is about 1 eV, a value sensibly higher than that of C_{60} pristine. This implies that the C atoms in the single layer have different chemical configurations (C-metal bonds) with respect to the C_{60}. Figure 17 left shows a set of IPES spectra for low coverages of C_{60} deposited on Pt(111) at RT and the single monolayer evolution after annealing at different temperatures. The bottom curves, related to the multilayer and the clean Pt(111), are shown for comparison. The top most spectrum is related to the chemisorbed (6x5) C_{60}/Au(110) that will be discussed in the next section. IPES spectra of the hexagonal phases of a single monolayer are substantially different from the C_{60}-Au(110) where was estimated by electron spectroscopies and an ionic character dominates interaction a charge transfer of 1±1 (see for discussion Hinterstein et al. 2008, Felici et al. 2009) electron. When the surface is annealed a 2-domain structure is formed rather dramatic changes in the spectrum IPES spectrum occur: the LUMO-derived feature appears centered at 0.5 eV and the spectral weight near EF is greatly increased, indicating that there is a redistribution of the empty states. This fact has been interpreted as a strong interaction between the substrate and the C_{60} overlayer when the covalent bond takes place. This result nicely confirms the formation of 12 Pt-C bonds, suggested by the SXRD analysis.

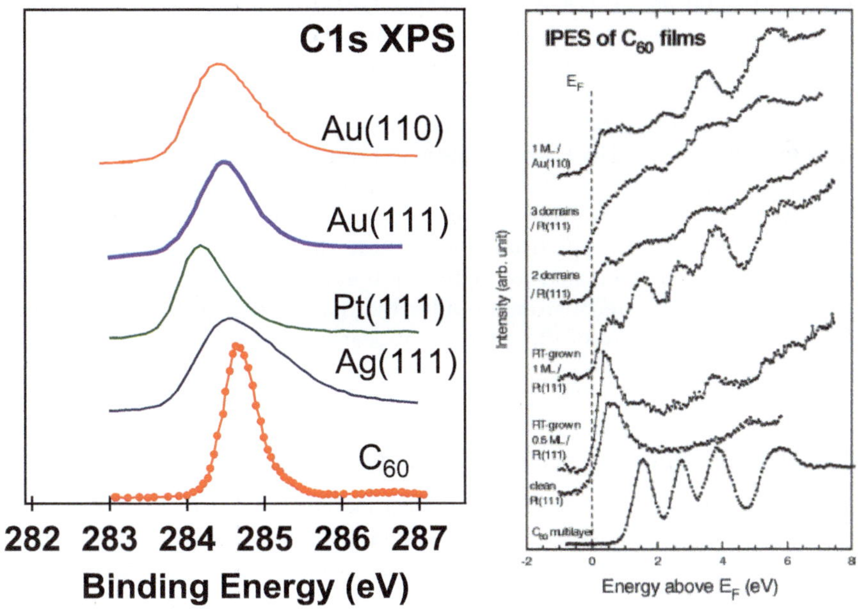

Fig. 17. a) C1s photoemission spectra of C_{60} deposited onto Ag(111), Pt(111), Au(110) and pristine C_{60}. b) Example of IPES spectra taken at normal incidence for different C_{60} overlayers on Pt(111). The spectra for the chemisorbed 1 ML C_{60}/Au(110) (top) and for the multilayer of C_{60} on the same Au(110) (bottom) are plotted for comparison (Pedio1999).

For the two-domain structure on Pt(111), vibrational studies have indicated no evidence for charge transfer. Actually in case of Pt substrate a further annealing of the system leads to a 3-domain structure and to fragmentation of the fullerene. The corresponding LEED pattern shows graphitic rings. Higher T and longer annealing produces changes in the IPES spectra. At this stage it has been suggested that the fragmentation products consist in pentagon-hexagon groups, still arranged in a pseudo-order related to graphitic-like features in diffraction, as discussed in ref. Pedio 1999.

Upon deposition of 1 ML of C_{60} on Au(110) it is possible to observe by STM a nice hexagonal pattern of the fullerene layer showing alternating brighter and darker lines. The uncovered regions of Au modify their reconstruction from 1x2 to a 1x5 surface cell. This 1x5 reconstruction is in perfect register with the brighter and darker lines in the STM images leading the authors to conclude that this kind of reconstruction was extending under the fullerene layer (Modesti et al. 1996 cited in Pedio et al. 2000). On clean 2x1-Au(110) deposition of few monolayers of C_{60} followed by thermal annealing leads to the formation of large terraces showing a (6x5) reconstruction. The evolution of the IPES spectra of C_{60} deposited on Au(110) for different molecular depositions at room temperature (RT) (Pedio et al. 1995 cited in Pedio et al. 2000) shows, for coverages < 1 ML, that all the molecular structures are broader and their energy separation is different with respect to the multilayer spectrum.

The first structure (LUMO) is practically not detectable and a clearly Fermi level emission is present. As soon as we reach the second layer, the LUMO feature appears clearly, though the peaks are still broader than in the multilayer system. For 1 ML deposition at about 450 °C we observed, in the LEED pattern, a (6x5) superstructure. The energy position of the features in the IPES spectra corresponds to the peaks of the 1ML C_{60} grown at RT, while their intensities resulr much more pronounced. The LUMO emission is confused with the Fermi. The topmost spectrum of Fig. 17 b) shows the 6x5 C_{60}/Au(110). During the formation of the first monolayer, when the system is annealed (either during C_{60} deposition or after) there is a strong deformation of *all* the molecular C_{60} localized states due to a charge transfer from the substrate to the C_{60} indicating a chemisorbed phase. SXRD data confirmed a strong redistribution of the substrate Au atoms induced by the formation of the 6x5 reconstruction after thermal annealing. When analyzed with x-ray diffraction the structure of the interface appeared to be much more complicated. The substrate reconstruction was not a "simple" 1x5 but a complex 6x5 reconstruction with no obvious symmetries. This makes the data analysis an almost impossible task. For solving this system we have applied direct methods which are able to provide hints on the electronic density maps of the surface unit cell (Pedio, 2000).

Starting from the model proposed by the direct methods, we have been able to refine the analysis and obtain a full 3D solution of the system (Hinterstein 2008). The peculiarity of this solution is the proof that fullerene adsorption induces a strong mass transport involving several layers of the substrate. The gold surface modifies itself in order to form a kind of calyx structure where the fullerene molecules fit. Using this proposed structure we have been able to calculate the electronic properties of the interface which can then be compared with the PES data (Fig. 18). The larger contact area between the C_{60} and the metal, resulting from the substrate rearrangement, allows for the formation of strong directional C-Au bonds. The theoretical simulations based on the SXRD model can single out the contributions of HOMO and LUMO of the spectra in PES and IPS (see Fig. 18), and clarify the LUMO contribution and the "ionicity" character of the bond.

Summarizing, in C_{60} deposited on (111) and (110) substrates the formation of directional bonds between C and substrate reconstructed atoms is the important factor of the high degree of metal surface deformation. These dimples formation is a thermally activated process that produce an anchoring of the molecule. The molecular states are results strongly hybridised.

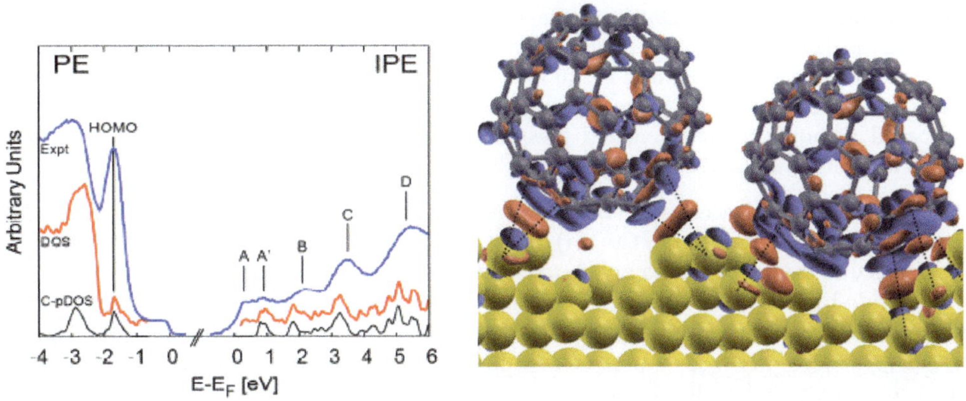

Fig. 18. Left: Comparison between experimental and calculated in-plane data corresponding to the C_{60}/Au(110)-p(6x5) surface reconstruction. The measured values and their associated uncertainties are proportional to the radii of the two empty semicircles. The filled semicircles are proportional to the calculated values using the final structure Right: Lateral views of the C_{60}/Au(110)-p(6x5) surface final structure. The observed corrugation has a height of one atomic level

4.2 Acenes on noble metals

Large acenes, C_{4n+2} H_{2n+4} (Figure 11 b) are conjugated organic molecules composed by *n* hexagonal rings, with H atoms, which saturate the frontier C orbitals. The π molecular states orientation is perpendicular to the molecular plane. Acenes form molecular crystals with triclinic structure showing polymorphism in the stacking d(001)-spacing (Mattheus 2001). In these crystals the electronic mobility is comparable with silicon (Schöhn 2000a in Mattheus 2001).

Thin films of tetracene (Tc, $C_{18}H_{10}$) and pentacene (Pn, $C_{22}H_{14}$), appear most promising candidates for the building of molecular organic semiconductors. Their transport properties in case of well-ordered thin films is highly anisotropic and is strongly influenced by the first layer structural, optical and electronic properties (Lukas, 2001, Söhnchen, 2004 Danişman, 2005, Käfer, 2007, Witte, 2004 and references therein).

The recent literature is quite rich on this topic. The bond of acenes on metals, have been revealed weaker than that of C_{60} on the same substrates. The presence of the H atoms at the molecular borders and their morphology reduce the possibility of hybridization with metal states. Different ordered phases are found for Tc and Pn adsorption on noble metals Pn/Au(111) (Witte 2004) (Schroder2003), Pn/Au(110) (Floreano2006, Badvek2008), Pn/Au(111) (Kafer 2007a, Duhm 2008), Pn/Ag(111),(Danişman 2005, Eremtchenko 2005, Pedio 2007, Kaefer2007b), Tc/Ag(111) (Langner et al. 2005) and in some cases the phase diagram appears complex.

As in other cases the growth mode of acenes is controlled by the nature of the substrate surface and the molecular orientation is determined by a competition of molecule-molecule vs molecule-substrate interactions. On noble metal surfaces acenes have a nearly flat orientation likely to allow the π conjugated molecular orbital to overlap at different degree with the metal electronic states, while on semiconductors acenes has the tendency to standing up. The orientation of pentacene molecules is controlled by the electronic structure of the substrate, namely the near-Fermi level density of states above the surface controls the interaction of the substrate with the pentacene π orbitals. A reduction of this density, as compared to noble metals, realized in semimetallic Bi(001) and Si(111)(5×2)Au surfaces, results in pentacene standing up (Thayer 2005).

The interaction of pentacene with metal surfaces has been studied extensively (Toyoda, 2010 and references therein) Pn/Cu(111) and Cu(110) (Lukas 2002, Lukas 2004). The intermolecular orientation onto noble metal surfaces, generally, adopts the preferred closed-packed side-by-side orientation Lukas 2002, Lukas 2004, but particular geometry of the substrate can induce a head-to head configuration (see for example Floreano, 2006). The film defects formation versus growth protocols and their effect on the subsequent grown layers, are widely discussed in the literature as well as the role of surface defects and stepped surfaces (Lukas 2001, Danişman 2005).

A recent DFT theoretical work relates the distance of the molecular plane of Pn from fcc (111) surfaces, with work function changes and the with hybridization of molecular states with substrate electronic properties. Calculation found that the Au(111) is rather "inert" to Pn layer, while pentacene interacting with Ag(111) and Cu(111) induces hybridization and a weak chemical bond. The sequence of the bond strength follows the Hammer Nørskov reactivity model (section 3). Similarly for the PTCDA molecule on (111) the strength of the bond follows the same trend (Duhm et al. 2008 b). On the other hand the interaction has been interpreted as mainly due to a redistribution of the metal substrate charge ("cushion" effect Witte 2005, Bredas 2010, Li 2009) drawn back by the deposition of weak adsorbates. This effect could lead to work function variations after acenes deposition, compatible with the experimental values on noble metals discussed by Baldacchini et al., 2006.

High Resolution photoemission of the Pn/Cu(119) and theoretical DFT results indicate that pentacene adsorbed on noble metals has the tendency to redistribute π states upon a weak "chemisorption" (Ferretti 2007).

For single layer of Pn/Cu(111) and Pn/Cu(110) the broadening of the NEXAFS features (Lukas, 2004) to was attributed to the electron coupling of the the molecular π orbital. Based on PES measurements, some kind of hybridization has been proposed for Pn/Cu(111) and Pn/Ag(111) (Jaeckel, 2008), while the molecule-substrate interaction for Pn/Au(111) is considered weaker.

Pn on Au(111) is an example of physisorptive organic/metal interface (Koch 2007). The interaction Pn_Au is weak, non-covalent and resulting from different contributions (Duhm2008). In case of Au(111) the Pn forms a variety of ordered phases, up to the completion of the single monolayer. The Au(111) herringbone reconstruction is still detectable by STM images (Schöder2003, Pong et al. 2009). The flat orientation (Käfer2007) of the first layer is related to a ΔW of 0.95 eV and to the formation of an intermolecular dipole modelled by an electrostatic model. The existence of a surface dipole built into molecular layers is conceptually different from the surface dipole at metal Surfaces. Its origin lies in the details of the molecular electronic structure and its magnitude depends on the orientation of molecules relative to the surface of an ordered assembly.

The analysis of the dependence by incident light polarization of C K-edge NEXAFS of the 1 ML Pn on Ag(111) indicate that the Pn molecule is chemisorbed with a tilt angle of 10°+/- 5°. Higher tilt angle 25°+/- 5° were found for submonolayer coverages. This deformation of the LUMO with respect to the LUMO transitions in gas phase Pn (Alagia, 2005) is present for coverages close to the single layer (Figure 19, Pedio2007). This framework in the Pn/Ag(111) is also suggested by (Kaefer 2007b). The electronic π molecular states changes across this orientation transition. XAS spectra show a redistribution of the oscillator strength in the C 1s LUMO excitations as a function of the Pn coverage, together with modifications of the density of states in the filled HOMO region.

Valence band UPS and NEXAFS spectroscopies applied to Tc and Pn adsorption on Ag(111) and Au(111), from subML to 1 ML coverages, show differences in the HOMO and LUMO states related to a redistribution of the molecular states (Pedio, 2008). From NEXAFS, LUMO results strongly modified in the cases of Tc and Pn deposited on Ag(111) substrate and less perturbed on Au(111).

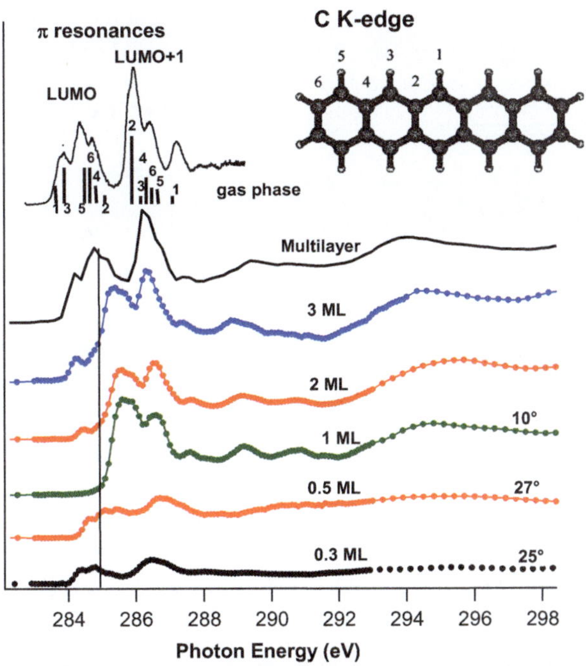

Fig. 19. C K edge XAS of Pentacene on Ag(111) at different molecular coverages in monolayer taken at the magic angle (Pedio 2007). Gas phase spectrum is shown (top); the bars indicate the calculated contributions from different inequivalent C atoms (see numbers in the sketch of pentacene molecule).

Further work is in progress to understand the hybridization degree and the dispersion of the molecular states of the various presented systems. In any case we can indicate a progression of the bond from the stronger Pn/Ag(111) to the weaker Tc/Au(111) (Pedio, 2008). The modification of the HOMO and LUMO molecular states confirms the trend found by DFT (Toyoda 2010) and supporting the modification of the π molecular states of adsorbed acenes on the Ag(111) substrate and the weaker interaction with Au(111).

A peculiar case is the Pn/Au(110). Even in this case the phase diagram is complex (Figures 20 and 21). A multitechnique approach permitted to determine the molecular orientation of the 6x3, 3x6 and 6x8 phases.

The phase diagram indicates that that two-dimensional commensurate growth only occurs in the monolayer range for a substrate temperature, *Ts*, higher than about 370 K. The highest coverage ordered phase displays a (6 x 8) symmetry and corresponds to the saturation coverage at *Ts* = 420 K. The (3 x 6) phase corresponds to the saturation coverage of the first layer at *Ts* =470 K. Helium scattering (HAS) experiments provide information similar to the S-XRD. Figure 20 shows HAS along the [001], ΓY, direction collected at RT after deposition at *Ts* = 420 K for the (6 x8) phase (filled triangles) and *Ts* = 470 K for the (6 x 3) and (3 x 6) phases (open circles and filled squares, respectively). The 2-fold symmetry pattern of the clean missing row reconstructed substrate is also reported at the bottom of the graphic (full line). As a guide to the eye, fractional peaks of the eighth order are indicated at the bottom x axis, while fractional peaks of the sixth order are reported at the top x axis.

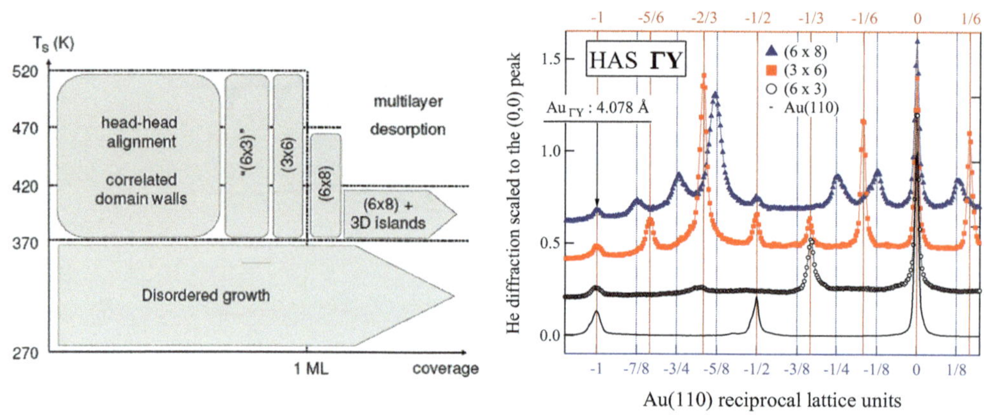

(a) (b)

Fig. 20. a) Phase diagram of pentacene films on Au(110). The (6 x 3) phase is quoted since it has not a well defined equilibrium coverage. The completion of the first monolayer is conventionally assigned between the (3 x 6) and (6 x 8) phases. In figure b are shown the peaks of the different ordered phase

The models of the systems were obtained by means of HAS (Floreano, 2006), NEXAFS and STM (Badvek, 2008). In the monolayer range, pentacene forms two commensurate phases with (3 x 6) and (6 x 8) symmetry. The (3 x 6) phase only contains equivalent, perfectly flat molecules that are aligned side-by-side along [001], thus forming crosstie chains extending along the $\left[1\bar{1}0\right]$ direction. These chains are also preserved in the high coverage (6 x 8) phase, but they display a wider separation since new chains are accommodated in between, with the additional molecules aligned head-to-tail along the $[1\bar{1}0]$ direction and tilted by approximately 90° around their molecular axis. This structure, formed by three molecules per unit cell (two flat and one tilted), forms large domains of nanorails.

In fig. 20 the HAS measurements shows the disappearing of the ½ peak of the gold reconstruction indicating a deconstruction of the Au(110) missing row caused by the molecular deposition. This has been also confirmed by preliminary SXRD results[4].

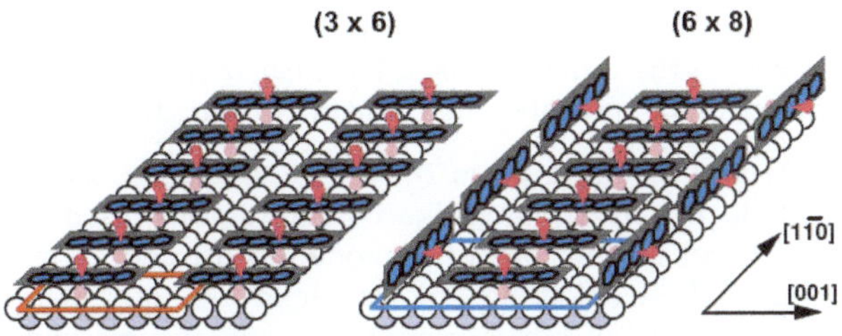

STM

Fig. 21. Models of the 3x6 and 6x8 Pn/Au(110) reconstruction, based on He scattering, STM and NEXAFS measurements. The directional pai states of the Pn are presented in red.

These results enlighten the local rearrangement of the charge at the interface. A possible depletion of electron charge in correspondence of the new chains, inducing an upstanding orientation of Pn, can be inferred.

4.3 Porphyrine derivatives on noble metals

Porphyrine derivatives constitute the essential functional units for many processes and Nature demonstrates their great functionality, variability and chemical stability in many cases. Metalloporphyrin are base components of biological molecules as chlorophyll (Mn porphyrin), emoglobin (Fe porphyrin) and are efficient in the Oxygen exchange.

They are formed by a macrocycle and surrounding groups (Fig. 11 right). The central porphine ring is composed by 4 Pyrrole rings linked by methine bridges (porphyrins) or N (Phtalocyanins). By varying the choice of the metal center, the bonding and characteristics of the metalloporphyrin or the phtalocyanine may be dramatically affected. This is reflected in spectroscopic measurements as XPS and XAS.

A clear description of the electronic properties of porphirinoids is still under debate in the literature, due to the high degree of variation in the simulations. The energy sequence of the valence and conduction electronic states show discrepancies, indicating that the energy ordering in metal-organic molecules crucially depends on the electronic solution employed and the functional theory adopted (Stenuit et al., 2010, Maron et al. 2008).

Porphyrins films grown by various methods (UHV vapour deposition, solutions etc) received a paramount interest and the literature is extremely rich. New interesting results have been achieved recently, as for example a Kondo effect related to the spin of the central atom in Co and Fe phtalocyanines on Au(111), the induced magnetic order of Mn

[4] L.Floreano privat communication

porphyrins on Co films as well as magnetic coupling effects (Wende et al. 2007 cited in Fanetti et al. 2011). Recently in the CoEOP deposited onto Ag(110) (Fanetti et al 2011), a charge transfer to the LUMO orbital has been found. DFT calculations indicate that the molecular interaction with the substrate leads to the electron transfer from the Ag substrate to the molecule. This can be responsible of the steric adsorption of the CoOEP, that results titled of 15°, at variance with other similar systems.

Porphyrins can be deposited in highly ordered single monolayers on metal surfaces, leading to ordered structures. Covalent interaction between the central metal atom and substrates, the "Cushion" effect induced at the noble metal surface, and the out-of-plane central atom dipole moments plays a role in the different cases. No clear trend, as a function of external group, central metal atom element and substrate element and structure, is still available.

On the noble metal (111) surfaces, and often on the (100) planes, both porphyrins and phtalocyanines are found to lay in a planar-like configuration presenting large terraces domains. In case of Au(111) substrates, the herringbone reconstruction remains detectable. The long range ordening of these "floating" molecular layers show dependency on the substrate element and on the kind of porphyrinoid and central metal atom. For example differences between single layers Cu and Ni octaethylporphyrins deposited on Au(111) have been detected in the diffraction patterns and in the electronic properties (Resta et al. 2010).

The interaction with (110) substrates appears different. Restructuring has been indicated by STM in case of Cu-phtalocyanine (CuPc) deposited on Ag(110) (ref. Bohringer et al. 1997 cited in Rosei et al. 2003). CuPc induces a transition from thermodynamically controlled 3-dimensional faceting at submonolayer coverage, to a kinetically dominated 2-dimensional step faceting process, when coverage approaches a full monolayer. Transition metal-phtalocyanine on Au(110) surfaces result regularly spaced. High Resolution PES show interface electronic states close to the Fermi level. Their energies are different from CuPc to FePc and CoPc molecules. Structurally CuPc on Au(110) present a complex phase diagram with strong mass transport of the substrate as a function of the molecular coverage from submonolayer to the 1 ML (Floreano 2008). Again the molecular deposition deconstructs the Au missing row and a set of higher periodicities (x5, x3, x7) are progressively detected along the [001] direction. In conclusion coverage-dependent transformation on fcc 110 substrates seems to be driven by the molecule–substrate interaction. This interaction seems to take place in other porphyrinoid deposited onto fcc (111) substrates but, in any case it appears such that no mass transport of these substrate takes place.

Due to the complexity of the roles played, in the balance among interactions, by the central metal atom in the porphyn ring, by the external groups and frontier orbitals, and by the interaction with the metal substrates (polarization, dipoles etc.), a complete understanding of these systems deserve further study.

Note, finally, that in some cases the molecular morphology results altered by STM (Donovan et al. 2010).

5. Conclusions

Noble metal surfaces offer a variety of structural/electronic moieties which can be used in the organic-inorganic engineering. The comparisons of the similarities and differences of the (111) and (110) surfaces show how the charge reservoir of isotropic or anisotropic electron density guides the interaction, from charge transfer to directional bonds, between molecules and substrate.

Noble metal surfaces are valuable substrates for building self assembled organic thin layers because of their:

- Low chemical reactivity together with high atomic surface mobility resulting in the capability of reordering after molecular deposition. Often order is thermally activated.
- Good degree of morphological order of the crystal surfaces
- Guide for the molecular ordering through the symmetry of the surface. If commensurable the overlayer must follow the substrate symmetry. In case the intermolecular interactions prevail, in strength, compared to the molecular-structural the layer can by uncommensurate.
- High electron density at the crystal surface (in case of (111) in Au, Ag and Cu the electrons form a quasi-two-dimensional gas at surface) providing a charge reservoir.

The details of interaction are crucial for the functionalization of the molecular films.

The anchoring of the molecules is due to a combination of effects (Rosei 2005): i) Adaptation of the surface geometry Larger interaction area, that allows to the molecule to fit into the local structure of substrates ; ii) Higher reactivity of the substrate (stronger bond) facilitates the creation of nanodimple sites underneath the molecules. A simple correlation exists between the bonding strength of a molecule and the metal coordination number of the adsorption site *d* -band model by Hammer and Nørskov. iii) The adsorption inducing restructuring is typically thermally activated: it provides an Energy gain in the adsorption energy ΔE_{ad} of the molecules on the reconstructed surface, balancing the energy cost ΔE_{metal} required to break the metal bonds

The formation of directional bonds seems the base mechanism for the nanopit formation, at least for strong chemisorbed species (as fullerene). C60 induced reconstruction of substrates is thermally activated, even on the close packed fcc (111) surfaces. The molecule results locally perturbed with the electronic properties (particularly the LUMO orbitals), that are modified by the interaction. The structural characterizationof the molecules onto the (110) substrate demonstrates the formation of NANOPITS and/or FACETTING of the metal surface.

In case of "flat" and conjugated molecules, as acenes and prophyrin derivatives, the interrelations of the various aspects (morphology, dipoles-electrostatic configuration at interfaces, distribution of the charge of the metal substrate etc…) are particularly evident. Molecular orientation, long range ordening, surface reconstruction and charge (re)distribution (interface dipole, molecular dipole, "cushion" effect) are all present at different but never negligible degree, depending on the chosen molecule and on the substrate morphology/surface energy/reactivity.

6. References

Africh C. 2011. Surface Structure and Reactivity laboratory CNR-IOM, TASC Available from: www.tasc.infm.it

Alagia, M.; Baldacchini, C. Betti, M. G. Bussolotti, F. Carravetta, V. Ekström, U.; Mariani, C. & Stranges S. (2005). Core-shell photoabsorption and photoelectron spectra of gas-phase pentacene: experiment and theory. *J Chem. Phys* Vol. 122, No. 12 (22 March 2005), pp. 124305

Ashcroft, N. W. & Mermin, N. D. (1976). Solid State Physics. Saunders College, HRW ISBN 0-03-083993-9

BALSAC Available from: http://www.fhiberlin.mpg.de/~hermann/Balsac/pictures.html

Bassani, F. & Altarelli M. (1983). Interaction of radiation with condensed matter, In *Handbook on Synchrotron Radiation,* Vol. I, E.E. Koch (Ed), 463-605, North-Holland Publishing Company Amsterdam (1983).

Baldacchini, C.; Mariani, C. Betti, M. G. Gavioli, L. Fanetti, M. & Sancrotti, M. (2006). Molecular gap and energy level diagram for pentacene adsorbed on fille d-band metal surfaces. *Appl. Phys. Lett* Vol. 89, 2006. pp. 152119 1 -3

Bavdek, G.; Cossaro, A. Cvetko, D. Africh, C. Blasetti, C. Esch, F. Morgante, A. & Floreano, L. (2008). Pentacene Nanorails on Au(110). *Langmuir* Vol. 24 (No. 3), (December 2007), pp 767–772.

Benfatto, M. & Felici, R. (2001). Resonant atomic scattering factor theory: A multiple scattering approach:*Phys. Rev. B* Vol. 64, No.11 (24 August 2001) 115410 (9 pages).

Brühwiler, P.; Karis, A. O. & Mårtensson N. (2002). Charge-transfer dynamics studied using resonant core spectroscopies. *Review of Modern Physics,* Vol. 74, (2002). pp. 703-740

Besenbacher, F. (1996). Scanning tunnelling microscopy studies of metal surfaces, *Rep. Prog. Phys. Vol.* 59 (1996), pp. 1737–1802

Chen C. (1993). *Introduction to Scanning Tunneling Microscopy* . Oxford University Press.

Cepek, C.; Goldoni, A. Modesti, S. (1996). Chemisorption and fragmentation of C_{60} on Pt(111) and Ni(110). *Phys. Rev. B* Vol. 53, (March 15, 1996)pp. 7466– 7472 Available from: http://link.aps.org/doi/10.1103/PhysRevB.53.7466

Cepek, C.; Giovanelli, L.; Sancrotti, M.; Costantini, G.; Boragno, C.& Valbusa U. (2000). Electronic structure and growth mode of the early stages of C60 adsorption at the Ag(001) surface. *Surf. Sci.* Vols. 454-456, pp. 766-770

Cepek, C.; Sancrotti, M. Greber T. & Osterwalder J. (2000). Electronic Structure of K doped C60 monolayers on Ag(001) *Surf. Sci.* Vols. 454-456, (2000) pp. 467–471

Cepek, C.; Vobornik, I. Goldoni, A. Magnano, E. Selvaggi, G. Kröger, J. Panaccione, G. Rossi, G. & Sancrotti M. (2001). Temperature-Dependent Fermi Gap Opening in the c(6×4)–C_{60}/Ag(100) Two-Dimensional Superstructure. *Phys. Rev. Lett.* Vol. 86, (April 2, 2001) pp. 3100-3102 Available from: http://link.aps.org/doi/10.1103/PhysRevLett.86.3100

Clark, D.E.; Unertl, W.N. & Kleban P.H. (1986). Specific-heat anomaly of Au(110) (1 × 2) studied by low-energy electron diffraction *Phys. Rev. B* Vol. 34 No. 6 (1986) pp. 4379-4381.

Danişman, M. F.; Casalis, L. & Scoles, G. (2005). Supersonic molecular beam deposition of pentacene thin films on two Ag(111) surfaces with different step densities. *Phys. Rev. B* Vol. 72 pp. 085404.

Donovan, P.; Robin, A.; Dyer, M. S. Persson, M. & Raval, R. (2010). Unexpected Deformations Induced by Surface Interaction and Chiral Self- Assembly of CoII-Tetraphenylporphyrin (Co-TPP) Adsorbed on Cu(110):A Combined STM and Periodic DFT Study. *Chem. Eur. J.* vol. 16, (2010) pp. 11641 – 11652 Available from: http://dx.doi.org/10.1002/chem.201001776

Dose, V. (1985). Momentum-resolved Inverse Photoemission. *Surf. Sci. Rep* Vol. 5, No. 8 (1985), pp. 337-378. Available from:

http://www.sciencedirect.com/science/article/pii/0167572985900068

Duhm, S.; Heimel, G.; Salzmann, I.; Glowatzki, H.; Johnson, R. L.; Vollmer, A.; Rabe, J. P. & Koch, N. (2008). Orientation-dependent ionization energies and interface dipoles in ordered molecular assemblies. *Nature Materials*, Vol. 7, pp. 326-330

Duhm, S; Gerlach, A.; Salzman, I.; Broeker, B.; Johnson, R.L.; Screiber, F. & Koch, N. (2008). b) PTCDA on Au(111), Ag(111) and Cu(111):Correlation of interface charge transfer to bonding distance. *Organic Electronics*. Vol. 9 pp. 111-118

Eremtchenko, M.; Temirov, R.; Bauer, D.; Schaefer, J. D. & Tautz F. S. (2005). Formation of molecular order on a disordered interface layer: Pentacene/Ag(111). *Phys. Rev. B* Vol. 72, No. 11 pp. 115430 (9 pages).

Fall, C.J.; Bingelli, N. & Baldereschi, A. (2000). Work-function anisotropy in noble metals: Contributions from d states and effects of the surface atomic structure. *Phys. Rev. B* Vol. 61, No. 12, pp. 8489–8495.

Feidenhans'l, R. (1989). Surface Structure Determination by X-ray Diffraction. *Surf.Sci Rep.*, Vol. 10, pp. 105-188.

Hammer, B. (2006). Special sites at Noble and late transition metal catalysts. *Chemistry and Material Science, Topics in Catalysis*, Vol. 37, No.1, pp. 3-16.

Hinterstein, M.; Torrelles, X. Felici, R. Rius, J. Huang, M. Fabris, S. Fuess, H. & Pedio, M. (2008). Looking underneath fullerenes on Au(110): Formation of dimples in the substrate, Phys Rev B Vol. 77 No. 15, (2008) pp. 153412 (4 pages), also selected in Virtual Journal of Nanoscale Science & Technology vol 17, May 12, 2008 issue 19.

Hüfner, S. (1995). *Photoelectron Spectroscopy Principles and applications*. Springer Berlin ISNB 3-540-19108-9

Käfer, D; Ruppel, L. & Witte, G. (2007). a). Growth of pentacene on clean and modified gold surfaces. *Phys. Rev. B*. Vol. 75 pp. 085309. Available from: http://link.aps.org/doi/10.1103/PhysRevB.75.08530

Käfer, D. & Witte, G. (2007). b). Evolution of pentacene films on Ag(111): Growth beyond the first monolayer. *Chem. Phys. Lett.* Vol. 442. (June 2007) pp. 376-383

Koch, N.; Vollmer, A. Duhm, S. Sakamoto, S. Suzuki, Y. (2007) The Effect of Fluorination on Pentacene/Gold Interface Energetics and Charge Reorganization Energy. *Adv. Mater. Vol.* 19, (January 2007) pp.112-116.

Fanetti, M. Calzolari, A. Vilmercati, P. Castellarin-Cudia, C. Borghetti, P. Di Santo, G. Floreano, L. Verdini, A. Cossaro, A. Vobornik, I. Annese, E. Bondino, F. Fabris, S. Goldoni, A. (2011). The Structure and the Molecule-Substrate Interaction in a Co-Octaethyl Porphyrin Monolayer on the Ag(110) Surface. *J. Physical Chemistry* C in press. Manuscript ID: jp-2011-011233

Felici, R.; Pedio, M. Borgatti, F. Iannotta, S. Capozi, M. Ciullo, G. Stierle, A. (2005). X-ray diffraction characterization of Pt(111) surface nanopatterning induced by C_{60} adsorption. *Nat. Mat.* Vol 4 (September2005) pp.688-692.

Felici R. & Pedio M. (2009). Nanostructure induces by the adsorption of fullerenes: structural and Electronic properties review paper in Proceedings of the Workshop Synchrotron Radiation and Nanostructures, Papers in Honour of Paolo Perfetti eds A. Cricenti and G. Margaritondo pp. 209-223 (2009), Word Scientific New Jersey, ISBN-13 978-981-4280-83-9

Ferretti, A.; Baldacchini, C. Calzolari, A. Di Felice, R. Ruini, A. Molinari, E. & Betti M. G. (2007). Mixing of Electronic States in Pentacene Adsorption on Copper. *Phys. Rev. Lett.* Vol. 99 No. 4, July 25, 2007 pp. 046802 (4 pages).

Floreano, L.; Cossaro, A. Cvetko, D. Bavdek, G. & Morgante A Phase Diagram of Pentacene Growth on Au(110). (2006). *J. Phys. Chem. B*, 110 (10), (February 17, 2006) pp 4908–4913.

Floreano, L.; Cossaro, A. Gotter, R. Verdini, A. Bavdek, G. Evangelista, F. Ruocco, A. Morgante, A. & Cvetko, D. (2008). Periodic arrays of Cu-Phthalocyanine chains on Au(110), *J. Phys. Chem. C* Vol. 112 (2008) pp.10794.2008.

Gavioli, L. & Cepek, C. (2008). Ultrathin Fullerene-Based films via STM and STS. In: Applied Scanning Probe Methods (Eds) Bhushan B; Fuchs, H & Tomitori,M. series Nanoscience and Technology, 2008 Sprinter-Verlag Berlin.

Gargiani, P.; Angelucci, M. Mariani, C. & Betti M. G. (2010). Metal-phthalocyanine chains on the Au(110) surface: Interaction states versus *d*-metal states occupancy. *Phys. Rev. B* Vol. 81, No. 8 pp. 085412 (7 pages).

Giudice, E.; Magnano, E. Rusponi, S. Boragno, C. & Valbusa, U. (1998). Morphology of C_{60} thin films grown on Ag(001) *Surf. Sci.* Vol. 405 No. 2-3 (5 May 1998) pp. L561-565.

Goldoni A. Cepek C. Magnano E. Laine A. D. Sancrotti M. (1998). Temperature dependence of the electronic structure near E_F and electron-phonon interaction in C60/Ag(100) single layers. *Phys. Rev. B* Vol. 58 No. 4, pp. 2228-2332

Li, H.; Duan, Y. Coropceanu, V. & Bredas J-L. (2009). Electronic structure of the pentacene-gold interface: a density-functional theory study. *Organic Electronics* Vol. 10 pp. 1571-1578.

Langner, A.; Hauschild, A. Fahrenholz, S. & Sokolowski, M. (2005). Structural properties of tetracene films on Ag(1 1 1) investigated by SPA-LEED and TPD. *Surf Sci* Vol. 574 (10 November 2004) pp. 153-165.

Li, H. L. ; Pussi, K. Hanna, K. J. Wang, L.-L. Johnson, D. D. Cheng, H.-P. Shin, H. Cuartarolo, S. Moritz, W. Smerdon, J. A. Mc.Grath, R. & Diehl R. D.. 2009. Surface Geometry of C_{60} on Ag(111). *Phys. Rev. Lett.* Vol. 103 No. 5, pp. 056101 (4 pages).

Lu, X.; Grobis, M. Khoo, K. H. Louie, S. G. & Cromie M. F. (2003). Spatially Mapping the Spectral Density of a Single C60 Molecule. *Phys. Rev. Lett.* Vol. 90 No.9, 096802 (4 pages).

Lukas, S.; Vollmer, S. Witte, G. &Woll Ch. (2001). Adsorption of acenes on flat and vicinal Cu(111) surfaces: Step induced formation on lateral order. *J.Chem. Phys* Vol. 114, No. 22, pp. 10123-10130.

Lukas, S.; Witte, G. & Wöll Ch. (2002) Novel Mechanism for Molecular Self-Assembly on Metal Substrates: Unidirectional Rows of Pentacene on Cu(110) Produced by a Substrate-Mediated Repulsion *Phys. Rev. Lett.* Vol. 88, No. 2 (December 28, 2002) pp. 028301 (4 pages).

Lukas, S.; Söhnchen, S. Witte, G. & Wöll Ch.. (2004). Epitaxial growth of pentacene films on metal surfaces. *Chem.Phys.Chem* Vol. 5 pp. 266-270.

Lüth, H. (1996) Surfaces and Interfaces of Solid Materials, Springer Berlin. ISBN 3-540-58576-1 3.

Marom N.; Hod, O. Scuseria, G. E. & Kronik L. 2008. Electronic structure of copper phthalocyanine: A comparative density functional theory study. *J. Chem. Phys.* Vol. 128, No. 16. pp. 164107-(6 pages).

Mattheus, C. C.; Dros, A. B. Baas, J. Meetsma, A. de Boer, J. L. & Palstra T.M. Polymorphysm in pentacene. (2001) *Acta Crystallographica Section C Crystal Structure Communications* Vol. C57 pp. 939-941. ISSN 0108-2701.

Maxwell, A. J.; Brühwiler, P. A. Nilsson, A. Mårtensson, N. & Rudolf P. (1994). Photoemission, autoionization, and x-ray-absorption spectroscopy of ultrathin-film C_{60} on Au(110). *Phys. Rev. B* Vol. 49 No. 15, (April 15, 1994) pp. 10717 -10725.

NIST X-ray Photoelectron Spectroscopy Database 20, version 3.4 (Web version), http://srdata.nist.gov/xps/

Nickel, B.; Fiebig, M. Schiefer, S. Göllner, M. Huth, M. Erlen, C. & Lugli P. (2008). Pentacene devices: Molecular structure, charge transport and photo response *Phys. Stat.Sol. (a)* pp. 205, No. 3, pp. 526–533.

Pedio, M.; Hevesi, K. Zema, N. Capozi, M. Perfetti, P. Gouttebaron, R. Pireaux, J.-J. Caudano R. & Rudolf, P. (1999). C_{60}/metal surfaces: adsorption and decomposition. *Surf. Sci.* Vol. 437, (June 1999) pp. 249- 260.

Pedio, M.; Felici, R. Torrelles, X. Rudolf, P. Capozi, M. Rius, J. & Ferrer S. (2000). Study of C_{60}/Au(110)-p(6×5) Reconstruction from In-Plane X-Ray Diffraction Data *Phys. Rev. Lett.* Vo. 85 No. 5, pp. 1040-1043.

Pedio, M.; Doyle, B. Mahne, N. Giglia, A. Borgatti, F. Nannarone, S. Johansson, S. Temirov, R. Tautz, S. Casalis, L. Hudej, R. Danisman, M.F. & Nickel B. (2007). Growth of Pentacene on Ag(111) surface: A NEXAFS study. *Applied Surface Science* Vol. 254, pp. 103-107

Pedio, M.; Doyle, B. Mahne, N. Giglia, A. Nannarone, S. Montecchi, M. Pasquali, L. (2008). *J. Phys.: Conf. Ser.* Molecular states of polyacenes grown on noble metal surfaces. Vol. 100 pp. 052072 1-4, IVC-17/ICSS-13 and ICN+T2007 Lund July 2007. Available from: http://iopscience.iop.org/1742-6596/100/5/052072/

Pong, I. F.; Yau, S. Huang, P-Y. & Chen, M.-C. (2009). In situ STM Imaging of the Structures of pentacene Molecules adsorbed on Au(111). *Langmuir*, Vol 25, No 17 (November 2009) pp. 9887-9893.

Resta, A.; Felici, R. Kumar, M. & Pedio M., (2010). Ni and Cu octaethyl porphyrins ordered monolayer on Au(111) surfaces. *Journal of Non-Crystalline Solids* Vol. 356 pp. 1951–1954.Robinson, I.K. (1986). Crystal truncation rods and surface roughness. *Phys. Rev. B* Vol. 33, No. 6, (march 1986), pp. 3830-3836.

Robinson, K. & Tweet D. J. (1992). Surface X-ray diffraction. *Rep. Prog. Phys.* Vol. 55, pp. 599-651. Available from: http://iopscience.iop.org/0034-4885/55/5/002.

Rosei, F.; Schunack, M. Naitoh, Y. Jiang, P. Gourdon, A. Laegsgaard, E. Stensgaard, I. Joachim, Ch. & Besenbacher F. 2003. Properties of Large organic molecules on metal surfaces. *Progress in Surface Science* Vol. 71, pp. 95-146.

Sandy, A.R.; Mochrie, S.G.J. Zehner, D.M. Huang, K.G. & Gibbs D. (1991). Structure and phases of the Au(111) surface: X-ray-scattering measurements. *Phys Rev B* Vol. 43 No. 6, (15 feb 1991) pp. 4667- 4687.

Schöder, P. G.; France, C. B. Park, J. B. & Parkinson B. A. (2003). Orbital Alignment and Morphology of Pentacene Deposited on Au(111) and SnS2 Studied Using Photoemission Spectroscopy . *J. Phys. Chem. B* Vol. 107. No. 10 (February 2003) pp. 2253-2261

Scudiero, L.; Barlow, D.E. & Hipps, K.W. (2002). Scanning Tunneling Microscopy, Orbital Mediated Tunneling Spectroscopy, and Ultraviolet Photoelectron Spectroscopy of Nickel(II) Octaethylporphyrins Deposited from vapor. *J. Phys. Chem. B*, Vol. 106, pp. 996-1003.

Singh-Miller, N. E. Marzari, N. (2009). Surface energies, work functions, and surface relaxations of low-index metallic surfaces from first principles. *Phys Rev. B* Vol. 80 No. 23 (2009) pp. 235407 (9 pages).

Smoluchowski, R. (1941). Anisotropy of the Electronic Work Function of Metals. *Phys. Rev.* Vol. 60, No. 9 pp. 661-674 . ISSN 0031899X.

Söhnchen, S.; Lukas, S. & Witte, G. (2004). Epitaxial growth of pentacene films on Cu(110). *J. Chem. Phys.* Vol. 121, pp. 525. Available from: http://jcp.aip.org/resource/1/jcpsa6/v121/i1/p525_s1

Stenuit, G.; Castellarin-Cudia, C. Plekan, O.K. Feyer, V. Prince, K.C. Goldoni, A. & Umari, P. (2010). Valence electronic properties of porphyrin derivatives. *Phys Chem Chem Phys* Vol. 12, No. 36, pp. 10812 (7 pages). Available from: http://pubs.rsc.org/en/content/articlelanding/2010/cp10.1039/c004332j

Stöhr, J. (1992). NEXAFS Spectroscopy, Springer-Verlag Berlin ISBN 3-540-54422-4

Stolze, P. (1994). Simulation of surface defects. *J. Phys::Condens. Matter* Vol 6 (1994) pp. 9495-9517

Toyoda, K.; Hamada, I. Lee, K. Yanagisawa, S. & Morikawa, Y. (2010). *J. Chem. Phys.* Vol. 132 pp. 134793 (9 pages).

Thayer, G.E.; Sadowski, J. T. Meyer zu Heringdorf, F. Sakurai, T. & Tromp R. M. (2005). *Phys. Rev. Lett.* Vol. 95, No.25 pp. 256106 (9 pages).

Triguero, L.; Pettersson, L.G. M. & Ågren, H. (1998). Calculations of near-edge x.ray absorption spectra of gas-phase and chemisorbed molecules by means of density-functional and transition-potential theory. *Phys. Rev. B* Vol. 58, No.12 pp. 8097-8110.

Vitos, L.; Ruban, A. V. Skiver, H.L. & Kollàr, J. (1998). The Surface energy of metal. *Surf. Sci.* Vol. 441, No. 1-2 (August 1998) pp. 186-202. Available from: http://www.sciencedirect.com/science/article/pii/S003960289800363X

van Hove, M. A.; Koestner, R. J. Stair, P. C. Biberian, J. P. Kesmodel, L. L. Bartos, I. & Somorjai, G. A. (1981). The surface reconstructions of the (100) crystal faces of iridium, platinum and gold: I. Experimental observations and possible structural models. *Surf. Sci.* Vol. 103, No.1. (1 feb 1981), pp. 189-217.

Vlieg, E.; (2000). ROD: a program for surface X-ray crystallography *J. Appl. Cryst.* Vol. 33, pp. 401- 405. ISSN 0021-8898

Witte G. & Wöll Ch. (2004). Growth of aromatic molecules on solid substrates for applications in organic electronics *J. Mater. Res.* Vol. 19, No. 7, pp. 1889- 1916 and references therein.

Witte, G.; Lukas, S. Bagus, P.S. & Wöll, Ch. (2005). Vacuum level alignment at organic/metal junctions:"Cushion" effect and the interface dipole. *Appl. Phys. Lett.*Vol. 87 pp. 263502 (3 pages).

Werner W. *Tutorial on Electron Spectroscopies*. 20.10.2011. Available from: http://eaps4.iap.tuwien.ac.at/~werner/qes_tut.html#main

Woodruff, D. P. & Delchar, T. A. (1988). *Modern Techniques of Surface Science*. Cambridge University Press ISBN 0 521 35719 5.

Zahn, D. R.T.; Gavrila, G. N. & Gorgoi M. (2006). The transport gap of organic semiconductors studied using the combination of direct and inverse photoemission. *Chemical Physics* Vol. 325 pp. 99–112

15

Organic Aqua Regia: Discovery, Fundamentals, and Potential Applications

Wei Lin
School of Materials Science & Engineering, Georgia Institute of Technology
USA

1. Introduction

Recently, the author and his colleagues reported the discovery of a series of organic mixtures that dissolve various noble metals such as gold (Au), silver (Ag), and palladium (Pd) efficiently at room temperature (Lin *et al.*, 2010). The author named these mixtures *organicus liquor regius* as the male counterpart of *aqua regia* (female) in *Latin*. For the readers to catch the term more easily, the author uses the term *organic aqua regia* (OAR) from here after. OAR is composed of thionyl chloride ($SOCl_2$) and an effective organic component. By varying the composition and reaction conditions, selective dissolution of noble metals was achieved. The discovery of OAR and their potential applications in many fields have attracted extensive attention (Yeston, 2010; NatureEditorial, 2010; Urquhart, 2010; Ritter, 2010). The chapter provides a brief description of our work on, and understanding of, OAR from three aspects: discovery, fundamental chemistry, and some of the potential applications that have been preliminarily demonstrated.

Section 2 describes how the author discovered OAR when studying chemical bonding of functionalized carbon nanotubes to an Au surface, and introduces some known properties (dissolution selectivity towards metals, and dissolution kinetics) of OAR, with a touch on some of the recipes of OAR that have been developed so far. A summary of the chemistry, strength, and issues of *aqua regia*—the ever most powerful solution to dissolve noble metals in history, and a tabulated comparison between OAR and *aqua regia* is presented. When one thinks about "dissolution" of sodium chloride in water, the word—dissolve—might sound somewhat misleading for describing the oxidation process of metallic Au by the reagent, the accompanying complexation, and the solvation processes in the solution; however, the author keeps using the terms such as "dissolve", "dissolution", and "solubility" in the chapter following the conventional description of oxidative dissolution of metals in hydrometallurgy. The author believes the readers have no problem distinguishing the dissolution of metals with that of salts/sugar.

A significant portion of the chapter is then devoted in Section 3 to some fundamental chemistry of OAR, with the focus on the Au-$SOCl_2$-pyridine system. $SOCl_2$ and its mixture with pyridine (py) have been widely used in organic chemistry for decades; however, the chemistry of $SOCl_2$-py is still now well understood. Our rigorous study on $SOCl_2$-py has provided some new fundamental understanding of the mixture. Also discussed is the possible mechanism for the Au dissolution in the $SOCl_2$-py mixture. A surface reaction mechanism is proposed. Section 4 discusses some potential applications of OAR such as etching of metals in microelectronics industry, and recovery of noble metals from catalysis industry and consumer products, *etc.* Our recent progress is presented.

2. Discovery of OAR

As many discoveries were, the discovery of OAR was accidental. In 2007, the author was involved in developing a chemical bonding process to anchor in situ functionalized vertically aligned carbon nanotubes to a modified Au surface. The basic chemistry was to assemble a thin layer of 4-mercaptobenzoic acid molecules on the Au surface, and then form the bonding between the functional groups (*e.g.*, hydroxyl groups) on the carbon nanotubes and the acid groups of the 4-mercaptobenzoic acid molecules via esterification (Fig. 1). However, there were some fundamental challenges to such a bonding process. First, solid-solid reaction at the interface was very unlikely to occur given the low functionalization degree of the carbon nanotubes and the irregular surface of the carbon nanotubes (at that time very few people talked about the irregular surface of the macroscopically well-aligned carbon nanotubes). Second, a wet chemical reaction at the interface was preferred, *e.g.*, a catalyzed esterification reaction in an aqueous solution. However, the wet chemical process—if not controlled well—would damage the vertical alignment of the carbon nanotubes (at that time very few people revealed the truth that a vertically aligned carbon nanotubes array/bundle could easily collapse in wettable liquids). Third, direct esterificaiton of benzoic acid with alcohol had been known to be very inefficient (Vulakh *et al.*, 1975; Zuffanti, 1948), and therefore, kinetically unlikely at the interface. To address these issues, the benzoic acid group was first transformed to the benzoic acid chloride group by the reaction with $SOCl_2$ in the presence of a certain concentration of py as the catalyst. A dilute solution of $SOCl_2$ (5~10 ppm) in acetonitrile (CH_3CN) with a trace amount of py (py:$SOCl_2$=1:2 in mole) was used (illustrated in Fig. 2).

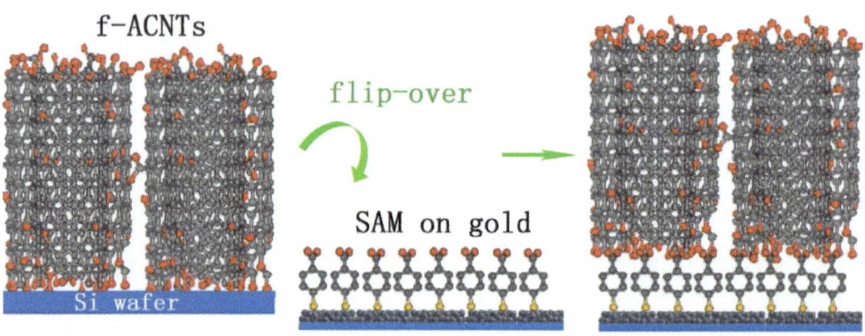

Fig. 1. Illustration of the idea of chemical bonding of in situ functionalized vertically aligned carbon nanotubes (f-ACNTs) to a functionalized Au surface. The red dots on the f-ACNTs represent the functional groups such as epoxide and hydroxyl groups (Lin *et al.*, 2008).

When the author was studying the functionalization (esterification) reaction, something unexpected but interesting happened one day: in the presence of py, a relatively high concentration of $SOCl_2$ resulted in efficient dissolution of Au in the organic solution at room temperature. Soon after the author confirmed that Au was not soluble at all in either $SOCl_2$

or py, or the above-mentioned very dilute $SOCl_2$-py/CH_3CN solution, the author realized the similarity between the organic mixtures and *aqua regia*. *Aqua regia* has been used for centuries as a powerful etchant to dissolve noble metals. The beauty of *aqua regia* is that the simple 1:3 mixture of concentrated nitric and hydrochloric acids can dissolve various noble metals while the noble metals are not soluble in either of the acids. The organic mixtures have exactly the same beauty with *aqua regia*: simple mixtures; therefore, the author names the mixtures *organic aqua regia* (OAR). My curiosity pushed me to try other materials in the $SOCl_2$-py mixture (with and without CH_3CN); qualitative results are listed in Table 1.

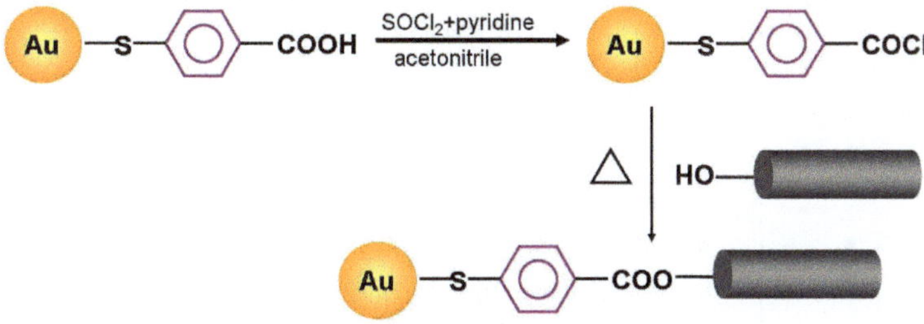

Fig. 2. Esterification process at the interface between the f-ACNTs and the modified Au surface.

material	dissolution (Y/N)
silver	Y
gold	Y
platinum	N
palladium	Y
copper	Y
iron	Y
nickel	Y
titanium	N
tungsten	N
tantalum	N
chromium	N
tin	Y
indium	Y
Teflon	N
silicon	N
silicon oxide	N

Table 1. Qualitative results of dissolution of various materials in the $SOCl_2$-py mixture.

The discovery promises more than the beauty and simplicity. The distinct difference between OAR and *aqua regia* is that OAR are non-aqueous, which is important. Compared with inorganic chemistry, organic chemistry provides a precise control over chemical reactivity. The ability to engineer organic reactions would probably enable selective dissolution of noble metals. A few non-aqueous solutions for dissolving noble metals have been preliminarily investigated (Senanayake, 2008; Raisanen *et al.*, 2007; Mortier *et al.*, 2005;

Cau *et al.*, 2003; Nakao, 1992), however, with relatively unsatisfactory solubility, selectivity, efficiency, stability, or/and simplicity. Soon after the discovery of OAR, the author and his colleagues started to work on the formulation of OAR for selective dissolution of noble metals. Some preliminary results are presented here. Fig. 3 shows the kinetic results of the dissolution of noble metals in the mixture of $SOCl_2$ and py with a volumetric ratio of 3:1. The mixture dissolves Au at a rate of 0.3 mol m^{-2} h^{-1} at room temperature, which is faster than Au dissolution in conventional cyanide leaching agents (<0.004 mol m^{-2} h^{-1}) and iodide etchants (< 0.16 mol m^{-2} h^{-1}) (Senanayake, 2008; Qi & Hiskey, 1991). Ag and Pd can also be dissolved at high dissolution rates (0.8 mol m^{-2} h^{-1} and 0.5 mol m^{-2} h^{-1}, respectively); platinum (Pt) is completely inert. In a comparison, Fig. 4 shows that a $SOCl_2$-DMF (N,N'-dimethylformamide) mixture dissolves Au at a rate of 0.3 mol m^{-2} h^{-1}; neither Pd nor Pt is apparently dissolved.

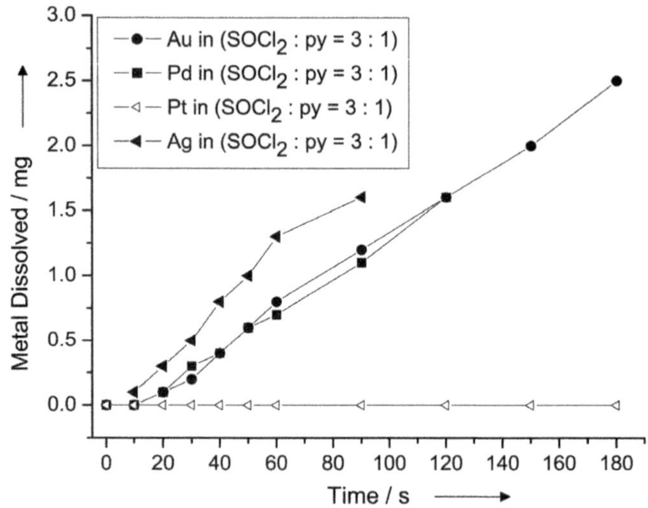

Fig. 3. Kinetic studies of the dissolution of Au, Pd, Ag, and Pt in a 3:1 $SOCl_2$-py mixture.

Such dissolution selectivity may find many important applications such as recovery of noble metals. The global energy crisis demands for green energy technologies, which undoubtedly requires increased noble metal resources. However, on the earth, noble metals are scarce. The ability to recover high-purity noble metals via recovery processes will be paramount to the sustainable development of human being. Among the noble metals, Pt is most widely used as a catalyst in many green technologies, in particular, proton-exchange membrane fuel cell. Pt recovery, however, has long been a challenging issue. Catalyst product stream starts with high-purity Pt, and combines Pt with other noble metal elements for various purposes. The conventional Pt recovery technologies are complicated, mainly relying on the dissolution of Pt in strong inorganic acids (*aqua regia*) and subsequent precipitation of the dissolved Pt from the solution (Awadalla *et al.*, 1992; Edwards & Nattrass, 1978; Lakewood & Arvada, 1977). However, *aqua regia* dissolves all the metals at the same time and cannot separate them out. It ends up with low-purity Pt which doesn't go back to the product stream directly; it has to be refined, but refinery is costly. Moreover, auqa regia is notoriously dangerous to work with; it is not recyclable; it is environmentally hazardous. In comparison, OAR may provide a route to improve the recycling quality and efficiency of Pt by a selective dissolution process. The selective dissolution process removes the impurity noble metals (Ag, Au, and Pd, *etc.*) before the final dissolution/separation of Pt. Schematic illustration of the recovery process using OAR is shown in Fig. 5; the preliminary results will be discussed in Section 4. Besides, OAR is recyclable (simply by distillation), and dilute OAR are relatively safe and easy to handle. A comparison between *aqua regia* and OAR is listed in Table 2.

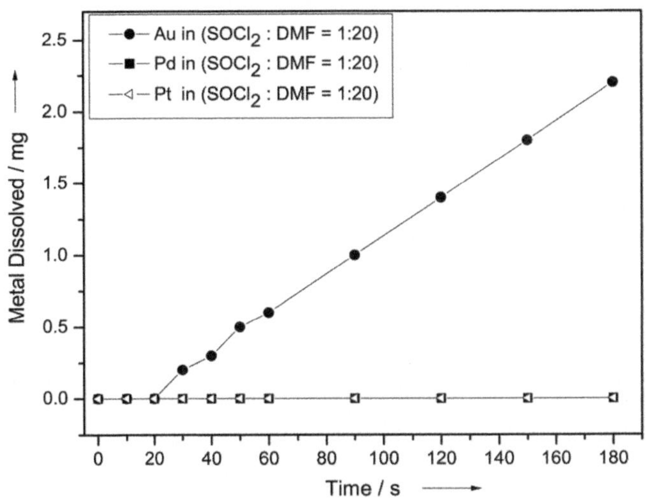

Fig. 4. Kinetic studies of the dissolution of Au, Pd, and Pt, respectively, in a 1:20 $SOCl_2$-DMF mixture.

So far, the chemicals that have been found to work as the effective basic components to compose OAR with $SOCl_2$ for dissolving Au include, but are by no means limited to, the following chemicals and their derivatives: pyrrole, pyrrolidine, pyrrolidone, isoxazole, isothiazole, pyrazole, imidazole, thiazole, oxazole, pyrazolone, bipyrazole, pyridine, pyridazine, pyrimidine, pyrazine, triazine, indole, quinoline, purine, pteridine, phthalocyanine, N,N'-dicyclohexylcarbodiimide, DMF, N,N'-dimethylbenzylamine, dodecyltrimethylammonium bromide, tri-p-tolyl-phosphine, *etc.* Effective dissolution of Au in a mixture of $SOCl_2$ with any of the following chemicals has not been observed: maleimide, azobisisobutyronitrile, aniline, polyaniline, phenanthroline, methylbenzyl cyanide, 2-acetyl-1-methylpyrrole, benzyltriethylammonium tetrafluoroborate.

3. Fundamental chemistry of the $SOCl_2$-py mixture and the Au-$SOCl_2$-py system

One key feature that all the aforementioned organic candidates have in common is that they more or less have charge-transfer interactions with $SOCl_2$ (Schenk & R. Steudel, 1963; Korshak *et al.*, 1971), where the sulfur in $SOCl_2$ is an electron acceptor, and the nitrogen (or phosphor) an electron donor. Let us focus on the Au-$SOCl_2$-py system.

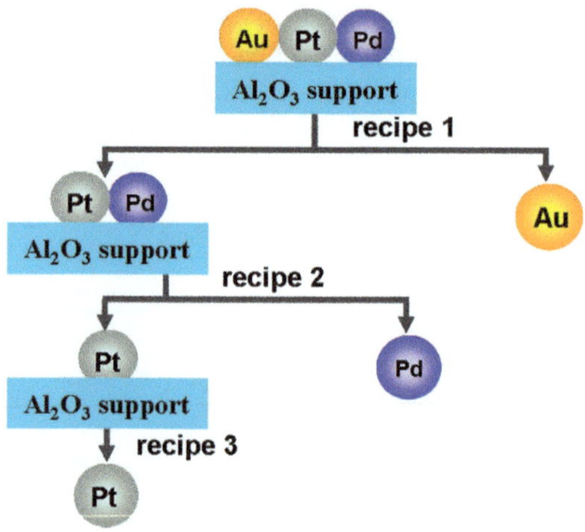

Fig. 5. Schematic illustration of the recovery of high-purity Pt on the basis of the dissolution selectivity of OAR toward noble metals.

	aqua regia	OAR
composition	nitric acid + hydrochloric acid	innumerous
concentration	concentrated	customized
safety	explosive and corrosive	corrosive
recyclable	N	Y
dissolution of noble metals	indiscriminate	selective

Table 2. A comparison between *aqua regia* and OAR

3.1 $SOCl_2$-py mixture

Py (structure a in Fig. 6) has been known for long to be able to catalyze the synthesis of acyl chlorides (RCOCl) by the reaction between organic acids and $SOCl_2$ (Cade & Gerrard, 1953; Gerrard & Thrush, 1953; Higashi *et al.*, 1986; Human & Mills, 1946). So has DMF. From a bromination reaction of py in the presence of $SOCl_2$, Garcia *et al.* proposed a reaction between py and $SOCl_2$ involving the probable equilibrium among 1-(chlorosulfinyl)-pyridinium chloride (structure b in Fig. 6), 1-(chlorosulfinyl)-4-chloro-4-hydropyridine (structure c in Fig. 6), and further reaction products (Garcia *et al.*, 1960). Such proposed reaction equilibrium was thought by Higashi *et al.* to be the mechanism for esterification reactions between carboxylic acids and $SOCl_2$ using py as the catalyst (Higashi *et al.*, 1986). However, no systematic spectroscopic study on the $SOCl_2$-py mixture has been reported so far. It is a fact that there is a strong interaction between py and $SOCl_2$, which can be seen by the "white fog" formed when either of them is added to the other at room temperature. In fact, immediate yellowing of the solution is observed when py and $SOCl_2$ are mixed fast at room temperature. Such a strong interaction is due to the formation of a charge-transfer complex between py and $SOCl_2$ (partial transfer of the electron density on the nitrogen to the sulphur). The charge transfer (structure shown in Fig. 7), in general, weakens the bonds within the acceptor molecule (R. Steudel & Y.

Steudel, 2007), which accounts for the observed redshifts of both the asymmetric (from 446 to 427~426 cm^{-1}) and the symmetric (from 497 to 494~474 cm^{-1} depending on the ratio of $SOCl_2$ to py) Cl-S-Cl stretching in the Raman spectra of the $SOCl_2$-py mixtures (Fig. 8). $SOCl_2$-py is observed directly in the mass spectrum of the $SOCl_2$-py mixture (Lin *et al.*, 2010). Nuclear magnetic resonance (NMR) spectra of the $SOCl_2$-py mixture are shown in Fig. 9. Table 3 summarizes the chemical shifts. We see the deshielding (low-field chemical shifts) of ^{14}N (ca, 5 ppm), ^{13}C (ca, 1.35 and 2.42 ppm for C-3 and C-4, respectively), and ^{1}H (ca, 0.35, 0.49 and 0.50 ppm for H-2, H-3 and H-4, respectively), and shielding (high-field chemical shift) of C-2 (ca, -1.77 ppm) relative to pure py. These chemical shifts are attributed to the charge-transfer interaction (Ricca & Severini, 1988). We also observe shifts and intensity changes of py-related vibration peaks in the FTIR of the $SOCl_2$-py mixture relative to pure py and $SOCl_2$ (Fig. 10). Such shifts are in agreement with our simulation results (Lin *et al.*, 2010).

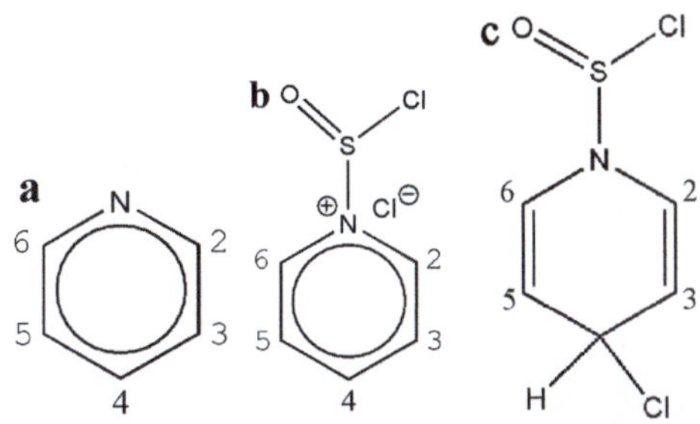

Fig. 6. Structures of py (a), 1-(chlorosulfinyl)-pyridinium chloride (b), and 1-(chlorosulfinyl)-4-chloro-4-hydropyridine (c). The numbering of atoms is applied to the NMR analysis.

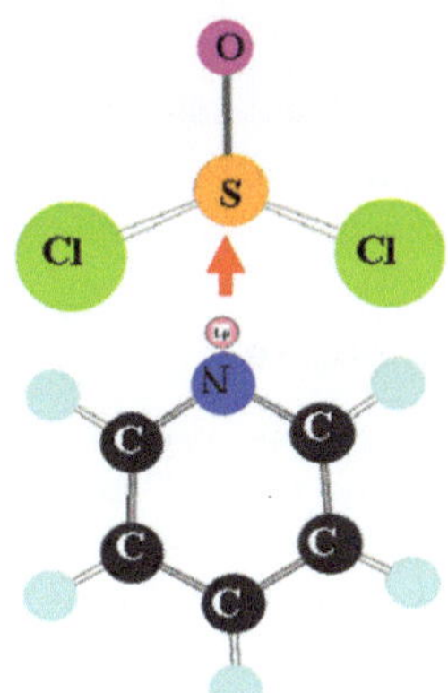

Fig. 7. The molecular illustration of the $SOCl_2$-py charge-transfer complex.

In the Raman and the FTIR spectra of the $SOCl_2$-py mixtures, almost all the vibration modes of pure py and pure $SOCl_2$ remain unchanged. This indicates that the parent molecules, although perturbed by the charge-transfer interaction, maintain their basic structural integrity, and that the dominating product is a molecular adduct rather than a rearrangement, elimination or dissociated product. Neither 1-(chlorosulfinyl)-pyridinium chloride nor 1-(chlorosulfinyl)-4-chloro-4-hydropyridine is identified in the NMR spectra or

FTIR spectra (Cade, & Gerrard, 1953; Higashi *et al.*, 1986; Human & Mills, 1946; Garcia *et al.*, 1960). However, the experimental data of electrical conductivity measurement of the mixtures suggest the presence of mobile ions due to the dissociation. Therefore, 1-(chlorosulfinyl)-pyridinium chloride, though being a weak electrolyte in $SOCl_2$ (Fig. 11), exists as the dissociated form of the $SOCl_2 \cdot$ py adduct in the solutions. Geometry optimization results show that: $SOCl_2$-py adduct (non-dissociated) is energetically more favorable than 1-(chlorosulfinyl)-pyridinium chloride, the dissociated form of the adduct, the latter being 16.8 kcal mol^{-1} higher in energy than the former (the energy difference will be lowered when put into a condense phase, depending on the dielectric constant of the solvent). This helps explain why the adduct is a weak electrolyte in $SOCl_2$.

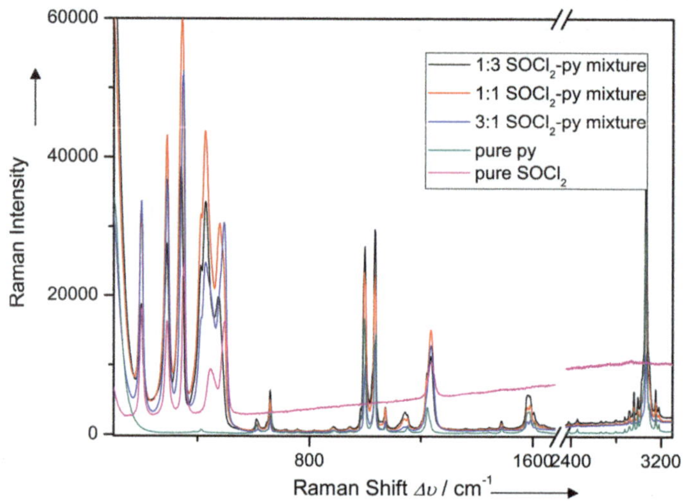

Fig. 8. 532-nm-laser-wavelength Raman spectra of py, $SOCl_2$, and their mixtures with varied ratios.

3.2 Au dissolution in the $SOCl_2$-py mixture

It has been confirmed that the valence state of gold after the dissolution is Au(III), in the form of $[AuCl_4]^-$ (Lin *et al.*, 2010). The Raman spectrum in Fig. 12 shows the $[AuCl_4]^-$ structure. From the view of thermodynamics, dissolution of noble metals usually requires a strong oxidant, and simultaneously ligands that coordinate with the noble metal ion to reduce the redox potential of the metal in the solution. Obviously, Cl^- is the ligand in this case. Although even chemists may not be aware of the strong oxidizing ability of $SOCl_2$, it should be noted that Au(I)→Au(III) oxidation by $SOCl_2$ has been demonstrated under certain conditions (Schmidbaur & Jandik, 1983; Bovio *et al.*, 1993). $SOCl_2$ oxidizes Cu and Ag readily at room temperature. Thus, $SOCl_2$ is a strong oxidant, and it can be the oxidant in our case. However, Au cannot be oxidized by $SOCl_2$ alone. No weight loss was detected of an Au film after it was immersed in $SOCl_2$ at room temperature or even at 70 °C (refluxing) for 1 week. Purging oxygen into the $SOCl_2$ bath doesn't oxidize Au, either. Py, undoubtedly, plays a very important role in the oxidative dissolution process. We believe that the charge-transfer interaction activates $SOCl_2$ to oxidize Au.

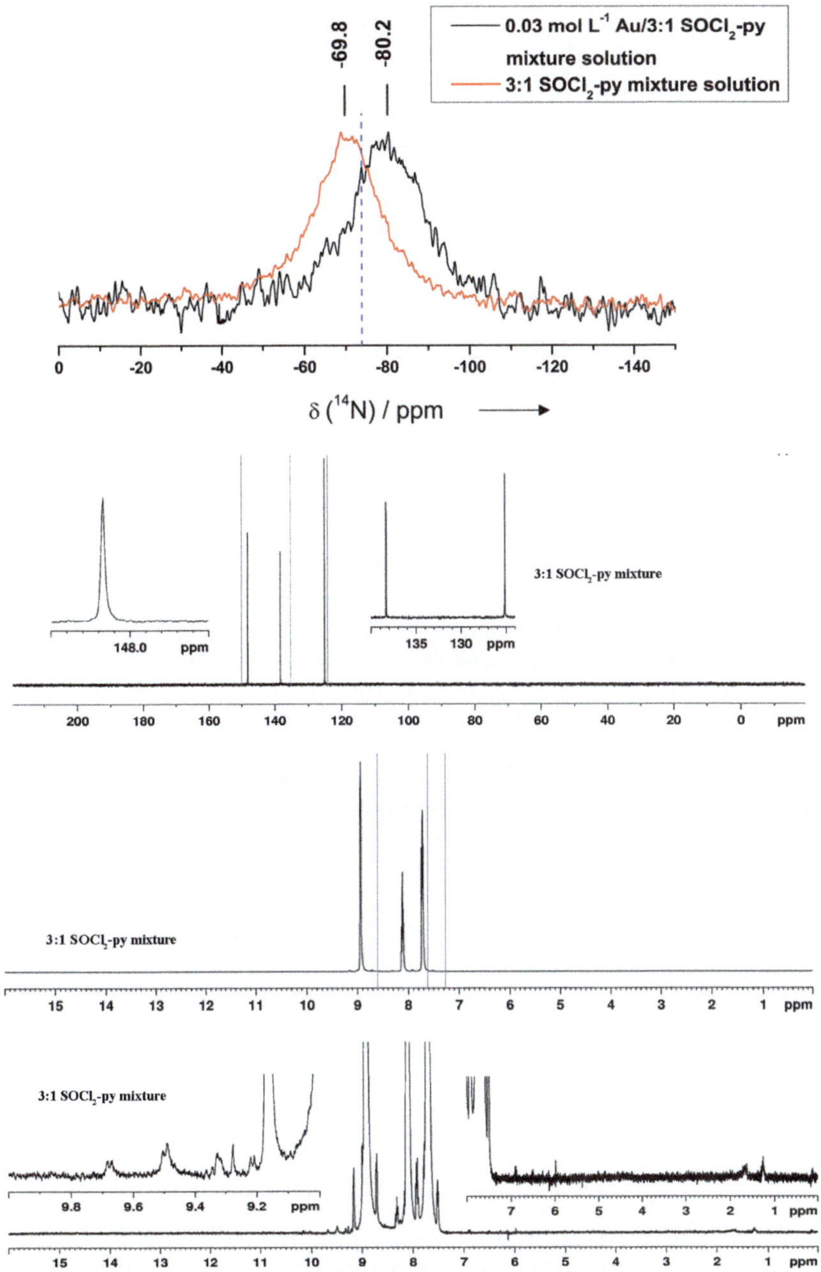

Fig. 9. ^{14}N, ^{13}C, and ^{1}H NMR spectra of a 3:1 $SOCl_2$-py mixture and its Au solution. The dashed lines represent the chemical shifts of δ=74 (^{14}N), 8.59 (H-2 and H-6), 7.23 (H-3 and H-5), 7.62 (H-4), 149.94 (C-2 and C-6), 123.75 (C-3 and C-5) and 135.89 (C-4) ppm for pure py. The insets are the magnified regions of the spectra.

δ (^1H) of 3:1 $SOCl_2$-py mixture	δ (^{13}C) of 3:1 $SOCl_2$-py mixture
9.68 (d, J=6.0)	148.17 (2.9)
9.50 (d, J=5.8)	138.32 (2.3)
9.28 (s)	125.15 (2.0)
9.21 (d, J=4.6)	
9.17 (d, J=4.1)	
8.94 (d, J=4.5)	
8.75 (d, J=4.2)	
8.12 (t, J=7.8)	
7.72 (t, J=6.6)	
7.52 (t, J=6.1)	
1.56 (s)	
1.18 (s)	

Table 3. Summary of solution NMR data. Left column: "s", "d" and "t" in the brackets represent "singlet", "doublet" and "triplet". The values in bold are the major chemical shifts. Right column: the values in the brackets are the half-height line-width ($\nu_{1/2}$).

One drawback of thermodynamics is that it doesn't tell the reaction kinetics (in most cases, thermodynamics doesn't relate to kinetics). For example, the coordinative oxidation mechanism may not be able to explain why the dissolution rate of Au is strongly dependent on the recipe of the OAR, especially the organic component such as py, DMF, pyrrolidine, pyrrolidone, isoxazole, isothiazole, pyrazole, imidazole, *etc.* The reaction rate in different OAR recipes can be orders of magnitude different! Moreover, Pd and Pt are less noble than Au, but are not oxidatively dissolved in many recipes of OAR. The fundamental view of the dissolution reaction, therefore, points toward an interfacial reaction mechanism as shown in Fig. 13. The proposed mechanism involves two important roles by the organic component, say py, in two major steps. First, py coordinates to the Au surface by putting the non-bonding electrons in py (according to molecular orbital theory) to the accessible and symmetry allowed atomic orbital of the metal element. As such, the work function of the noble metal is effectively reduced (this has been known for quite a while), and therefore, the surface noble metal atoms are more liable to oxidation by the oxidant, *i.e.*, activated $SOCl_2$. Second, py forms the charge-transfer complex with $SOCl_2$ to activate the oxidant; upon capture of the electrons from Au, Cl-S-Cl dissociates to generate the Cl^- as the ligand. The author proposes that the surface coordination between Au and py, and later on the desorption of py after the electron transfer from Au to sulfur, determine the reaction rate. As such, the electron structure and physical properties of the organic component determine its interactions with $SOCl_2$ and the gold surface, and consequently, determine the dissolution kinetics. The two sets of experiment below may shed some light on the interfacial reaction mechanism.

In one set of experiment, Au is put in $SOCl_2$; then a small amount of water is added to generate free Cl^- (caution: $SOCl_2$ reacts with water rigorously). The presence of Cl^- can be very easily picked up by electrical conductivity measurement. Then anhydrous py is added as usual. It is found that the dissolution rate of Au is reduced. When DMF is used rather than py, the dissolution sometimes doesn't occur. For comparison, in the other set of experiment, Au is put in the anhydrous py before adding the $SOCl_2$ (a small amount of water is added to the $SOCl_2$

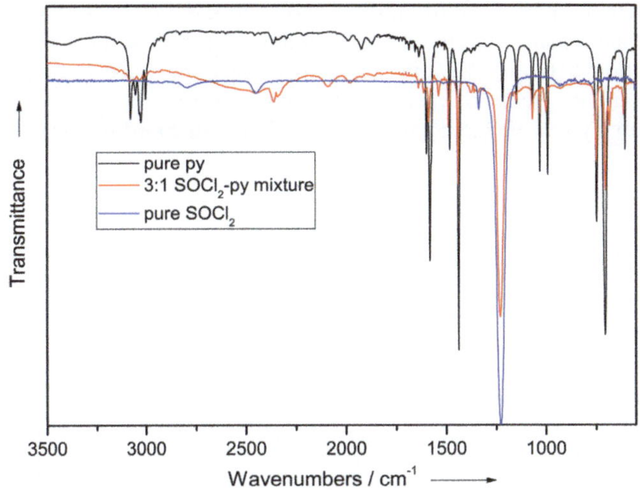

Fig. 10. FTIR spectra of py, $SOCl_2$, and their mixture.

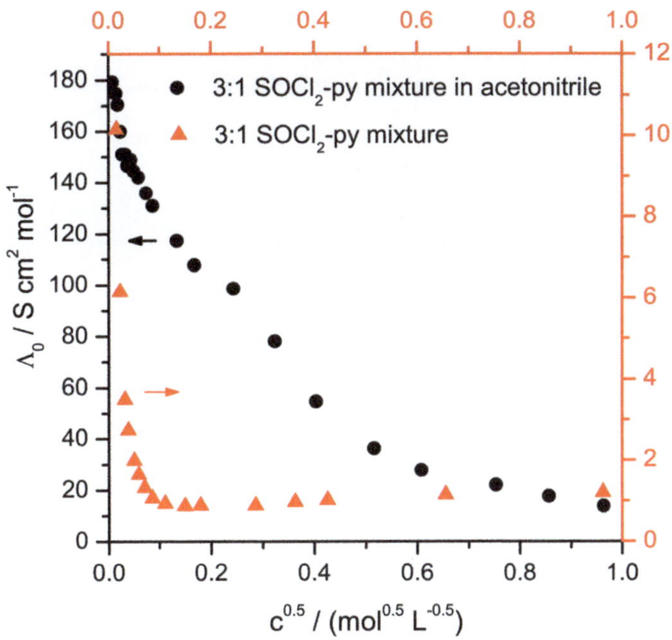

Fig. 11. Conductivity measurement results (III): Λ-$c^{0.5}$ plot, where c is the concentration of $SOCl_2 \cdot$ py adduct in the solution, and Λ the molar conductivity of the electrolyte. It is clear that $SOCl_2 \cdot$ py is a weak electrolyte in $SOCl_2$ matrix; its dissociation degree increases in acetonitirle. This is in line with our prediction because $SOCl_2$ has a lower dielectric constant (9.3) and higher viscosity (0.6 cP) than CH_3CN (36.2 and 0.3442 cP, respectively) (Geary, 1971; Harkness & Daggett Jr., 1965).

first). No distinct change in the dissolution rate of Au is found. When DMF is used rather than py, the dissolution rate is somewhat reduced. From the experiments, the author postulates that when the coordination between Au and py is interrupted by Cl^- (Cl^- adsorption on metal surfaces is notorious in electrochemistry), the reaction rate is greatly affected. Py-Au coordination is relatively less affected by the Cl^- than DMF-Au coordination.

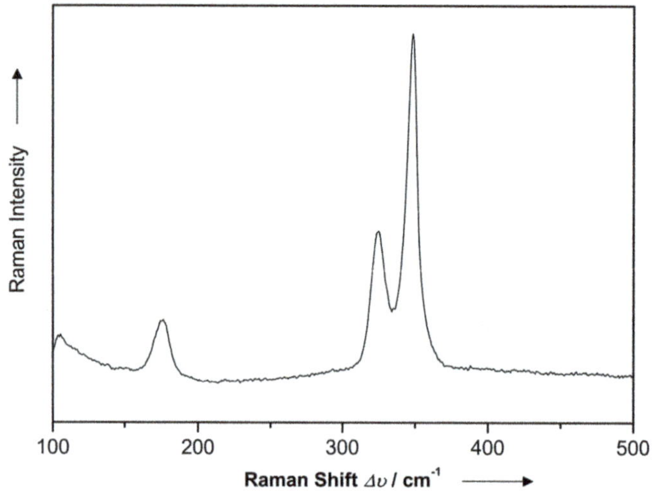

Fig. 12. Raman spectrum of a vapor-etched Au/Si surface. Vibrations attributed to $[AuCl_4]^-$ structure are observed at 170, 324, and 348 cm^{-1}.

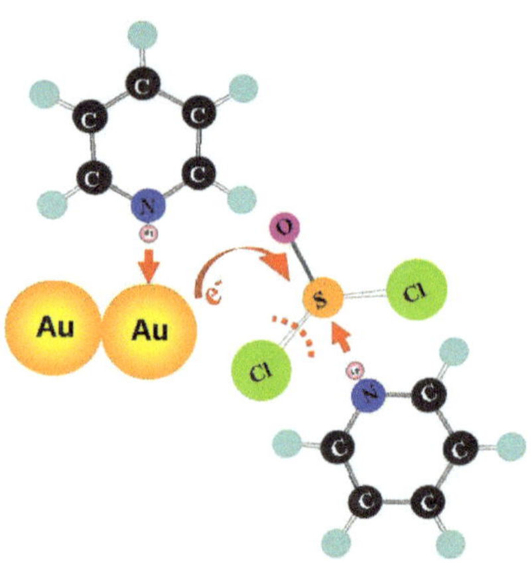

Fig. 13. A proposed oxidative dissolution mechanism for Au in OAR.

A more direct proof comes from the fact that the Au dissolution rate is distinctly reduced in the mixture of $SOCl_2$ and fully deuterated py (py-d_5). The charge transfer interaction between $SOCl_2$ and py-d_5 is evidenced in the Raman spectrum (Fig. 14); deuteration doesn't alter the chemical properties. However, due to the larger mass of py-d_5 than py, the surface adsorption and desorption of py-d_5 are thought to be slower than its undeuterated counterpart.

It is noted that no known reduction products of $SOCl_2$ such as sulfur have been confirmed so far. Therefore, the following possible reactions have been proposed:

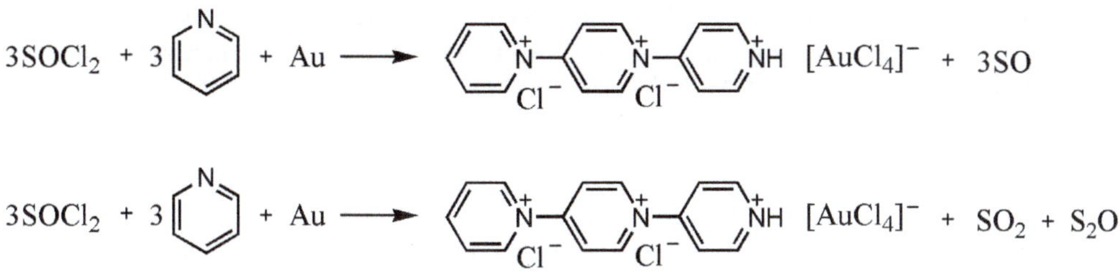

As for Pt and Pd in OAR, the author believes their solubilities are related to surface passivation. It is well known in the research field of "corrosion science" that pure Pt and Pd are thermodynamically less noble than Au but, in reality, more noble than Au. The reason is that Pt and Pd surfaces passivate against oxidants over a wide range of pH values in aqueous solutions, while Au surface doesn't. This also explains why none of titanium, tantalum, chromium, and tungsten can be dissolved in OAR. The passivation prevents the ligand and the oxidant from accessing the metal atoms. To get clearer about the mechanism for the dissolution selectivity, electrochemical characterizations and simulations are needed.

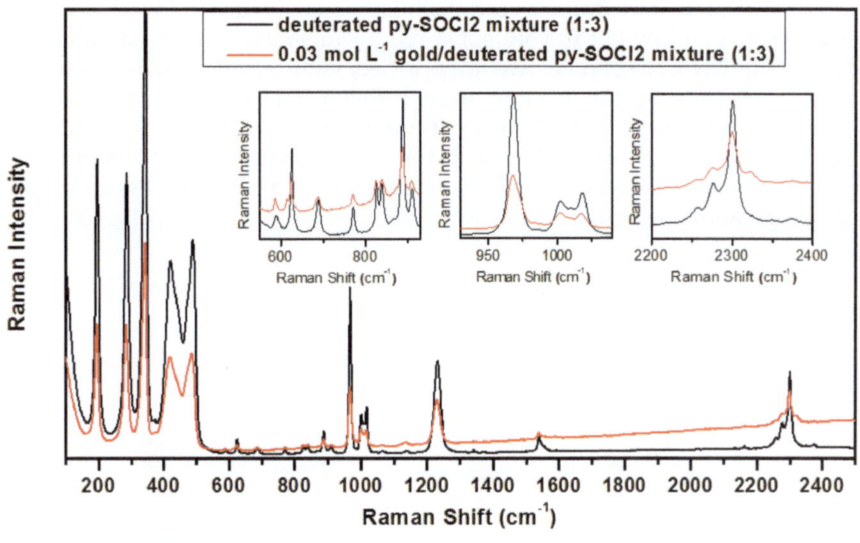

Fig. 14. Raman spectra of deuterated OAR and Au/OAR solution. The Raman shifts of the asymmetric and the symmetric Cl-S-Cl stretching are at 422 and 490 cm^{-1}, respectively, indicating the charge-transfer interaction.

4. Potential applications

There are many potential applications of OAR in metallurgy, metal etching for integrated circuit fabrication in electronics, recovery of noble metals from catalysis industry, and synthesis of noble metal nano particles, *etc.* Two preliminary results are discussed here.

4.1 Recovery of Pt from bimetallic catalyst nanoparticles

Bimetallic nanoparticles (NPs) of noble metal elements represent a new class of functional materials, and have been investigated intensively in the past fifteen years. Among the various combinations of bimetallic NPs, Au-Pt core-shell NPs (sometimes called Pt-decorated Au NPs, Pt-layered Au NPs, or Au-Pt dendritic hetero-aggregate NPs to account for the morphology of the the non-uniform and porous Pt shell) is an important category that has demonstrated improved catalytic activities (Zhou *et al.*, 2006; Zhang *et al.*, 2009; Lu *et al.*, 2004; Wang *et al.*, 2009). Recently, the author demonstrated an efficient leaching-extraction process to recover Pt directly from Pt-Au bimetallic nanoparticles using OAR (Lin, 2011). The purities of the Pt recovered from a mixture of Au and Pt nanoparticles, and from Pt-Au core-shell nanoparticle catalyst were as high as 99.49±0.22 %, and 95.02±0.08 %, respectively. The novel recovery process promises applications in catalysis industry. Here we show the example of leaching-extraction of Pt from Pt-Au core-shell nanoparticle catalyst. Au core was formed by reducing 24 mg $HAuCl_4 \cdot 3H_2O$ in 10 mL oleylamine at 120 °C for 3 h under an Ar blanket. 25 mg platinum acetylacetonate was dissolved in 10 ml oleylamine at 70 °C, and then added into the dark purple Au colloidal solution at 120 °C under vigorous stirring. The solution was fast heated up to 235 °C, and kept for 3 h. The whole synthesis process was under Ar protection. The core-shell NPs were collected by precipitation with ethanol and centrifugation, and washed with mixtures of acetone and hexane. 10 mg of the dark brown core-shell NPs were added into organic *aqua regia* composed of $SOCl_2$ and py (3:1 in volume) at room temperature, and then diluted into 400 mL CH_3CN. The CH_3CN solution was then mixed with 50 mL hexane under vigorous stirring. After the mixing, the mixture went through a fast phase separation at room temperature (Fig. 15). The dissolved Au went into the bottom CH_3CN phase (yellow), and can be recovered simply by calcination (Fig. 16). Pt NPs were extracted out of the CH_3CN solution, and re-dispersed in the upper hexane phase (brown). The Pt NPs can be recovered simply by evaporating the hexane.

Fig. 15. A photograph showing the extraction result of Pt NPs from the dissolved Au after the leaching process: the bottom yellow phase is the CH_3CN solution with dissolved Au; the upper brown phase is the extracted Pt NPs dispersed in hexane. The two layers can be easily separated (inset).

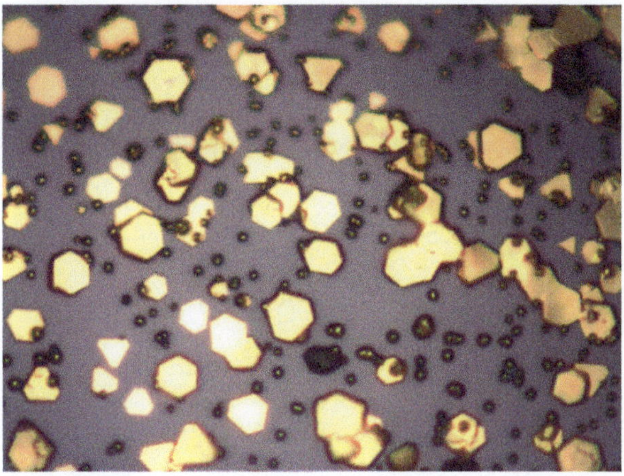

Fig. 16. An optical microscope image showing the recovered Au particles by calcinations.

4.2 Fast etching of copper for applications in electronic industry

Besides recovery of noble metals by OAR from waste electronic products/processing, etching of metals by OAR may find an important application in subtractive patterning of copper (Cu) connection/conduction paths in electronic packaging. Various Cu etching processes have been developed for the fabrication of printed circuit board, and Cu/low-K interconnects associated with back-end-of-line (Bryce & Berk, 1995; Cakir, 2006; Cakir *et al.*, 2005; Halpern, 1953). However, the etching rates are still not satisfactory (Table 4); etchants of higher etching rates are desired. Recently, the author reported fast etching of Cu in $SOCl_2/CH_3CN$ solutions (Lin & Wong, 2011). The etching rate can be tuned over a wide range by varying the concentration of the etchant, and the stirring rate of the liquid. For example, the etching rate reaches 36 mg min^{-1} cm^{-2} in 1 mol L^{-1} $SOCl_2/CH_3CN$ under stirring

Reference	Etchant	Concentration (mol L^{-1})	Temperature (^{o}C)	Etching rate (mg min^{-1} cm^{-2})
Cakir *et al.*, 2005	$FeCl_3$	3.76	50	3.5
Cakir *et al.*, 2005	$CuCl_2$	2.33	50	1.7
Halpern, 1953	$FeCl_3$	2	50	27
Halpern, 1953	$FeCl_3$	2	30	17
Bryce & Berk, 1995	$O_2/NH_3\cdot H_2O$	0.9 atm/0.74	26	0.03
Cakire, 2006	Alkaline	N/A	30-60	13-27
Cakire, 2006	$CuCl_2$	>2	50-54	11-22
Cakire, 2006	$FeCl_3$	>2	43-49	11-22
Lin & Wong, 2011	$SOCl_2/CH_3CN$	1	20	36
Lin & Wong, 2011	$SOCl_2/CH_3CN$	2	20	53
Lin & Wong, 2011	$SOCl_2/CH_3CN$	1	20	320
Lin & Wong, 2011	$FeCl_3$	2	50	19

Table 4. A brief comparison of etchants for Cu.

at room temperature, which is much faster than any currently used etchant for Cu. With the assistance of sonication, the etching rate reaches 320 mg min^{-1} cm^{-2}. The reactions involved are shown below. Based on the selectivity of OAR, Au and Pt are used as the effective masks on Cu and Au, respectively, for the subtractive patterning of Cu and Au conduction paths. Preliminary studies showed acceptable and controllable undercuts of the etched conduction paths by the OAR.

$$6Cu + 2SOCl_2 \rightarrow Cu_2S + 4CuCl + SO_2 \quad (1)$$

$$4CuCl + 2SOCl_2 \rightarrow 4CuCl_2 + S + SO_2 \quad (2)$$

$$Cu_2S + 2SOCl_2 \rightarrow 2CuCl_2 + 2S + SO_2 \quad (3)$$

Overall reaction:

$$2Cu + 2SOCl_2 \rightarrow 2CuCl_2 + S + SO_2 \quad (4)$$

5. Acknowledgment

The author acknowledges Prof. C. P. Wong for financial support of part of the research work.

6. References

Awadalla, F. T.; Molnar, R. E.; & Riteey, G. M. (1992). Recovery of Platinum Group Metals (PGM) from Acidic Solutions by Reduction Precipitation with Sodium Borohydride. US Patent, 5,304,233

Bovio, B.; Burini, A. & Pietroni, B. R. (1993). Reactions of Trimeric 1-benzyl-2-gold(I) imidazole Leading to Au(I) Carbene Complexes – Crystal-Structure of [1-benzyl-3-benzoyl-imidazolin-2-yliden] chloro Gold(I). *Journal of Organometallic Chemistry*, Vol.452, pp. 287-291

Bryce, C. & Berk, D. (1995). Kinetics of the Dissolution of Copper in Iron(III) Chloride Solutions. *Industrial & Engineering Chemistry Research*, Vol.34, pp. 1412-1418

Cade, J. A. & Gerrard, W. (1953). Action of Thionyl Chloride on Carboxylic Acids in Presence of Pyridine. *Nature*, Vol.172, pp. 29

Cakir, O.; Temel, H. & Kiyak, M. (2005). Chemical Etching of Cu-ETP Copper. *Journal of Materials Processing Technology*, Vol.162, pp. 275-279

Cakir, O. (2006). Copper Etching with Cupric Chloride and Regeneration of Waste Etchant. *Journal of Materials Processing Technology*, Vol.175, pp. 63-68

Cau, L.; Deplano, P.; Marchio, L.; Mercuri, M. L.; Pilia, L.; Serpe, A. & Trogu, E. F. (2003). New Powerful Reagents Based on Dihalogen/N,N '-dimethylperhydrodiazepine-2,3-dithione Adducts for Gold Dissolution: the IBr Case. *Dalton Transactions*, Vol.10, pp. 969-1974

Edwards, R. I. & Nattrass, M. J. (1978). Separation and Purification of Platinum and Palladium. US Patent, 4,105,742

Garcia, E. E.; Greco, C. V. & Hunsberger, I. M. (1960). Facile Bromination of Pyridine-Type Heterocycles at the Beta-position. *Journal of the American Chemical Society*, Vol.82, pp. 4430-4431

Geary, W. J. (1971). The Use of Conductivity Measurements in Organic Solvents for the Characterization of Coordination Compounds. *Coordination Chemistry Reviews* Vol.7, pp. 81-122

Gerrard, W. & Thrush, A. M. (1953). Reactions in Carboxylic Acid Thionyl Chloride Systems. *Journal of the Chemical Society*, Vol., pp. 2117-2120

Halpern, J. (1953). Kinetics of the Dissolution of Copper in Aqueous Ammonia. *Journal of the Electrochemical Society*, Vol.100, pp. 421-428

Harkness, A. C. & Daggett Jr., H. M. (1965). The Electrical Conductivities of Some Tetra-n-alkylammonium Salts in Acetonitrile. *Canadian Journal of Chemistry*, Vol.43, pp. 1215-1221

Higashi, F.; Mashimo, T. & Takahashi, I. (1986). Preparation of Aromatic Polyesters by Direct Polycondensation with Thionyl Chloride Pyridine. *Journal of Polymer Science Part a-Polymer Chemistry*, Vol.24, pp. 97-102

Human, J. P. E. & Mills, J. A. (1946). Action of Thionyl Chloride on Carboxylic Acids in Presence of Pyridine. *Nature*, Vol.158, pp. 877

Korshak, V. V.; Vinograd, Sv. & Pankrato, Va. (1971). New Ways to Synthesize Polybenzimidazoles and Polybenzoxazoles. *Vysokomolekulyarnye Soedineniya Section B*, Vol.13, pp. 550-551

Lakewood, J. B. & Arvada, C. (1977). Separation and Selective Recovery of Platinum and Palladium by Solvent Extraction. US Patent, 4,041,126

Lin, W.; Xiu, Y. H.; Jiang, H. J.; Zhang, R. W.; Hildreth, O.; Moon, K. S. & Wong, C. P. (2008). Self-Assembled Monolayer-Assisted Chemical Transfer of In Situ Functionalized Carbon Nanotubes. *Journal of the American Chemical Society*, Vol.130, pp. 9636-9637

Lin, W.; Zhang, R. W.; Jang, S. S.; Wong, C. P. & Hong , J. I. (2010). "Organic Aqua Regia"-Powerful Liquids for Dissolving Noble Metals. *Angewandte Chemie-International Edition*, Vol.49, pp. 7929-7932

Lin, W. (2011). Recovery of High-purity Pt from Pt-Au Bimetallic Nanoparticles Using Organic Aqua Regia. *Rare Metals*, accepted

Lin, W. & Wong, C. P. (2011). Fast Etching of Copper in Thionyl Chloride/Acetonitrile Solutions. *Corrosion Science*, Vol. 53, pp. 3055-3057

Lu L. H.; Sun, G. Y.; Zhsng, H. J.; Wang, H. S.; Xi, S. Q.; Hu, J. Q.; Tian, Z. Q. & Chen, R. (2004). Fabrication of Core-Shell Au-Pt Nanoparticle Film and Its Potential Application as Catalysis and SERS Substrate. *Journal of Materials Chemistry*, Vol.14, pp. 1005-1009

Mortier, T.; Persoons, A. & Verbiest, T. (2005). Oxidation of Solid Gold in Chloroform Solutions of Cetyltrimethylammonium Bromide. *Inorganic Chemistry Communications*, Vol.8, pp. 1075-1077

Nakao, Y. (1992). Dissolution of Noble Metals in Halogen Halide Polar Organic Solvent Systems. *Journal of the Chemical Society-Chemical Communications*, Vol.5, pp. 426-427

NatureEditorial (2010). Dissolving Precious Metals. *Nature*, Vol.467, pp. 503

Qi, P. H. & Hiskey, J. B. (1991). Dissolution Kinetics of Gold in Iodide Solutions. *Hydrometallurgy*, Vol.27, pp. 47-62

Raisanen, M. T.; Kemell, M.; Leskela, M. & Repo, T. (2007). Oxidation of Elemental Aold in Alcohol Solutions. *Inorganic Chemistry*, Vol.46, pp. 3251-3256

Ricca, G. & Severini, F. (1988). C-13 and H-1-NMR Study of the Pyridine-initiated Oligomerization of Maleic-anhydride and of the Polymer Structure. *Polymer*, Vol.29, pp. 880-882

Ritter S. K. A New Twist on Aqua Regia, In: *C&EN*, March 31, 2011, Available from: http://pubs.acs.org/cen/news/89/i14/8914notw4.html

Schenk, P. W. & Steudel, R. (1963). Addukte Des Dischwefelmonoxyds. *Angewandte Chemie*, Vol.75, pp. 793

Schmidbaur, H.; & Jandik, P. (1983). Au(I)-Au(II) AND Au(I)-Au(III) Oxidation in Ylide Complexes with Chlorinating Agents. *Inorganica Chimica Acta-Articles*, Vol.74, pp. 97-99

Senanayake, G. (2008). A Review of Effects of Silver, Lead, Sulfide and Carbonaceous Matter on Gold Cyanidation and Mechanistic Interpretation. *Hydrometallurgy*, Vol.90, pp. 46-73

Steudel, R. & Steudel, Y. (2007). Charge-Transfer Complexes between the Sulfur Molecules SO2, S2O, S-3, SONH, and SOCl2 and the Amine Donors NH3 and NMe3 - A Theoretical Study. *European Journal of Inorganic Chemistry*, Vol.27, pp. 4385-4392

Urquhart, J. Challenging Aqua Regia's Throne, In: *Chemistry World*, September 23, 2010, Available from: <http://www.rsc.org/chemistryworld/News/2010/September/23091001.asp>

Vulakh, E. L.; Freidlin, E. G. & Gitis, S. S. (1975). Reaction of Substituted Benzoic Acids with Thionyl Chloride. *Zhurnal Organicheskoi Khimii*, Vol.11, pp. 1481-1486

Wang, S. Y.; Kristian, N.; Jiang, S. P. & Wang, X. (2009). Controlled Synthesis of Dendritic Au@Pt Core-Shell Nanomaterials for Use as An Effective Fuel Cell Electrocatalyst. *Nanotechnology*, Vol.20.

Yeston, J. (2010). Golden Selection. *Science*, Vol.330, pp. 153

Zhang, W.; Li, L.; Du, Y. K.; Wang, X. M. & Yang, P. (2009). Gold/Platinum Bimetallic Core/Shell Nanoparticles Stabilized by a Fr,chet-Type Dendrimer: Preparation and Catalytic Hydrogenations of Phenylaldehydes and Nitrobenzenes. *Catalysis Letters*, Vol.127, pp. 429-436

Zhou, S. G.; McIlwrath, K.; Jackson, G. & Eichhorn, B. (2006). Enhanced CO Tolerance for Hydrogen Activation in Au-Pt Dendritic Heteroaggregate Nanostructures. *Journal of the American Chemical Society*, Vol.128, pp. 1780-1781

Zuffanti, S. (1948). Preparation of acyl chlorides with thionyl chloride. *Journal of Chemical Education*, Vol.25, pp. 481

16

Alloying Effect in Low Loaded Rh Catalysts Supported on High Surface Area Alumina on Their Activity in CH_4 and NO Decomposition

Mieczysława Najbar** et al*
Faculty of Chemistry
Jagiellonian University
Poland

1. Introduction

1.1 The activity of Rh sites in Al-Rh alloys nanocrystallites in the reaction based on the electron transfer to the adsorbed molecules

In the alloys of transition metals (TM) and main group ones (MGM) the valence electrons from MGM can fill the d band of the TMs entirely and slightly increase the number of electrons in their s band (Azaroff, 1960; Kittel, 2005). Such Rh atoms enriched in electrons, present in Al-Rh alloys, were found to be very active electron donors (Pietraszek et al., 2007). It was revealed that they transfer the electrons to the antibonding Л orbital of adsorbed nitric oxide molecules causing NO decomposition to dinitrogen and dioxygen (Pietraszek et al., 2007).

Due to the great difference in the Al and Rh electro-negativity, the alloys are sometimes considered as nonstoichiometric chemical compounds. Because of the large range of nonstoichiometry those compounds should rather be estimated as intermediate phases (IP) (Azaroff, 1960).

The presence of the Al-Rh alloy nanocrystallites isostructural with Al_9Rh_2 (Bostrom et al., 2005) was earlier revealed in the freshly prepared Rh/δAl_2O_3 catalysts containing 0.06 and 1.5 wt.% Rh (Pietraszek et al., 2007; Zimowska et al., 2006).

It is well known that in bimetallic alloy crystallites (Gasser, 1985) the metal with the lower melting point segregates in their surface layers. The melting point of aluminum is much lower than that of rhodium (Lide, 2004-2005). The enrichment in aluminum of the surface of

* Patrick Da Costa[1], Jarosław Dutkiewicz[2], Valerio Choque[3], Narcis Homs[3], Paweł Kornelak[2], Agnieszka Pietraszek[2], Pilar Ramirez de la Piscina[3] and Janusz Sobczak[4]
[1]*The Jean le Rond d'Alembert Institute, University Pierre et Marie Curie, France*
[2]*Faculty of Chemistry, Jagiellonian University, Poland*
[3]*Department of Inorganic Chemistry, University of Barcelona, Spain*
[4]*Institute of Physical Chemistry PAS, Poland*
**Corresponding Author

the nanocrystallites of the Al-Rh intermediate phase at temperature 873K, close to Al melting point, should be very distinct.

The electron transfer from Rh atoms enriched in electrons to the antibonding orbitals of such molecules as O_2 and CH_4, leading to dissociation, can also be expected. The comparison of the enthalpy of the formation of NO, O_2 and CH_4 (Lide, 2004-2005) allows to expect the following sequence of the initial temperature of the dissociation: $T_{CH4} > T_{O2} > T_{NO}$.

Methane dissociation is the main step in one of the most intensively investigated process of the partial methane oxidation (PMO). Thus, it was interesting to compare the activity of the Rh active sites in the Al-Rh nanocrystallites which formed on the surface of the low-loaded Rh/Al_2O_3 catalysts with that of the Rh active sites in the rhodium clusters present on the surface of the high-loaded catalysts.

The study of the influence of the concentration of the atomic oxygen surface species on the Al-Rh alloy nanocrystallites on the composition of the PMO products is necessary to understand the role of the CH_4: O_2 ratio in the PMO process.

The investigations of the Al segregation in the Al-Rh alloy nanocrystallites allow one to understand the evolution of the surface structure during the catalyst use at temperatures higher than the synthesis temperature.

The simultaneous studies of the NO direct decomposition and CH_4 oxidation may confirm the expected sequence of the temperatures of the initial dissociation of CH_4, O_2 and NO. The knowledge of this sequence would help to estimate the temperature range of the effective direct NO decomposition in the off gases containing methane and an excess of oxygen.

1.2 The methane conversion in the oxygen presence to the synthesis gas over alumina supported rhodium catalysts

CO and H_2 mixture (the synthesis gas - syn-gas) is applied in the Fischer-Tropsch and methanol syntheses. Hydrogen for ammonia synthesis and the food industry is also obtained from the syn-gas.

At present the synthesis gas is produced mainly by the highly endothermic, and thus unfriendly for environment, methane steam reforming. The exothermic partial methane oxidation would be more economical and therefore more friendly for environment.

Since a C-H bond can be broken on Rh sites at relatively low temperatures (Wang et al., 1996), the endothermic methane dissociation to C atoms and dihydrogen, followed by carbon oxidation, is most frequently considered as a plausible mechanism of the partial methane oxidation to the synthesis gas over the Rh/Al_2O_3 catalysts (Enger et al., 2008; Hofstad et al., 1998; Mallens et al., 1997; Nakagawa et al., 1999; Wang et al., 1996). At PMO temperatures hydrogen practically does not adsorb on the Rh surface and thus the possibility of its interaction with surface oxygen or OH groups is rather low (Rieck & Bell, 1985).

The high price of rhodium limits the use of the high loaded Rh/Al_2O_3 catalysts. Therefore it is of interest to investigate the possibility of the use of the low-loaded ones in PMO. In such catalysts Al-Rh alloys can be formed, besides clusters of metallic Rh, as a result of the strong metal-support interaction (SIMS) (Pietraszek et al., 2007;Zimowska et al., 2006).

As it was discussed elsewhere (Zimowska et al., 2006), calcination of the precursors of the low loaded Rh/δAl_2O_3 catalysts (with the specific surface area ca 287m^2/g) causes Rh incorporation into the near-to-surface layers of the alumina nanocrystallites resulting in a Rh^{x+}/Al_2O_3 solid solution (Rh^{x+}/Al_2O_3 s.s.) formation. The reduction of such a s.s. leads to Al-Rh alloys.

1.3 The NO direct decomposition in the oxygen presence over alumina supported precious metal catalysts

Nitrogen oxides NO_x in the flue gases from fuel combustion contain more then 90% of NO. It is formed from air in the endothermic N_2 oxidation at temperatures above 1273K. NO is a thermodynamically unstable molecule with a high, positive formation enthalpy $\Delta H^o_{f\ (298)}$ = 90,2 kJ/mol. Therefore it could decompose to N_2 and O_2 at low temperatures (between 293 and 973K) (Chuang & Tan, 1997; Garin, 2001; Tonetto et al., 2003).

The NO decomposition seems to be the best way of NO removal from the flue gases of stationary sources of emission from a practical, environmental as well as an economical point of view. This process does not require any reducer that allows the avoidance of secondary pollutants like oxygenated hydrocarbons, CO, CO_2, NH_3 or even cyanate and isocyanate (Chuang & Tan, 1997; Parvulescu et al., 1998; Tonetto et al., 2003). However, in the case of the direct NO decomposition N_2O can be formed aside from N_2 in one of the two possible paths of the process (Parvulescu et al., 1998):

$$2NO \rightarrow N_2 + O_2 \quad (1)$$

$$2NO \rightarrow N_2O + \tfrac{1}{2}\, O_2 \quad (2)$$

Oxide supported Pt, Rh and Pd high-loaded catalysts are known to be the most active in the direct NO decomposition among the metal catalysts supported on metal oxides (Almusaiteer et al., 2000; Garin, 2001; Gorte et al., 1981; Ishii et al., 2002; Papp & Sabde, 2005; Pietraszek et al., 2007; Rahkamaa & Salmi, 1999; Root et al., 1983; Sugisawa et al., 2001; X. Wang et al., 2004). However, the NO dissociation proceeds on Rh active species in rhodium clusters with a satisfactory rate only above 623K and it is greatly suppressed by the oxygen presence. The reactive atomic species at such temperatures favor NO oxidation. On the other hand, the direct NO decomposition on the Rh active sites in Al-Rh alloy nanocrystallites (Pietraszek et al., 2007) formed on the low loaded alumina supported catalysts may proceed with a relatively high rate at 473K, below the temperature of the oxygen dissociation. The investigation of the alloying effect on the NO direct decomposition in the presence of oxygen and methane on a 0.18 wt.% Rh/δAl_2O_3 catalyst with rhodium present mostly in the Al-Rh nanocrystallites should give relevant information about the direct NO decomposition in off gasses formed during methane combustion.

2. Results and discussion

2.1 Experimental

Two 0.18 wt.% Rh/δAl_2O_3 catalysts were obtained by one- or three-step Rh^{3+} deposition on δ-alumina support. It was expected that such a procedure would produce catalysts with Rh sites of different coordination.

The structure of the Rh sites was determined by the FTIR spectroscopy of CO adsorbed and by XPS.

The δAl_2O_3 support ($287 m^2/g$) was obtained by the sol-gel method from aluminum tri-sec-butylate (Zimowska et al., 2006).

The one-step (1-S) and three-step (3-S) Rh depositions were performed from a $RhCl_3 \cdot 2H_2O$ aqueous solution of the same concentration.

The details of the Rh deposition as well as those of oxidising and reducing thermal treatments were described elsewhere (Pietraszek et al., 2007; Zimowska et al. 2006).

The first step of the 3-S synthesis of the 0.18 wt.% Rh/δAl_2O_3 catalyst consists of:

i/ the impregnation of the δAl_2O_3 support with the $RhCl_3 \cdot 2H_2O$ aqueous solution, ii/ the drying of a wet precursor, iii/ the calcination of a dry precursor at 773K and reduction at 623K and 773K of a calcined precursor. The used procedure yields the 0.06 wt.% Rh/δAl_2O_3 catalyst.

The next step of the 3-S synthesis of the 0.18wt.% Rh/δAl_2O_3 catalyst differs from the previous one only in the first stage; the 0.06 wt.% Rh/δAl_2O_3 catalyst (not the δAl_2O_3 support) is impregnated by the $RhCl_3 \cdot 2H_2O$ aqueous solution. As a result of the next drying, calcination and reduction the 0.12wt.%Rh/δAl_2O_3 catalyst is obtained.

In the third step of the 3-S synthesis the 0.12 wt.% Rh/δAl_2O_3 catalyst is impregnated by the same $RhCl_3 \cdot 2H_2O$ aqueous solution. The next drying, calcinations and reduction yields the 0.18 wt. % Rh/ δAl_2O_3 catalyst.

The 1-S synthesis of the 0.18 wt.% Rh/δAl_2O_3 catalyst consists of: i/ the impregnation of the δAl_2O_3 support with a triple portion of the $RhCl_3 \cdot 2H_2O$ aqueous solution, ii/ the drying of a wet precursor, iii/ the calcination of a dry precursor and the reduction of a calcined precursor.

The FTIR spectroscopy of CO adsorbed and XPS measurements were performed for the fresh one-step (F-1-S) and three-step (F-3-S) catalysts as well as for the one-step and three-step catalysts subjected to the interaction with six methane pulses (U-1-S) and (U-3-S).

CO was adsorbed on the surface of the catalysts at room temperature and the spectra were taken also at room temperature after outgasing at 373K using a Nicolet 5700 Nexus spectrometer at a resolution of 4 cm^{-1}.

The XPS measurements for F-1-S, F-3-S, U-1-S, and U-3-S catalysts were carried out using a scanning photoelectron spectrometer PHI 5000 VersaProbe (Physical Electronics USA/ULVAC Japan) with monochromatic Al Kα radiation (1486.6eV). The signal was collected from 200 x200µm area using an x-ray beam focused to 100µm diameter with 25 W power. For the charge shift compensation all the measurements were made with neutralization by low energy electrons and argon ions. The survey spectra were obtained with a pass energy of 117.4eV in 0.4eV increments. The detailed spectra of the Rh3d, and the Al2p regions were measured with a pass energy of 23.5eV in 0.1eV increments. The binding energy scale (BE) was calibrated based on the Al2p in a Al_2O_3 support signal (BE = 74.3eV). Signal components were fitted using a non-linear least squares fitting Casa XPS software. The quantitative analysis was done after a Shirley-type background subtraction using the Scofield sensitivity factors.

The investigation of the interaction of the catalysts with methane pulses was carried out at 873K in an acid-resistant steel tubular reactor with a 6mm internal diameter and 30cm length, under the pressure of 280 kPa and at GHSV =9000h^{-1}. The 700 mm^3 methane pulses were introduced into the Ar stream flowing through the reactor, the Carboxen 1000 packed column and the TCD detector of a Hewlett-Packard 5890 chromatograph. The time between consecutive methane pulses was ca. 32min.

The direct NO decomposition in methane and the oxygen presence was investigated under the steady-state conditions in 473-673K temperature range. The reaction mixture of 150ppm NO 1500ppm CH_4, 7%O_2 and Ar as a balance passed over the catalyst placed on quartz-wool in a quartz reactor with GHSV = 20 000h^{-1}. Before the steady-state experiment the catalyst was heated in the reaction mixture from the r.t. to 773K and consecutive measurements were performed stepwise starting from 673K to 473K. A FID detector was used to follow the total concentration of hydrocarbons and the N_2 formation was checked by a micro-GC.

2.2 The physicochemical characterisation of 0.18wt.%Rh/δAl_2O_3 catalysts by FTIR spectroscopy of CO adsorbed and XPS

Fig. 1. presents the FTIR spectra of CO adsorbed on the F-3-S 0.18wt.%Rh/δAl_2O_3 catalyst (a) and on the F-3-S 0.18 wt.% Rh/δAl_2O_3 one (b). In agreement with the data extensively reported in the literature (Finocchio et al., 2007 ; Hadjiivanov &Vayssilov, 2002; Kraus et al., 1989; Lavalley et al., 1990; Paul et al., 1999; Yates et al., 1979) two main bands at 2087 and 2013 cm^{-1}, present in the spectra of both the catalysts, can be assigned to symmetric and asymmetric stretching vibrations of gem-bicarbonyls adsorbed on the Rh atomically dispersed

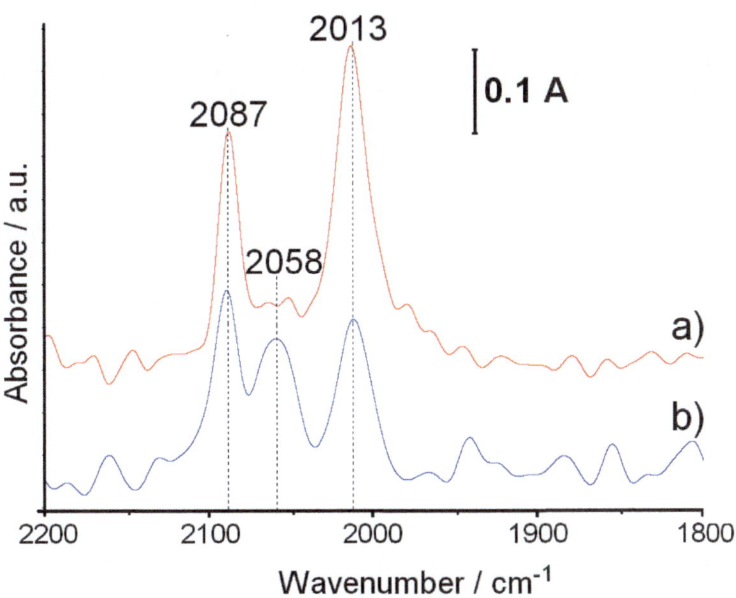

Fig. 1. DRIFT spectra CO adsorbed on the surface of the 0.18 wt.% Rh /δAl_2O_3 catalysts obtained by: a/three-step rhodium deposition, b/ one-step rhodium deposition

on the alumina surface. However, the exact position of the atomically dispersed Rh in the structure of catalyst has yet to be determined (Finocchio et al., 2007 ; Hadjiivanov &Vayssilov, 2002; Kraus et al., 1989; Lavalley et al., 1990; Paul et al., 1999; Yates et al., 1979). According to above cited papers the peak at 2058cm^{-1}, clearly seen in the spectrum of the F-1-S catalyst, results from C-O stretching vibrations in CO linearly adsorbed on the Rh clusters.

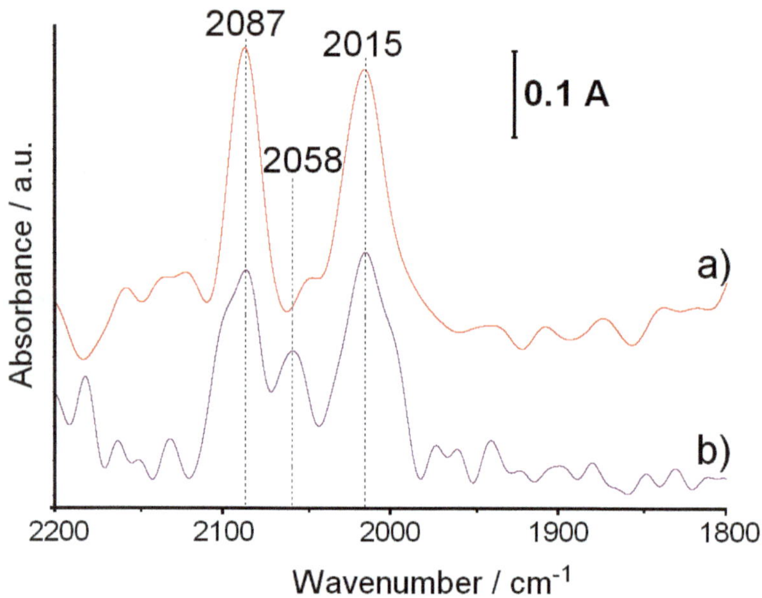

Fig. 2. DRIFT spectra of CO adsorbed on the surface of the 0.18 wt.% Rh /δAl_2O_3 catalysts: a/ obtained by a three-step rhodium deposition and subjected to interaction with six methane pulses, b/ obtained by a one-step rhodium deposition and subjected to interaction with six methane pulses.

Thus one may conclude that on the surface of the F-1-S catalysts the atomically dispersed Rh coexists with the Rh clusters. However, on the surface of the F-3-S catalyst atomically dispersed species are hardly seen.

Fig. 2 presents the FTIR spectra of CO adsorbed on the three-step (a) and one-step (b) 0.18wt.% Rh/δAl_2O_3 catalysts subjected to interaction with 6 methane pulses at 873K.

The distinct decrease of the ratio of Rh in clusters to Rh in alloy nanocrystallites as a result of the catalysts interaction with methane pulses is observed. The increase of the width of the gem-bicarbonyl peaks reveals that during the pulse experiment atomically dispersed Rh with a coordination slightly different than in the fresh catalysts is formed.

The comparison of XPS results for the fresh and used catalysts gives further insight in the nature of atomically dispersed rhodium species and in the mechanism of their creation as well as their interaction with the clustered metallic rhodium species.

In Table 1 the BEs of peaks obtained by the deconvolution of the Al2p$_{3/2}$ and Rh3d$_{5/2}$ bands as well as the percentage of the particular Al and Rh species (($Me^x/\Sigma Rh^x+\Sigma Al^x)\cdot 100$) in F-1-S, U-1-S, F-3-S and U-3-S catalysts are presented.

Catalyst	Peak	BE [eV]	$(Me^x/\Sigma Rh^x+\Sigma Al^x)\cdot 100$ [at.%]*
F-1-S	Rh3$d_{5/2}$	310.0 307.4	0.08 0.05
	Al2$p_{3/2}$	74.3	99.87
U-1-S	Rh3$d_{5/2}$	307.5 305.0	0.003 0.003
	Al2$p_{3/2}$	75.4 74.3	15.69 84.30
F- 3-S	Rh3$d_{5/2}$	310.2 307.8	0.06 0.09
	Al2$p_{3/2}$	75.5 74.3	9.72 90.13
U- 3-S	Rh3$d_{5/2}$	309.6 307.2	0.003 0.003
	Al2$p_{3/2}$	75.4 74.3	14.21 85.70

M^x - metallic or cationic Rh and/or Al species

Table 1. The XPS results for the fresh one-step (F-1-S) and the three-step (F-3-S) 0.18 wt.% Rh/ δ Al_2O_3 catalysts and for those catalysts subjected to the interaction with six methane pulses (U-1-S and U-3-S).

The Al2$p_{3/2}$ bands in the spectra of the particular catalysts were deconvoluted into two peaks. The major peaks were ascribed to Al in alumina - BE=74.3eV (Wagner et al., 2007).

The BE of the less intensive Al2$p_{3/2}$ peaks as well as those of the Rh 3$d_{5/2}$ peaks were determined with respect to this one. The less intensive Al2$p_{3/2}$ peaks with exceptionally high BE (75.4-75.5eV) were ascribed to Al in the Al-Rh alloy. The transfer of the valence Al electrons mainly to Rh unoccupied d-orbitals (Azaroff, 1960; Kittel, 2005) in the alloy should cause the increase of the Al2$p_{3/2}$ BE and decrease of Rh3$d_{5/2}$ one with respect to those in separate metals (Wagner, 2007).

The Rh3$d_{5/2}$ bands in the XPS spectra of the F-1-S and F-3-S catalysts were deconvoluted into the peaks of cationic species with BE equal to 310.0 and 310.2eV and of Al-Rh alloy ones with BE equal to 307.4 and 307.8 eV, a little lower than that of metallic Rh (Wagner, 2007).

The atomically dispersed Rh species (Finocchio et al., 2007; Hadjiivanov &Vayssilov, 2002; Kraus et al., 1989; Lavalley et al., 1990; Paul et al., 1999; Yates et al., 1979) demonstraded in the FTIR spectra of CO adsorbed on the F-1-S and the F-3-S catalysts (Fig.1) by bands at 2087 and 2013 cm^{-1}, occur probably in the surface lattice position of the Al-Rh alloy nanocrystallites.

The presence of only one $3d_{5/2}$ peak of the Rh species with BE =307.4 eV in the spectrum of the F-1-S catalyst reveals an electronic interaction between the Al-Rh intermediate phase nanocrystallites and the Rh clusters. It shows that the Rh clusters occur in the immediate proximity of the Al-Rh nanocrystallites. Such an electronic interaction was earlier observed by Thiam et al. (Thiam et al., 2004) between aluminum foil and a rhodium overlayer.

A higher percentage of the metallic Rh species in the F-3-S catalyst (60 %) than in the F-1-S one (ca 40%) could easily be explained by the differences in the syntheses. The one-step Rh deposition is thought to have occurred preferably on the smallest nanocrystallites with the highest surface/bulk ratio, possibly by an Rh^{3+} cation exchange with protons of the accessible acidic OH groups on alumina. Thus, the Rh deposition calculated per volume unit of alumina crystallites is higher the smaller their dimensions are. Rhodium clusters are formed from an excess of the aqueous Rh^{3+} ions solution present in pores.

On the other hand, the three-step Rh deposition occurs mostly by exchange of the Rh cations with the acidic OH group protons on the surface of the smallest δ alumina crystallites, which is renovated in each synthesis step in the course of the calcinations. It results in the exceptional enrichment of these crystallites in Rh. An incorporation of rhodium into δ alumina during the thermal treatments causes the formation of the Rh^{x+}/Al_2O_3 solid solution. The reduction of the s.s. nanocrystallites, resulting in inter-metallic Al-Rh alloy formation, is easier the higher the rhodium concentration is.

It is obvious that the enrichment of Rh atoms in electrons is the greater the higher their dispersion is in the aluminium matrix. Earlier this problem in particular was discussed regarding the Cu-Ni alloys (Kittel, 2005).

The interaction of the F-1-S catalyst with six CH_4 pulses causes a complete disappearance of the cationic Rh species and a great decrease in the content of the metallic Rh species. Simultaneously the Rh species with $3d_{5/2}$ BE equal to 305.0 eV and the Al species with an exceptionally high $2p_{3/2}$ BE (75.4eV) do appear.

The disappearance of the cationic species reveals the formation of the new nanocystallites of the Al-Rh alloy by the reduction of the Rh^{x+}/Al_2O_3 s.s. of low Rh content.

The major decrease of the content of Rh species with $3d_{5/2}$ BE close to 307 eV and the appearance of the species with $3d_{5/2}$ BE = 305 eV and the Al ones with $2p_{3/2}$ BE = 75.4 eV could be the result of the Al surface segregation in the alloy nanocrystallites formed at 773K. However, Rh species with $3d_{5/2}$ BE = 305 eV and the Al ones with $2p_{3/2}$ BE = 75.4 eV could also be formed as a result of the reduction of the Rh^{x+}/Al_2O_3 s.s. with a low Rh content.

The heating of the F-3-S catalyst and its interaction with 6 methane pulses causes a great decrease in the total contents of the Rh species in the surface nanolayers, the disappearance of the species with $3d_{5/2}$ BE = 307.8 eV, the appearance of the ones with $3d_{5/2}$ BE = 307.2 eV and an increase of the content Al species with $2p_{3/2}$ BE = 75.4 eV.

The major decrease in the content of the cationic and metallic Rh species, the the decrease in the $3d_{5/2}$ BE of the Rh metallic species as well as the increase of the content of the Al ones with $2p_{3/2}$ BE = 75.4 eV reveal the reduction of the Rh/Al_2O_3 s.s. areas with a small Rh content and Al segregation in the Al-Rh alloy nanocrystallites formed at 773K. The great extent of the surface Al segregation can be caused by great difference in Al and Rh melting points and the proximity of the used temperature and the temperature of Al melting (Lide, 2004-2005).

2.3 Methane dissociation on 1-S and 3-S 0.18 wt.% Rh/δAl_2O_3 catalysts

In Fig 3. the methane conversion (C_{CH4}), selectivity to hydrogen (S_{H2}), selectivity to CO (S_{CO}) and selectivity to CO_2 (S_{CO2}) in six consecutive CH_4 pulses over F-1-S catalyst are presented.

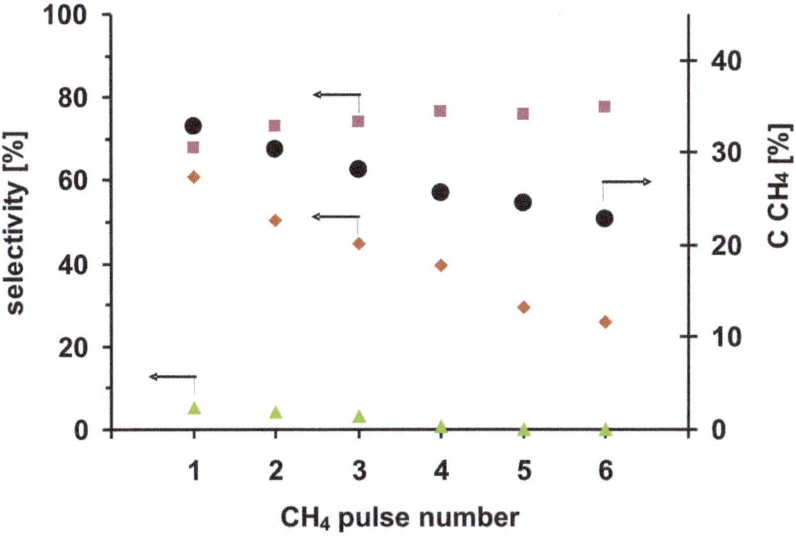

Fig. 3. CH_4 conversion (•) and selectivity to H_2(■), CO (♦) and CO_2 (▲) in six consecutive CH_4 pulses over the one-step 0.18 wt.% Rh/δAl_2O_3 catalyst (873K, 280kPa, 9000 h^{-1}).

The CH_4 conversion in the first pulse over the 1-S catalyst is equal to 33%, the selectivity to H_2 - 68%, the selectivity to CO - 61% and to CO_2 - 10%. In the following methane pulses C_{CH4} continuously decreases down to 23% in the sixth pulse. The S_{H2} increases up to 78 % in the sixth pulse, S_{CO} decreases, almost linearly, down to 22% in the sixth pulse and the S_{CO2} drops to 0% in the fourth pulse.

The simultaneous decrease of the C_{CH4}, the S_{CO} and the S_{CO2} and the increase of S_{H2} in the consecutive methane pulses can be explain by the decreasing concentration of the atomic species of oxygen adsorbed on the surface Rh species in Al-Rh alloy nanocrystallites and in Rh clusters present on the surface of the 1-S catalyst.

The initial S_{H2} increase accompanied by a decrease in S_{CO2} reveals that further formation of alloy nanocrystallites with a low concentration of Rh sites have a greater ability to transfer electrons to methane molecules. The 2.8 fold decrease in the S_{CO} in the sixth methane pulse with respect to that in the first methane pulse with a simultaneous increase of the selectivity to hydrogen, suggesting deeper methane dissociation, clearly shows a carbon deposition on the catalyst surface as a result of the deficit of oxygen species adsorbed.

In Fig. 4 the methane conversion, selectivity to hydrogen, selectivity to CO and the selectivity to CO_2 in six consecutive CH_4 pulses introduced to the reactor containing the F-3-S catalyst are presented.

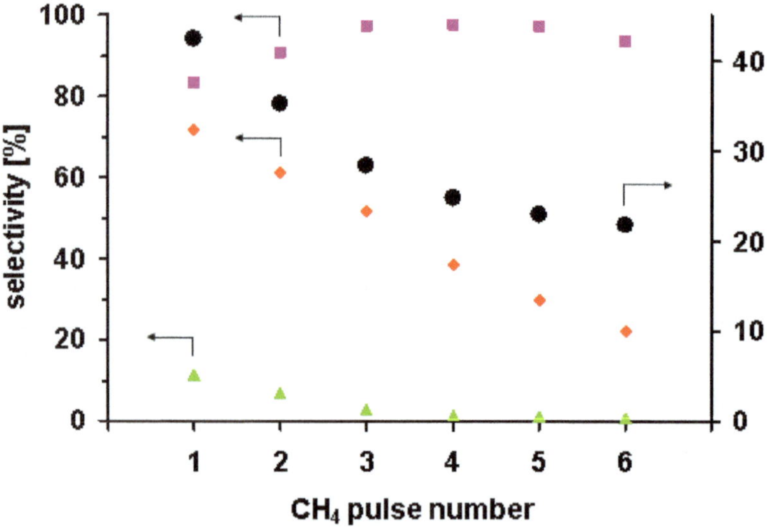

Fig. 4. CH_4 conversion (•) and selectivity to H_2(■), CO (♦) and CO_2 (▲) in six consecutive CH_4 pulses over the 3-S 0.18wt.% Rh/δAl_2O_3 catalyst (873K, 280kPa, 9000 h^{-1}).

The C_{CH4} in the first methane pulse is equal to 42%, the S_{H2} - 83%, the S_{CO} - 72% and the S_{CO2} - 11%. In the next methane pulses the C_{CH4} decreases down to 22% in the sixth pulse. The S_{H2} increases up to 98% in the third pulse and slowly decreases to 94% in the sixth pulse. The S_{CO} decreases almost linearly down to 22% in the sixth pulse and S_{CO2} drops to 0% in the fourth pulse.

The simultaneous decrease in the S_{CO2} and increase of the S_{H2} in the consecutive methane pulses up to the forth one could be ascribed to the formation of the new Al-Rh alloy nanocrystallites of a low content of Rh sites with a great ability to transfer the electrons to the methane molecules adsorbed.

There is a 50% decrease in the C_{CH4} and approximately a 70% decrease in the S_{CO} in the sixth pulse in comparison with the first one, not accompanied by the decrease in the S_{H2} which reveals carbon deposition on the Al-Rh alloy nanocrystallites.

The formation of the new alloy nanocrystallites by the reduction of the Rh/Al_2O_3 s.s. of the relatively low rhodium concentration causes the creation of a very active Rh species.

The stronger S_{CO} decrease on the 3-S catalyst (3.3 fold) of a higher activity than on the 1-S one (2.8 fold) of lower activity reveals that carbon is preferably deposited on the Al-Rh nanocrystallites.

The higher conversion of the first methane pulse over the F-3-S catalyst than over the F-1-S reveals that the Rh active species in the Al-Rh alloy nanocrystallites are more active in the methane dissociation than those in the Rh clusters (Figs.1 and 2). The decrease of the activity of both the catalysts in the consecutive pulses corresponds to the decrease of the total content of the Rh species accessible for methane adsorption.

The slower the decrease of the methane conversion over the 1-S catalyst than that over the 3-S one may be assigned to a slower carbon deposition on the Rh clusters than on the Al-Rh nanocrystallites.

On the other hand, the higher selectivity to hydrogen over the 3-S catalyst than over the 1-S one clearly shows a beneficial influence of the Al proximity on the Rh sites ability to transfer electrons to antibonding orbitals of methane.

Fig. 5 presents CH_4 conversion and selectivity to H_2, CO and CO_2 in 41 consecutive CH_4 pulses over the 3-S 0.18wt.% Rh/δAl_2O_3 catalyst (873K, 280kPa, 9000 h^{-1}). The methane pulses were introduced into the reactor mostly in time intervals close to 30min. Less regular time intervals (30-37min.) were applied between the 11th and 17th pulses as well as the 25th and 33rd. The C_{CH4} and S_{CO} decrease and S_{H2} increases in consecutive pulses up to the pulse 11. Irregular intervals between the 11 - 17 pulses and the 25 – 33 ones cause fluctuations in the C_{CH4}, the S_{CO} and the S_{H2} that can be ascribed to the changes in the amount of the carbon deposit oxidized by the oxygen from the gas phase. The 75 min. time interval between 34th and 35th pulses resulted in the C_{CH4} increase of 11% and of the S_{CO} of 110%. The increase of the time between the 40 and 41 pulses to 110 min. results in the increase of the C_{CH4} of 25% and of the S_{CO} of 230%. However, the selectivity to hydrogen does not change distinctly in any case. Thus, one can conclude that a long term interaction of the surface of the Al-Rh nanocrystallites, containing the carbon deposit, with the gas phase containing traces of oxygen causes both a carbon deposit removal by oxidation and oxygen adsorption. The oxidation of the carbon deposit causes the recovery of the Rh sites active in the methane dissociation that results in the increase of the C_{CH4}. The surface oxygen species cause oxidation of the carbon formed due to methane dissociation, that leads to the S_{CO} increase. However, their population is not high enough to have an effect on the S_{H2}. Thus, to perform the PMO with the high methane conversion and the high selectivity to CO and H_2 the proper CH_4:O_2 ratio is needed.

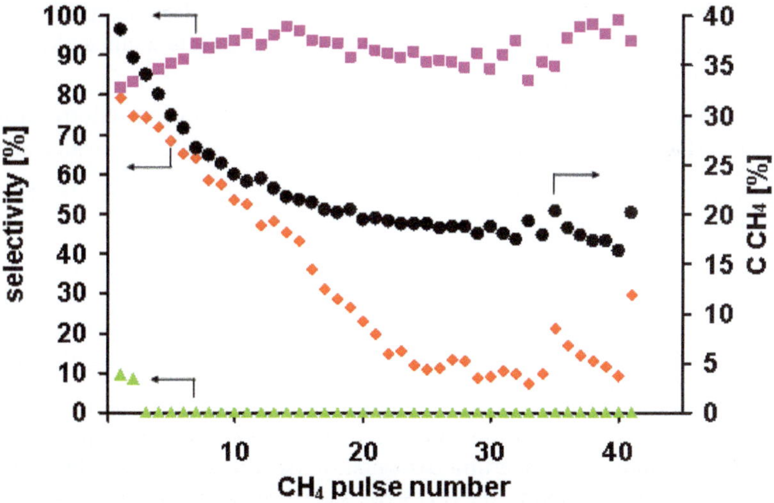

Fig. 5. CH_4 conversion (•) and selectivity to H_2(■), CO (♦) and CO_2 (▲) in 41 consecutive CH_4 pulses over the 3-S 0.18wt.% Rh/δAl_2O_3 catalyst (873K, 280kPa, 9000 h^{-1}).

2.4 Direct NO decomposition over the three-step (3-S) 0.18 wt.% $Rh/\delta Al_2O_3$ catalyst

NO conversion (C_{NO}), selectivity to N_2 (S_{N2}) calculated from the results of the steady-state experiments over the 3-S 0.18wt.%Rh/Al_2O_3 catalysts, are presented in Figure 6.

The steady-state experiment was performed with the use of the reaction gas containing 150ppm NO, 1500ppm CH_4, 7%O_2 and Ar as a balance.

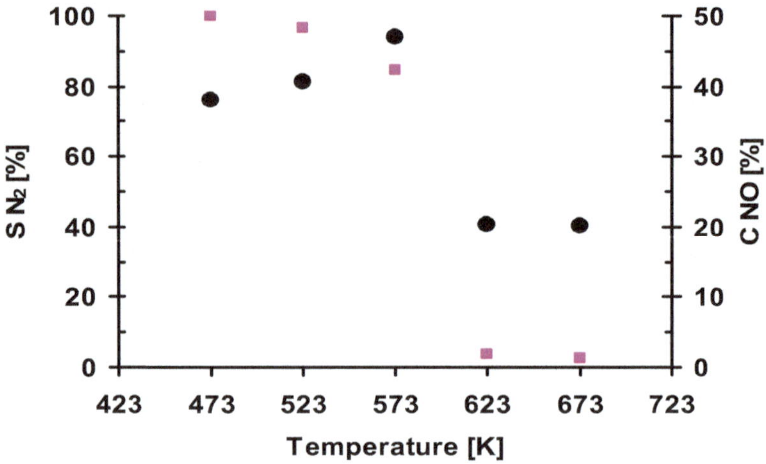

Fig. 6. NO conversion (•) and selectivity to NO (■) over three-step a 0.18 wt.% $Rh/\delta Al_2O_3$ catalyst (150ppmNO +1500ppm CH_4 + 7%O_2 /He , 20000h^{-1}, data from steady-state experiment)

The highest NO conversion, increasing with a temperature increase, from 38 to 47%, is observed at a temperature range 473K - 573K. The highest but decreasing from 100% to 85% selectivity to N_2 is achieved in the same temperature. A simultaneous sudden drop of the S_{N2} and a deep decrease in C_{NO} are observed at 623K.

The direct NO decomposition over the high-loaded rhodium catalysts supported on alumina was earlier observed only at temperatures higher than 623K (Garin, 2001).

Thus, the low-temperature NO decomposition on the low loaded 0.18wt.%$Rh/\delta Al_2O_3$ catalyst confirms exceptional activity of the Rh sites in Al-Rh alloy.

A simultaneous increase of the C_{NO} and the decrease of the S_{N2} with the temperature increase in the 473K - 573K range could be explained by the competitive dissociative O_2 adsorption on the Rh active sites, increasing with temperature. The atomic oxygen species block the Rh sites active in the direct NO decomposition and facilitate the NO oxidation.

The simultaneous sudden drop of the S_{N2} and a great decrease in the C_{NO} at 623K shows that blocking the Rh sites-active in the NO dissociation becomes much stronger. Such blocking may originate in a methane dissociation on the same sites. In Fig. 7, the CH_4 conversion (C_{CH4}) and selectivity to CO_2 (S_{CO2}), measured during the thermo-programmed catalyst heating at the rate of 3K/min in the reaction mixture containing 150ppm NO, 1500ppm CH_4, 7%O_2 and Ar as a balance, are presented.

The temperature of the sudden drop of the S_{N2} and C_{NO} (Fig. 6) corresponds to the initial temperature of the total methane oxidation to CO_2 (Fig. 7). This confirms competitive methane adsorption on the Rh active sites leading to the formation of the carbon species undergoing oxidation to CO_2.

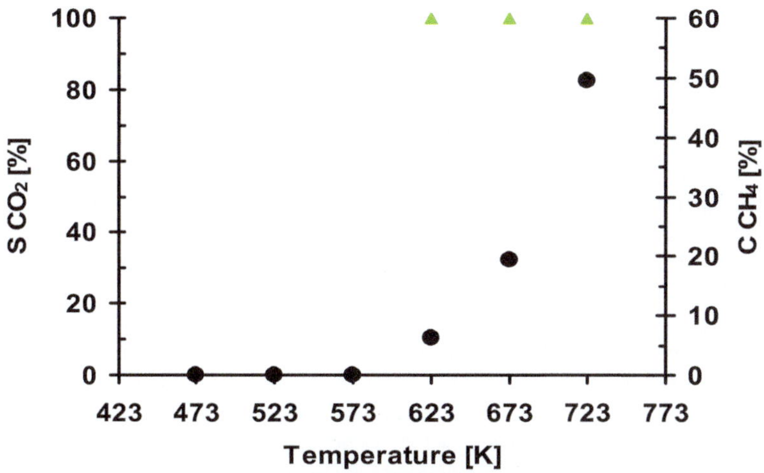

Fig. 7. CH_4 conversion (•) and selectivity to CO_2 (▲) over three-step 0.18 wt.% Rh/δAl_2O_3 catalyst (150ppmNO +1500ppm CH_4 + 7%O_2 /He , 20000h^{-1}, data from TPSR experiment)

3. Conclusions

The DRIFT spectroscopy of the adsorbed CO and XPS were used to show the formation of the Al-Rh alloy nanocrystallites on the surface of the 0.18 wt % Rh/δAl_2O_3 catalysts obtained by the one-step or three-step impregnation of the high surface area alumina support with the $RhCl_3$ aqueous solution (1-S and 3-S catalysts).

The effect of Al-Rh alloying on the catalyst activity in the methane dissociation and the direct NO decomposition was investigated.

The activity and selectivity to dihydrogen and to carbon monooxide of both the catalysts were determined at 873K. The results showed unambiguously that alloying distinctly enhances the Rh site activity in the electron donation to antibonding orbitals of methane molecules adsorbed. The observed enhancement of the Rh site activity was ascribed to the Al valence electrons transfer to the Rh 4d shell and slightly to the Rh 5s shell. The direct NO decomposition in the presence of oxygen and methane was investigated in the temperature range of 673-473K over a 3-S catalyst. The NO conversion with 100% selectivity to dinitrogen on the Rh active sites was revealed at an unexpectedly low temperature (473K). Simultaneous increase of the activity and decrease in the selectivity to dinitrogen with a temperature increase up to 573K clearly revealed a competitive dissociative oxygen adsorption on the same active sites. However, the sudden decrease at 623K of the catalyst activity in the NO decomposition and the selectivity to dinitrogen, accompanied by an initial methane oxidation to CO_2, undoubtedly showed competitive methane dissociation.

4. Acknowledgment

The polish authors (M. N, J. D. and P. K.) acknowledge the financial support by the European Regional Development Fund under the Innovate Economy Operational Programme 2007-2013, POIG.01.01.02-12-112/09 project.

5. References

Almusaiteer, K.; Krishnamurthy, R. & Chuang, S.S.C. (2000). In situ infrared study of catalytic decomposition of NO on carbon-supported Rh and Pd catalysts. *Catalysis Today*, Vol.55, No.3, pp. 291-299, ISSN 0920-5861

Azaroff, L.V. (1960). *Introduction to Solids*, McGRAW-Hill Book Cmpany, New York

Bostrom, M.; Rosner, H.; Prots, Y.; Burkhardt, U. & Grin,Y. (2005). The Co_2Al_9 Structure Type Revisited. *Zeitschrift für anorganische und allgemeine Chemie*,Vol.631, No.2-3, pp.534-541, ISSN 1521-3749

Chuang, S.S.C. & Tan, C.–D. (1997). Promotion of oxygen desorption to enhance direct NO decomposition over Tb−Pt/Al_2O_3 catalyst. *The Journal of Physical Chemistry B*, Vol.101, No.15, pp. 3000-3004, ISSN 1520-5207

Enger, B.C.; Lodeng, R. & Holmen, A. (2008). A review of catalytic partial oxidation of methane to synthesis gas with emphasis on reaction mechanisms over transition metal catalysts. *Applied Catalysis A: General*, Vol.346, No.1-2, pp.1-27, ISSN 0926-860X

Finocchio, E.; Busca, G.; Forzatti, P.; Groppi, G. & Beretta, A. (2007). State of supported rhodium nanoparticles for methane catalytic partial oxidation (CPO): FT-IR studies. *Langmuir*, Vol.23, No.20, pp. 10419–10428, ISSN 1520-5827

Garin, F. (2001). Mechanism of NO_x decomposition. *Applied Catalysis A: General*, Vol.222, No.1-2, pp. 183-219, ISSN 0926-860X

Gasser, R.P.H. (1985). An introduction to chemisorption and catalysis by metals, Oxford University Press, ISBN 0-19-855163-0, New York

Gorte, R.J.; Schmidt, L.D. & Gland, J. L. (1981). Binding states and decomposition of NO on single crystal planes of Pt. *Surface Science*, Vol.109, No.2, pp. 367-380, ISSN 0039-6028

Hadjiivanov, K.I. & Vayssilov, G.N. (2002). Characterization of oxide surfaces and zeolites by carbon monoxide as an IR probe molecule. *Advances in Catalysis*, Vol.47, pp. 307-511, ISBN 0-12-007844-9

Hofstad, K.H.; Hoebink, J.H.B.J.; Holmen, A. & Marin, G.B. (1998). Partial oxidation of methane to synthesis gas over rhodium catalysts. *Catalysis Today*, Vol.40, No. 2-3, pp. 157-170, ISSN 0920-5861

Ishii, M.; Hayashi, T. & Matsumoto, S. (2002). Adsorption, desorption and decomposition of nitrogen monoxide on Rh(1 0 0) studied by electron-stimulated desorption, Auger electron spectroscopy and temperature-programmed desorption. *Applied Catalysis A: General*, Vol.225, No.1-2, pp. 207-213, ISSN 0926-860X

Kittel, Ch. (2005). *Introduction to Solid State Physics* John Willey & Sons. Inc, ISBN 0-471-41526-x, USA

Kraus, L.; Zaki, M.I.; Knözinger, H. & Tesche, B. (1989). Support and additive effects on the state of rhodium catalysts. *Journal of Molecular Catalysis*, Vol.55, No.1, pp. 55-69

Lavalley, J.C.; Saussey, J.; Lamotte, J.; Breault, R.; Hindennann, J.P. & Kiennemann, A. (1990). Infrared study of carbon monoxide hydrogenation over rhodium/ceria and rhodium/silica catalysts. *The Journal of Physical Chemistry*, Vol.94, No.15, pp. 5941–5947

Lide, D.R. (2004-2005). *Handbook of Chemistry and Physics*, CRC Press, ISBN 0-8493-0485-7, Boca Raton

Mallens, E. P. J.; Hoebink, J.H.B.J. & Marin, G.B. (1997). The reaction mechanism of the partial oxidation of methane to synthesis gas: A transient kinetic study over rhodium and a comparison with platinum. *Journal of Catalysis*, Vol.167, No.1, pp. 43-56, ISSN 0021-9517

Nakagawa, K.; Ikenaga, N.; Kobayashi, T. & Suzuki, T. (1999). Transient response of catalyst bed temperature in the pulsed reaction of partial oxidation of methane to synthesis gas over supported rhodium and iridium catalysts. *Journal of Catalysis*, Vol.186, No.2, pp. 405-413, ISSN 0021-9517

Papp, H. & Sabde, D.P. (2005). An investigation on the mechanism of NO decomposition over Rh/SiO_2 catalysts in presence of pulse injected H_2. *Applied Catalysis B: Environmental*, Vol.60, No.1-2, pp. 65-71, ISSN 0926-3373

Parvulescu, V.I.; Grange, P. & Delmon, B. (1998). Catalytic removal of NO. *Catalysis Today*, Vol.46, No.4, pp. 233-316, ISSN 0920-5861

Paul, D.K.; Marten, C.D. & Yates, J.T.Jr. (1999). Control of $Rh^I(CO)_2$ formation on Rh/Al_2O_3 catalysts by complexation of surface –OH groups using NH_3. *Langmuir*, Vol.15, No.13, pp. 4508–4512, ISSN 1520-5827

Pietraszek, A.; Da Costa, P.; Marques, R.; Kornelak, P.; Hansen, T.W.; Camra, J. & Najbar, M. (2007). The effect of the Rh–Al, Pt–Al and Pt–Rh–Al surface alloys on NO conversion to N_2 on alumina supported Rh, Pt and Pt–Rh catalysts. *Catalysis Today*, Vol.119, No.1-4, pp. 187-193, ISSN 0920-5861

Rahkamaa, K. & Salmi, T. (1999). Investigation of the catalytic decomposition of NO and N_2O on supported Rh with transient techniques. *Chemical Engineering Science*, Vol.54, No.20, pp. 4343-4349, ISSN 0009-2509

Rieck, J.S. & Bell, A.T. (1985). Studies of the interactions of H_2 and CO with silica- and lanthana-supported palladium. *Journal of Catalysis*, Vol.96, No.1, pp. 88-105, ISSN 0021-9517

Root, T.W.; Schmidt, L.D. & Fisher, G.B. (1983). Adsorption and reaction of nitric oxide and oxygen on Rh(111). *Surface Science*, Vol.134, No.1, pp. 30-45, ISSN 0039-6028

Sugisawa, T.; Shiraishi, J.; Machihara, D.; Irokawa, K.; Miki, H.; Kodama, C.; Kuriyama, T.; Kubo, T. & Nozoye, H. (2001). Adsorption and decomposition of NO on Pt (112). *Applied Surface Science*, Vol.169-170, pp. 292-295, ISSN 0169-4332

Thiam, M.M.; Hrčiř , T.; Matolin, V. & Nehasil, V. (2004). EELS and AES investigation of Rh thin film growth on polycrystalline Al substrate. *Vacuum*, Vol.74, No.2, pp. 141-145, ISSN 0042-207X

Tonetto, G.M.; Ferreira, M.L. & Damiani, D.E. (2003). A combined theoretical and experimental study of NO decomposition on Pd and Pd-Mo catalysts. *Journal of Molecular Catalysis A: Chemical*, Vol.193, No.1-2, pp. 121-137, ISSN 1381-1169

Wagner, Ch.D.; Naumkin, A.V.; Kraut-Vass, A.; Allison, J.W.; Powell, C.J. & Rumble, J.R.Jr. (2007). NIST X-ray Photoelectron Spectroscopy Database NIST Standard Reference Database 20, Version 3.5, Available from: http://srdata.nist.gov/xps/

Wang, D.; Dewaele, O.; Groote, A.M.D. & Froment, G.F. (1996). Reaction mechanism and role of the support in the partial oxidation of methane on Rh/Al_2O_3. *Journal of Catalysis*, Vol.159, No.2, pp. 418-426, ISSN 0021-9517

Wang, X.; Sigmon, S.M.; Spivey, J.J. & Lamb, H.H. (2004). Support and particle size effects on direct NO decomposition over platinum. *Catalysis Today*, Vol.96, No.1-2, pp. 11-20, ISSN 0920-5861

Yates, J.T.; Duncan, T.M.; Worley, S.D. & Vaughan, R.W. (1979). Infrared spectra of chemisorbed CO on Rh. *The Journal of Chemical Physics*, Vol.70, pp. 1219-1224, ISSN 1089-7690

Zimowska, M.; Wagner, J. B.; Dziedzic, J.; Camra, J.; Borzęcka-Prokop, B. & Najbar, M. (2006). Some aspects of metal-support strong interactions in Rh/Al_2O_3 catalyst under oxidising and reducing conditions. *Chemical Physics Letters*, Vol.417, No.1-3, pp.137-142, ISSN 0009-2614

17

Electrooxidation as a Pretreatment Process Before Cyanidation

Fatma Arslan
Istanbul Technical University, Mining Faculty
Mineral Processing Engineering Department, Istanbul
Turkey

1. Introduction

The poor extraction of refractory gold ores is an old and still existing problem. Gold losses in cyanidation are mainly due to the presence of sulfide and carbonaceous materials (1,2,3). The decomposition of sulfide minerals during the cyanidation of refractory gold ores creates two types of impurities: metal cations and sulfur compounds (HS^-, $S_2O_3^{2-}$, and SO_3^{2-}). The metal cations form complexes with cyanide ion and consume reagent. The sulfide ion is considered to be a powerful poison during the cyanidation and may adsorb on gold, causing the inhibition of gold dissolution reaction by passivation. Sulfur compounds form thiocyanate by consuming cyanide ion and oxygen.

Carbonaceous gold ores contain (a) an activated component which is capable of adsorbing gold complexes from the solution, (b) a mixture of hydrocarbons usually associated with active carbon components (they formed a gold complex during the deposition that was not attacked by cyanide), and (c) an organic acid similar to humic acid, which contains functional groups capable of interacting with gold complexes to form organic gold compounds which might possibly be formed by chelation (1,2,3).

Various procedures had been investigated in an attempt to enhance gold recovery from different ores, including roasting, kerosene treatment, flotation, and aqueous oxidation. If roasting or aqueous oxidation is carried out before cyanidation, the sulfur species tend to oxidize to the less harmful sulfate form. There are several aqueous oxidation techniques prior to cyanidation to treat refractory gold ores: chemical (chlorine, hypochlorite, chlorates, perchlorates, ozone, oxygen, and sulfuric acid) or electrochemical oxidation, pressure leaching, and bacterial leaching.

Oxidation of carbon and sulfide minerals by sodium hypochlorite is a possible method to render the ore amenable to cyanidation, and earlier work by the USBM indicates that electrochemical generation of hypochlorite is promising. An electro-oxidation process was developed by the U.S. Bureau of Mines for generating oxidizing conditions in situ in a pulp prepared from finely ground carbonaceous gold ore and sodium chloride solution (4). Gold extraction by cyanidation after oxidation increased almost in every case compared to that obtained without pretreatment. Electro-oxidation of carbonaceous gold ores, sulfidic gold and silver ores containing pyrite, pyrrhotite, and arsenopyrite has been extensively studied in a laboratory and pilot scale. There are also several studies carried out on the oxidation of pyrite, pyrrhotite, chalcopyrite, sphalerite, arsenopyrite, gersdorffite (NiAsS) in chlorine saturated water.

Arsenopyritic gold is a major source of refractory gold not amenable to roasting. Bio-oxidation is a popular treatment option but is slow and needs careful control of the As content of the feed. Electrochemical slurry oxidation in-situ with chlorine in acidified salt water can provide a more environmentally friendly process flow-sheet than carbon-in-pulp/leach processing and can be applied to a small or large resource. Chlorine and hypochlorite are not cheap and not fully used in oxidation, electrooxidation process may be alternative and economical to those since all hypochlorite produced in situ are fully used in oxidation.

In order to explain the mechanism of electrooxidation of refractory (carbonaceous and sulfide type) gold ores it is better to understand the production of hypochlorite that is given as follows.

2. Hypochlorite production

Hypochlorite solutions are easy to handle and relatively safe compared to chlorine, but purchased liquid hypochlorite has some disadvantages. Its transportation cost is high due to having concentrations normally range up to 12-15% of available chlorine. Even these concentrations tend to be unstable and decompose easily. A sodium hypochlorite solution generated in situ from readily available salt, on the other hand, is safe and relatively stable (5). Hypochlorite solutions are prepared commercially by electrolyzing a stirred sodium chloride solution. The electrolysis of sodium chloride solutions gives Cl_2 molecules and OH^- ions, and stirring ensures that these species react with each other. The resulting solution is a hypochlorite ion that is a strong reagent (6). Two types of cell design are used to produce hypochlorite, the diaphragm type and non-diaphragm type electrolytic cells.

2.1 Diaphragm type electrolytic cells

Chlorine is produced in an electrolyzer where a diaphragm separates the anodic compartment from the cathodic (7) A solution of sodium chloride flows into the anodic compartment and the reaction at the anode is:

$$2Cl^- \rightarrow Cl_2 + 2e^- \qquad (1)$$

A gaseous chlorine dissolves to some extent in the brine and the following equilibria are established in the diaphragm:

$$Cl_2 + H_2O \rightarrow ClO^- + Cl^- + H^+ \qquad (2)$$

$$H^+ + ClO^- \rightarrow HClO \qquad (3)$$

The cathodic reaction is:

$$2H_2O + 2e^- \rightarrow H_2 + 2OH^- \qquad (4)$$

The OH^- ions diffuse into the diaphram, where they combine with H^+ ions to form water:

$$H^+ + OH^- \rightarrow H_2O \quad (5)$$

In addition, chemical and electrochemical chlorate formation reactions take place according to the following reactions:

$$2HClO + ClO^- \rightarrow ClO_3^- + 2Cl^- + 2H^+ \quad (6)$$

$$6OCl^- + 6OH^- \rightarrow 2ClO_3^- + 4Cl^- + 3H_2O + 3/2O_2 + 6e^- \quad (7)$$

Reactions (3) and (5) proceed much more rapidly than mass transport due to convection and diffusion. Reaction (6) proceeds relatively slowly with respect to those proceedings. The anodic side becomes acidic and the cathodic side is alkaline. The neutralization reaction in the diaphragm results in the formation of salt and water. At low pH, chlorate formation is mainly due to the homogeneous chemical reaction (6), as the electrochemical reaction (7) occurs at high pH (39). Reaction (6) is competitive with the main anodic reaction (1); therefore, it is inefficient with regard to chlorate formation. For chlorate formation according to reaction (6), both pH and temperature controls are important. The stoichiometric ratio of hypochlorous acid to hypochlorite ion required by reaction (3) is achieved by having a pH of about 6 and the rate of reaction increased by about two orders of magnitude by raising the temperature from 20°C to 80°C (8). At elevated temperatures, however, the solubility of chlorine decreases and the amount of undissociated hypochlorous acid escaping from the reactor increased, both of which lead to process inefficiencies. Figure 1 shows the temperature effect on the solubility of chlorine gas (9). The saturation concentration for chlorine is given as 0.06 mole/liter (10).

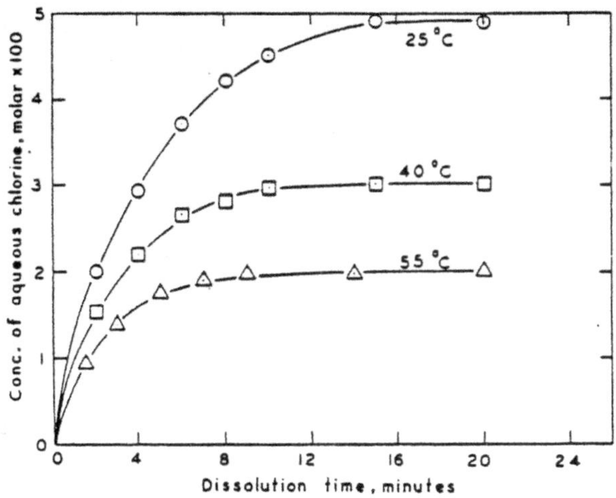

Fig. 1. Temperature effect on the solubility of chlorine gas (9). (Experimental conditions: 0.1 M HCl solution and 400 rpm.)

Cell components, such as electrodes and separators, affect the design and the selection of the best electrolyzer. The design of the anode structure of a diaphragm type chlorine cell using a graphite plate anode used to be limited by the mechanical properties and the machinability of graphite. Gradual degradation during electrolysis was also a problem (11). Therefore, carbon electrodes were substituted nearly everywhere (chlor-alkali electrolysis, hypochlorite and chlorate productions) by oxide coated Ti and dimensionally stable anodes (DSA), and the coatings are reported to have life times of 3 to 5 years (11). The reasons for using titanium anodes are summarized as follows: (11, 12)

- The excellent stability of titanium against general and pitting corrosion in acidic and slightly basic aqueous solutions.
- Application of RuO_2 coatings or other coatings containing metal oxides or platinum group metals allows the reduction of the over potential for anodic chlorine and anodic oxygen evolution (13). These coatings are particularly stable on a titanium base metal.
- The price of titanium has decreased greatly during the last decade so that costs are no longer prohibitive for a more extensive use of the Ti anode.

Noble metals, in particular Pt metals, are too expensive to be used as bulk materials for electrodes. It is more appropriate to apply them as a coating to a suitable support. Now, because of using metal anodes, cell designs are more flexible. Due to corrosion, the cathodic compartment of a diaphragm type chlorine cell must be fabricated with corrosion resistant materials such as stainless steel (11).

Production of chlorate was studied with a divided laboratory cell equipped with the oxide coated (RuO_2-TiO_2, Pt-IrO_2, and PdO) Ti anodes and Pt foil cathode in the respective compartments (14). Figure 2 shows an example of the concentration changes of the chemical species existing in the anolyte as functions of electrolysis time. The concentrations of ClO^-, HCIO, and ClO_2^- increased for the first three hours and then became constant. On the other hand, the ClO_2^- concentration increased almost constantly with time after the electrolytic cell reached the steady state. The rates were found to be independent of anode material and are a function of temperature and solution pH. Only the PdO coated anode showed slightly lower values than the others.

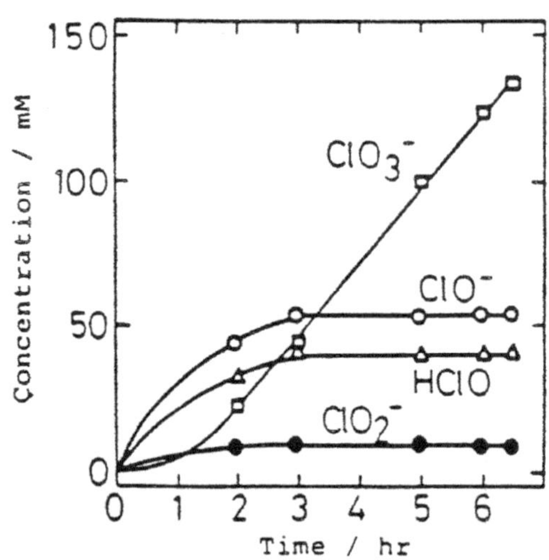

Fig. 2. Solution composition change during electrolysis of NaCl solution (15). (Experimental conditions: 4.0 M NaCl, pH=7.0, T=50oC, E=1.2V vs SCE)

Ibl and Landolt have also studied the anodic chlorate formation in dilute NaCl solutions in a diaphragm type cell using graphite electrodes (10). The diaphragm was porous polyethylene and the solution temperature was 7±0.5°C. Hypochlorite and chlorate concentrations were measured as a function of time and the results are illustrated in Figure 3 (10). In the each run, the change of hypochlorite concentration with time was followed until a steady state with a constant hypochlorite concentration was reached. Concentration profiles near the anode are shown in Figure 4 (10). The hypochlorite concentration increases with the distance from the anode where chlorine concentration shows a decreasing trend.

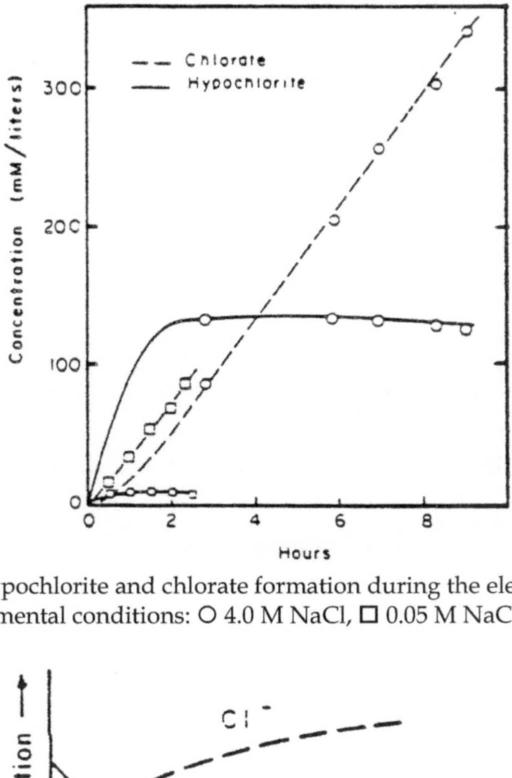

Fig. 3. Variation of hypochlorite and chlorate formation during the electrolysis of NaCl solution (10). (Experimental conditions: O 4.0 M NaCl, □ 0.05 M NaCl)

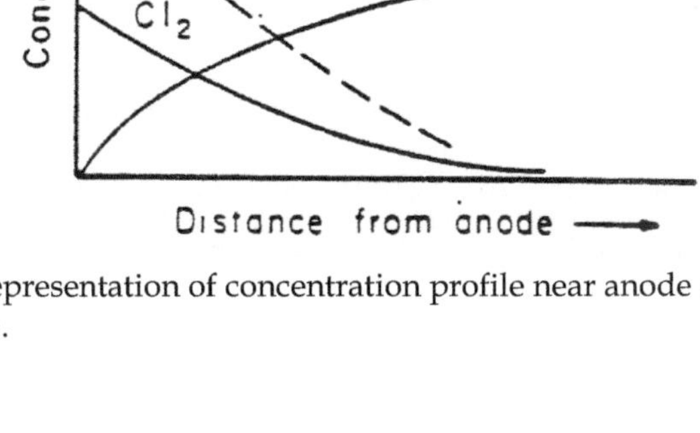

Fig. 4. Schematical representation of concentration profile near anode during the electrolysis of NaCl solution (10).

Separators are another component in the cell design which should be considered. Na^+, Cl^-, and OH^- are the principal ionic species in the cell (11). The sodium ion is the principal electrolytic current carrier across the separator. Hydroxyl ions also carry some current which reduces the current efficiency of hypochlorite production process. Their transport into anolyte causes chlorate formation with the chlorine and unacceptable losses in reduced chlorine evolution. Therefore, the separator should be designed in such a way that the transportation of the OH^- ions from the catholyte across the anolyte could be prevented. A diaphragm should be operated such that the bulk flow through the pores is sufficient to keep the OH^- ions from migrating across the separator towards the anode in order to prevent reaction (7).

2.2 Non-diaphragm type electrolytic cells

Chemical reactions which occur in this type of cell are summarized in the following. In the electrolysis of NaCI solutions, chlorine is generated at the anode according to the following reaction (5,10):

$$2Cl^- \rightarrow Cl_2 + 2e^- \quad (8)$$

The chlorine generated immediately undergoes hydrolysis according to the reactions:

$$Cl_2 + H_2O \rightarrow HClO + Cl^- + H^+ \quad (9)$$

$$HClO \rightarrow ClO^- + H^+ \quad (10)$$

In non-diaphragm hypochlorite cells, solution pH is allowed to range from 7 to 9, insuring complete hydrolysis of the chlorine generated. The pH is balanced by the cathodic evolution of hydrogen as follows:

$$2H+ + 2e^- \rightarrow H_2 \quad (11)$$

Equations (8) to (11) describe the ideal behavior of a non-diaphragm hypochlorite cell. There are, however, several competing reactions which contribute to cell inefficiencies. Chlorate may be formed by either anodic oxidation of hypochlorite,

$$6OCl^- + 3H_2O \rightarrow 2ClO_3^- + 6H^+ + 4Cl^- + 3H_2O + 3/2O_2 + 6e^- \quad (12)$$

or by the chemical reaction between the hypochlorite ion and hypochlorous acid:

$$2HClO + ClO^- \rightarrow ClO_3^- + 2Cl^- + 2H^+ \quad (13)$$

Both of these reactions (12 and 13) have been studied extensively because they are significant in the electrolytic production of chlorate. The chemical chlorate formation, equation (14), proceeds very slowly at room temperature and a basic pH, and when the solution is warmed the disproportionate reaction of hypochlorite occurs rapidly (16). This reaction is, therefore, not significant because cells do not operate over 40°C, and normally equilibrate near pH 9 when making 1 % available chlorine solutions.

Current efficiency may also be lost due to the anodic discharge of hydroxyl ions:

$$2OH^- \rightarrow 2H^+ + O_2 + 4e^- \quad (14)$$

This reaction competes with anodic chlorine evolution. Another reaction which is well documented in the literature (17) because of its importance in chlorate production cells is the cathodic reduction of hypochlorite:

$$ClO^- + H_2O + 2e^- \rightarrow Cl^- + 2OH^- \quad (15)$$

Finally, a small contribution to inefficiency can result from the chemical reaction:

$$2ClO^- \rightarrow O_2 + 2Cl^- \quad (16)$$

This reaction is normally significant only in the presence of electrolyte impurities such as iron, or in the presence of active catalytic surfaces.

In summary, a non-diaphragm hypochlorite cell functions through equations (8) to (11). Major current efficiency losses result from equations (12), (13), and (14), with minor losses from equations (15) and (16).

In the basic solutions, the following reaction occurs instead of reaction (9) to produce hypochlorite (18):

$$1/2Cl_2 + 2OH^- \rightarrow OCl^- + H_2O + e^- \quad (17)$$

In reaction (8), neither H^+ nor OH^- is involved, so the reduction potential is independent of pH. However, chlorine does not exist in aqueous solution at pH's higher than 5. Hypochlorous acid is a weak acid ($pK_a = 7.4$), and therefore it predominates at any pH below 7.4 and OCl^- is predominant in the basic solutions. The influence of pH on the dissociation of hypochlorous acid is shown in Figure 5 (19).

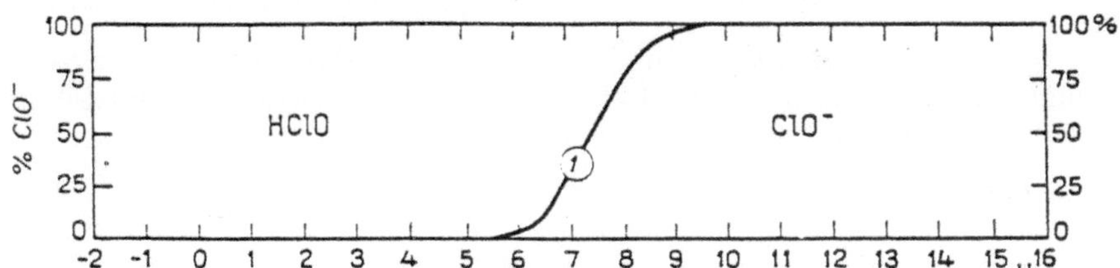

Fig. 5. Effect of pH on the dissociation of hypochlorous acid (19).

ClO_3^- is produced with the electrolysis of hot saturated NaCl solution with the addition of a little dichromate to modify the cathode behavior. Its influence is the formation of a thin diaphragm of insoluble chromium chromate on the cathode which is produced by the reduction of dichromate to chromic ions (20). The diaphragm prevents contact of the bulk of the solution with the electrode and thus with the hydrogen. The electrodes employed were graphite anode and iron cathode (18). In another study, a smooth platinum anode and a wrought iron cathode were used as electrodes (20). If the solution were kept cold and if the electrolysis were stopped at an earlier point, the main product would be hypochlorite.

The relative effects of operating parameters on cell performance were studied by Bennett in batch electrolysis (5). Experiments were performed in a simple compartment cell using DSA anodes (widely used by the chlorine industry), bare titanium cathodes, 1 A/in^2 current density, and 30 g/l NaCI feed. Anodic current efficiency for producing hypochlorite is found to be approximately 40%. Of the 60% current efficiency lost, approximately 35% is due to the cathodic reduction of hypochlorite, 10% to the anodic oxidation of hypochlorite, and 15% to the anodic evolution of oxygen from hydroxyl ion. According to these results, current density had a significant effect on the hypochlorite production. When 0.5 A/in^2 current density was used, anodic current efficiency was near zero due to the sharp increase in the tendency for the cathodic reduction of hypochlorites. Later, a multi-stage cell with continuous flow was developed. Hypochlorite production was gradually increased and total current efficiency in this case could be near 65%, or 25% higher than that for a single compartment.

At lower concentrations, salt utilization was found to be poor, and at higher concentrations, the loss of current efficiency made the operation economically unattractive. A multi-stage electrolytic hypochlorite cell (SANILAC, developed by Diamond Shamrock Corp.) developed as a result of this study, using DSA anodes, is capable of producing 50 lbs/day of chlorine (5). The cell feed consists of a 3% NaCI solution. At available chlorine concentrations of 9.0-9.5 g/l, this cell uses 2.3 to 2.4 kWh(AC) and approximately 3 lb of salt per lb of chlorine. Several sizes of cells producing up to 1000 lb/day of chlorine are commercially available. Similar technology is also being used to produce chlorine economically from seawater.

2.3 Electrooxidation of refractory gold ores

Electrolysis was investigated by the USBM as a means of generating oxidizing conditions in situ in the ore pulp prepared from finely ground carbonaceous ore and NaCl brine (4). This would result in (a) the potential reduction of costs incident to the purchase of hypochlorite and chlorine reagents as compared: (a) the costs of the power and salt, and (b) producing the required oxidant at the desired rate so that excess oxidant would not be consumed by the ore.

The initial research was conducted on the Carlin type carbonaceous gold ores. Later, molybdenum and rhenium reeovery from molybdenite concentrates, and silver and mercury recovery from tailings and Cinnabar ores were studied (21-26).

There are several important parameters in the investigation of electrooxidation, such as salt concentration, electrolysis time, temperature, current density, rate of electrolysis, types of electrodes, electrode spacing, and particle size of the ore.

The effect of salt concentration in the pulp on gold extraction was determined in a series of experiments in which the oxidation time was 7 hours and pulp density was maintained at 40%. Gold extraction increased linearly with respect to NaCl concentration (27). The concentration of salt maintained in the pulp is a principal variable since it affects the conductivity of the pulp and, consequently, the efficieney of hypochlorite production. The use of salt concentration in excess of 200 lb/ton ore resulted in a 95% gold extraction. The corresponding power consumption decreased from 100 kWh/ton ore to 65 kWh/ton ore at the 95% gold extraction range (4).

The effect of temperature on gold extraction was determined at 30, 40, and 50°C. The maximum gold extraction was obtained at 40°C. Generally, the electrode spacing should be as close as is consistent with good pulp flow between the electrodes since the resistance between the electrodes inereases with an inerease in electrode spacing. Conductivity is a function of pulp density and electrode spacing; and pulp density must be adjusted to give the desired temperature. Generally, the use of lower pulp densities enables the use of closer electrode spacing, which in turn allows the use of lower salt concentrations. However, this is limited by lower concentrations of the oxidant agent because of slow reaction rates (28).

Current density is another factor which affects the voltage and the temperature. As current density increases, the voltage required also increases. The result of high current density is excessive power consumption and increase of heat input to the system. High current densities also cause undesirable electrode reactions that produce $NaClO_3$ in the electrolytic cell. According to the results, it was found that a current density of 0.5 to 0.75 A/in^2 of anode surface was acceptable (4).

Grinding is an important part of any hydrometallurgical process beeause it is necessary to liberate the mineral from the host rock so that the minerals can come in contact with the reactants. Gold extraction increases as the partide size becomes smaller. It was found that 70% -200 mesh was satisfactory for this ore; however, every ore must be evaluated in order to find the optimum partide size for reaction (28).

Several electrode systems were investigated by the USBM (4, 28, 29). The first electrode design utilized was tubular electrodes, as shown in Figure 6. Preliminary experiments with this electrode indicated that considerable resistance to pulp flow was encountered, resulting in a buildup of pulp inside the cathode pipe. The circulation of pulp through this cell was satisfactory during the initial experiments, but with time a deposit of slimes was observed to build up on the cathode, restricting pulp flow.

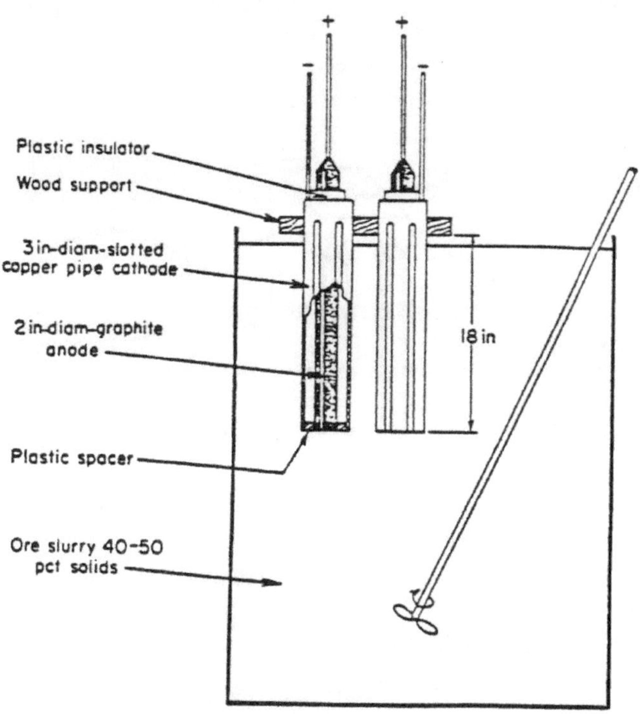

Fig. 6. Tubular type electrode design (from ref. 28)

The second electrode system was a plate-type electrode system consisting of the series of copper cathodes and graphite anodes, as shown in Figure 7. Pulp flow through the eleetrodes was excellent, but a cathode deposit was still observed to build up with the continued use of the electrode. An alternative arrangement was to substitute graphite for the copper in the plate-type electrodes so that the deposit could be removed by reversing the current. The plate-type arrangement of the cell also offered much less resistance to the pulp flow through the 0.5 inch spacing between the electrodes.

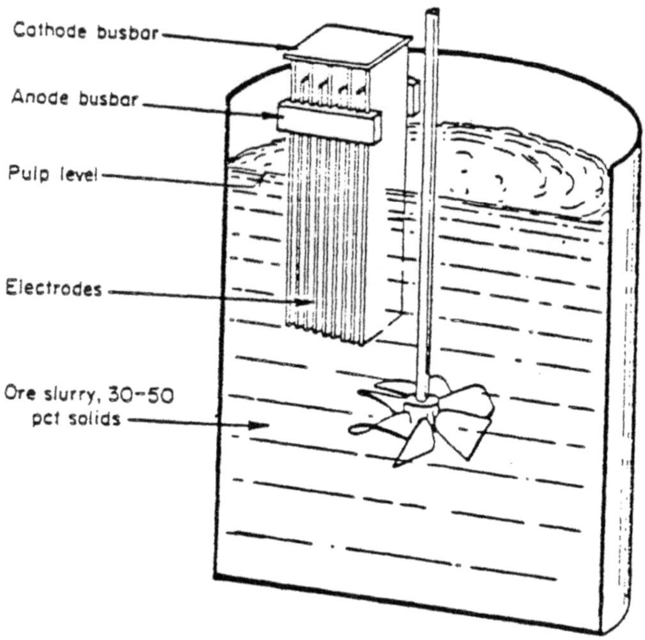

Fig. 7. Plate type electrode design (from ref. 28).

The other electrode design, utilizing a PbO_2 coated titanium anode and an iron cathode, was investigated to determine its effectiveness in generating that required oxidizing conditions at lower salt concentrations. 90% gold extraction by subsequent cyanidation was achieved with 4% salt solution, but because of the lower conductivity of the pulp, an additional 20% power was required.

Some pilot plant studies were also done. The data from the larger scale experiments indicated that scale-up to commercial plants should not present any problems.

In silver and mercury recovery from tailings and Cinnabar ores, the plate-type electrode design which is similar to that used in the treatment of carbonaceous gold ores has been used (21-23).

The bipolar cell design shown in Figure 8 has been used in the recovery of Mo and Re from molybdenite ores (26). The spacing between electrodes was 5/16 in and the cell consists of ten graphite electrodes held in a plexiglass container. 0.5 A/in^2 was applied. The operating temperature was maintained at 35-40°C by passing the pulp through water-cooled heat exchangers positioned between the stirred vessel and the cell. The pH was adjusted between 6 and 8. The capacity of the electrolytic oxidation was 1.5 lb/hr of Mo and the concentrate was fed at a rate of between 3.4 and 4.7 lb /h in order to determine the effects of process variables on Mo extraction (26).

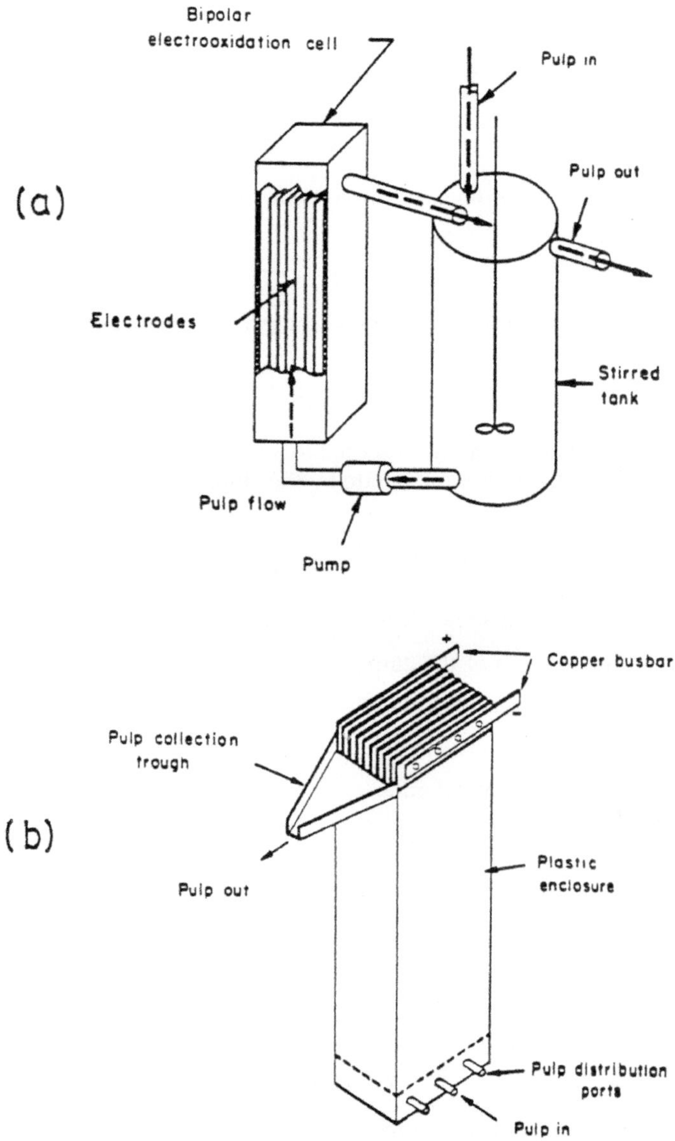

Fig. 8. Pulp flow sequence through agitator and bipolar cell (a), and electrooxidation electrode design (b) (from ref. 26)

Chemical reactions which occur during the electrooxidation can be summarized as follows, (23) the chloride ion is converted to chlorine at the anode:

$$2Cl^- \rightarrow Cl_2 + 2e^- \quad (18)$$

At the cathode water is hydrolyzed to produce hydroxyl ion and hydrogen:

$$2H_2O + 2e- \rightarrow H_2 + 2OH^- \quad (19)$$

The reaction between the chlorine and the hydroxyl ion produces hypochlorite according to the following reaction:

$$Cl_2 + 2OH^- \rightarrow OCl^- + Cl^- + H_2O \quad (20)$$

Hypochlorite then reacts with the carbonaceous material in the ore, passivating activated carbon:

$$C + 4OCl^- \rightarrow MCl_2 + SO_4^{2-} + 2Cl^- \quad (21)$$

and converting the metal sulfides to chlorides that can hydrolyze to a hydroxide depending the pH of the slurry (30):

$$MS + 4OCl^- \rightarrow CO_2 + 2Cl^- \quad (22)$$

The dissolution of the cinnabar can be observed as a two-step process involving oxidation of the sulfide followed by dissolution of sulfate by the chloride ion as shown in the following reactions (23):

$$HgS + 4OCl^- \rightarrow HgCl_4^{2-} + SO_4^{2-} \quad (23)$$

$$HgSO_4 + \text{excess } Cl^- \rightarrow HgCl_4^{2-} + SO_4^{2-} \quad (24)$$

Also, direct oxidation of some of the cinnabar apparently occurs at the anode to produce mercuric sulfate:

$$HgS + 4O^{2-} \rightarrow HgSO_4 + 8e^- \quad (25)$$

The hypochlorite reacts with molybdenite:

$$MoS_2 + 9OCl^- + OH^- \rightarrow MoO_4^{2-} + 9Cl^- + 2\,SO_4^{2-} \quad (26)$$

The reaction for the oxidation of rhenium with hypochlorite is:

$$Re_2S_7 + 28OCl^- + 16OH^- \rightarrow 2ReO_4^- + 28Cl^- + 7SO_4^{2-} + H_2O \quad (27)$$

When the technique was used for carbonaceous gold ores, the electrolyte performance was compared with the results obtained by chlorine on the same ore sample (30). The comparison showed that 1.3 to 1.5 kWh of power was required to replace around equivalent of chlorine. At that time, a monopolar cell was used which has the disadvantage of requiring a high amperage rectifier and a large bus bar. In 1978, the cell had been improved, and the new design (a bipolar cell) had overcome the disadvantages. This cell is capable of producing a pound equivalent of chlorine for approximately 1 kWh. Depending on the chlorine prices and the electrical power supply capacity in northeastern Nevada, the electrooxidation technique may provide an economical treatment for carbonaceous ores.

The mechanism of electrooxidation of carbonecous and sulfide type ores was investigated by Arslan and Duby (31-34) using the two compartment cell as shown in Figure 9, by separating the anolyte and catholyte parts in NaCl solutions. Laboratory work was carried out on the production of hypochlorite and the oxidation of pyrite in an electrochemical cell with a 10% sodium chloride electrolyte at 35–40°C. The production rate of hypochlorite was higher when the pH was not controlled than when it was maintained at the initial value of about 6.5.

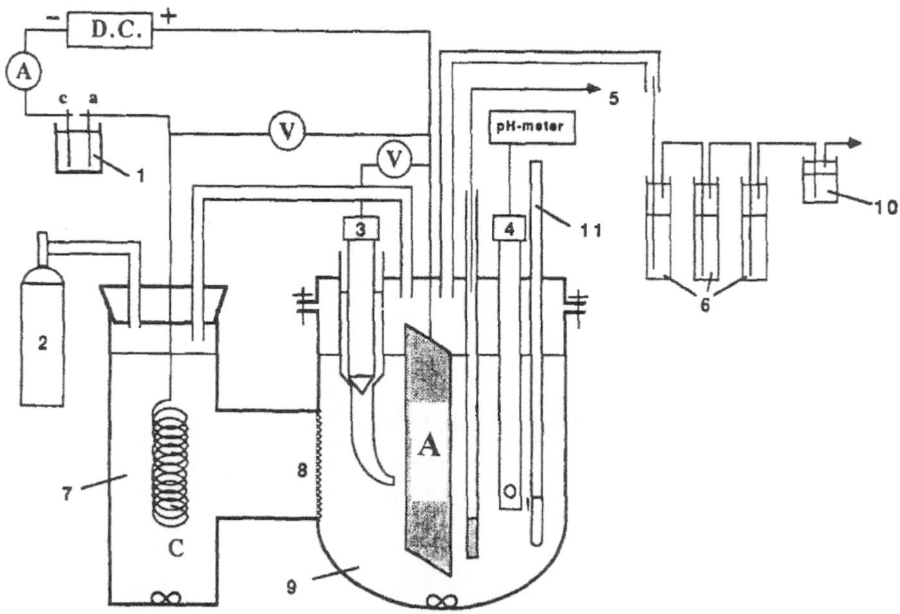

Fig. 9. Experimental arrangement: C=cathode, A=anode, 1=copper coulometer, 2=nitrogen gas tank, 3=saturated calomel electrode, 4=pH electrode, 5=sampling, 6=0.5M NaOH solution, 7=catholyte, 8=glass frit, 9=anolyte, 10=saturated CaO solution, 11=thermometer. (31)

First of all hypohlorite production is studied without any addition of carbon, pyrite or sulfide type gold ore (31,32). Similarly to the early ones, the chemical reactions which occur during the electrolysis of a NaCl solution can be expressed as follows. The chloride ion is converted to chlorine at the anode:

$$2Cl^- \rightarrow Cl_2 + 2e^- \quad (28)$$

At the cathode, water is hydrolyzed to produce hydroxyl ion and hydrogen:

$$2H_2O + 2e^- \rightarrow H_2 + 2OH^- \quad (29)$$

At the anodic side, hypochlorite is produced according to the following reactions. Chlorine reacts with water to produce hypochlorous acid:

$$Cl_2 + H_2O \rightarrow HClO + H^+ + Cl^- \quad (30)$$

then, the dissociation of hypochlorous acid occurs depending on the solution pH:

$$HClO \rightarrow ClO^- + H^+ \quad (31)$$

and hydroxyl ions produced at the cathode react with hydrogen ions to form water:

$$H^+ + OH^- \rightarrow H_2O \quad (32)$$

After a certain time, hypochlorite production levels off and this is attributed to the production of chlorates, either chemically,

$$2HClO + ClO^- \rightarrow ClO_3^- + 2H^+ + 2Cl^- \quad (33)$$

or electrochemically,

$$6ClO^- + 3H_2O \rightarrow 2ClO_3^- + 6H^+ + 4Cl^- + 3/2O_2 + 6e^- \quad (34)$$

Electrochemical reduction of hypochlorites at the cathode is also possible according to the following reaction:

$$ClO^- + 2H_2 + 2e^- \rightarrow H_2O + Cl^- \quad (35)$$

Another reaction is the anodic discharge of hydroxyl ions:

$$2H_2O \rightarrow 4H^+ + O_2 + 4e^- \quad (36)$$

During the experiments the potential and pH measurements are also supported these reactions according to the potential-pH diagrams drawn for the system as hon in Figure 10 and 11. (31)

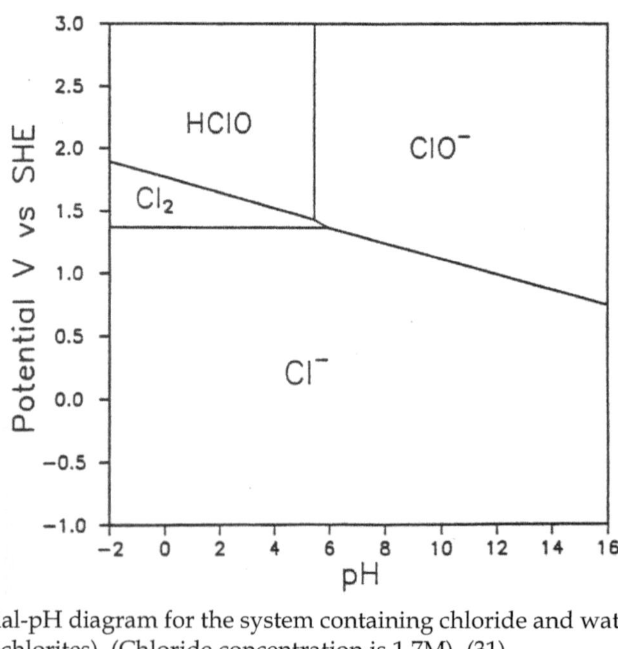

Fig. 10. A potential-pH diagram for the system containing chloride and water at 40°C (stability of hypochlorites). (Chloride concentration is 1.7M). (31)

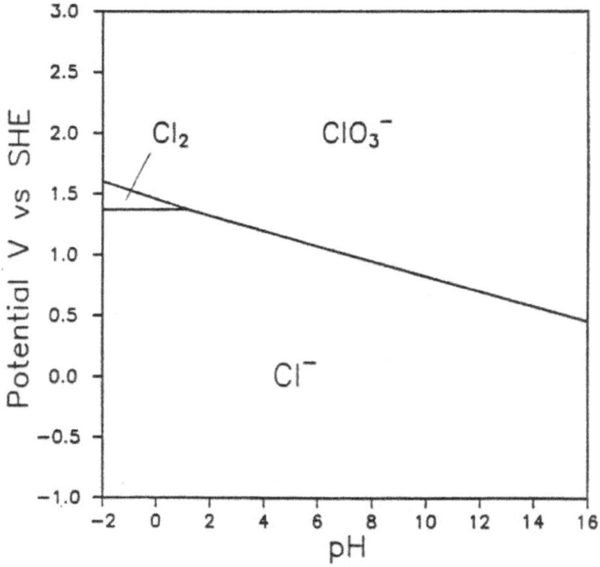

Fig. 11. A potential-pH diagram for the system containing chloride and water at 40°C (stability of chlorates). (Chloride concentration is 1.7M). (31)

During the electrooxidation of carbon, when the graphite anode was used, carbon oxidation was observed but it was difficult to determine the amount of CO_2 attributed to the oxidation of the graphite anode itself (31-32). Therefore, experiments were run using a Pt/Ir coated Ti anode. Carbon is oxidized by hypochlorite which is produced electrochemically in the cell, according to the following equations:

$$C + 2ClO^- \rightarrow CO_2 + 2Cl^- \quad (37)$$

$$C + 2HClO \rightarrow CO_2 + 2Cl^- + 2H^+ \quad (38)$$

According to the equations (37 and 38) two moles of HClO (or OCl^-) are required for the oxidation of one mole of carbon to CO_2, corresponding to 4 electrons. Potential and ph measurements taken during the tests were in agreement with those reactions and the potential-ph diagrams of C-Water system shown in Figure 12.

Electrooxidation of pyrite with hypochlorite generated electrochemically in the cell can be expressed as (31,33):

$$2\,FeS_2 + 15ClO^- + H_2O + 2H^+ \rightarrow 2Fe^{3+} + 4HSO_4^- + 15Cl^- \quad (39)$$

$$2\,FeS_2 + 15HClO^- + H_2O \rightarrow 2Fe^{3+} + 4HSO_4^- + 13H^+ + 15Cl^- \quad (40)$$

Pyrite oxidation with hypochlorite is a chemical reaction as given in reactions (39 and 40), but the production of chlorine that results in hypochlorite production is an electrochemical process. When the overall process is considered, 7.5 moles of HClO are required per mole of pyrite corresponding to 15 electrons.

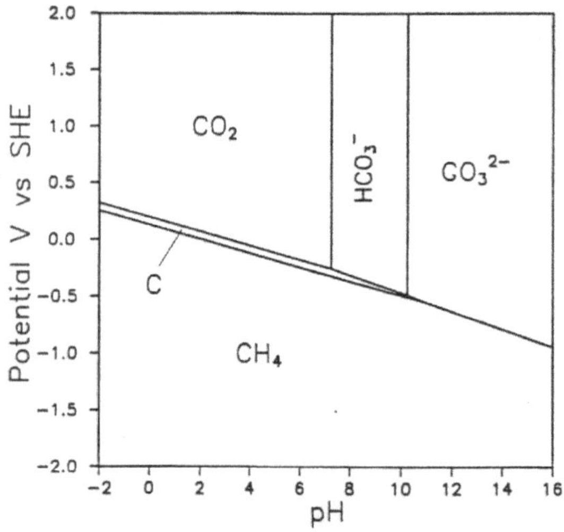

Fig. 12. A potential-pH diagram for the system containing carbon and water at 40°C (for dissolved substances). (Carbon concentration is 0.2M). (31)

Total iron in the solution increased with time at the beginning but leveled off after 4 hours and the current efficiency was about 97%. The oxidation rate was directly related to the current and it was not affected by the pulp density.The solution pH was 1.7 at the end of a 7 hour run. This shows pyrite oxidation according to reaction (40). It was reported in the literature (26) that the oxidation of concentrates high in sulfur or pyrite causes a sharp drop in the pH value as observed in our study. The potential and pH measurements are also agreed with those findings by comparing the results with the potential-pH diagrams given in Figures 12 and 13.

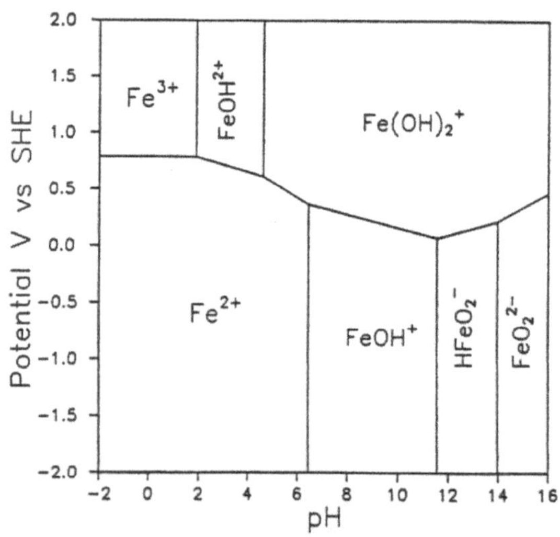

Fig. 13. A potential-pH diagram for the system containing iron and water at 40°C (for dissolved substances). (Iron concentration is 0.05M). (31)

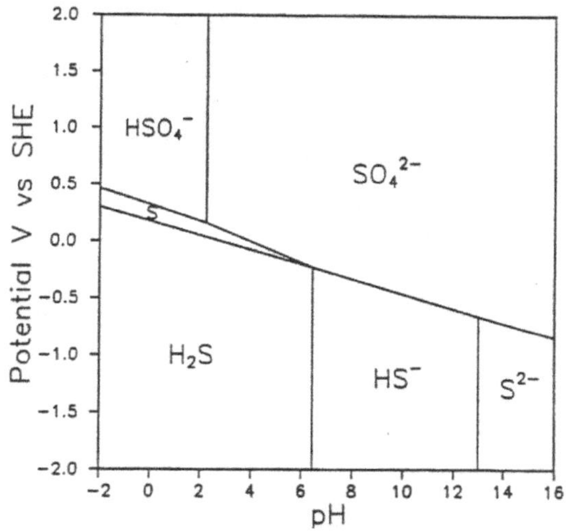

Fig. 14. A potential-pH diagram for the system containing sulfur and water at 40°C (for dissolved substances). (Sulfur concentration is 0.1M). (31)

In tests of electrooxidation of pyritic gold flotation concentrate, stoichiometric calculations show that 23% of total iron in the concentrate is in the form of pyrite, 71% as pyrrhotite, and 6% as arsenopyrite (31, 34). Therefore, the total iron dissolved in the solution may come from these minerals. Pyrite (as given above), pyrrhotite, and arsenpyrite are oxidized according to the following reactions:

$$2FeS_2 + H_2O + 15HClO \rightarrow 2Fe^{3+} + 15Cl^- + 4HSO_4^- + 13H^+ \quad (41)$$

$$2FeS + 9HClO \rightarrow 2Fe^{3+} + 9Cl^- + 2HSO_4^- + 5H^+ + H_2O \quad (42)$$

$$FeAsS + 7HClO + H_2O \rightarrow Fe^{3+} + 7Cl^- + H_3AsO_4 + HSO_4^- + 5H+ \quad (43)$$

According to reactions (41-43), 15 electrons are needed for pyrite oxidation, 9 electrons for pyrrhotite oxidation and 14 electrons for arsenopyrite oxidation. Thus, the weighted average number of electrons is approximately 10.5 and the current efficiencies are calculated by using n=10.5 in all cases. At the end of each run the anolyte pH was around 1 and the catholyte pH was about 12. Analysis of anolyte solutions showed no Fe^{2+}, therefore, all iron in the solution was in the form of Fe^{+3}. This agrees with the oxidation reactions given above and can be seen from Figure 14.

Abrantes and Costa is also studied electro-oxidation as a potential alternative process to conventional pre-treatment of a refractory gold ore was investigated. (35) A particulate bed anode of a Portuguese concentrate containing gold was electroleached in a two-compartment electrolytic cell. The effectiveness of the electrochemical treatment is emphasized by comparison with the results of direct leaching under similar conditions. The influence of the operative parameters (e.g., temperature, electrolysis time, current and slurry density) on the overall gold extraction are studied. Up to 90% gold extraction was obtained after electrolytic pre-treatment. A comparison between chemical and electrochemical oxidation using chloride and sulphate solutions was also carried out. The results demonstrate the better performance of chloride media and that the electrolytic route enhances the amount of precious metal dissolved.

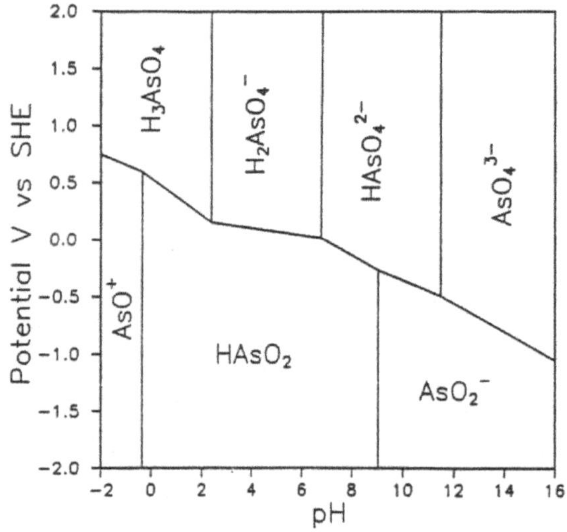

Fig. 15. A potential-pH diagram for the system containing arsenic and water at 40°C (for dissolved substances). (Arsenic concentration is 1.0×10^{-3}M). (31)

Since arsenopyritic gold is a major source of refractory gold not amenable to roasting and bio-oxidation is a popular treatment option but is slow and needs careful control of the As content of the feed, H. G. In Linge and N. J. Welham studied, electrochemical slurry oxidation in-situ with chlorine in acidified salt water in a membrane cell reactor has been evaluated at laboratory-scale level as a potential processing route (36). A refractory mine concentrate (cyanideable gold content ~7%) containing arsenopyrite (FeAsS), pyrite (FeS_2) and gersdorffite (NiAsS) has been used to demonstrate that gold recovery > 90% could be achieved. Reaction rates are but sale of the co-produced NaOH can be a valuable process credit. This route can provide a more environmentally friendly process flow-sheet than carbon-in-pulp/leach processing and can be applied to a small or large resource.

The silver mine of the Phoenix Mountain in Guangxi is complicated silver mine with higher arsenic, antimony and copper content (37). The electroleaching of silver is studied and it was found that electrolysis conditions are controlled easy. The recurrence of the process of scaling-up from small to large-size plant is very good, and the silver extraction rate is over 95%.

The electrochemical leaching technology for complex refractory ore was also studied for its high flexibility, easy operation and little pollution (38). It is pointed out that out-field intensification and multifield coupling technology was the development trend for ore processing.

3. References

[1] Osseo-Asare K., Afenya P.M., and Abotski G.M., "Carbonaceous matter in gold ores: isolation, characterization, and adsorption behavior in aurocyanide solutions",

Precious Metals: Mining, Extraction, and Processing, Ed. By V. Kudryk, D.A. Corrigan, and W.W. Liang, 1984, TMS Publication, pp. 125-144.
[2] Guay W.J. and Gross M.A., "The treatment of refractory gold ores containing carbonaceous material and sulfides", The AIME annual meeting, Chicago, Illinois, Feb. 22-26, 1981.
[3] Demopoulos G.P., "Mineralogical and technological aspects of gold extraction from refractory ores", presented at the Professional Development Seminar in "Mineral Processing of Gold Ores" held at the Dept. of Mining and Metallurgical Eng., McGill Univ., April 30-May 2, 1987.
[4] Scheiner B.J., Lindstrom R.E., Guay W.J., and Peterson D.G., "Extraction of gold from carbonaceous ores: pilot plant studies", Bureau of Mines RI 7597, 1972.
[5] Bennett J.E., "Non-diaphragm electrolytic hypochlorite generators", Chemical Engineering Progress, vol 70, no 12, Dec. 1974, pp 60-63.
[6] Masterton W.l., Slowinski E.J., and Stanitski C.L., Chemical Principles, 5th ed., Saunders College Pub., Philadelphia, 1981, pp 530-533.
[7] Rousar I., Micka K., and Kimla A., Electrochemical Engineering I, Elsevier Science Pub. Co. Inc., New York, 1986, pp 326-330.
[8] Pickett D.J., Electrochemical Reactor Design, Elsevier Science Pub. Co. Inc., New York, 1977, pp 362-363.
[9] Cho E.H., "Coal desulfurization with aqueous chlorine", Met. Trans. B., Oct 1989, vol 20B, pp 567-571.
[10] Ibl N. and Landolt D., "On the mechanism of anodic chlorate formation in dilute NaCl solutions", J. Electrochem. Soc., vol 115, no 7, July 1968, pp 713-720.
[11] Ismail M.I., Electrochemical Reactors –Their Science and Technology, Part A, Elsevier Science Pub. Co. Inc., New York, 1989, pp 73-75.
[12] de Nora V., "From DSA to NCA", proc. of the symp. on Performance of Electrodes for Industrial Electrochemical Processes, vol 89-10, The Electrochem. Soc. Inc., 1989, pp 1-13.
[13] Muranaga T., Kanaya Y., and Yokota N., "Characteristic of platinum group metal anode for chlor-alkali cells", proc. of the symp. on Performance of Electrodes for Industrial Electrochemical Processes, vol 89-10, The Electrochem. Soc. Inc., 1989, pp 77-86.
[14] Chemistry of Cyanidation, Mineral Dressing Notes, American Cyanamid Co., AIME annual meeting, New York, Feb. 13-16, 1950.
[15] Tasaka A. et al., "Activation energy of chlorate production with the oxide coated anodes", proc. of the symp. on Electrochemical Engineering in the Chlor-Alkali and Chlorate Industries, The Electrochem. Soc. Inc., 1988, pp 147-156.
[16] Segal B.G., Chemistry –Experiment and Theory, John Wiley and Sons Inc., New York, 1985, pp 570-572.
[17] Wu J., "Kinetics of the reduction of hypochlorite ion", J. Electrochem. Soc., vol 134, no 6, 1987, pp 1462-1467.
[18] Selly N.J., Experimental Approach to Electrochemistry, John Wiley and Sons Inc., New York, 1977, pp 99-101, 123.
[19] Pourbaix M., Atlas of Electrochemical Equilibria in Aqueous Solutions, NACE, 1974, pp 591-603.
[20] Knibbs N.V.S. and Palfreeman H., "The theory of Electrochemical Chlorate and perchlorate formation", Trans. of Faraday Soc., 1920, no 16, pp 402-433.

[21] Scheiner B.J., Pool D.L., and Lindstrom R.E., "Recovery of silver and mercury from mill tailings by electrooxidation", Bureau of Mines RI 7660, 1972.
[22] Shedd E.S., Scheiner B.J., and Lindstrom R.E., "Recovery of mercury from cinnabar ores by electrooxidation–extraction plant amenability tests", Bureau of Mines RI 8083, 1975.
[23] Scheiner B.J., Lindstrom R.E., and Shanks D.E., "Recovery of mercury from cinnabar ores", Bureau of Mines RI 7350, 1973.
[24] Scheiner B.J., Pool D.L., Lindstrom R.E., and McClelland G.E., "Prototype commercial electrooxidation cell for the recovery of molybdenum and rhenium from molybdenite concentrates", Bureau of Mines RI 8357, 1979.
[25] Lindstrom R.E. and Scheiner B.J., "Extraction of molybdenum and rhenium from concentrates by electrooxidation", Bureau of Mines RI 7802, 1973.
[26] Scheiner B.J., Lindstrom R.E., and Pool D.L., "Extraction and recovery of molybdenum and rhenium from molybdenite concentrates by electrooxidation: process demonstration", Bureau of Mines RI 8145, 1976.
[27] Scheiner B.J., Lindstrom R.E., and Henrie T.A., "Processing refractory carbonaceous ores for gold recovery", Journal of Metals, March 1971, pp 37-40.
[28] Scheiner B.J., Lindstrom R.E., and Henrie T.A., "Oxidation process for improving gold recovery from carbon-bearing gold ores", Bureau of Mines RI 7573, 1971.
[29] Scheiner B.J., Lindstrom R.E., and Henrie T.A., "Process for recovering gold from carbonaceous ores", U.S. Patent no. 3,574,600, April 13, 1971.
[30] Scheiner B.J., "Relation of mineralogy to treatment methods for carbonaceous gold ores", The SME annual meeting, Denver, Colorado, Feb. 24-27, 1987.
[31] F. Arslan, Electrooxidation of refractory gold ores. Ph.D. Thesis, Columbia Univ. (1991).
[32] F. Arslan, P.F. Duby , Mechanism Of Electrooxidation Of Refractory Gold Ores, SME, 1991, Preprint Number 91-38 9 pages.
[33] F. Arslan, P.F. Duby, Electro-oxidation of pyrite in sodium chloride solutions, Hydrometallurgy, Volume 46, Issues 1-2, August 1997, pp 157-169.
[34] F. Arslan, P.F. Duby, Electrooxidation of gold-bearing sulfide concentrate, Minerals & Metallurgical Processing, Feb 1, 2003. vol. 20, no 1, pp. 10-14.
[35] L. M. Abrantes and M. C. Costa, Electro-oxidation as a pretreatment for gold recovery, Hydrometallurgy, Volume 40, Issues 1-2, January 1996, pp 99-110.
[36] H. G. Linge and N. J. Welham, Gold recovery from a refractory arsenopyrite (FeAsS) concentrate by in-situ slurry oxidation, Minerals Engineering, Volume 10, Issue 6, June 1997, Pages 557-566.
[37] WANG Wei-xi, LUO Jian-ping, LI Li-jun, Study on Extracting Silver from Silver Concentrate in Guangxi by Slurry Electrolysis Process. Nonferrous Metals (Extractive Metallurgy), 2006-05.
[38] Z.H. Liu, Y. Li, X.X. Zhou, J. Du, C.Y. Tao, Research Progress of Electro-oxidation of Intensification leaching for Refractory Ore, Advance Materials Research (volumes 236-238), Vol. Application of Chemical Engineering, May 2011, pp 775-780.

Part 5

Application in Biosystem

18

Nobel Metal Nanoparticles in Bio-LED

Yen-Hsun Su[1], Sheng-Lung Tu[2] and Wei-Min Zhang[3]
[1]*Department of Material Science and Engineering*
National Dong Hwa University, Tainan
[2]*Department of Resource Engineering, National Cheng Kung University, Tainan*
[3]*Department of Physics, National Cheng Kung University, Tainan*
Taiwan

1. Introduction

Nanomaterials are well known for their interesting physical chemical properties [1-7]. Recently, noble metal nanoparticles have attracted attention [9-15] due to their optical property of surface plasmon resonance. Surface plasmon resonance produces hot light spots due to a stronger electromagnetic field. Because of their low toxicity, ease of delivery, and strong and tunable surface plasmon resonance, noble metal nanoparticles are used in the field of the photothermal ablation therapy to treat cancer [10]. Plants help people confront the global warming by absorbing CO_2 in the photosynthesis process. If plants were able to do the photosynthesis process day and night, absorbing CO_2 will act in day and night.

In this paper, we investigate the life time of noble metal nanoparticle colloids on plants specifically. The buds of Floppers, Bryophyllum pinnatum (Lam.) Kurz., are watered the noble metal cluster colloids. The buds are recorded for six weeks. The results show that noble metal nanoparticle colloids exit plant growth, with the Pt nanoparticles aggregating on the leaf. We expect that this research has potential applications in the fields of biosensing, bio-LED, bioimaging, nanophotonics, photothermal ablation therapy, and nanotoxicity.

2. Experimental produre

Chemicals and materials. We prepared 10 L standard solution by adding MgCl= 10.0 ppm, KCl=0.5 ppm, NaCl= 8 ppm, and $FeCl_3$= 0.04 ppm in Milli-Q water. Trisodium citrate (99%) and $HAuCl_4$, $AgNO_3$, and H_2PtCl_6 were purchased from Showa Chemical Co. (Tokyo, Japan) and Alfa Aesar Chemical Co. (Lancashire, England) and used as supplied, respectively. Milli-Q water was obtained from the Milli-Q Element system (Massachusetts, USA), which employs ultra-clean materials and a succession of optimized water purification technologies to produce 18.2 MΩ/cm resistivity ultrapure water (at 25 °C), ideal for trace analysis methods. The solution used in this work was freshly prepared and deoxygenated by bubbling nitrogen gas for at least 30 min prior to use. All materials and apparatus were sterile.

Preparation of noble nanoparticles. Au NPs were prepared by a chemical-reduction method with Au, Pt, Ag-salt and Na-cit standard solutions. 1.5 ml of freshly prepared 0.34 M trisodium $citrate_{(aq)}$ solution was added to 15 ml of 0.5 mM $HAuCl_{4(aq)}$ solution, $AgNO_{3(aq)}$

solution, and $H_2PtCl_{6(aq)}$ solution at room temperature and under vigorous stirring for 48 hours. These solutions were stirred for 30 min to control the reaction rate of Au, Ag, and Pt nanoparticles, respectively. All materials and apparatus were sterile.

Preparation of Floppers, Bryophyllum pinnatum (Lam.) Kurz. We cut 40 leaves from the plant Bryophyllum pinnatum (Lam.) Kurz., and these were then placed in a 50 ml glass bottom and watered with standard solution. The environment was controlled at 300 K, with 100mW/square-cm radiation with sun lamps and 60% humidity. Buds grew at the edges of the leaves, and when the length of a bud reached 5.980 mm, it was cut, and put at the bottom of a 3 ml glass bottom. The cut buds were hold in a 300 K environment with 100mW/square-cm radiation with sun lamps and 60% humidity. A total of 697 buds were tested using either standard solution or noble metal colloids. The buds were watered with 0.5 ml of the standard solution or noble metal colloids. And all materials and apparatus were sterile.

Instruments and Measurements. Dark-field spectroscopy, zetasizer, and TEM were used in the analysis. The morphology of the noble metal nanoparticles was observed by TEM. The as-grown product was observed with a Hitachi model HF-2000 transmission electron microscope operating at 200 kV. The sample was directly dispersed onto a 200-mesh holey carbon film on a Cu grid for TEM analysis. After being watered with standard solution or noble metal colloids for six weeks, the leaves and roots of the buds were observed via dark-field spectroscopy.

3. Results

3.1 Morphology of Au, Ag, and Pt nanoparticles

The concentration of the citratic acidic group modifies the noble metal nanoparticles to control their size. The average size for Au, Ag, and Pt nanoparticles was measured with a zetasizer, and was 6.3 nm, 3.2 nm, and 5.8 nm, respectively. From the TEM observation in Fig. 1, the size of the Au, Ag, and Pt nanoparticles is in good agreement with the zetasizer measurements.

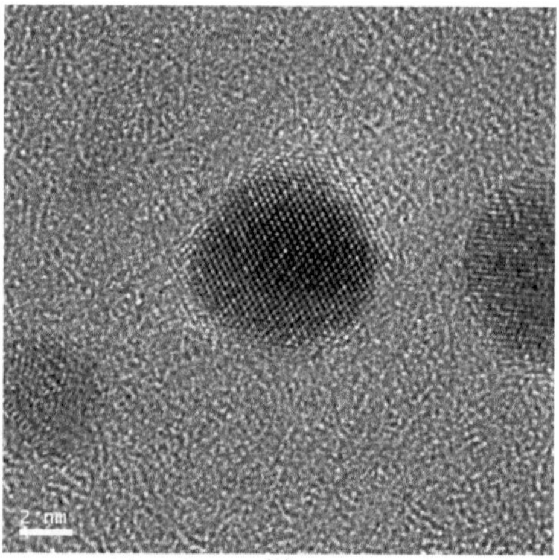

(a)

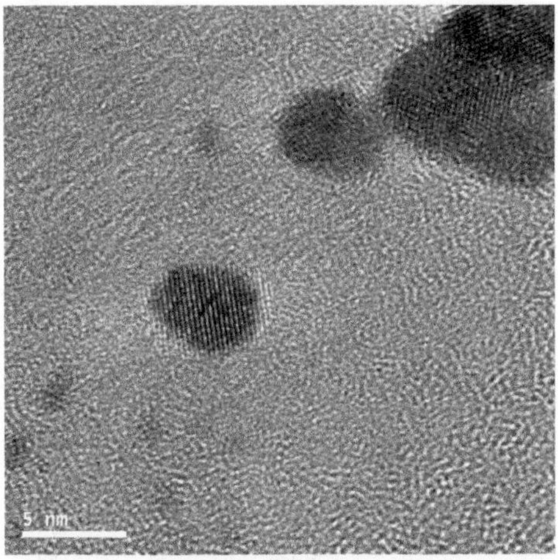

(b)

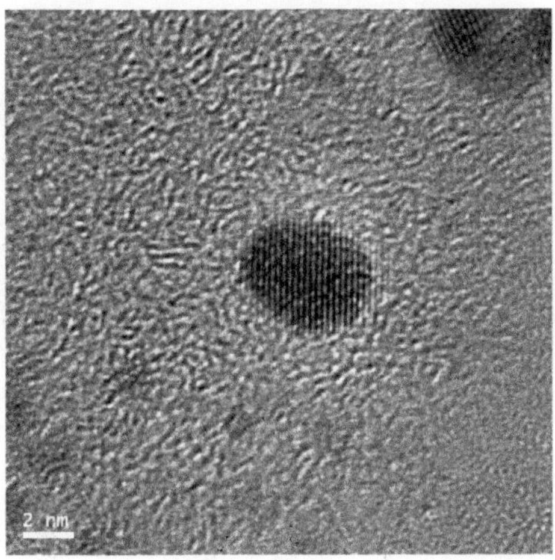

(c)

Fig. 1. Morphology of (a) Au, (b) Ag, and (c) Pt nanoparticles under TEM observation.

3.2 Buds watered with standard solution or noble metal colloids for six weeks

After being watered with standard solution or noble metal colloids for six weeks, dark-blue masses on the leaves are seen under the dark-field microscopic observation in Fig. 2. The leaves scatter green light, as shown in Fig. 2 (a). Metal clusters usually scatter strong light due to surface plasmon resonance [9,14]. And as the size of the particles (clusters) increases, so does the damping factor and the absorption. On the other hand, the light scattering gets weaker, as the size of particles (clusters) increases [14,15]. Thus, the metal cluster presents the dark-blue masses. In addition, the nanoparticles aggregate on the surface of leaves, and Fig. 2 shows that Pt nanopartcles aggregate together the most.

(a)

(b)

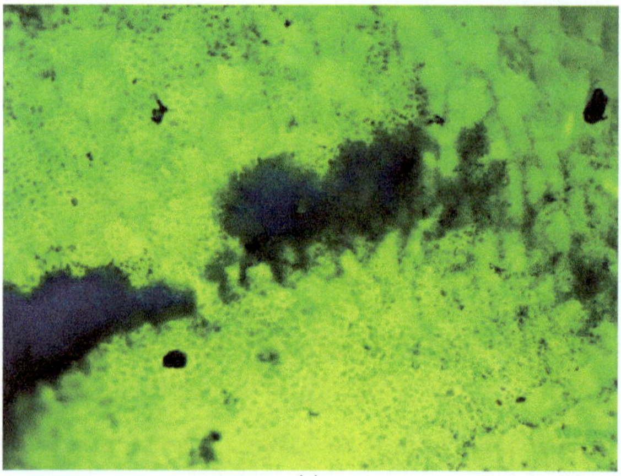

(c)

(d)

Fig. 2. Dark-field microscopic image of leaf on the bud watered with (a) standard solution, (b) Au colloids, (c) Pt colloids, and (d) Ag colloids for six weeks.

In Fig. 3, the dark-field microscopic observation shows the roots scatter the weak light, while the metal clusters scatter stronger light. As the size of the metal clusters decreases, the surface plasmon resonance increases, as does the scattering of light [14,15]. Therefore, the particles (clusters) size on the roots are smaller than these on the leaves. However, the light scattering of Pt clusters is weaker than that of the Au or Ag clusters, and Pt nanopartcles also aggregate together more seriously than the others on the roots.

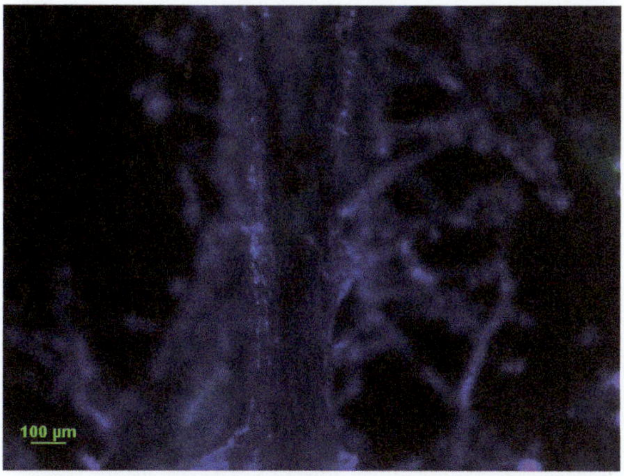

(a)

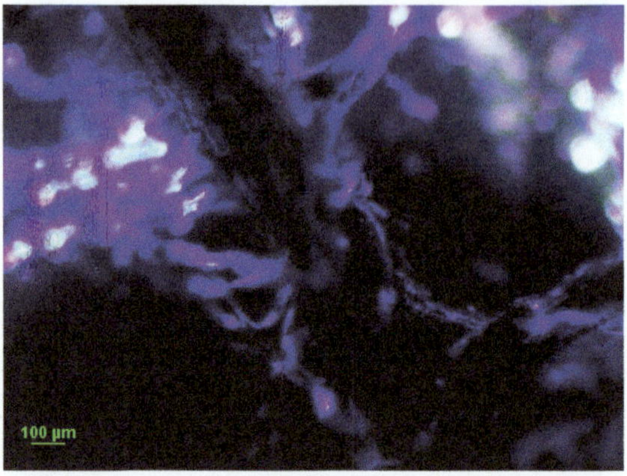

(b)

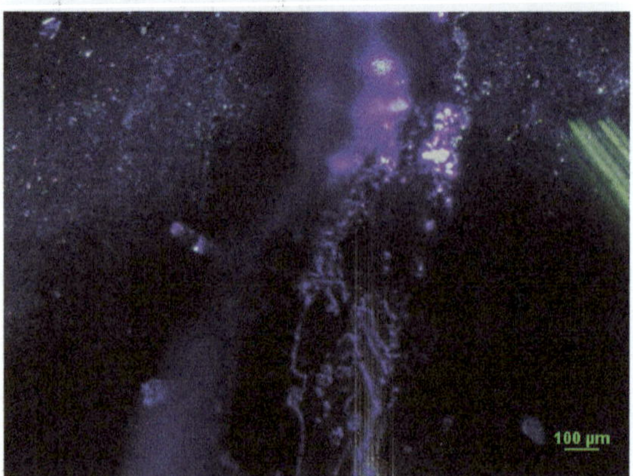

(c)

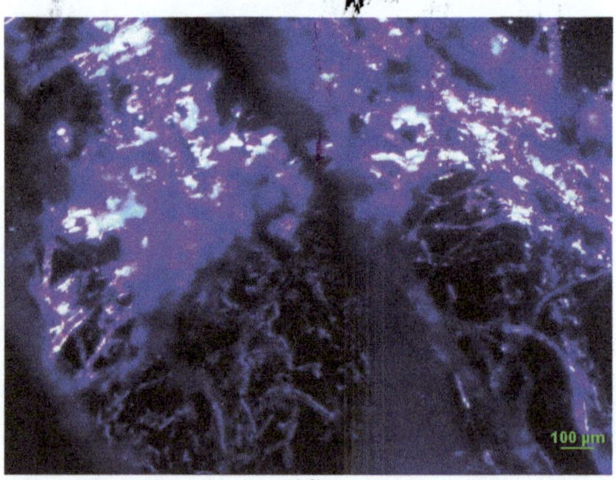

(d)

Fig. 3. Dark-field microscopic image of roots on the bud watered with (a) standard solution, (b) Au colloids, (c) Pt colloids, and (d) Ag colloids for six weeks.

According to Fig. 2~3, we observe the noble will not be corroded in leaves and roots. Install, noble metal nanoparticle will aggregate together in vein of leaves or the surface of roots.

4. Conclusions

We found that noble metal colloids significantly exit Floppers, Bryophyllum pinnatum (Lam.) Kurz. for several weeks. Future detailed studies should be performed to understand how the mechanism of nanoparticle leaves the plant body.

5. Acknowledgments

This work was financially supported by the National Science Council of Taiwan, No. 100-2218-E-259-003-MY3, which is gratefully acknowledged.

6. References

[1] Pedano ML, Li S, Schatz G C, Mirkin CA 2010 Angew. Chem. Int. Ed. 49 78-82.
[2] Camden J P, Dieringer J A, Wang Y, Masiello DJ, Marks LD, Schatz GC, Van Duyne RP 2008 J. Am Chem. Soc. 130 12616.
[3] Khlebtsov B, Khlebtsov N 2008 Nanotechnology 19 435703.
[4] Hao F, Sonnefraud Y, Van Dorpe P, Maier SA, Halas N J, Nordlander P 2008 Nano Lett. 8 983.
[5] Longo A, Pepe GP, Carotenuto G., Ruotolo A, De Nicola S, Belotelov VI, Zvezdin, AK 2007 Nanotechnology 18 365701.
[6] Maly J, Lampova H, Semeradtova A, Stofik M, Kovacik L 2009 Nanotechnology 20 385101.
[7] Mutombo P, Kiss AM, Berko A, Chab V 2006 Nanotechnology 17 4112.
[8] Bouhelier A, Bachelot R, Lerondel G, Kostcheev S, Royer P, Wiederrecht1 GP 2005 Phys. Rev. Lett. 95 267405.
[9] Schatz G.C 2010 J. Phys. Chem. Lett. 1 802.
[10] Zhang J Z 2010 J. Phys. Chem. Lett. 1 686.
[11] Guo PF, Wu S, Ren QJ, Lu J, Chen Z, Xiao SJ, Zhu YY 2010 J. Phys. Chem. Lett. 1 315.
[12] Wang C, Zhang M, Wang H, Zou S 2010 J. Phys. Chem. Lett. 1 79.
[13] Su YH, Chang SH, Teoh LG, Tu ST, Hon MH 2009 J. Nanosci. Nanotechnol. 9 1181.
14. Su YH, Teoh LG, Lai WH, Chang SH, Yang HY, Hon MH 2007 Appl. Spectrasc. 61, 1007.
[14] Su YH, Chang SH, Teoh LG, Chu WH, Tu SL 2009 J. Phys. Chem. C, 113 3923.
[15] Skirtach AG, Javier AM, Kreft O, Kohler K, Alberola AP, Mohwald H, Parak WJ, Sukhorukov GB 2006 Angew. Chem.. Int. Ed. 45 4612.
[16] Troutman TS, Barton JK, Romanowski M 2008 Adv. Mater. 20 2604.
[17] Alkilany AM, Nagaria PK, Hexel CR, Shaw TJ, Murphy CJ, Wyatt MD 2009 Small 5 701.
[18] Male KB, Lachance B, Hrapovic S, Sunahara G., Luong JHT 2008 Anal. Chem. 80 5487.
[19] Gobin AM, Moon JJ, West J, Ephrin L, Int. 2008 J. Nanomed. 3 351.
[20] Bernardi RJ, Lowery AR, Thompson PA, Blaney SM, West J L 2008 J. Neuro-Oncol. 86 165.
[21] Lal S, Clare SE, Halas NJ 2008 Acc. Chem. Res. 41 1842.

[22] Huang XH, El-Sayed IH, Qian W, El-Sayed MA 2006 J. Am. Chem. Soc. 128 2115.
[23] von Maltzahn G, Park JH, Agrawal A, Bandaru NK, Das SK, Sailor MJ, Bhatia SN 2009 Cancer Res. 69 3892.
[24] Letfullin RR, Joenathan C, George TF, Zharov V P 2006 Nanomedicine 1 473.
[25] Mironava T, Hadjiargyrou M, Simon M, Jurukovski V, Rafailovich MH 2010 Nanotoxicology 4 120.

19

Green Synthesis of Noble Metal (Au, Ag, Pt) Nanoparticles, Assisted by Plant-Extracts

Victor Sanchez-Mendieta[1] and Alfredo Rafael Vilchis-Nestor[2]
[1]*College of Chemistry, Autonomous University of the State of Mexico*
[2]*Research Center for Sustainable Chemistry UAEMex-UNAM*
México

1. Introduction

The physicochemical and optoelectronic properties of metallic nanoparticles are strongly dependent on the size and size-distribution, but also nanoparticles shape contributes significantly to the control of their properties. Wide varieties of physical and chemical procedures have been developed in order to synthesize nanoparticles of different compositions, sizes, shapes and controlled polydispersity. Nevertheless, the routinely physicochemical techniques for nanoparticle production such as photochemical reduction [1], laser ablation [2], electrochemistry [3], lithography [4] or high energy irradiation [5], either remain expensive or employ hazardous substances, such as organic solvents, and toxic reducing agents like sodium borohydride and N,N-dimethylformamide. In addition, due to the high surface energy of the nanoparticles, these tend to form aggregates; therefore, surface passivating and capping reagents are frequently added to the reaction systems to avoid coalescence. The development of reliable, eco-friendly processes for the synthesis of nanomaterials is an important aspect of nanotechnology. Nanotechnology also requires the synthesis of nanomaterials of different chemical compositions, sizes and morphology with an excellent control over these characteristics.

With the growing need to minimize or eliminate the use of environmental-risk substances, as the green chemistry principles describe [6], the synthesis of nanoparticles using biological entities has received increasing attention in the last decade [7]. The biosynthetic procedures involve either living organisms such as bacteria [8], fungi [9] and plants [10] or biomass, like plant extracts [11-14]. Biological synthetic processes have emerged as a simple and viable alternative to more complex physicochemical approaches to obtain nanomaterials with adequate control of size and shape [15].

The use of the highly structured physical and biosynthetic activities of microbial cells for the synthesis of nanosized materials has recently emerged as a novel approach for the synthesis of metal nanoparticles. The interactions between microorganisms and metals have been well documented [16] and the ability of microorganisms to extract and/or accumulate metals is already employed in biotechnological processes such as bioleaching and bioremediation [17]. A general scheme for nanoparticles synthesis in microorganisms is illustrated in figure 1, reported by Das-Marsili [18], where is easily observed the scheme of biomineralization process for nanoparticles synthesis assisted by microorganisms. However more detailed studies are required to fully understand the difference in nanoparticles morphology and size between different metals when the same kind of microorganism is employed. Furthermore, the specific mechanism of reduction, nucleation and grow model of metallic ions and nanoparticles respectively, when interact with metabolic process of the microorganism and biomolecules, remains as a big challenge to be reached.

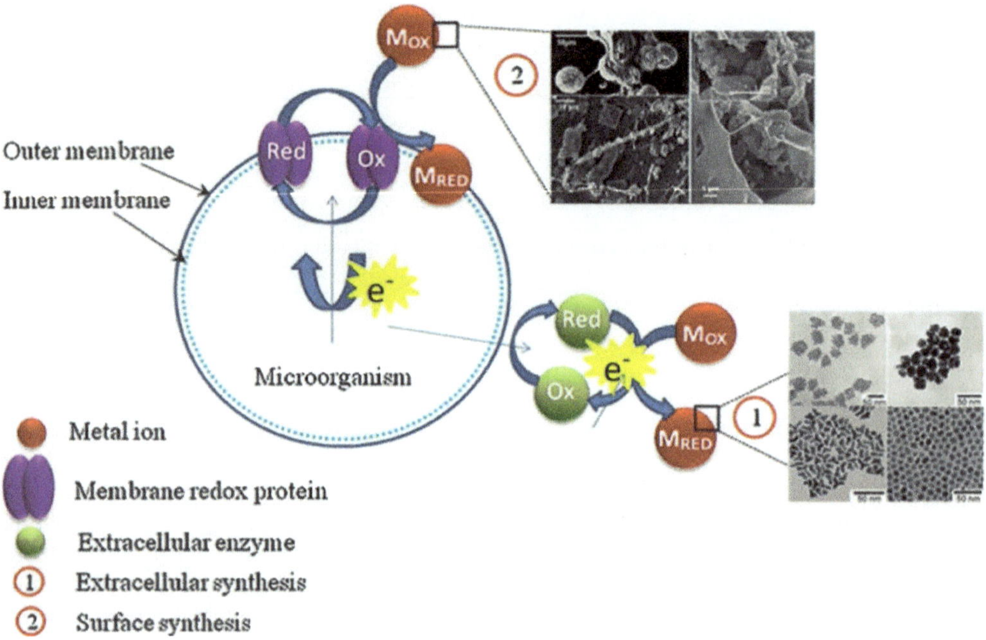

Fig. 1. Scheme of nanoparticles formation by microorganism. (Das-Marsili [18])

1.1 Noble metals nanoparticles synthesis assisted by bacteria

Many organisms, both unicellular and multicellular, are known to produce inorganic materials either intracellularly or extracellularly. One approach that shows immense potential is based on the biosynthesis of nanomaterials using bacteria.

Microorganisms, particularly bacteria, are often exposed to extreme environmental conditions and an ability to resist those stresses is essential for their survival. The specific defense mechanisms help the bacteria to quell such stresses, including the toxicity due to high concentration of foreign metal ions or metals. Such mechanisms include efflux systems, alterations of solubility and toxicity by changes in the redox state of the metal ions, extracellular complexation or precipitation of the metals intracellularly and the lack of specific metal transport systems. [19] One major advantage of having prokaryotes as nanoparticles synthesizers is that they can be easily modified using genetic engineering techniques for overexpression of specific enzymes, apart from the ease of handling.

1.2 Noble metals nanoparticles synthesis assisted by fungus

The use of fungi in the synthesis of nanoparticles is a relatively recent addition and holds promise for large scale nanoparticles production. The use of eukaryotes, especially fungi, is potentially exciting since they secrete large amounts of proteins and enzymes, thus increasing productivity, and are simple to deal with both in the laboratory and at industrial scale. Moreover the process can be easily scaled up, economically viable with the possibility of easily covering large surface areas by suitable growth of mycelia. Furthermore, downstream processing would be much simpler using fungi. [19]

Further information about state of the art of the biosynthesis of metallic nanoparticles assisted by microorganisms, can be revised in some reviews available in literature. [7, 18-20]

1.3 Noble metals nanoparticles synthesis assisted by plants

Among the use of living organisms for nanoparticle synthesis, plants have found application particularly in metal nanoparticle synthesis. Use of plants for synthesis of nanoparticles could be advantageous over other environmentally benign biological processes as this eliminates the elaborate process of maintaining cell cultures. Biosynthetic processes for nanoparticles would be more useful if nanoparticles were produced extracellularly using plants or their extracts and in a controlled manner according to their size, dispersity and shape. Plant use can also be suitably scaled up for large-scale synthesis of nanoparticles. [21] Noble metals, especially Au and Ag, have been extensible tested for the biosynthetic process assisted by plants, in order to obtain metallic nanoparticles with control over shape and size. A list of the metallic nanoparticles obtained by biosynthesis employing plants biomass or extracts can be checked in recent reviews, reported by Yadav *et al.* [21] and Bali *et al.* [22].

In the case of the plants and yeast, the possibility to obtain metallic particles with nanometric dimension were explored just after, these organism were employed for the remediation of metal-contaminated water [23] due to a growing necessity to develop environmentally friendly systems to retrieve metals. It has been shown that many plants can actively uptake and bioreduce metal ions from soils and solutions during the detoxification process, thereby forming insoluble complexes with the metal ion in the form of nanoparticles.

The first successfully report of synthesis of nanoparticles assisted by living plants appeared in 2002 when it was shown that gold nanoparticles, ranging in size from 2 to 20 nm, could form inside alfalfa seedlings [10]. Subsequently it was shown that alfalfa also could form silver nanoparticles when exposed to a silver rich solid medium. [24]

The research group of Gardea-Torresdey from the University of Texas at El Paso, is the major responsible of the experiments that involve the formation of metallic nanoparticles (primarily gold and silver) by different living plant systems as can be observed in table 1.

A virtual circle has been achieved following the biomass method when the catalytic function of the nanoparticles rich biomass was substantiated by the reduction of aqueous 4-nitrophenol, the first report of gold nanoparticle-bearing biomatrix directly reducing a toxic pollutant [25].

The use of plants in metal extraction (phytoremediation) has appeared as a promising alternative in the removal of heavy metal excess from soil and water, and in the process a new method to produce metallic nanoparticles was developed. In order to understand how the phytoremediation technology works, researchers have used X-ray absorption spectroscopy. X-ray absorption spectroscopy (XAS) consists of two complimentary techniques X-ray absorption near edge structure (XANES) and extended X-ray absorption fine structure (EXAFS), which provide invaluable information about the coordination chemistry of metals and toxic element interactions with phytoremediation systems [27].

Using X-ray absorption spectroscopy (XAS) analysis, XANES, and EXAFS, Gardea-Torresdey group has tried to elucidate the process involved during the nanoparticles formation inside living plants and the biomolecules associated with the particles in different parts of the plant. Actually it can be determined the oxidation state, coordination environment, and the average radii of the nanoparticles bound to the hops biomass [28]. The studies suggest that the carboxyl or other oxygen containing ligands on the biomass are the functional groups responsible for the binding and the reduction site for gold on the biomass.

Plant	Metal nanoparticle	Shape	Size	Year/ Reference
Sesbania drummondii	Gold	Spherical	6-20 nm	2007- [25]
Avena sativa	Gold	Fcc: tetrahedral, decahedral, hexagonal, icosahedral multitwinned, irregular, and rod shape	5-20 nm and 25-85 nm, depending on pH	2004-[26]
Medicago sativa (alfalfa sprouts)	Silver	Mostly spherical	2-20 nm	2003-[24]
Medicago sativa	Gold	fcc twinned and icosahedron structure	2-20 nm	2002-[23]

Table 1. Nanoparticles obtained by bioreduction process, assisted by living plants.

From the first reports on the formation of gold and silver nanoparticles by living plants, it was opened up new and exciting ways to fabricate nanoparticles. It showed how it is possible to link materials science and biotechnology in the new emerging field of nanobiotechnology.

The next methodology associated with plants to produce metallic nanoparticles employs the dried biomass of the plants and a metallic salt, as bioreducing agent and precursor, respectively. The general procedure can be described in a general way as follow, the plant is dried (typically sun-dried), then is suspended in water and placed in an ultrasonic bath to homogenize the suspension. Finally the suspension is allowed to rest with the metallic ion aqueous solutions followed by a homogenization process.

Silver or gold precursors at ambient temperature produces both silver nanoparticles (55–80 nm) and triangular or spherical gold nanoparticles, from sun-dried biomass of *Cinnamomum camphora* leaf when are incubated together [13]. This report evidence that the mechanism concerned to the gold and silver nanoparticles formation is different despite of the same reducing agent employed. For gold nanoparticles there is strongly evidence that suggests that these particles are the result of coalescence of smaller particles, because there is no reason to expect a rapid growth in the gold particles and the formation of nanotriangles. The author´s indicate the marked difference of shape control between gold and silver nanoparticles was attributed to the comparative advantage of protective biomolecules. The polyol components and the water-soluble heterocyclic components were mainly responsible for the reduction of silver ions or chloroaurate ions and the stabilization of the nanoparticles, respectively. The process were scaled-up successfully for biological production of silver nanoparticles by lixivium of sundried *Cinnamomum camphora* leaf in continuous-flow tubular microreactors [29]

Silver nanoparticles (NPs) were rapidly synthesized by treating silver ions with a *Capsicum annuum* L. extract [14]. The reaction process was simple and convenient to handle, and was monitored using ultraviolet-visible spectroscopy (UV-vis). The reduction of silver ions and stabilization of the silver NPs was thought to occur through the participation of proteins.

From the bioreduction method, originally proposed by Gardea-Torresday [30], where bionanotechnolgy shows a huge potencial to produce metallic nanoparticles with different shapes (figure 2), nevertheless the most important results employing biomass have been reached by J. Ascencio research group, who have developed a procedure based on the use of the tannins of the biomass of *Medicago sativa* (alfalfa) to obtain Au nanorods [31], bimetallic nanoparticles [32, 33] and even lanthanide clusters [34, 35], and lately has also been demonstrated effective for the synthesis Zn nanoparticles [36], and iron oxide [wuestite ($Fe_{0.902}O$) and magnetite (Fe_3O_4)] clusters [37]. The method is based on the reduction of metal ions through biomass at controlled pH conditions to improve the size control of nanoparticles. (See table 2)

Plant	Metal nanoparticle	Shape	Size	Year/ Reference
Medicago sativa	Gold	decahedral multiple twinned, icosahedral multiple twinned particles, irregular shaped particles, Fcc tetrahedral particles,and hexagonal platelet particles	From 20-100 nm depending on pH	1999-[30]
Cinnamomum camphora	Gold and Silver	Triangular or spherical	55–80 nm	2007-[13]
Capsicum annuum L	Silver	Spherical	10-40 nm. Depending on reduction time	2007-[14]
Medicago sativa	gold	nanorods	20 nm diameter	2001-[31]
Medicago sativa	Eu-Au bimetallic	spherical	From 2-30 nm, depending on pH	2003-[32]

Plant	Metal nanoparticle	Shape	Size	Year/ Reference
Medicago sativa	Ti/Ni bimetallic, core-shell structure with the Ti covering a Ni core	Mainly spherical	From 2-6 nm, depending on pH	2006-[33]
Medicago sativa	Sm	spherical	From 2-8 nm at pH 4	2005-[35]
Medicago sativa	Zn	Spherical	2-5.6 nm depending on pH	2006-[36]
Medicago sativa	Yb	Spherical	2-10 nm	2004-[34]
Medicago sativa	wuestite (Fe0.902O) and magnetite (Fe3O4)	spherical	3.1 nm average size	2007-[37]

Table 2. Nanoparticles obtained by bioreduction process, assisted by plant biomass.

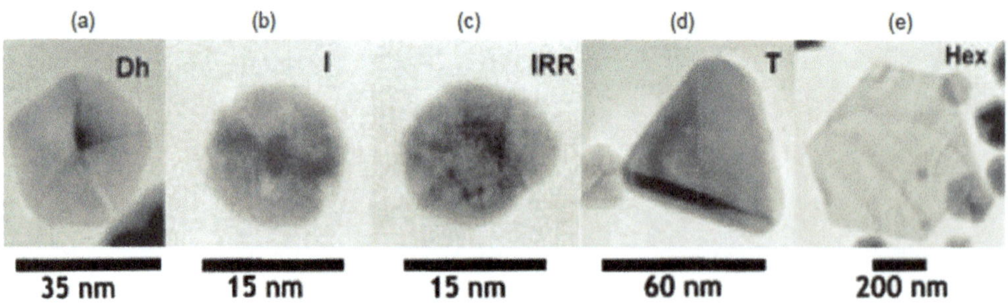

Fig. 2. Morphologies of gold nanoparticles obtained with *Medicago sativa* biomass. Enlargement showing the different kinds of gold particles observed. (a) decahedral multiple twinned; (b) icosahedral multiple twinned; (c) irregular shaped; (d) FCC tetrahedral; (e) FCC hexagonal platelets. *Yacaman et al.* [30]

Here we present a simple green synthetic methodology of silver, gold and platinum nanoparticles, which involves the in-situ reduction of aqueous Ag(I), Au(III) and Pt(IV) ions employing *Camellia sinensis*, and *Opuntia ficus indica* aqueous extract as a reducing and capping agents.

2. Experimental section

2.1 Materials

For the green synthesis, water soluble salts ($HAuCl_4$, $AgNO_3$ and H_2PtCl_6) were employed as precursor of nanoparticles, all purchased from Sigma-Aldrich Chemicals. The above reagents were of analytical purity and were used without further purification.

2.2 Extract preparation

In general manner, different portions of plants: 30g of fresh cladodes of *Opuntia ficus indica* were rinsed with de-ionized water and finely cut. In the case of *Camellia sinensis* 1.5 g of dried leaves were employed to prepare the extract. Afterwards, it was boiled in 100 mL of de-ionized water for five minutes. The mixture was cooled and vacuum filtered. The resulting extract was used for further experiments.

2.3 Synthesis of metallic nanoparticles

Noble metals (Ag, Au and Pt) nanoparticles were prepared by adding different volumes of the plants extracts in a range between 1- 10 mL to 5 mL of 10^{-3} M $AgNO_3$, $HAuCl_4$ and H_2PtCl_6 solutions, respectively. Then, the volume was mixed with de-ionized water until to reach a final volume of 20 mL. This mixture was allowed to stand for 24 h. No additional reducing agent or surfactants were needed for the synthesis of gold, silver or platinum nanoparticles.

2.4 UV-Vis absorbance spectroscopy

The bioreduction of Ag(I), Au (III), and Pt(IV) in aqueous solution was monitored by following the UV-Vis absorbance of the reaction mixture during 4 hours as a function of time, and after 24 hours of reaction. UV-Vis analysis was performed on a CARY 5000 spectrophotometer operated at a resolution of 1 nm at room temperature.

2.5 Transmission electron microscopy (TEM)

Samples for TEM studies were prepared by placing a drop of the silver, gold or platinum nanoparticles colloidal suspension obtained by bioreduction on lacy carbon-coated grids. TEM observations were performed on a transmission electron microscope JEOL 2010.

3. Results

3.1 Camellia sinensis system

It is well known that *Camellia sinensis* contains polyphenols and terpenoids, such as β-cariophyllene, linalool, cis-jasmone, α-terpineol, δ-cadinene, indole, geraniol, among the major bio-components, which have bactericidal and antioxidant activity, and several other useful properties. [38] As stated in previously reported bio-reduction methods, [12, 13, 14] these type of compounds contribute to the metal ion reduction processes, and can also control the size and stabilize the nanostructures formed. For instance, gallic acid, a secondary metabolite present in plants like green tea that also has powerful antioxidant properties, has been successfully used to obtain water soluble Au and Ag nanostructures, and has proved to be an efficient capping agent of these noble-metal nanoparticles.[38, 39]

Addition of *Camellia sinensis* extract to aqueous $HAuCl_4$ solutions, at ambient temperature, quickly turned the initial yellowish solution into pink, grey-blue and pale violet, respectively (inset of figure 3a). In case of formation of silver nanoparticles, the solution turned from yellowish to bright yellow and to dark brown (inset of figure 3b). In the case of platinum nanoparticles, when the time reaction increased, the color of the solution gradually changed from yellow brown to dark brown.

UV-Vis spectra of Au and Ag nanoparticles in aqueous solution are shown in Figure 3a and 3b, respectively. [11] The absorption peaks at around 550 nm for Au and around 430 nm for Ag are characteristic of these noble-metal particles. Colloidal solutions of gold and silver nanoparticles show bright colors, as mentioned before, their origin is attributed to the collective oscillation of the free conduction electrons induced by an electromagnetic field. These resonances are also denoted as surface plasmons. These oscillations depend on the particle size, and with increasing size the plasmon absorption maximum is shifted to longer wavelength and the bandwidth increases. Red shift are observed in both spectra (figure 3), which could be due to an increment in the particle size. In other words, the wavelength shift observed in the plasmon bands is a consequence of the different volume of *Camellia sinensis* extract added to the metal ions solution; however, the initial metal ion concentrations and the reaction time play also a crucial role in the optical properties and size obtained for these nanostructures. Meanwhile, stable gold nanoparticles are obtained until 24 h of reaction (figure 3a). A good control of optical properties of gold nanotriangles, synthesized with lemongrass extract, was also achieved with a simple variation of the extract concentrations. [48]

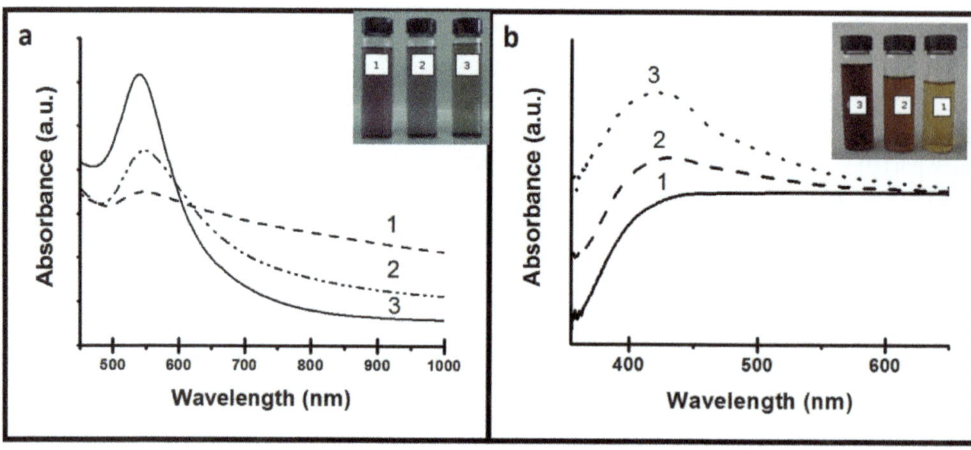

Fig. 3. a) UV-Vis-NIR absorption spectra of gold nanoparticles formed after 24 hours of the reaction of different amounts of *Camellia sinensis* extract with 5 mL of 10^{-3} M aqueous solution $HAuCl_4$ (4, 2 and 1 mL of *Camellia sinensis* extract, curves 1-3 respectively). **b)** UV-Vis absorption spectra of silver nanostructures formed after 4 hours of the reaction of different amounts of *Camellia sinensis* extract with 5 mL of 10^{-3} M aqueous solution $AgNO_3$ (1, 5 and 10 mL of *Camellia sinensis* extract, curves 1-3 respectively).

One of the most convenient techniques for characterization of Pt nanoparticles was UV–vis spectroscopy. Due to the ligand-to-metal charge transfer transition of the $[PtCl_4]^{-2}$, it exhibited a peak at 215 nm in its UV–Vis spectrum. The peak at 215 nm disappeared after the reaction, indicating that the $[PtCl_4]^{-2}$ ions were completely reduced. The Pt nanoparticles had absorption in all ranges of the UV–Vis spectrum and the absorption increased with the decrease of wavelength [40]. Unfortunately, the remnant biomolecules that act as capping and reducing agents, present a strong absorption band in the region between 210 and 260 nm, causing interference with the UV-Vis band of the $[PtCl_4]^{-2}$.

TEM images of poly-disperse gold nanoparticles are shown in figure 4 depending on the volume extract used during the synthesis. Some anisotropic nanoparticles with irregular contours could be observed in figure 4b, with an average particle size of around 25 nm, in contrast with the smaller average size (2.5 nm) of the gold nanoparticles obtained from reaction that employs 0.2 mL of camellia sinensis extract. TEM studies also revealed that the concentration of tea extract plays the main role in the morpholgy of the metal nanoparticles. When 0.2 mL of green tea extract was used for formation of Au nanoparticles, more anisotropic particles were obtained (figure 4a). However, by increasing the amount of *Camellia sinensis* extract from 0.2 to 1 mL, the resulted nanoparticles are slightly bigger and more spherical. It is believed that, when a larger concentration of bio-reducing agent is used, a relatively fast nucleation process occurs, which is followed by a slower growing stage due to stronger interactions between protective bio-molecules and growing Au nanocrystals, leading to the formation of more isotropic particles. [13]

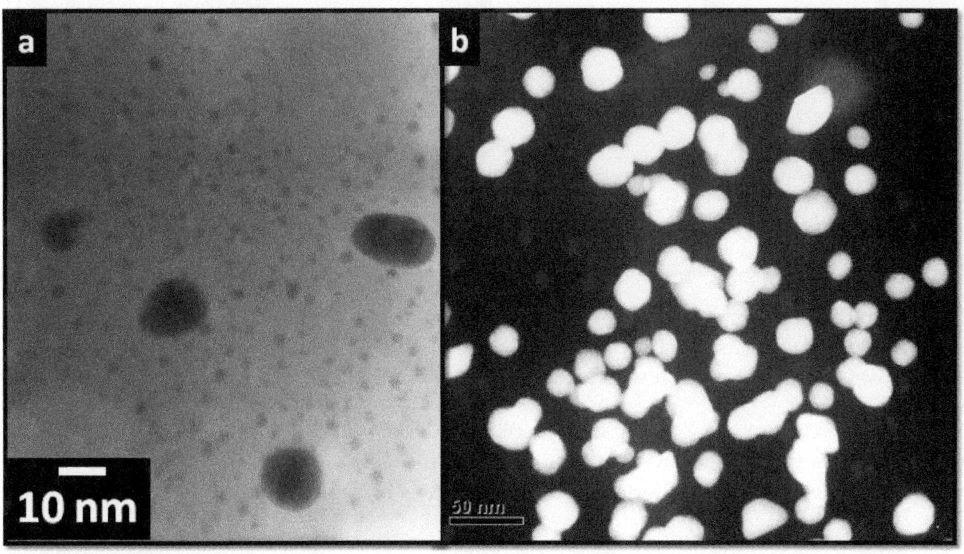

Fig. 4. **a**) Bright-field TEM micrograph of gold nanoparticles formed after 24 hours of the reaction, using 0.2 mL of *Camellia sinensis* extract; and **b**) Dark-field TEM micrograph of gold, nanoparticles formed after 24 hours of the reaction 1mL of *Camellia sinensis* extract, respectively. In both cases 5 mL of 10^{-3} M aqueous solution $HAuCl_4$ were employed for the synthesis.

Typical TEM image of silver nanoparticles can be observed in figure 5. The inset in figure 5 corresponds to SAED pattern, associate to face-centered cubic (FCC) silver. The diffraction spots suggest that grow as a single crystal. On the contrary, when a larger volume of *Camellia sinensis* extract is added to a silver ions solution, the resultant nanocrystals exhibit anisotropy, which is mainly attributed to both, a rapid reduction of the Ag ions and fast sinter of growing crystals into anistropic nanostructures (figure 5). This last outcome can be explained by weaker interactions of Ag growing nanoparticles and protective bio-molecules compared to those with Au nanoparticles. This also explains the different rate of formation of stable Au and Ag nanoparticles using this methodology.

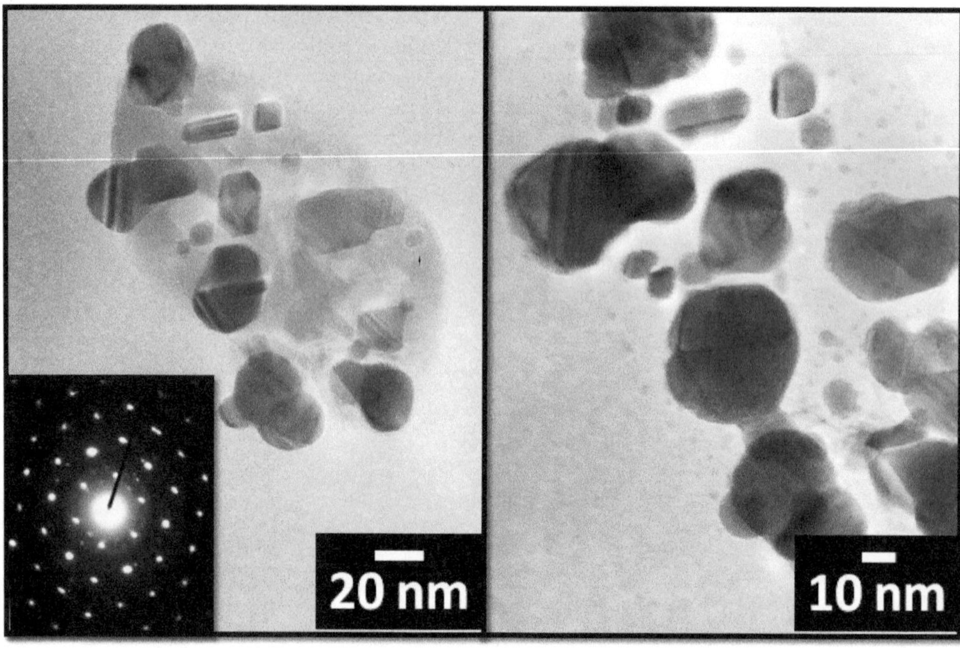

Fig. 5. Typical TEM images of silver nanostructures formed after 24 hours of the reaction, using 2 mL of *Camellia sinensis* extract with 5 mL of 10^{-3} M aqueous solution $AgNO_3$. The inset shows the SAED pattern which correspond to a FCC structure.

In case of platinum, the nanoparticles obtained by bioreduction with *Camellia sinensis* aqueous extract, are shown in the micrographs of figure 6. The morphology of the nanoparticles was mainly spherical with a size between 2-10 nm.

Control of the size and morphology of the resultant nanostructures can be related to the interactions between biomolecules (e.g., terpenoids, polyphenols and phenolic acids) and metal atoms. Therefore, it is thought that phenolic acid type biomolecules present in *Camellia sinensis* extract, are responsible for the reduction of silver, chloroaurate and chloroplatinate ions, and also for the stabilization of the nanoparticles throughout electrostatic interactions. Although good efforts have been done to elucidate these reduction and stabilization mechanisms, more studies are required.

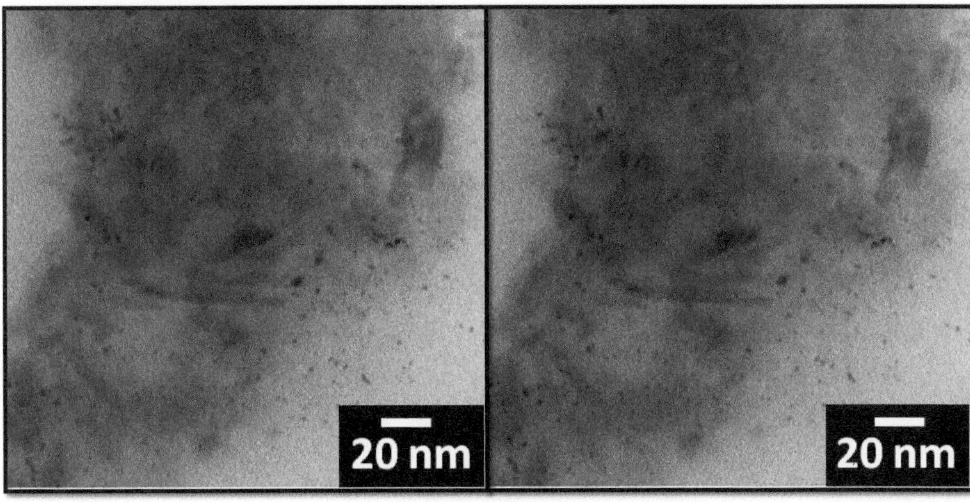

Fig. 6. Typical TEM images of platinum nanoparticles formed after 24 hours of the reaction, using 5 mL of *Camellia sinensis* extract with 5 mL of 10^{-3} M aqueous solution H_2PtCl_6.

3.2 Opuntia ficus indica system

Chemical composition of *Opuntia ficus indica* cladodes are mainly water (92%), carbohydrates (~4-6%), proteins (~1%), vegetable fats (~0.2%), minerals (~1%) such as calcium and iron predominantly, and vitamins mainly ascorbic acid (vitamin C) [41,42]. On another hand, it has been demonstrated by Sun et al. [43] that gold nanoparticles can be obtained directly and simply by reduction of Au(III) ions with ascorbic acid as reducing agent.

The carbohydrate fraction of the *Opuntia ficus indica* can be attributed mainly to the starch and soluble sugars contents. The cladodes showed higher starch contents (from 7 to 13/100 g of dry weight) than soluble sugars (from 6 to 2/100 g of dry weight), in contrast, protein and fats contents showed poorer quantities. Besides, cladodes exhibited also low pH values due to the presence of many organic acids such as: malic, citric and oxalic acids [44]. Carboxylic moieties show a great affinity towards the surface of the nanoparticles, for this reason nanoparticles can be stabilized through electrostatic interactions with carboxylic groups, as reported previously by K Yoosaf et al. [38].

Opuntia ficus indica is considered as a great source of natural antioxidant compounds due to the content of polyphenols, higher than 900 mg/100g of dry matter, which are mainly responsible for an antioxidant activity [44]. This type of compounds play an important role during metal ion reduction processes in biosynthetic methods, as stated in previously reported [13]. Although a more detailed study is required to establish a detailed mechanism of formation and stabilization of the metallic nanoparticles, we can assumed that hydroxyl and carboxylic groups present in the biomolecules of the *Opuntia ficus indica* extract play an important role in the Ag(I) and Au(III) ions reduction and can also control the size and the stability of the nanostructures formed.

When *Opuntia ficus indica* aqueous extract is combined with 10^{-3} M aqueous $AgNO_3$ and $HAuCl_4$, solutions color changes from pale yellow to dark yellow and from yellowish to red due to the formation of silver and gold nanoparticles respectively, as a consequence of the excitation of surface plasmon vibrations.

The UV-vis absorption spectra of silver nanoparticles, formed after 1 hour of reaction using different quantities of *Opuntia ficus indica* extract, are shown in Fig. 7(a) Spectra 1, 2 and 3 correspond respectively to 10, 5 and 1 ml of *Opuntia ficus indica* extract used as bioreducing agent. Spectra 2 and 3 show a single absorption band centered at 398 nm. In contrast, spectra 1 exhibits an insignificant intensity band at 404 nm width, which can be attributable to an insufficient amount of reductive biomolecules for the Ag^{+1} reduction ; consequently fewer silver nanoparticles are formed.

As described by Huang et al. [13], weaker binding of biomolecules with the nascent silver nanoparticles could lead to isotropic growth of the crystals and further formation of spherical nanoparticles.

In plant-mediated synthesis, the control of the size of silver nanoparticles has been proposed to be time-reaction dependent [45]. Basically, the longer the reaction time, the larger the sizes and the nanoparticles change from polycrystalline to single crystalline. Fig. 7(b) shows UV-vis spectra after 1 and 65 hours or reaction time. The fact that the spectra are very similar indicates that the particles have essentially the same size and shape (spheroids). This behavior takes place only when 5 ml of *Opuntia ficus indica* extract are used as the reducing agent. However, when this quantity is doubled, the corresponding UV-Vis band disappears after 65 hours, and silver micro-particles are deposited in the reaction solution as a result of a coalescence process of nanoparticles and probably due to the poor efficacy of *Opuntia ficus indica* extract capping biomolecules in stabilizing the silver nanoparticles through time, thus increasing the size of Ag crystals.

Addition of *Opuntia ficus indica* extract to 10^{-3} M aqueous $HAuCl_4$ solution led to the appearance of a blue color in solution after about 1 h of reaction, indicating the formation of gold nanoparticles (inset in figure 9). The UV-vis absorption spectrum recorded from this solution shows the characteristic surface plasmon resonance (SPR) band of gold nanoparticles centered at 540 nm (Figure 8). The kinetics of formation of gold nanoparticles was followed by UV-Vis spectroscopy, and the spectra obtained are shown in Figure 8. It is observed that with the progress of the reaction the absorbance intensity at 540 nm increases monotonically with time when 1 mL of *Opuntia ficus indica* extract is employed, while with time, the band centered at 560 nm for the system that used 2 mL of *Opuntia ficus indica* extract for the reduction undergoes a further red shift before stabilizing at 640 nm after completion of reaction. These time-dependent features in the UV-Vis spectra are characteristic of aggregated spherical nanoparticles or anisotropic nanostructures whose dimensions change with time. Figure 8b shows the UV-Vis absorbance spectra of gold nanoparticles synthesized 2 mL of *Opuntia ficus indica* extract recorded from 05 to 4 h of the reaction. It can be observed in figure 9 a clearly band shift to longer wavelengths and strong intensity with the time. It is well known that rod-shaped and flat gold nanoparticles absorb in the NIR region of the electromagnetic spectrum [12, 46]. Such nanostructures exhibit two well-separated absorption bands wherein the low wavelength band centered at ca. 540 nm corresponds to the transverse surface plasmon vibration while the long wavelength component corresponds to the longitudinal surface plasmon absorption. [47] The first signal could be enclosed by the second one, causing that the spectra profile non symmetric of figure 8b.

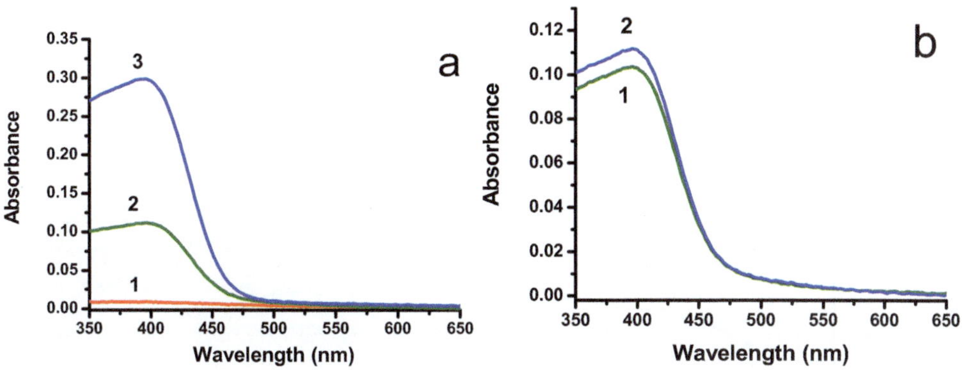

Fig. 7. **a)** UV-Vis absorption spectra of silver nanoparticles formed after 1 hour of the reaction of different amounts (1, 5 and 10 ml, curves 1-3 respectively) of *Opuntia ficus indica* extract with 5 ml of 10^{-3} M aqueous solution of Ag NO_3. **b)** UV-Vis absorption spectra of silver nanostructures formed with 5 ml of *Opuntia ficus indica* extract and 5 ml of 10^{-3} M aqueous solution of $AgNO_3$ after 1 and 65 hours of the reaction (curves 1 and 2, respectively).

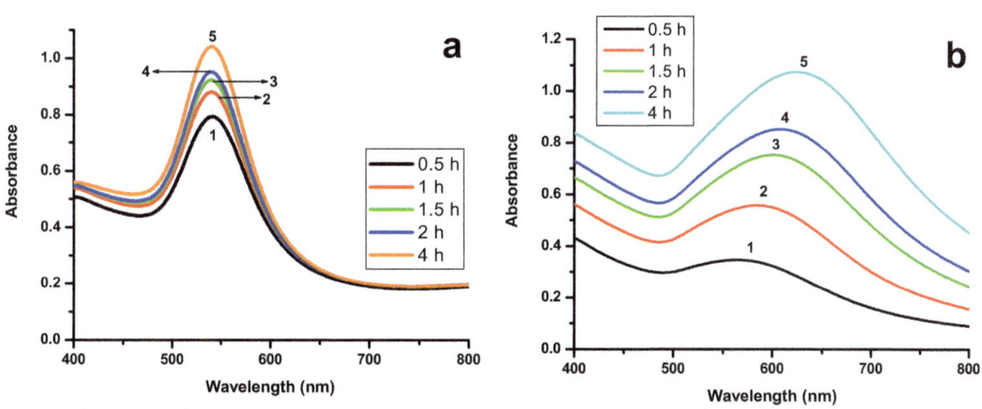

Fig. 8. UV-Vis absorption spectra of gold nanoparticles measured during the reaction of **a)** 1 mL and **b)** 2ml of *Opuntia ficus indica* aqueous extract with 5 ml of 10^{-3} M solution of $HAuCl_4$, respectively (final volume of reaction mixture adjuste to 10 mL using deionized water); after 0.5, 1, 1.5, 2 and 4 h of reaction (curves 1-5, respectively)

Figure 9 presents a plot of the plasmon intensity at 540 nm against reaction time, from the graph in figure 8a. It can be seen easily that the plasmon intensity at the reaction time of 2 h is near to that at 4 h, meaning completion of the reaction. The inset of figure 9 presents the final color in the mixture solution after 4 h of reaction time. The extract was light green before reaction with gold ions and changed to pale blue and blue after 4 hours of reaction time, with 1 mL and 2 mL of *Opuntia ficus indica* extract, respectively, and then the color did not change any more with reaction time. The characteristic blue of gold nanoparticles solutions provided a convenient spectroscopic signature to indicate their formation.

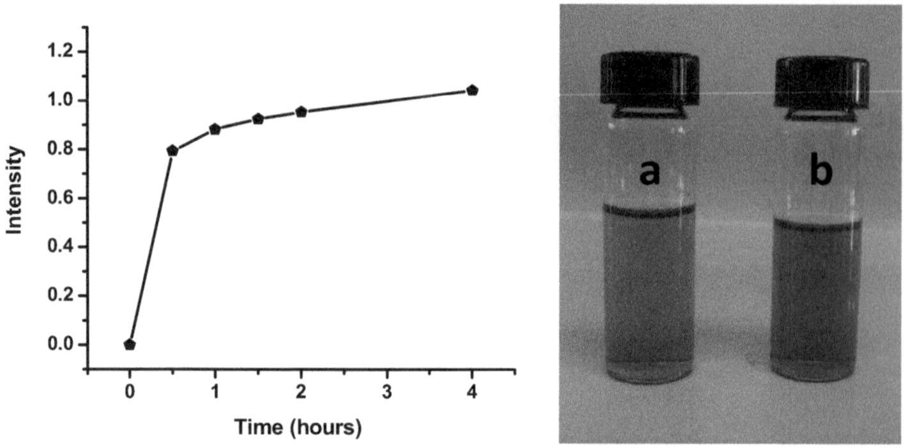

Fig. 9. Plot of the intensity of surface plasmon resonance at 540 nm against the reaction time, for gold nanoparticles obtained from 1 mL of *Opuntia ficus indica* aqueous extract with 5 ml of 10^{-3} M solution of $HAuCl_4$. The inset shows photos of the different nanoparticle solutions after 4 h of the reaction whose labels correspond to spectra **a)** and **b)** shown in the figure 8.

TEM analysis reveals that silver nanoparticles are predominately ellipsoids, as can be observed in Fig. 10(a). Micrograph 10b and histogram 10b correspond to silver nanoparticles formed after 24 hours of reaction, using 5 ml of 10^{-3} M of aqueous solution $AgNO_3$. Nanoparticles sizes are between 8 and 50 nm, with an average of 23.5 nm ± 5 nm. A typical selected area diffraction pattern is shown in the inset of Fig. 10a. Diffraction rings can be indexed as (200), (220), (311) and (222) reflections (indicated by numbers 1, 2, 3 and 4 respectively), corresponding to a FCC structure of silver. For the 1 and 10 ml of biomass extract, the average particle size was 8.6 and 53 nm, respectively. Figure 11(a) shows silver nanoparticles after 24 hours of the bioreduction process, with the same amount of silver ions and volume of reducing agent used previously in Figure 10(a) but kept at 60°C instead of at room temperature. Average particle size is reduced to 8.1 nm, and the particles are quite spherical.

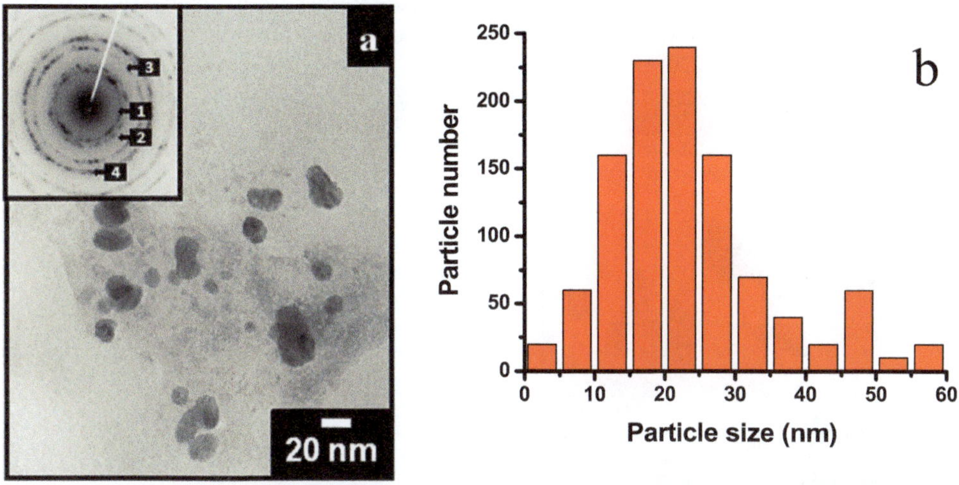

Fig. 10. **a)** Representative TEM image of silver nanoparticles synthesized with 5 ml of *Opuntia ficus indica* extract when reacting with 5 ml of 10^{-3} M aqueous solution of $AgNO_3$. **Inset 10 a)** Electron diffraction pattern of **(a)**, which can be indexed on the basis of the FCC structures of silver. **b)** Size distribution of Ag nanoparticles synthesized with 5 ml of *Opuntia ficus indica* extract when reacting with 5 ml of 10^{-3} M aqueous solution of $AgNO_3$.

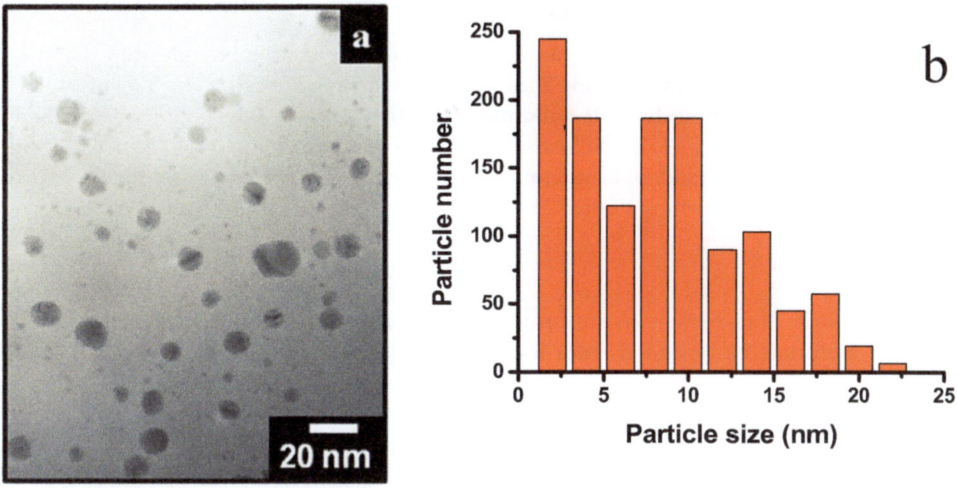

Fig. 11. **a)** Representative TEM image of Ag nanoparticles formed after 24 hours of the bioreduction, with 5 ml of *Opuntia ficus indica* extract and 5 ml of 10^{-3} M aqueous solution of $AgNO_3$ at 60°C. **b)** Histogram showing the Ag nanoparticles size distribution obtained at 60°C.

Figure 12 shows typical TEM micrograph of gold nanoparticles synthesized using of *Opuntia ficus indica* extract. The morphology of the nanoparticles observed was icosahedral, multiple-twined particles, but also irregular shapes were formed, with an average size of 40 nm. The inset SAED pattern revealed that the diffraction rings from inner to outer, could be indexed as (111), (200), (220), and (311) reflections, respectively, corresponding to face-centered cubic (FCC) gold. The diffraction rings also suggested the NPs were polycrystalline.

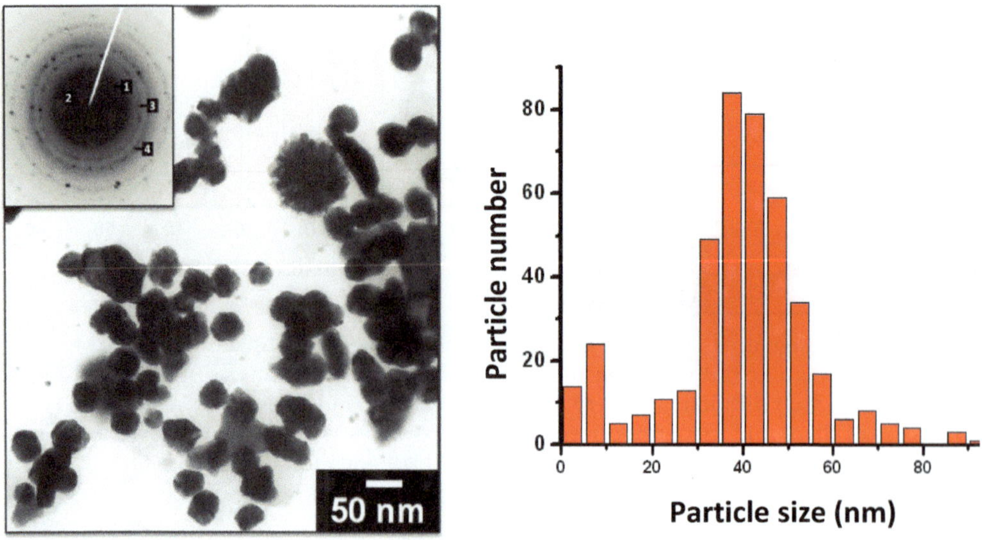

Fig. 12. **a)** Transmission electron micrograph of gold nanoparticles after bioreduction. Gold nanoparticles were obtained by exposing 2 mL of *Opuntia ficus indica* extract to 5 ml of 10^{-3} M solution of $HAuCl_4$. The inset shows the electron diffraction pattern of the gold particles in the main part of the image. Rings 1, 2, and 3 arise due to reflections from the (111), (200), (220) and (311) lattice planes of FCC gold. **b)** Histogram of size distribution of gold nanoparticles synthesized by the experiment of figure 12a. The average particle size is 40 nm.

4. Summary

If nanotechnology is to realize its full promise and potential, it is essential that these next-generation materials be designed from the outset to be benign to human health and the environment.

Nanoparticles of gold, silver and platinum have been successfully prepared using a simple and efficient green nanochemistry methodology. Furthermore, we have demonstrated that use of natural, renewable and low cost biological reducing agents, such as *Camellia sinensis* and *Opuntia ficus indica* can produce noble metal nanostructures in aqueous solution at ambient conditions, avoiding, in consequence, the presence of hazardous and toxic solvents and waste.

The colloidal systems of silver and gold nanoparticles, stabilized by the bio-molecules present in the *Camellia sinensis* and *Opuntia ficus indica* aqueous solutions, are currently under evaluation as potential useful bio-imaging agents for medical applications due to their characteristic optical properties exhibited.

5. Acknowledgements

Authors are grateful to Francisco Ruiz (Centro de Nanociencia y Nanotecnología Universidad Nacional Autónoma de México) for TEM measurements.

6. References

[1] S. Eustis, H. Y. Hsu, MA. El-Sayed, *J. Phys. Chem. B.* (2005), 109, 4811
[2] F. Mafune, J. Kohno, Y. Takeda, T. J. Kondow, *Phys. Chem. B.* (2002), 106, 7575.
[3] L. Rodríguez-Sánchez, M. C. Blanco, M. A. López-Quintela, *J. Phys, Chem. B.* (2002) 104, 9683.
[4] G. Zhang, D.J. Wang, *J. Am. Chem. Soc.* (2008) 130, 5616.
[5] M. Treguer, C. Cointet, H. Remita, J. Khatouri, M. Mostafavi, J. Amblard, J. J. Belloni, *Phys. Chem. B.* (1998) 102, 4310.
[6] PT. Anastas, MM. Kirchhoff, *Acc. Chem. Res.* (2002) 35, 686.
[7] P. Mohanpuria, N. K. Rana, S. K. Yadav, *J. Nanopart. Res.* (2008) 10, 507.
[8] R. Joerger, T. Klaus, C. G. Granqvist, *Adv. Mater.* (2000) 12, 407.
[9] K. C. Bhainsa, S. F. D'Souza, *Colloids and Surfaces B: Biointerfaces.* (2006) 47, 160.
[10] J. L. Gardea-Torresday, J. G. Parsons, E. Gomez, J. Peralta-Videa, H. E. Troiani, P. Santiago, J. Yacaman, *Nano. Lett.* (2002) 2, 397.
[11] AR. Vilchis-Nestor, V. Sánchez-Mendieta, Marco A. Camacho-López, Miguel A. Camacho-López, JA. Arenas-Alatorre, *Mater. Lett.* 62, 3103 (2008)
[12] S. Shankar, A. Ahmad, M. Sastry, *Biotechnol Prog.* 19, 1627 (2003)
[13] J. Huang , Q. Li, D. Sun, Y. Lu, Y. Su, X. Yang, H. Wang, Y. Wang, W. Shao, N. He, J. Hong, C. Chen, *Nanotechnology* (2007) 105104, 18.
[14] S. Li, Y. Shen, A. Xie, X. Yu, L. Qui, L. Zhang, Q. Zhang. *Green Chem.* (2007) 9, 852.
[15] S.S. Shankar, A. Rai, B. Ankamwar, A. Singh, A. Ahmad, M. Sastry, *Nature Materials* (2004) 3, 482.
[16] T.J. Beveridge, M.N. Hughes, H. Lee, K.T. Leung, R.K. Poole, I. Savvaidis, S. Silver, J.T. Trevors. *Advances in Microbial Physiology*, 38 (1997), 177–243.
[17] Jean-Marc Bollag, Tawna Mertz, and Lewis Otjen. Bioremediation through Rhizosphere Technology, Chapter 1: Role of Microorganisms in Soil Bioremediation, 1994, *ACS Symposium Series*, Volume 563, 2-10.
[18] Sujoy K. Das, Enrico Marsili. A green chemical approach for the synthesis of gold nanoparticles: characterization and mechanistic aspect. *Rev Environ Sci Biotechnol* (2010) 9, 199–204
[19] D. Bhattacharya and R. K. Gupta. Nanotechnology and Potential of Microorganisms. *Critical Reviews in Biotechnology*, 25, 199–204, (2005).
[20] M. Gericke, A. Pinches. *Hydrometallurgy* (2006) 83, 132–140.
[21] V. Kumar and S.K. Yadav. *Journal of Chemical Technology and Biotechnology* (2009) 84-2, 151–157.
[22] R. Bali, N. Razak, A. Lumb and A.T. Harris. The synthesis of metallic nanoparticles inside live plants. *Proceedings of the 2006 International Conference on Nanoscience and Nanotechnology* (ICONN'2006), 224-227.
[23] G. Gamez, K. Dokken, I. Herrera, J. G. Parsons, K. J. Tiemann, and J. L. Gardea-Torresdey. Chemical processes involved in Au(III) binding and bioreduction by alfalfa biomass. *Proceedings of the 2000 Conference on Hazardous Waste Research.*

[24] Jorge L. Gardea-Torresdey, E. Gomez, J. R. Peralta-Videa, Jason G. Parsons, H. Troiani, and M. Jose-Yacaman. *Langmuir* 2003, 19, 1357-1361.
[25] N. C. Sharma, S. Sahi, J. G. Parsons, Jorge L. Gardea-Torresdey, Tarasankar Pal. *Environ. Sci. Technol.* 2007, 41, 5137-5142.
[26] V. Armendariz, I. Herrera, J. R. Peralta-Videa, Miguel Jose-Yacaman, H. Troiani, P. Santiago and Jorge L. Gardea-Torresdey. *Journal of Nanoparticle Research* (2004) 6, 377–382,.
[27] Jorge L. Gardea-Torresdeya, Jose R. Peralta-Videa, G. de la Rosaa, J.G. Parsons. *Coordination Chemistry Reviews* (2005) 249, 1797–1810.
[28] M.L. Lopez, J.G. Parsons, T, J.R. Peralta Videa, J.L. Gardea-Torresdey. *Microchemical Journal* (2005) 81, 50–56.
[29] J. Huang, L. Lin, Q. Li, D. Sun, Y. Wang, Y. Lu, N. He, K. Yang, X. Yang, H. Wang, W. Wang and W. Lin. *Ind. Eng. Chem. Res.* (2008) 47, 6081–6090.
[30] J.L. Gardea-Torresdey, K.J. Tiemann, G. Gamez, K. Dokken, S. Tehuacanero and M. José-Yacamán. *Journal of Nanoparticle Research* 1 (1999), 397–404.
[31] J. Gardea-Torresday, K. Tiemman, G. Gamez, K. Dokken, S. Tehuacanero, M. Jose-Yacaman. *J. Nanoparticles Res.* 3 (2001) 475.
[32] J.A. Ascencio, Y. Mejia, H.B. Liu, C. Angeles, G. Canizal. *Langmuir* (2003) 19, 5882.
[33] P.S. Schabes-Retchkiman, G. Canizal, R. Herrera-Becerra, C. Zorrilla, H.B. Liu, J.A. Ascencio. *Optical Materials* (2006), 29, 95–99.
[34] J.A. Ascencio, A.C. Rodríguez-Monroy, H.B. Liu, G. Canizal. *Chem.Lett.* (2004), 33 1056.
[35] Jorge A. Ascencio, Ana C. Rincon, and Gerardo Canizal. J. Phys. Chem. B (2005), 109, 8806-8812.
[36] G. Canizal, P.S. Schabes-Retchkiman, U. Pal, Hong Bo Liu, J.A. Ascencio. *Materials Chemistry and Physics* (2006) 97, 321–329.
[37] Raúl Herrera-Becerra, Cristina Zorrilla, and Jorge A. Ascencio. *J. Phys. Chem. C* (2007) 111, 16147-16153.
[38] Yoosaf K, Ipe BI, Suresh CH, Thomas G. *J Phys Chem C* (2007) 12839, 111.
[39] Scampicchio M, Wang J, Blasco AJ, Arribas AS, Mannino S, Escarpa A. *Anal Chem.* (2006) 2060, 78.
[40] W. Yang, Y. Maa, J. Tang, X. Yang. *Colloids and Surfaces A: Physicochem. Eng. Aspects* 302 (2007) 628–633.
[41] D. Guzmán-Loayza, J. Chávez, *Rev. Soc. Quím.* Perú, 73, 41 (2007).
[42] N. Salim, C. Abdelwaheb, C. Rabah, B. Ahcene, *African Journal of Biotechnology.* 8, 1623, (2009).
[43] K. Sun, J. Qiu, J. Liu, Y. Miao, *J. Mater. Sci*, 44, 754, (2009).
[44] MA. Ayadi, W. Abdelmaksoud, M. Ennouri, H. Attia, *Industrial Crops and Products* 30, 40, (2009).
[45] V. Kumar, S. Kumar, J. Chem. *Technol. Biotechnol.* 84, 151 (2008).
[46] Cecilia Noguez. *J. Phys. Chem. C* 2007, *111*, 3806-3819.
[47] Luis M. Liz-Marzán, *Langmuir* 2006, *22*, 32-41.
[48] S. Shiv Shankar, Akhilesh Rai, Absar Ahmad, and Murali Sastry. *Chem. Mater.* 2005, *17*, 566-572.

Permissions

The contributors of this book come from diverse backgrounds, making this book a truly international effort. This book will bring forth new frontiers with its revolutionizing research information and detailed analysis of the nascent developments around the world.

We would like to thank Yen-Hsun Su, for lending his expertise to make the book truly unique. He has played a crucial role in the development of this book. Without his invaluable contribution this book wouldn't have been possible. He has made vital efforts to compile up to date information on the varied aspects of this subject to make this book a valuable addition to the collection of many professionals and students.

This book was conceptualized with the vision of imparting up-to-date information and advanced data in this field. To ensure the same, a matchless editorial board was set up. Every individual on the board went through rigorous rounds of assessment to prove their worth. After which they invested a large part of their time researching and compiling the most relevant data for our readers. Conferences and sessions were held from time to time between the editorial board and the contributing authors to present the data in the most comprehensible form. The editorial team has worked tirelessly to provide valuable and valid information to help people across the globe.

Every chapter published in this book has been scrutinized by our experts. Their significance has been extensively debated. The topics covered herein carry significant findings which will fuel the growth of the discipline. They may even be implemented as practical applications or may be referred to as a beginning point for another development. Chapters in this book were first published by InTech; hereby published with permission under the Creative Commons Attribution License or equivalent.

The editorial board has been involved in producing this book since its inception. They have spent rigorous hours researching and exploring the diverse topics which have resulted in the successful publishing of this book. They have passed on their knowledge of decades through this book. To expedite this challenging task, the publisher supported the team at every step. A small team of assistant editors was also appointed to further simplify the editing procedure and attain best results for the readers.

Our editorial team has been hand-picked from every corner of the world. Their multi-ethnicity adds dynamic inputs to the discussions which result in innovative outcomes. These outcomes are then further discussed with the researchers and contributors who give their valuable feedback and opinion regarding the same. The feedback is then collaborated with the researches and they are edited in a comprehensive manner to aid the understanding of the subject.

Apart from the editorial board, the designing team has also invested a significant amount of their time in understanding the subject and creating the most relevant covers. They scrutinized every image to scout for the most suitable representation of the subject and create an appropriate cover for the book.

The publishing team has been involved in this book since its early stages. They were actively engaged in every process, be it collecting the data, connecting with the contributors or procuring relevant information. The team has been an ardent support to the editorial, designing and production team. Their endless efforts to recruit the best for this project, has resulted in the accomplishment of this book. They are a veteran in the field of academics and their pool of knowledge is as vast as their experience in printing. Their expertise and guidance has proved useful at every step. Their uncompromising quality standards have made this book an exceptional effort. Their encouragement from time to time has been an inspiration for everyone.

The publisher and the editorial board hope that this book will prove to be a valuable piece of knowledge for researchers, students, practitioners and scholars across the globe.

List of Contributors

Shigeki Matsunaga
Nagaoka National College of Technology, Japan

Takahiro Koishi
University of Fukui Graduate School of Engineering, Japan

Masatoshi Saito and Shigeru Tamaki
Niigata University, Japan

Mauro C. Geraldes
Department of Geology, Rio de Janeiro State University, Rio de Janeiro-RJ, Brazil

José R. Parga and José A. Díaz
Department of Metallurgy and Materials Science Institute Technology of Saltillo, Saltillo Coahuila, México

Jesús L. Valenzuela
Department of Chemical Engineering and Metallurgy, University of Sonora, Hermosillo, Sonora, México

Leandro A. Voisin
University of Chile, Department of Mining Engineering, Chile

Dragos-Pinzaru Oana-Georgiana, Dumitru-Daniel Herea and Horia Chiriac
National Institute of Research & Development for Technical Physics, Romania

Yong Yang, Lixin Zhao and Song Hu
Institute of Optics and Electronics, Chinese Academy of Sciences, Chengdu, China

Guanxiao Cheng
College of Electronic Science and Technology, Shenzhen University, Shenzhen, China

Shaolin Zhou
School of Electronic and Information Engineering, South China University of Technology, Guangzhou, China

Sherif A. El-Safty
Materials Research Laboratory for Environmental and Energy, National Institute for Materials Science (NIMS), Tsukuba-shi Ibaraki, Japan
Graduate School for Advanced Science and Engineering, Waseda University, Shinjuku-ku, Tokyo, Japan

Nguyen Duc Hoa
Materials Research Laboratory for Environmental and Energy, National Institute for Materials Science (NIMS), Tsukuba-shi Ibaraki, Japan

Vaso Bojanić
University of Banja Luka, Faculty of Agriculture, B&H

Miomir Pavlović
University of Eastern Sarajevo, Faculty of Technology Zvornik Republic of Srpska, B&H

Junling Lu, Yu Lei and Jeffrey W. Elam
Argonne National Laboratory, Argonne, IL, USA

Yasser A. Abdelaziz
National Institute for Standards (NIS), Giza, Egypt

Takashiro Muroi
Industrial Catalysts Laboratory, Japan

Sachiko Ono and Hidetaka Asoh
Department of Applied Chemistry, Faculty of Engineering, Kogakuin University, Tokyo, Japan

Ryo Watanabe, Yasushi Sekine, Masahiko Matsukata and Eiichi Kikuchi
Department of Applied Chemistry, Waseda University, Japan

Maddalena Pedio and Cinzia Cepek
IOM CNR, Laboratorio TASC Area Science Park, Basovizza Trieste, Italy

Roberto Felici
ESRF Grenoble, France

Wei Lin
School of Materials Science & Engineering, Georgia Institute of Technology, USA

Mieczysława Najbar
Faculty of Chemistry, Jagiellonian University, Poland

Patrick Da Costa
The Jean le Rond d'Alembert Institute, University Pierre et Marie Curie, France

Jarosław Dutkiewicz, Paweł Kornelak and Agnieszka Pietraszek
Faculty of Chemistry, Jagiellonian University, Poland

Valerio Choque, Narcis Homs and Pilar Ramirez de la Piscina
Department of Inorganic Chemistry, University of Barcelona, Spain

Janusz Sobczak
Institute of Physical Chemistry PAS, Poland

Fatma Arslan
Istanbul Technical University, Mining Faculty Mineral Processing Engineering Department, Istanbul, Turkey

Yen-Hsun Su
Department of Material Science and Engineering, National Dong Hwa University, Tainan, Taiwan

Sheng-Lung Tu
Department of Resource Engineering, National Cheng Kung University, Tainan, Taiwan

Wei-Min Zhang
Department of Physics, National Cheng Kung University, Tainan, Taiwan

Victor Sanchez-Mendieta
College of Chemistry, Autonomous University of the State of Mexico, México

Alfredo Rafael Vilchis-Nestor
Research Center for Sustainable Chemistry UAEMex-UNAM, México